Autonomous Mobile Robots

Sensing, Control, Decision Making and Applications

CONTROL ENGINEERING

A Series of Reference Books and Textbooks

Editor

FRANK L. LEWIS, PH.D.

Professor
Applied Control Engineering
University of Manchester Institute of Science and Technology
Manchester, United Kingdom

1. Nonlinear Control of Electric Machinery, *Darren M. Dawson, Jun Hu, and Timothy C. Burg*
2. Computational Intelligence in Control Engineering, *Robert E. King*
3. Quantitative Feedback Theory: Fundamentals and Applications, *Constantine H. Houpis and Steven J. Rasmussen*
4. Self-Learning Control of Finite Markov Chains, *A. S. Poznyak, K. Najim, and E. Gómez-Ramírez*
5. Robust Control and Filtering for Time-Delay Systems, *Magdi S. Mahmoud*
6. Classical Feedback Control: With MATLAB®, *Boris J. Lurie and Paul J. Enright*
7. Optimal Control of Singularly Perturbed Linear Systems and Applications: High-Accuracy Techniques, *Zoran Gajif and Myo-Taeg Lim*
8. Engineering System Dynamics: A Unified Graph-Centered Approach, *Forbes T. Brown*
9. Advanced Process Identification and Control, *Enso Ikonen and Kaddour Najim*
10. Modern Control Engineering, *P. N. Paraskevopoulos*
11. Sliding Mode Control in Engineering, *edited by Wilfrid Perruquetti and Jean-Pierre Barbot*
12. Actuator Saturation Control, *edited by Vikram Kapila and Karolos M. Grigoriadis*
13. Nonlinear Control Systems, *Zoran Vukić, Ljubomir Kuljača, Dali Donlagič, and Sejid Tesnjak*
14. Linear Control System Analysis & Design: Fifth Edition, *John D'Azzo, Constantine H. Houpis and Stuart Sheldon*
15. Robot Manipulator Control: Theory & Practice, Second Edition, *Frank L. Lewis, Darren M. Dawson, and Chaouki Abdallah*
16. Robust Control System Design: Advanced State Space Techniques, Second Edition, *Chia-Chi Tsui*
17. Differentially Flat Systems, *Hebertt Sira-Ramírez and Sunil Kumar Agrawal*

Autonomous Mobile Robots

Sensing, Control, Decision Making and Applications

Shuzhi Sam Ge
The National University of Singapore

Frank L. Lewis
Automation and Robotics Research Institute
The University of Texas at Arlington

CRC Press
Taylor & Francis Group
Boca Raton London New York

CRC Press is an imprint of the
Taylor & Francis Group, an **informa** business

A TAYLOR & FRANCIS BOOK

CRC Press
Taylor & Francis Group
6000 Broken Sound Parkway NW, Suite 300
Boca Raton, FL 33487-2742

First issued in paperback 2019

© 2006 by Taylor & Francis Group, LLC
CRC Press is an imprint of Taylor & Francis Group, an Informa business

No claim to original U.S. Government works

ISBN-13: 978-0-8493-3748-2 (hbk)
ISBN-13: 978-0-367-39089-1 (pbk)

Library of Congress Cataloging-in-Publication Data

Catalog record is available from the Library of Congress

**Visit the Taylor & Francis Web site at
http://www.taylorandfrancis.com**

**and the CRC Press Web site at
http://www.crcpress.com**

Preface

The creation of a truly autonomous and intelligent system — one that can sense, learn from, and interact with its environment, one that can integrate seamlessly into the day-to-day lives of humans — has ever been the motivating factor behind the huge body of work on artificial intelligence, control theory and robotics, autonomous (land, sea, and air) vehicles, and numerous other disciplines. The technology involved is highly complex and multidisciplinary, posing immense challenges for researchers at both the module and system integration levels. Despite the innumerable hurdles, the research community has, as a whole, made great progress in recent years. This is evidenced by technological leaps and innovations in the areas of sensing and sensor fusion, modeling and control, map building and path planning, artificial intelligence and decision making, and system architecture design, spurred on by advances in related areas of communications, machine processing, networking, and information technology.

Autonomous systems are gradually becoming a part of our way of life, whether we consciously perceive it or not. The increased use of intelligent robotic systems in current indoor and outdoor applications bears testimony to the efforts made by researchers on all fronts. Mobile systems have greater autonomy than before, and new applications abound — ranging from factory transport systems, airport transport systems, road/vehicular systems, to military applications, automated patrol systems, homeland security surveillance, and rescue operations. While most conventional autonomous systems are self-contained in the sense that all their sensors, actuators, and computers are onboard, it is envisioned that more and more will evolve to become open networked systems with distributed processing power, sensors (e.g., GPS, cameras, microphones, and landmarks), and actuators.

It is generally agreed that an autonomous system consists primarily of the following four distinct yet interconnected modules:

(i) Sensors and Sensor Fusion
(ii) Modeling and Control
(iii) Map Building and Path Planning
(iv) Decision Making and Autonomy

These modules are integrated and influenced by the system architecture design for different applications.

This edited book tries for the first time to provide a comprehensive treatment of autonomous mobile systems, ranging from related fundamental technical issues to practical system integration and applications. The chapters are written by some of the leading researchers and practitioners working in this field today. Readers will be presented with a complete picture of autonomous mobile systems at the systems level, and will also gain a better understanding of the technological and theoretical aspects involved within each module that composes the overall system. Five distinct parts of the book, each consisting of several chapters, emphasize the different aspects of autonomous mobile systems, starting from sensors and control, and gradually moving up the cognitive ladder to planning and decision making, finally ending with the integration of the four modules in application case studies of autonomous systems.

The first part of the book is dedicated to sensors and sensor fusion. The four chapters treat in detail the operation and uses of various sensors that are crucial for the operation of autonomous systems. Sensors provide robots with the capability to perceive the world, and effective utilization is of utmost importance. The chapters also consider various state-of-the art techniques for the fusion and utilization of various sensing information for feature detection and position estimation. Vision sensors, RADAR, GPS and INS, and landmarks are discussed in detail in Chapters 1 to 4 respectively.

Modeling and control issues concerning nonholonomic systems are discussed in the second part of the book. Real-world systems seldom present themselves in the form amenable to analysis as holonomic systems, and the importance of nonholonomic modeling and control is evident. The four chapters of this part, Chapters 5 to 8, thus present novel contributions to the control of these highly complicated systems, focusing on discontinuous control, unified neural fuzzy control, adaptive control with actuator dynamics, and the control of car-like vehicles for vehicle tracking maneuvers, respectively.

The third part of the book covers the map building and path planning aspects of autonomous systems. This builds on technologies in sensing and control to further improve the intelligence and autonomy of mobile robots. Chapter 9 discusses the specifics of building an accurate map of the environment, using either single or multiple robots, with which localization and motion planning can take place. Probabilistic motion planning as a robust and efficient planning scheme is examined in Chapter 10. Action coordination and formation control of multiple robots are investigated in Chapter 11.

Decision making and autonomy, the highest levels in the hierarchy of abstraction, are examined in detail in the fourth part of the book. The three chapters in this part treat in detail the issues of representing knowledge, high level planning, and coordination mechanisms that together define the cognitive capabilities of autonomous systems. These issues are crucial for the development of intelligent mobile systems that are able to reason and manipulate available information. Specifically, Chapters 12 to 14 present topics pertaining to knowledge representation and decision making, algorithms for planning under uncertainties, and the behavior-based coordination of multiple robots.

In the final part of the book, we present a collection of chapters that deal with the system integration and engineering aspects of large-scale autonomous systems. These are usually considered as necessary steps in making new technologies operational and are relatively neglected in the academic community. However, there is no doubt that system integration plays a vital role in the successful development and deployment of autonomous mobile systems. Chapters 15 and 16 examine the issues involved in the design of autonomous commercial robots and automotive systems, respectively. Chapter 17 presents a hierarchical system architecture that encompasses and links the various (higher and lower level) components to form an intelligent, complex system.

We sincerely hope that this book will provide the reader with a cohesive picture of the diverse, yet intimately related, issues involved in bringing about truly intelligent autonomous robots. Although the treatment of the topics is by no means exhaustive, we hope to give the readers a broad-enough view of the various aspects involved in the development of autonomous systems. The authors have, however, provided a splendid list of references at the end of each chapter, and interested readers are encouraged to refer to these references for more information. This book represents the amalgamation of the truly excellent work and effort of all the contributing authors, and could not have come to fruition without their contributions. Finally, we are also immensely grateful to Marsha Pronin, Michael Slaughter, and all others at CRC Press (Taylor & Francis Group) for their efforts in making this project a success.

Editors

Shuzhi Sam Ge, IEEE Fellow, is a full professor with the Electrical and Computer Engineering Department at the National University of Singapore. He earned the B.Sc. degree from the Beijing University of Aeronautics and Astronautics (BUAA) in 1986, and the Ph.D. degree and the Diploma of Imperial College (DIC) from the Imperial College of Science, Technology and Medicine in 1993. His current research interests are in the control of nonlinear systems, hybrid systems, neural/fuzzy systems, robotics, sensor fusion, and real-time implementation. He has authored and co-authored over 200 international journal and conference papers, 3 monographs and co-invented 3 patents. He was the recipient of a number of prestigious research awards, and has been serving as the editor and associate editor of a number of flagship international journals. He is also serving as a technical consultant for the local industry.

Frank L. Lewis, IEEE Fellow, PE Texas, is a distinguished scholar professor and Moncrief-O'Donnell chair at the University of Texas at Arlington. He earned the B.Sc. degree in physics and electrical engineering and the M.S.E.E. at Rice University, the M.S. in Aeronautical Engineering from the University of West Florida, and the Ph.D. at the Georgia Institute of Technology. He works in feedback control and intelligent systems. He is the author of 4 U.S. patents, 160 journal papers, 240 conference papers, and 9 books. He received the Fulbright Research Award, the NSF Research Initiation Grant, and the ASEE Terman Award. He was selected as Engineer of the Year in 1994 by the Fort Worth IEEE Section and is listed in the Fort Worth Business *Press Top 200 Leaders in Manufacturing*. He was appointed to the NAE Committee on Space Station in 1995. He is an elected guest consulting professor at both Shanghai Jiao Tong University and South China University of Technology.

Contributors

Martin Adams
School of Electrical and Electronic
  Engineering
Nanyang Technological University
Singapore

James S. Albus
National Institute of Standards
  and Technology
Gaithersburg, Maryland

Alessandro Astolfi
Electrical and Electronics
  Engineering Department
Imperial College London
London, UK

Stephen Balakirsky
Intelligent Systems Division
National Institute of Standards
  and Technology
Gaithersburg, Maryland

Anthony Barbera
National Institute of Standards
  and Technology
Gaithersburg, Maryland

José A. Castellanos
Instituto de Investigación en
  Ingeniería de Aragón
Universidad de Zaragoza
Zaragoza, Spain

Luiz Chaimowicz
Computer Science Department
Federal University of Minas
Gerais, Brazil

Jingrong Cheng
Department of Electrical Engineering
University of California
Riverside, California

Peng Cheng
Department of Computer Science
University of Illinois
Urbana-Champaign, Illinois

Sesh Commuri
School of Electrical & Computer
  Engineering
University of Oklahoma
Norman, Oklahoma

Jay A. Farrell
Department of Electrical Engineering
University of California
Riverside, California

Rafael Fierro
MARHES Laboratory
School of Electrical & Computer
  Engineering
Oklahoma State University
Norman, Oklahoma

Shuzhi Sam Ge
Department of Electrical and
  Computer Engineering
National University of Singapore
Singapore

Héctor H. González-Baños
Honda Research Institute USA, Inc.
Mountain View, California

Fan Hong
Department of Electrical and
  Computer Engineering
National University of Singapore
Singapore

David Hsu
Department of Computer Science
National University of Singapore
Singapore

Huosheng Hu
Department of Computer Science
University of Essex
Colchester, UK

Chris Jones
Computer Science Department
University of Southern California
Los Angeles, California

Ebi Jose
School of Electrical and Electronic
  Engineering
Nanyang Technological University
Singapore

Vijay Kumar
Department of Mechanical
Engineering and Applied
  Mechanics
University of Pennsylvania
Philadelphia, Pennsylvania

Jean-Claude Latombe
Department of Computer Science
Stanford University
Palo Alto, California

Steven M. LaValle
Department of Computer Science
University of Illinois
Urbana-Champaign, Illinois

Tong Heng Lee
Department of Electrical and
  Computer Engineering
National University of Singapore
Singapore

Frank L. Lewis
Automation and Robotics Research
  Institute
University of Texas
Arlington, Texas

Yu Lu
Department of Electrical Engineering
University of California
Riverside, California

Maja J. Matarić
Computer Science Department
University of Southern California
Los Angeles, California

Elena Messina
Intelligent Systems Division
National Institute of Standards and
  Technology
Gaithersburg, Maryland

Mario E. Munich
Evolution Robotics Inc.
Pasadena, California

José Neira
Instituto de Investigación en
  Ingeniería de Aragón
Universidad de Zaragoza
Zaragoza, Spain

Jason M. O'Kane
Department of Computer Science
University of Illinois
Urbana-Champaign, Illinois

James P. Ostrowski
Evolution Robotics Inc.
Pasadena, California

Michel R. Parent
IMARA Group
INRIA-Rocquencourt
Le Chesnay, France

Stéphane R. Petti
Aisin AW Europe
Braine-L'Alleud, Belgium

Minhtuan Pham
School of Electrical and Electronics
 Engineering
Nanyang Technological University
Singapore

Paolo Pirjanian
Evolution Robotics Inc.
Pasadena, California

Julian Ryde
Department of Computer Science
University of Essex
Colchester, UK

Andrew Shacklock
Singapore Institute of Manufacturing
 Technology
Singapore

Jiali Shen
Department of Computer Science
University of Essex
Colchester, UK

Chun-Yi Su
Department of Mechanical
 Engineering
Concordia University
Montreal, Quebec, Canada

Juan D. Tardós
Instituto de Investigación en
 Ingeniería de Aragón
Universidad de Zaragoza
Zaragoza, Spain

Elmer R. Thomas
Department of Electrical Engineering
University of California
Riverside, California

Benjamín Tovar
Department of Computer Science
University of Illinois
Urbana-Champaign, Illinois

Danwei Wang
School of Electrical and Electronics
 Engineering
Nanyang Technological University
Singapore

Han Wang
School of Electrical and Electronics
 Engineering
Nanyang Technological University
Singapore

Zhuping Wang
Department of Electrical and
 Computer Engineering
National University of Singapore
Singapore

Jian Xu
Singapore Institute of
 Manufacturing Technology
Singapore

Abstract

As technology advances, it has been envisioned that in the very near future, robotic systems will become part and parcel of our everyday lives. Even at the current stage of development, semi-autonomous or fully automated robots are already indispensable in a staggering number of applications. To bring forth a generation of truly autonomous and intelligent robotic systems that will meld effortlessly into the human society involves research and development on several levels, from robot perception, to control, to abstract reasoning.

This book tries for the first time to provide a comprehensive treatment of autonomous mobile systems, ranging from fundamental technical issues to practical system integration and applications. The chapters are written by some of the leading researchers and practitioners working in this field today. Readers will be presented with a coherent picture of autonomous mobile systems at the systems level, and will also gain a better understanding of the technological and theoretical aspects involved within each module that composes the overall system. Five distinct parts of the book, each consisting of several chapters, emphasize the different aspects of autonomous mobile systems, starting from sensors and control, and gradually moving up the cognitive ladder to planning and decision making, finally ending with the integration of the four modules in application case studies of autonomous systems.

This book is primarily intended for researchers, engineers, and graduate students involved in all aspects of autonomous mobile robot systems design and development. Undergraduate students may also find the book useful, as a complementary reading, in providing a general outlook of the various issues and levels involved in autonomous robotic system design.

Contents

I

Sensors and Sensor Fusion

Mobile robots participate in *meaningful and intelligent* interactions with other entities — inanimate objects, human users, or other robots — through sensing and perception. Sensing capabilities are tightly linked to the ability to *perceive*, without which sensor data will only be a collection of meaningless figures. Sensors are crucial to the operation of autonomous mobile robots in unknown and dynamic environments where it is impossible to have complete a priori information that can be given to the robots before operation.

In biological systems, visual sensing offers a rich source of information to individuals, which in turn use such information for navigation, deliberation, and planning. The same may be said of autonomous mobile robotic systems, where vision has become a standard sensory tool on robots. This is especially so with the advancement of image processing techniques, which facilitates the extraction of even more useful information from images captured from mounted still or moving cameras. The first chapter of this part therefore, focuses on the use of visual sensors for guidance and navigation of unmanned vehicles. This chapter starts with an analysis of the various requirements that the use of unmanned vehicles poses to the visual guidance equipment. This is followed by an analysis of the characteristics and limitations of visual perception hardware, providing readers with an understanding of the physical constraints that must be considered in the design of guidance systems. Various techniques currently in use for road and vehicle following, and for obstacle detection are then reviewed. With the wealth of information afforded by various visual sensors, sensor fusion techniques play an important role in exploiting the available information to

1

further improve the perceptual capabilities of systems. This issue is discussed, with examples on the fusion of image data with LADAR information. The chapter concludes with a discussion on the open problems and challenges in the area of visual perception.

Where visual sensing is insufficient, other sensors serve as additional sources of information, and are equally important in improving the navigational and perceptual capabilities of autonomous robots. The use of millimeter wave RADAR for performing feature detection and navigation is treated in detail in the second chapter of this part. Millimeter wave RADAR is capable of providing high-fidelity range information when vision sensors fail under poor visibility conditions, and is therefore, a useful tool for robots to use in perceiving their environment. The chapter first deals with the analysis and characterization of noise affecting the measurements of millimeter wave RADAR. A method is then proposed for the accurate prediction of range spectra. This is followed by the description of a robust algorithm, based on target presence probability, to improve feature detection in highly cluttered environments.

Aside from providing robots with a view of the environment it is immersed in, certain sensors also give robots the ability to analyze and evaluate its own state, namely, its position. Augmentation of such information with those garnered from environmental perception further provides robots with a clearer picture of the condition of its environment and the robot's own role within it. While visual perception may be used for localization, the use of internal and external sensors, like the Inertial Navigation System (INS) and the Global Positioning System (GPS), allows refinement of estimated values. The third chapter of this part treats, in detail, the use of both INS and GPS for position estimation. This chapter first provides a comprehensive review of the Extended Kalman Filter (EKF), as well as the basics of GPS and INS. Detailed treatment of the use of the EKF in fusing measurements from GPS and INS is then provided, followed by a discussion of various approaches that have been proposed for the fusion of GPS and INS.

In addition to internal and external explicit measurements, landmarks in the environment may also be utilized by the robots to get a sense of where they are. This may be done through triangulation techniques, which are described in the final chapter of this part. Recognition of landmarks may be performed by the visual sensors, and localization is achieved through the association of landmarks with those in internal maps, thereby providing position estimates. The chapter provides descriptions and experimental results of several different techniques for landmark-based position estimation. Different landmarks are used, ranging from laser beacons to visually distinct landmarks, to moveable landmarks mounted on robots for multi-robot localization.

This part of the book aims to provide readers with an understanding of the theoretical and practical issues involved in the use of sensors, and the important role sensors play in determining (and limiting) the degree of autonomy mobile robots possess. These sensors allow robots to obtain a basic set of observations upon which controllers and higher level decision-making mechanisms can act upon, thus forming an indispensable link in the chain of modules that together constitutes an intelligent, autonomous robotic system.

1 Visual Guidance for Autonomous Vehicles: Capability and Challenges

Andrew Shacklock, Jian Xu, and Han Wang

CONTENTS

1.1   INTRODUCTION

1.1.1   Context

Current efforts in the research and development of visual guidance technology for autonomous vehicles fit into two major categories: unmanned ground vehicles (UGVs) and intelligent transport systems (ITSs). UGVs are primarily concerned with off-road navigation and terrain mapping whereas ITS (or automated highway systems) research is a much broader area concerned with safer and more efficient transport in structured or urban settings. The focus of this chapter is on visual guidance and therefore will not dwell on the definitions of autonomous vehicles other than to examine how they set the following roles of vision systems:

- Detection and following of a road
- Detection of obstacles
- Detection and tracking of other vehicles
- Detection and identification of landmarks

These four tasks are relevant to both UGV and ITS applications, although the environments are quite different. Our experience is in the development and testing of UGVs and so we concentrate on these specific problems in this chapter. We refer to achievements in structured settings, such as road-following, as the underlying principles are similar, and also because they are a good starting point when facing complexity of autonomy in open terrain.

    This introductory section continues with an examination of the expectations of UGVs as laid out by the Committee on Army Unmanned Ground Vehicle Technology in its 2002 road map [1]. Next, in Section 1.2, we give an overview of the key technologies for visual guidance: two-dimensional (2D) passive imaging and active scanning. The aim is to highlight the differences between various options with regard to our task-specific requirements. Section 1.3 constitutes the main content of this chapter; here we present a visual guidance system (VGS) and its modules for guidance and obstacle detection. Descriptions concentrate on pragmatic approaches adopted in light of the highly complex and uncertain tasks which stretch the physical limitations of sensory systems. Examples are given from stereo vision and image–ladar integration. The chapter ends by returning to the road map in Section 1.4 and examining the potential role of visual sensors in meeting the key challenges for autonomy in unstructured settings: terrain classification and localization/mapping.

1.1.2 Classes of UGV

The motivation or driving force behind UGV research is for military application. This fact is made clear by examining the sources of funding behind prominent research projects. The DARPA Grand Challenge is an immediate example at hand [2]. An examination of military requirements is a good starting point, in an attempt to understand what a UGV is and how computer vision can play a part in it, because the requirements are well defined. Another reason is that as we shall see the scope and classification of UGVs from the U.S. military is still quite broad and, therefore, encompasses many of the issues related to autonomous vehicle technology. A third reason is that the requirements for survivability in hostile environments are explicit, and therefore developers are forced to face the toughest problems that will drive and test the efficacy of visual perception research. These set the much needed benchmarks against which we can assess performance and identify the most pressing problems. The definitions of various UGVs and reviews of state-of-the-art are available in the aforementioned road map [1]. This document is a valuable source for anyone involved in autonomous vehicle research and development because the future requirements and capability gaps are clearly set out. The report categorizes four classes of vehicles with increasing autonomy and perception requirements:

Teleoperated Ground Vehicle (TGV). Sensors enable an operator to visualize location and movement. No machine cognition is needed, but experience has shown that remote driving is a difficult task and augmentation of views with some of the functionality of automatic vision would help the operator. Fong [3] is a good source for the reader interested in vehicle teleoperation and collaborative control.

Semi-Autonomous Preceder–Follower (SAP/F). These devices are envisaged for logistics and equipment carrying. They require advanced navigation capability to minimize operator interaction, for example, the ability to select a traversable path in A-to-B mobility.

Platform-Centric AGV (PC-AGV). This is a system that has the autonomy to complete a task. In addition to simple mobility, the system must include extra terrain reasoning for survivability and self-defense.

Network-Centric AGV (NC-AGV). This refers to systems that operate as nodes in tactical warfare. Their perception needs are similar to that of PC-AGVs but with better cognition so that, for example, potential attackers can be distinguished.

TABLE 1.1
Classes of UGV

Class	kph	Capability gaps	Perception tasks	TRL 6
Searcher (TGV)		All-weather sensors	Not applicable	2006
Donkey (SAP/F)	40	Localization and mapping algorithms	Detect static obstacles, traversable paths	2009
Wingman (PC-AGV)	100	Long-range sensors and sensors for classifying vegetation	Terrain assessment to detect potential cover	2015
Hunter-killer (NC-AGV)	120	Multiple sensors and fusion	Identification of enemy forces, situation awareness	2025

The road map identifies perception as the priority area for development and defines increasing levels of "technology readiness." Some of the requirements and capability gaps for the four classes are summarized and presented in Table 1.1. Technology readiness level 6 (TRL 6) is defined as the point when a technology component has been demonstrated in a relevant environment.

These roles range from the rather dumb donkey-type device used to carry equipment to autonomous lethal systems making tactical decisions in open country. It must be remembered, as exemplified in the inaugural Grand Challenge, that the technology readiness levels of most research is a long way from meeting the most simple of these requirements. The Challenge is equivalent to a simple A-to-B mobility task for the SAP/F class of UGVs. On a more positive note, the complexity of the Grand Challenge should not be understated, and many past research programs, such as Demo III, have demonstrated impressive capability. Such challenges, with clearly defined objectives, are essential for making progress as they bring critical problems to the fore and provide a common benchmark for evaluating technology.

1.2 VISUAL SENSING TECHNOLOGY

1.2.1 Visual Sensors

We first distinguish between passive and active sensor systems: A passive sensor system relies upon ambient radiation, whereas an active sensor system illuminates the scene with radiation (often laser beams) and determines how this is reflected by the surroundings. Active sensors offer a clear advantage in outdoor applications; they are less sensitive to changes in ambient conditions. However, some applications preclude their use; they can be detected by the enemy in military scenarios, or there may be too many conflicting sources in a civilian setting. At this point we also highlight a distinction between the terms "active vision" and "active sensors." Active vision refers to techniques in which (passive) cameras are moved so that they can fixate on particular features [4]. These have applications in robot localization, terrain mapping, and driving in cluttered environments.

1.2.1.1 Passive imaging

From the application and performance standpoint, our primary concern is procuring hardware that will acquire good quality data for input to guidance algorithms; so we now highlight some important considerations when specifying a camera for passive imaging in outdoor environments.

The image sensor (CCD or CMOS). CMOS technology offers certain advantages over the more familiar CCDs in that it allows direct access to individual blocks of pixels much as would be done in reading computer memory. This enables instantaneous viewing of regions of interest (ROI) without the integration time, clocking, and shift registers of standard CCD sensors. A key advantage of CMOS is that additional circuitry can be built into the silicon which leads to improved functionality and performance: direct digital output, reduced blooming, increased dynamic range, and so on. Dynamic range becomes important when viewing outdoor scenes with varying illumination: for example, mixed scenes of open ground and shadow.

Color or monochrome. Monochrome (B&W) cameras are widely used in lane-following systems but color systems are often needed in off-road (or country track) environments where there is poor contrast in detecting traversable terrain. Once we have captured a color image there are different methods of representing the RGB components: for example, the RGB values can be converted into hue, saturation, and intensity (HSI) [5]. The hue component of a surface is effectively invariant to illumination levels which can be important when segmenting images with areas of shadow [6,7].

Infrared (IR). Figure 1.1 shows some views from our semi-urban scene test circuit captured with an IR camera. The hot road surface is quite distinct as are metallic features such as manhole covers and lampposts. Trees similarly contrast well against the sky but in open country after rainfall, different types of vegetation and ground surfaces exhibit poor contrast. The camera works on a different transducer principle from the photosensors in CCD or CMOS chips. Radiation from hot bodies is projected onto elements in an array that heat up, and this temperature change is converted into an electrical signal. At present, compared to visible light cameras, the resolution is reduced (e.g., 320×240 pixels) and the response is naturally slower. There are other problems to contend with, such as calibration and drift of the sensor. IR cameras are expensive

FIGURE 1.1   A selection of images captured with an IR camera. The temperature of surfaces gives an alternative and complementary method of scene classification compared to standard imaging. Note the severe lens distortion.

and there are restrictions on their purchase. However, it is now possible to install commercial night-vision systems on road vehicles: General Motors offers a thermal imaging system with head-up display (HUD) as an option on the Cadillac DeVille. The obvious application for IR cameras is in night driving but they are useful in daylight too, as they offer an alternative (or complementary) way of segmenting scenes based on temperature levels.

Catadioptric cameras. In recent years we have witnessed the increasing use of catadioptric[1] cameras. These devices, also referred to as omnidirectional, are able to view a complete hemisphere with the use of a parabolic mirror [8]. Practically, they work well in structured environments due to the way straight lines project to circles. Bosse [9] uses them indoors and outdoors and tracks the location of vanishing points in a structure from motion (SFM) scheme.

1.2.1.2   Active sensors

A brief glimpse through robotics conference proceedings is enough to demonstrate just how popular and useful laser scanning devices, such as the ubiquitous SICK, are in mobile robotics. These devices are known as LADAR and are

---

[1] Combining reflection and refraction; that is, a mirror and lens.

available in 2D and 3D versions but the principles are essentially similar: a laser beam is scanned within a certain region; if it reflects back to the sensor off an obstacle, the time-of-flight (TOF) is measured.

2D scanning. The majority of devices used on mobile robots scan (pan) through 180° in about 13 msec at an angular resolution of 1°. Higher resolution is obtained by slowing the scan, so at 0.25° resolution, the scan will take about 52 msec. The sensor thus measures both range and bearing $\{r, \theta\}$ of obstacles in the half plane in front of it. On a moving vehicle the device can be inclined at an angle to the direction of travel so that the plane sweeps out a volume as the vehicle moves. It is common to use two devices: one pointing ahead to detect obstacles at a distance (max. range ~80 m); and one inclined downward to gather 3D data from the road, kerb, and nearby obstacles. Such devices are popular because they work in most conditions and the information is easy to process. The data is relatively sparse over a wide area and so is suitable for applications such as localization and mapping (Section 1.4.2). A complication, in off-road applications, is caused by pitching of the vehicle on rough terrain: this creates spurious data points as the sensor plane intersects the ground plane. Outdoor feature extraction is still regarded as a very difficult task with 2D ladar as the scan data does not have sufficient resolution, field-of-view (FOV), and data rates [10].

3D scanning. To measure 3D data, the beam must be steered though an additional axis (tilt) to capture spherical coordinates $\{r, \theta, \phi$: range, pan, tilt$\}$. There are many variations on how this can be achieved as an opto-electromechanical system: rotating prisms, polygonal mirrors, or galvono-metric scanners are common. Another consideration is the order of scan; one option is to scan vertically and after each scan to increment the pan angle to the next vertical column. As commercial 3D systems are very expensive, many researchers augment commercial 2D devices with an extra axis, either by deflecting the beam with an external mirror or by rotating the complete sensor housing [11].

It is clear that whatever be the scanning method, it will take a protracted length of time to acquire a dense 3D point cloud. High-resolution scans used in construction and surveying can take between 20 and 90 min to complete a single frame, compared to the 10 Hz required for a real-time navigation system [12]. There is an inevitable compromise to be made between resolution and frame rate with scanning devices. The next generation of ladars will incorporate flash technology, in which a complete frame is acquired simultaneously on a focal plane array (FPA). This requires that individual sensing elements on the array incorporate timing circuitry. The current limitation of FLASH/FPA is the number of pixels in the array, which means that the FOV is still small, but this can be improved by panning and tilting of the sensor between subframes, and then creating a composite image.

In summary, ladar offers considerable advantages over passive imaging but there remain many technical difficulties to be overcome before they can meet the tough requirements for vehicle guidance. The advantages are:

- Unambiguous 3D measurement over wide FOV and distances
- Undiminished night-time performance and tolerance to adverse weather conditions

The limitations are:

- Relatively high cost, bulky, and heavy systems
- Limited spatial resolution and low frame rates
- Acquisition of phantom points or multiple points at edges or permeable surfaces
- Active systems may be unacceptable in certain applications

The important characteristics to consider, when selecting a ladar for a guidance application, are: angular resolution, range accuracy, frame rate, and cost. An excellent review of ladar technology and next generation requirements is provided by Stone at NIST [12].

1.2.2 Modeling of Image Formation and Calibration

1.2.2.1 The ideal pinhole model

It is worthwhile to introduce the concept of projection and geometry and some notation as this is used extensively in visual sensing techniques such as stereo and structure from motion. Detail is kept to a minimum and the reader is referred to standard texts on computer vision for more information [13–15]. The standard pinhole camera model is adopted, while keeping in mind the underlying assumptions and that it is an ideal model. A point in 3D space $\{\tilde{X} \in \mathbb{R}^3\}$ projects to a point on the 2D image plane $\{\tilde{x} \in \mathbb{R}^2\}$ according to the following equation:

$$\mathbf{x} = \mathbf{P}X \colon \mathbf{P} \in \mathbb{R}^{3 \times 4} \tag{1.1}$$

This equation is linear because we use homogeneous coordinates by augmenting the position vectors with a scalar ($X = [\tilde{X}^\mathrm{T} \ 1]^\mathrm{T} \in \mathbb{R}^4$) and likewise the image point ($\mathbf{x} = [x \ y \ w]^\mathrm{T} \in \mathbb{R}^3 \colon \tilde{x} = \mathbf{x}/w$). A powerful and more natural way of treating image formation is to consider the ray model as an example of projective space. $\mathbf{P}$ is the projection matrix and encodes the position of the camera and its intrinsic parameters. We can rewrite (1.1) as:

$$x = K[R \quad T]\tilde{X}: K \in \mathbb{R}^{3 \times 3}, \quad R \in SO(3), \quad T \in \mathbb{R}^3 \qquad (1.2)$$

Internal (or intrinsic) parameters. These are contained in the calibration matrix K, which can be parameterized by: focal length (f), aspect ratio (α), skew (s), and the location of the offset of the principal point in the image $\{u_0, v_0\}$.

$$K = \begin{pmatrix} f & s & u_0 \\ 0 & \alpha f & v_0 \\ 0 & 0 & 1 \end{pmatrix} \qquad (1.3)$$

External (or extrinsic) parameters. These are the orientation and position of the camera with respect to the reference system: R and T in Equation 1.2.

1.2.2.2 Calibration

We can satisfy many vision tasks working with image coordinates alone and a projective representation of the scene. If we want to use our cameras as measurement devices, or if we want to incorporate realistic dynamics in motion models, or to fuse data in a common coordinate system, we need to upgrade from a projective to Euclidean space: that is, calibrate and determine the parameters. Another important reason for calibration is that the wide-angle lenses, commonly used in vehicle guidance, are subject to marked lens distortion (see Figure 1.1); without correction, this violates the assumptions of the ideal pinhole model.

A radial distortion factor is calculated from the coefficients $\{k_i\}$ and the radial distance (r) of a pixel from the center $\{x_p, y_p\}$.

$$\delta(r) = 1 + k_1 r^2 + k_2 r^4: r = ((\tilde{x}_d - x_p)^2 + (\tilde{y}_d - y_p)^2)^{0.5} \qquad (1.4)$$

The undistorted coordinates are then

$$\{\tilde{x} = (\tilde{x}_d - x_p)\delta + x_p, \tilde{y} = (\tilde{y}_d - y_p)\delta + y_p\} \qquad (1.5)$$

Camera calibration is needed in a very diverse range of applications and so there is wealth of reference material available [16,17]. For our purposes, we distinguish between two types or stages of calibration: linear and nonlinear.

1. Linear techniques use a least-squares type method (e.g., SVD) to compute a transformation matrix between 3D points and their 2D projections. Since the linear techniques do not include any lens distortion model, they are quick and simple to calculate.

2. Nonlinear optimization techniques account for lens distortion in the camera model through iterative minimization of a determined function. The minimizing function is usually the distance between the image points and modeled projections.

In guidance applications, it is common to adopt a two-step technique: use a linear optimization to compute some of the parameters and, as a second step, use nonlinear iteration to refine, and compute the rest. Since the result from the linear optimization is used for the nonlinear iteration, the iteration number is reduced and the convergence of the optimization is guaranteed [18–20]. Salvi [17] showed that two-step techniques yield the best result in terms of calibration accuracy.

Calibration should not be a daunting prospect because many software tools are freely available [21,22]. Much of the literature originated from photogrammetry where the requirements are much higher than those in autonomous navigation. It must be remembered that the effects of some parameters, such as image skew or the deviation of the principal point, are insignificant in comparison to other uncertainties and image noise in field robotics applications. Generally speaking, lens distortion modeling using a radial model is sufficient to guarantee high accuracy, while more complicated models may not offer much improvement.

A pragmatic approach is to carry out much of the calibration off-line in a controlled setting and to fix (or constrain) certain parameters. During use, only a limited set of the camera parameters need be adjusted in a calibration routine. Caution must be employed when calibrating systems *in situ* because the information from the calibration routine must be sufficient for the degrees of freedom of the model. If not, some parameters will be confounded or wander in response to noise and, later, will give unpredictable results. A common problem encountered in field applications is attempting a complete calibration off essentially planar data without sufficient and general motion of the camera between images. An *in situ* calibration adjustment was adopted for the calibration of the IR camera used to take the images of Figure 1.1. The lens distortion effects were severe but were suitably approximated and corrected by a two-coefficient radial distortion model, in which the coefficients (k_1, k_2) were measured off-line. The skew was set to zero; the principal point and aspect ratio were fixed in the calibration matrix. The focal length varied with focus adjustment but a default value (focused at infinity) was measured. Of the extrinsic parameters, only the tilt of the camera was an unknown in its application: the other five were set by the rigid mounting fixtures. Once mounted on the vehicle, the tilt was estimated from the image of the horizon. This gave an estimate of the camera calibration which was then improved given extra data. For example, four known points are sufficient to calculate the homographic mapping from ground plane to the image. However, a customized calibration routine was used that enforced the constraints and the physical degrees of freedom of the camera, yet was stable enough to work from data on the ground plane alone. As a final note on calibration: any routine should also provide quantified estimates of the uncertainty of the parameters determined.

1.3 VISUAL GUIDANCE SYSTEMS

1.3.1 Architecture

The modules of a working visual guidance system (VGS) are presented in Figure 1.2. So far, we have described the key sensors and sensor models. Before delving into task-specific processes, we need to clarify the role of VGS within the autonomous vehicle system architecture. Essentially, its role is to capture raw sensory data and convert it into model representations of the environment and the vehicle's state relative to it.

1.3.2 World Model Representation

A world model is a hierarchical representation that combines a variety of sensed inputs and a priori information [23]. The resolution and scope at each level are designed to minimize computational resource requirements and to support planning functions for that level of the control hierarchy. The sensory processing system that populates the world model fuses inputs from multiple sensors and extracts feature information, such as terrain elevation, cover, road edges, and obstacles. Feature information from digital maps, such as road networks, elevation, and hydrology, can also be incorporated into this rich world model. The various features are maintained in different layers that are registered together to provide maximum flexibility in generation of vehicle plans depending on mission requirements. The world model includes occupancy grids and symbolic object representations at each level of the hierarchy. Information at different hierarchical levels has different spatial and temporal resolution. The details of a world model are as follows:

Low resolution obstacle map and elevation map. The obstacle map consists of a 2D array of cells [24]. Each cell of the map represents one of the following situations: traversable, obstacle (positive and negative), undefined (such as blind spots), potential hazard, and so forth. In addition, high-level terrain classification results can also be incorporated in the map (long grass or small bushes, steps, and slopes). The elevation contains averaged terrain heights.

Mid-resolution terrain feature map. The features used are of two types, smooth regions and sharp discontinuities [25].

A priori information. This includes multiple resolution satellite maps and other known information about the terrain.

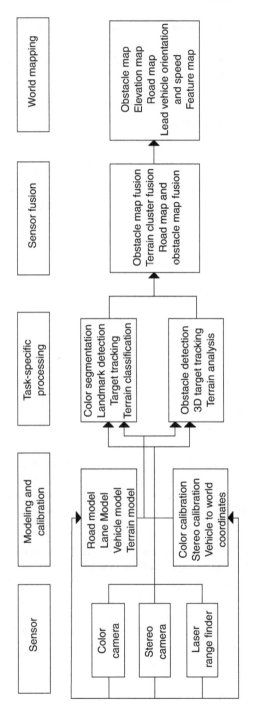

FIGURE 1.2   Architecture of the VGS.

Model update mechanism. As the vehicle moves, new sensed data inputs can either replace the historical ones, or a map-updating algorithm can be activated.

We will see real examples of occupancy grids in Section 1.5.3 and Section 1.3.6 (Figure 1.8 and Figure 1.9).

1.3.3 Physical Limitations

We now examine the performance criteria for visual perception hardware with regards to the classes of UGVs. Before we even consider algorithms, the physical realities of the sensing tasks are quite daunting. The implications must be understood and we will demonstrate with a simple analysis. A wide FOV is desirable so that there is a view of the road in front of the vehicle at close range. The combination of lens focal length (f) and image sensor dimensions (H, V) determine the FOV and resolution. For example, a 1/2" sensor has image dimensions ($H = 6.4$ mm, $V = 4.8$ mm). The angle of view (horizontally) is approximated by

$$\theta_H = 2\arctan\frac{H}{2f} \tag{1.6}$$

and it is easily calculated that a focal length of 5 mm will equate to an angle of view of approximately 65° with a sensor of this size. It is also useful to quote a value for the angular resolution; for example, the number of pixels per degree. With an output of 640 × 480 pixels, the resolution for this example is approximately 10 pixels per degree (or 1.75 mrad/pixel).

Now consider the scenario of a UGV progressing along a straight flat road and that it has to avoid obstacles of width 0.5 m or greater. We calculate the pixel size of the obstacle, at various distances ahead, for a wide FOV and a narrow FOV, and also calculate the time it will take the vehicle to reach the obstacle. This is summarized in Table 1.2.

TABLE 1.2

Comparison of Obstacle Image Size for Two Fields-of-View and Various Distances to the Object

Distance, d (m)	Obstacle size (pixel)		Time to cover distance (sec)		
	FOV 60°	FOV 10°	120 kph	60 kph	20 kph
8	35	113	0.24	0.48	1.44
20	14	45	0.6	1.2	3.6
50	5.6	18	1.5	3	9
300	0.9	3	9	18	54

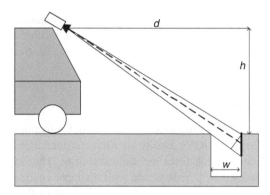

FIGURE 1.3   The ability of a sensor to image a negative obstacle is affected by the sensor's height, resolution, and the size of the obstacle. It is very difficult to detect holes until the vehicle is within 10 m.

It can be observed from Table 1.2 that:

1. The higher the driving speed, the further the camera lookahead distance should be to give sufficient time for evasive action. For example, if the system computation time is 0.2 sec and the mechanical latency is 0.5 sec, a rough guideline is that at least 50 m warning is required when driving at 60 kph.
2. At longer lookahead distances, there are fewer obstacle pixels in the image — we would like to see at least ten pixels to be confident of detecting the obstacle. A narrower FOV is required so that the obstacle can be seen.

A more difficult problem is posed by the concept of a negative obstacle: a hole, trench, or water hazard. It is clear from simple geometry and Figure 1.3 that detection of trenches from imaging or range sensing is difficult. A trench is detected as a discontinuity in range data or the disparity map. In effect we only view the projection of a small section of the rear wall of the trench: that is, the zone bounded by the rays incident with the forward and rear edges.

We conclude from Table 1.3 that with a typical camera mounting height of 2.5 m, a trench of width 1 m will not be reliably detected at a distance of 15 m, assuming a minimum of 10 pixels are required for negative obstacle detection. This distance is barely enough for a vehicle to drive safely at 20 kph. The situation is improved by raising the camera; at a height of 4 m, the ditch will be detected at a distance of 15 m. Alternatively, we can select a narrow FOV lens. For example, a stereo vision system with FOV ($15°H \times 10°V$) is able to

TABLE 1.3
Influence of Camera Height on Visibility of Negative Obstacles

Distance, d (m)	Visibility of negative obstacle (pixels) trench width $w = 1$ m	
	Camera height $h = 2.5$ m	Camera height $h = 4$ m
8	21 (0.31 m)	35 (0.5 m)
15	6.8 (0.17 m)	11 (0.27 m)
25	2.5 (0.1 m)	4 (0.16 m)

cover a width of 13 m at distance 25 m and possibly detect a ditch $\{w = 1$ m, $h = 4$ m$\}$ by viewing 8 pixels of the ditch.

There are several options for improving the chances of detecting an obstacle:

Raising the camera. This is not always an option for practical and operational reasons; for example, it makes the vehicle easier to detect by the enemy.

Increasing focal length. This has a direct effect but is offset by problems with exaggerated image motion and blurring. This becomes an important consideration when moving over a rough terrain.

Increased resolution. Higher-resolution sensors are available but they will not help if a sharp image cannot be formed by the optics, or if there is image blur.

The trade-off between resolution and FOV is avoided (at extra cost and complexity) by having multiple sensors. Figure 1.4 illustrates the different fields-of-view and ranges of the sensors on the VGS. Dickmanns [26,27], uses a mixed focal system comprising two wide-angle cameras with divergent axes, giving a wide FOV (100°). A high-resolution three-chip color camera with greater focal length is placed between the other cameras for detecting objects at distance. The overlapping region of the cameras' views give a region of trinocular stereo.

1.3.4 Road and Vehicle Following

1.3.4.1 State-of-the-art

Extensive work has been carried out on road following systems in the late 1980s and throughout the 1990s; for example, within the PROMETHEUS Programme which ran from 1987 until 1994. Dickmanns [28] provides a comprehensive review of the development of machine vision for road vehicles. One of the key tasks is lane detection, in which road markings are used to monitor the position

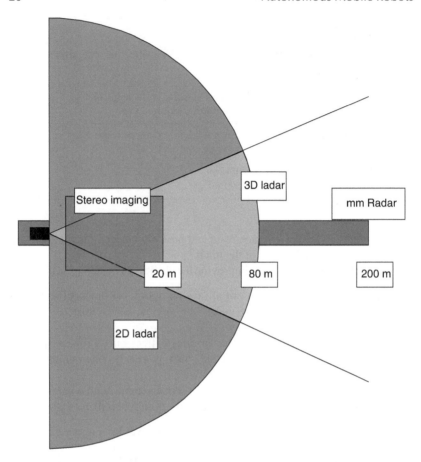

FIGURE 1.4 Different subsystems of the VGS provide coverage over different field-of-view and range. There is a compromise between FOV and angular resolution. The rectangle extending to 20 m is the occupancy grid on which several sensory outputs are fused.

of the vehicle relative to the road: either for driver assistance/warning or for autonomous lateral control. Lane detection is therefore a relatively mature technology; a number of impressive demonstrations have taken place [29], and some systems have achieved commercial realization such as Autovue and AssistWare. There are, therefore, numerous sources of reference where the reader can find details on image processing algorithms and details of practical implementation. Good places to start are at the PATH project archives at UCLA, the final report of Chauffeur II programme [30], or the work of Broggi on the Argo project [29].

The Chauffeur II demonstration features large trucks driving in convoy on a highway. The lead vehicle is driven manually and other trucks equipped with the system can join the convoy and enter an automatic mode. The system incorporates lane tracking (lateral control) and maintaining a safe distance to the vehicle in front (longitudinal control). This is known as a "virtual tow-bar" or "platooning." The Chauffeur II demonstration is highly structured in the sense that it was implemented on specific truck models and featured inter-vehicle communication. Active IR patterns are placed on the rear of the vehicles to aid detection, and radar is also used. The PATH demonstration (UCLA, USA) used stereo vision and ladar. The vision system tracks features on a car in front and estimates the range of an arbitrary car from passive stereo disparity. The ladar system provides assistance by guiding the search space for the vehicle in front and increasing overall robustness of the vision system. This is a difficult stereo problem because the disparity of features on the rear of car is small when viewed from a safe driving separation. Recently much of the research work in this area has concentrated on the problems of driving in urban or cluttered environments. Here, there are the complex problems of dealing with road junctions, traffic signs, and pedestrians.

1.3.4.2 A road camera model

Road- and lane-following algorithms depend on road models [29]. These models have to make assumptions such as: the surface is flat; road edges or markings are parallel; and the like. We will examine the camera road geometry because, with caution, it can be adapted and applied to less-structured problems. For simplicity and without loss of generality, we assume that the road lies in the plane $Z = 0$. From Equation 1.1, setting all Z coordinates of X to zero is equivalent to striking out the third column of the projection matrix $\mathbf{P}$ in Equation 1.2. There is a homographic correspondence between the points of the road plane and the points of the image plane which can be represented by a 3×3 matrix transformation. This homography is part of the general linear group GL_3 and as such inherits many useful properties of this group. The projection Equation 1.1 becomes

$$\mathbf{x} = HX\colon \; H \in \mathbb{R}^{3 \times 3} \qquad (1.7)$$

As a member of the group, a transformation H must[2] have an inverse, so there will also be one-to-one mapping of image points (lines) to points (lines) on the road plane. The elements of H are easily determined (calibration) by finding at least four point correspondences in general position on

---

[2] The exception to this is when the road plane passes through the camera center, in which case H is singular and noninvertible (but in this case the road would project to a single image line and the viewpoint would not be of much use).

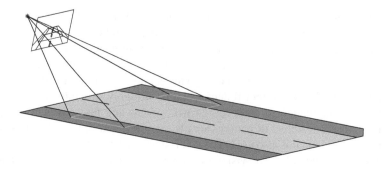

FIGURE 1.5   The imaging of planar road surface is represented by a one-to-one invertible mapping. A rectangular search region projects to a trapezoidal search region in the image.

the planes.[3] The homography can be expressed in any valid representation of the projective space: that is, we can change the basis to match the camera coordinate system. This means that the road does not have to be the plane $Z = 0$ but can be an arbitrary plane in 3D; the environment can be modeled as a set of discrete planes Π_i each with a homography H_i that maps it to the image plane.

In practice we use the homography to project a search region onto the image; a rectangular search space on the road model becomes a trapezoid on the image (Figure 1.5). The image is segmented, within this region, into road and nonroad areas. The results are then projected onto the occupancy grid for fusion with other sensors. Care must be taken because 3D obstacles within the scene may become segmented in the image as driveable surfaces and because they are "off the plane," their projections on the occupancy grid will be very misleading. Figure 1.6 illustrates this and some other important points regarding this use of vision and projections to and from the road surface. Much information within the scene is ignored; the occupancy gird will extend to about 20 m in front of the vehicle but perspective effects such as vanishing points can tell us a lot about relative direction, or be used to anticipate events ahead. The figure also illustrates that, due to the strong perspective, the uncertainty on the occupancy grid will increase rapidly as the distance from the vehicle increases. (This is shown in the figure as the regular spaced [2 m] lane markings on the road rapidly converge to a single pixel in the image.) Both of these considerations suggest that an occupancy grid is convenient for fusing data but

---

[3] Four points give an exact solution; more than four can reduce the effects of noise using least squares; known parameters of the projection can be incorporated in a nonlinear technique. When estimating the coefficients of a homography, principles of calibration as discussed in Section 4.2.2.2 apply. Further details and algorithms are available in Reference 13.

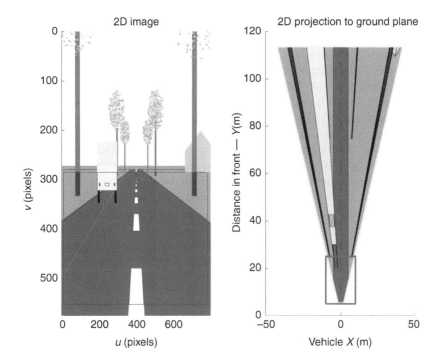

FIGURE 1.6   The image on the left is of a road scene and exhibits strong perspective which in turn results in large differences in the uncertainty of reprojected measurements. The figure on the right was created by projecting the lower 300 pixels of the image onto a model of the ground plane. The small box $(20 \times 20 \text{ m}^2)$ represents the extent of a typical occupancy grid used in sensor fusion.

transformation to a metric framework may not be the best way to represent visual information.

1.3.5   Obstacle Detection

1.3.5.1   Obstacle detection using range data

The ability to detect and avoid obstacles is a prerequisite for the success of the UGV. The purpose of obstacle detection is to extract areas that cannot or should not be traversed by the UGV. Rocks, fences, trees, and steep upward slopes are some typical examples. The techniques used in the detection of obstacles may vary according to the definition of "obstacle." If "obstacle" means a vehicle or a human being, then the detection can be based on a search for specific patterns, possibly supported by feature matching. For unstructured terrain, a more general definition of obstacle is any object that can obstruct the vehicle's driving path or, in other words, anything rising out significantly from the road surface.

Many approaches for extracting obstacles from range images have been proposed. Most approaches use either a global or a local reference plane to detect positive (above the reference plane) or negative (below the reference plane) obstacles. It is also possible to use salient points detected by an elevation differential method to identify obstacle regions [31]. The fastest of obstacle detection algorithms, range differencing, simply subtract the range image of an actual scene from the expected range image of a horizontal plane (global reference plane). While rapid, this technique is not very robust, since mild slopes will result in false indications of obstacles. So far the most frequently used and most reliable solutions are based on comparison of 3D data with local reference planes. Thorpe et al. [22] analyzed scanning laser range data and constructed a surface property map represented in a Cartesian coordinate system viewed from above, which yielded the surface type of each point and its geometric parameters for segmentation of the scene map into traversable and obstacle regions. The procedure includes the following.

Preprocessing. The input from a 2D ladar may contain unreliable range data resulting from surfaces such as water or glossy pigment, as well as the mixed points at the edge of an object. Filtering is needed to remove these undesirable jumps in range. After that, the range data are transformed from angular to Cartesian (x-y-z) coordinates.

Feature extraction and clustering. Surface normals are calculated from x-y-z points. Normals are clustered into patches with similar normal orientations. Region growth is used to expand the patches until the fitting error is larger than a given threshold. The smoothness of a patch is evaluated by fitting a surface (plane or quadric).

Defect detection. Flat, traversable surfaces will have vertical surface normals. Obstacles will have surface patches with normals pointed in other directions.

Defect analysis. A simple obstacle map is not sufficient for obstacle analysis. For greater accuracy, a sequence of images corresponding to overlapping terrain is combined in an extended obstacle map. The analysis software can also incorporate color or curvature information into the obstacle map.

Extended obstacle map output. The obstacle map with a header (indicating map size, resolution, etc.) and a square, 2D array of cells (indicating traversability) are generated for the planner.

1.3.5.2 Stereo vision

Humans exploit various physiological and psychological depth cues. Stereo cameras are built to mimic one of the ways in which the human visual system

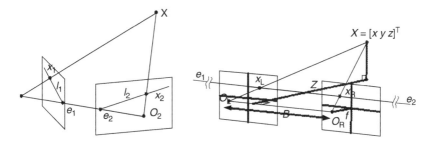

FIGURE 1.7 Epipolar geometry is valid for general positions of two views. The figure on the left illustrates the epipolar lines for two frames (1 and 2). However, if the optical axes are parallel and the camera parameters are similar, stereo matching or the search for corresponding features is much easier. The figure on the right illustrates the horizontal and collinear epipolar lines in a left–right configuration with fixed baseline B.

(HVS) works to obtain depth information [32]. In a standard configuration, two cameras are bound together with a certain displacement (Figure 1.7). This distance between the two camera centers is called the baseline B. In stereo vision, the disparity measurement is the difference in the positions of two corresponding points in the left and right images. In the standard configuration, the two camera coordinate systems are related simply by the lateral displacement B ($X_R = X_L + B$). As the cameras are usually "identical" ($f_L = f_R = f$) and aligned such that ($Z_L = Z_R = Z$) the epipolar geometry and projection equation ($x = f X/Z$) enable depth Z to be related to disparity d:

$$d = x_R - x_L = f\frac{X_L + B}{Z} - f\frac{X_L}{Z} = f\frac{B}{Z} \qquad (1.8)$$

where f is the focal length of the cameras. Since B and F are constants, the depth z can be calculated when d is known from stereo matching ($Z = fB/d$).

1.3.5.2.1 Rectification
As shown in Figure 1.7, for a pair of images, each point in the "left" image is restricted to lie on a given line in the "right" image, the epipolar line — and vice versa. This is called the epipolar constraint. In standard configurations the epipolar lines are parallel to image scan lines, and this is exploited in many algorithms for stereo analysis. If valid, it enables the search for corresponding image features to be confined to one dimension and, hence, simplified. Stereo rectification is a process that transforms the epipolar lines so that they are collinear, and both parallel to the scan line. The idea behind rectification [33] is to define two new perspective matrices which preserve the optical centers but with image planes parallel to the baseline.

1.3.5.2.2 Multi-baseline stereo vision

Two main challenges facing a stereo vision system are: mismatch (e.g., points in the left image match the wrong points in the right image) and disparity accuracy. To address these issues, multiple (more than two) cameras can be used. Nakamura et al. [34] used an array of cameras to resolve occlusion by introducing occlusion masks which represent occlusion patterns in a real scene. Zitnick and Webb [35] introduced a system of four cameras that are horizontally displaced and analyze potential 3D surfaces to resolve the feature matching problem.

When more than two cameras or camera locations are employed, multiple stereo pairs (e.g., cameras 1 and 2, cameras 1 and 3, cameras 2 and 3, etc.) result in multiple, usually different baselines. In the parallel configuration, each camera is a lateral displacement of the other. For a given depth, we then calculate the respective expected disparities relative to a reference camera (say, the leftmost camera) as well as the sum of match errors over all the cameras. The depth associated with a given pixel in the reference camera is taken to be the one with the lowest error. The multi-baseline approach has two distinctive advantages over the classical stereo vision [36]:

- It can find a unique match even for a repeated pattern such as the cosine function.
- It produces a statistically more accurate depth value.

1.3.5.2.3 General multiple views

During the 1990s significant research was carried out on multiple view geometry and demonstrating that 3D reconstruction is possible using uncalibrated cameras in general positions [14]. In visual guidance, we usually have the advantage of having calibrated cameras mounted in rigid fixtures so there seems little justification for not exploiting the simplicity and speed of the algorithms described earlier. However, the fact that we can still implement 3D vision even if calibration drifts or fixtures are damaged, adds robustness to the system concept. Another advantage of more general algorithms is that they facilitate mixing visual data from quite different camera types or from images taken from arbitrary sequences in time.

1.3.5.3 Application examples

In this section we present some experimental results of real-time stereo-vision-based obstacle detection for unstructured terrain. Two multi-baseline stereo vision systems (Digiclops from Pointgrey Research, 6 mm lens) were mounted at a height of 2.3 m in front and on top of the vehicle, spaced 20 cm apart. The two stereo systems were calibrated so that their outputs were referred to

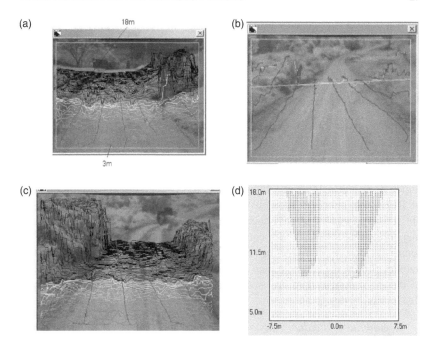

FIGURE 1.8 (a) Isodisparity profile lines generated from the disparity map using a LUT method. (b) A single isodisparity line (curved line), its reference line (straight) and detected obstacle pixels. (c) Detected obstacle points. (d) Obstacle map.

the same vehicle coordinate system. A centralized triggering signal was generated for the stereo systems and other sensors to synchronize the data capturing. The stereo systems were able to generate disparity maps at a frequency of 10 frames/sec. To detect obstacles, an isodisparity profile-based obstacle detection method was introduced [37], which converted the 3D obstacle detection into 1D isodisparity profile segmentation. The system output was an obstacle map with 75 × 75 elements, each representing a 0.2 m × 0.2 m area within 15 m × 15 m in front of the vehicle. Seventy-five isodisparity profiles were generated from the disparity map using a look-up-table method (Figure 1.8a). The name isodisparity comes from the fact that points in each profile line have the same disparity value. Regardless of the size of the disparity map (usually 320 × 240 pixels), the method was able to identify 75 × 75 points from the disparity image, which exactly matched the elements of the obstacle map. By processing these 75 × 75 points using reference-line-based histogram classification, obstacle points were detected with subpixel accuracy. Figure 1.8a shows the profiles of a typical test terrain with road and bushes. Figure 1.8b shows the calculated reference lines. It is noteworthy that the reference lines form a curved surface instead of a planar surface used by other approaches. The final obstacle detection result and map are displayed in Figure 1.8c and d, respectively.

1.3.6 Sensor Fusion

The most important task of a VGS is to provide accurate terrain descriptions for the path planner. The quality of terrain maps is assessed by miss rate and false alarm. Here, the miss rate refers to the occurrence frequency of missing a true obstacle while a false alarm is when the VGS classifies a traversable region as an obstacle region. Imaging a stereo vision system with a frame rate of 10 Hz will generate 3000 obstacle maps in 5 min. Even with a successful classification rate of 99.9%, the system may produce an erroneous obstacle map three times of which may cause an error in path planning. The objective of sensor fusion is to combine the results from multiple sensors, either at the raw data level or at the obstacle map level, to produce a more reliable description of the environment than any sensor individually. Some examples of sensor fusion are:

N-modular redundancy fusion: Fusion of three identical radar units can tolerate the failure of one unit.

Fusion of complementary sensors: Color terrain segmentation results can be used to verify 3D terrain analysis results.

Fusion of competitive sensors: Although both laser and stereo vision perform obstacle detection, their obstacle maps can be fused to reduce false alarms.

Synchronization of sensors: Different sensors have different resolutions and frame rates. In addition to calibrating all sensors using the same vehicle coordinates, sensors need to be synchronized both temporally and spatially before their results can be merged. Several solutions can be applied for sensor synchronization.

An external trigger signal based synchronization: For sensors with external trigger capability such as IR, color, and stereo cameras, their data capturing can be synchronized by a hardware trigger signal from the control system of the UGV. For laser or ladar, which do not have such capability, the data captured at the time nearest to the trigger signal are used as outputs. In this case, no matter how fast a laser scanner can scan (usually 20 frames/sec), the fusion frame rate depends on the slowest sensor (usually stereo vision, around 10 frames/sec).

A centralized time stamp for each image from each sensor: In this case sensors capture data as fast as they can. Since each sensor normally has its own CPU for data processing, a centralized control system will send out a standardized time stamp signal to all CPUs regularly (say, every 1 h) to minimize the time stamp drift.

When sensor outputs are read asynchronously, certain assumptions such as being Linear Time Invariant (LTI) [38] can be made to propagate asynchronized data to the upcoming sample time of the control system. Robl [38] showed examples of using first-order hold and third-order hold methods to predict sensor values at desired times. When different resolution sensors are to be fused at the data level (e.g., fusion of range images from ladar and stereo vision), down-sampling of sensor data with higher spatial resolution by interpolation is performed. For sensor fusion at the obstacle map level, spatial synchronization is not necessary since a unique map representation is defined for all sensors.

Example: Fusion of laser and stereo obstacle maps for false alarm suppression

Theoretically, pixel to pixel direct map fusion is possible if the calibration and synchronization of the geometrical constraints (e.g., rotation and translation between laser and stereo system) remain unchanged after calibration. Practically, however, this is not realistic, partially due to the fact that sensor synchronization is not guaranteed at all times: CPU loading, terrain differences, and network traffic for the map output all affect the synchronization. Feature-based co-registration sensor fusion, alternatively, addresses this issue by computing the best-fit pose of the obstacle map features relative to multiple sensors which allows refinement of sensor-to-sensor registration. In the following, we propose a localized correlation based approach for obstacle-map-level sensor fusion. Assuming the laser map L_{ij} and stereo map S_{ij} is to be merged to form F_{ij}. A map element takes the value 0 for a traversable pixel, 1 for an obstacle, and anything between 0 and 1 for the certainty of the pixel to be classified as an obstacle. We formulate the correlation-based sensor fusion as

$$F_{ij} = \begin{cases} L_{ij} & S_{ij} = \text{undefined} \\ S_{ij} & L_{ij} = \text{undefined} \\ (a_1 L_{ij} + a_2 S_{i+m,j+n})/(a_1 + a_2) & \max(\text{Corr}(L_{ij}S_{i+m,j+n})) \quad m,n \in \Omega \\ \text{undefined} & S_{ij}, L_{ij} = \text{undefined} \end{cases}$$

$$(1.9)$$

where Ω represents a search area and $\{a_1, a_2\}$ are weighting factors. $\text{Corr}(L, S)$ is the correlation between L and S elements with window size w_c:

$$\text{Corr}(L_{ij}S_{i+m,j+n}) = \sum_{p=-w_c/2}^{w_c/2} \sum_{q=-w_c/2}^{w_c/2} L_{i+p,j+q}S_{i+m+p,j+n+q} \qquad (1.10)$$

The principle behind the localized correlation sensor fusion is: instead of directly averaging L_{ij} and S_{ij} to get F_{ij}, a search is performed to find the best match within a small neighborhood. The averaging of the center pixel at a matched point produces the final fusion map.

In case an obstacle map only takes three values: obstacle, traversable, and undefined; the approach above can be simplified as

$$F_{ij} = \begin{cases} L_{ij} & S_{ij} = \text{undefined} \\ S_{ij} & L_{ij} = \text{undefined} \\ 1 & L_{ij} = 1, C_{so} > T_1, D < T_2 \\ 1 & S_{ij} = 1, C_{lo} > T_1, D < T_2 \\ 0 & \text{otherwise} \end{cases} \qquad (1.11)$$

where T_1 and T_2 are preset thresholds that depend on the size of the search window. In our experiments a window of size 5×5 pixels was found to work well. The choice of size is a compromise between noise problems with small windows and excessive boundary points with large windows. C_{so} and C_{lo} are obstacle pixel counts within the comparison window w_c, for L_{ij} and S_{ij}, respectively, D is the minimum distance between L_{ij} and S_{ij} in Ω:

$$D = \min \left(\sum_{p=-w_c/2}^{w_c/2} \sum_{q=-w_c/2}^{w_c/2} |S_{i+m+p,j+n+q} - L_{i+p,j+q}| \right) \quad (m,n) \in \Omega \quad (1.12)$$

$$C_{so} = \sum_{p=-w_c/2}^{w_c/2} \sum_{q=-w_c/2}^{w_c/2} S_{i+m+p,j+n+q} \qquad (1.13)$$

The advantage of implementing correlation-based fusion method is two-fold: it reduces false alarm rates and compensates for the inaccuracy from laser and stereo calibration/synchronization. The experimental results of using above mentioned approach for laser and stereo obstacle map fusion are shown in Figure 1.9.

The geometry of 2D range and image data fusion. Integration of sensory data offers much more than a projection onto an occupancy grid. There exist multiple view constraints between image and range data analogous to those between multiple images. These constraints help to verify and disambiguate data from either source, so it is useful to examine the coordinate transformations and the physical parameters that define them. This will also provide a robust framework for selecting what data should be fused and in which geometric representation.

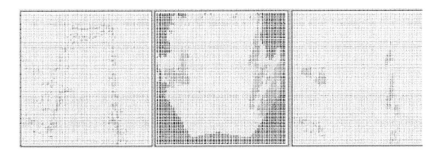

FIGURE 1.9 Sensor fusion of laser and stereo obstacle maps. False alarm in laser obstacle map (left image, three laser scanning lines at the top of the map), is suppressed by fusion with the stereo vision obstacle map (middle image), and a more reliable fusion result is generated (right image).

First, consider the relationship between a data point from the ladar and a world coordinate system. We can transform $\{r, \theta\}$ to a point X in a Cartesian space. A 3D point X will be detected by an ideal ladar if it lies in the plane $\Pi_{Z=0}$ expressed in the sensor's coordinate system. (This is neglecting the range limits, and the finite size and divergence of the laser beam). If the plane, in the world coordinate system, is denoted as Π_L, the set of points that can be detected satisfy

$$\Pi_L^T X = 0 \tag{1.14}$$

Alternatively we expand the rigid transformation equation and express this as a constraint (in sensor coordinates)

$$X_L = G_L^W X \qquad G_L^W = \begin{pmatrix} R_L^W & T \\ \mathbf{0} & 1 \end{pmatrix} \tag{1.15}$$

Only the third row of G $[r_{3i} \; T_Z]$ plays any part in the planar constraint on the point $\{\mathbf{X} = [X \; Y \; Z \; 1]^T\}$. The roles of the parameters are then explicit:

$$r_{31}X + r_{32}Y + r_{33}Z + T_Z = 0 \tag{1.16}$$

However, if the vehicle is moving over tough terrain there will be considerable uncertainty in the instantaneous parameters of R and T. We therefore look at a transformation between ladar data and image data without reference to any world coordinate system. Assuming there are no occlusions, X will be imaged as x on the image plane Π_I of the camera. As X lies in a plane Π_L, there exists a homography H (abbreviated from H_I^L ladar to image) that maps X to x.

$$x = HX: H \in R^{3 \times 3} \qquad\qquad (1.17)$$

This mapping is unambiguous and is parameterized by the geometry between the two sensors which is less uncertain than the geometry with reference to a world coordinate system. H can be solved from point correspondences and if required it can be decomposed into the geometric parameters relating the two planes.

The reverse mapping is not unambiguous: a point x is the image of the ray passing through x and the optical center O_C. We can map x (with H^{-1}) to a single point $p\ \{r,\ \theta\}$ in the laser parameter space but there is no guarantee that the true 3D point that gave rise to x in the image came from this plane. Another consideration is that image-ladar correspondences are rarely point-to-point but line-to-point. (ladar data rarely comes from a distinct point in 3D; it is more likely to have come from a set of points such as a vertical edge or the surface of a tree.) Consider the image of the pole shown in Figure 1.10; the pre-image of this is a plane, and so the image line could be formed from an infinite set of lines (a pencil) in this plane. However, knowledge of the laser point p, constrains the 3D space line to the pencil of lines concurrent with X. Furthermore, assuming that the base of the image line corresponds to the ground plane is sufficient to define a unique space line. There are various ways to establish mappings between the two types of sensors without reliance on *a priori* parameters with their associated uncertainties.

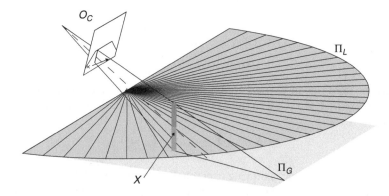

FIGURE 1.10   There is ambiguity in both ladar and imaging data. There are geometric constraints between the sets of data that will assist in disambiguation and improving reliability of both systems.

One of the key problems in processing ladar data is data association. For example, consider capturing data from a tree. The points that are detected depend on the viewpoint: that is, surface features are not pose invariant [10]. This problem becomes easier with the use of a putative model of the tree whose 2D position is determined by a centroid, which is invariant. Such a model is easier to initiate if image data provides the evidence that the data points match image features with the correct "tree-like" attributes. Once we have a model we can anticipate where to search for features to match data points and vice-versa. In this case we want to compare the real data with a model prediction but this has to be very efficient given the large amount of data and hypotheses that will occur. A typical problem is to test if a model patch will be detected by a sensor, and how many data points to expect. Range detection is equivalent to ray intersection and is more easily solved after projection into a 2D space: a cylindrical projection is sufficient and preserves the topology.

To summarize, in isolation there is much ambiguity in either sensor, and exchanging information using image constraints can reduce this problem. The difficulty is how to implement this practically as the concept of "being like a tree" is more abstract than the neat formulation of raw data fusion as seen in Section 1.3.6. This lack of precise mathematical formulation and reliance on heuristic rules deters many researchers. However, recent advances and increased processing speeds have made probabilistic reasoning techniques tractable and worthy of consideration in real-time problems such as visual guidance and terrain assessment.

1.4 Challenges and Solutions

The earlier sections have detailed many of the practical difficulties of visually based guidance and presented pragmatic techniques used during field demonstrations. To be realistic, autonomous vehicles represent a highly complex set of problems and current capability is still at the stage of the SAP/F "donkey" engaged in A-to-B mobility. To extend this capability, researchers need to think further along the technology road map [1] and tackle perception challenges such as: terrain mapping, detection of cover, classification of vegetation, and the like.

1.4.1 Terrain Classification

Obstacle detection based only on distance information is not sufficient. Long grass or small bushes will also be detected as obstacles because of their height. However, the vehicle could easily drive through these "soft" obstacles. Alternatively, soft vegetation can cover a dangerous slope but appear as a traversable surface. To reduce unnecessary avoidance driving, detected obstacles need to be classified as "dangerous" or "not dangerous." Color cameras can be used to perform terrain classification. Color segmentation relies on having a complete training set. As lighting changes, due to time of day or weather conditions, the appearance of grass and obstacle change as well. Although color normalization methods have been successfully applied to the indoor environment, they, to our knowledge, fail to produce reasonable results in an outdoor environment. Similarly, color segmentation can classify flat objects, such as fallen leaves, as obstacles, since their color is different from grass.

If dense range measurements in a scene are available (e.g., using ladar), they can be used, not only to represent the scene geometry, but also to characterize surface types. For example, the range measured on bare soil or rocks tends to lie on a relatively smooth surface; in contrast, in the case of bushes, the range is spatially scattered. While it is possible — although by no means trivial — to design algorithms for terrain classification based on the local statistics of range data [39–41], the confidence level of a reliable classification is low. Table 1.4 lists the most frequently encountered terrain types and possible classification methods.

1.4.2 Localization and 3D Model Building from Vision

Structure from motion (SFM) is the recovery of camera motion and scene structures — and in certain cases camera intrinsic parameters — from image

---

TABLE 1.4
Terrain Types and Methods of Classification

Terrain type	Sensors	Classification methods	Confidence level
Vegetable	IR/Color camera	Segmentation	Medium
Rocks	IR/Color camera	Segmentation	Medium
Walls/fence	Camera, stereo, laser	Texture analysis, obstacle detection	High
Road (paved, gravel, dirt)	IR/Color camera	Segmentation	Medium
Slope	Stereo, ladar	Elevation analysis, surface fit	High
Ditch, hole	Stereo, ladar		Low
Sand, dirt, mud, gravel	IR/Color camera	Segmentation	Medium
Water	Polarized camera, laser scanner	Feature detection, sensor fusion	Medium
Moving target	Camera, stereo	Optical flow, obstacle detection, pattern matching	High

sequences. It is attractive because it avoids the requirement for a priori models of the environment. The techniques are based on the constraints that exist between the multiple views of features. This is a mature area of computer vision that has attracted intensive research activity in the previous decade, prompted by the breakthroughs in multiple view geometry in the early 1990s. Much of the original work was motivated by mobile robotics but soon found more general application such as: the generation of special effects for cinema, scene recovery for virtual reality, and 3D reconstruction for architecture. Here, the theoretical drive has been inspired by the recovery of information from recorded sequences such as camcorders where the motion is general and little can be assumed regarding the camera parameters. These tasks can be accomplished off-line and the features and camera parameters from long sequences solved as a large-scale optimization in batch mode. As such, many would regard this type of SFM as a solved problem but the conditions in vehicle navigation are specific and require separate consideration:

- The motion is not "general," it may be confined to a plane, or restricted to rotations around axes normal to the plane.
- Navigation is required in real-time and parameters require continuous updating from video streams as opposed to the batch operations of most SFM algorithms.
- Sensory data, from sources other than the camera(s), are usually available.
- Many of the camera parameters are known (approximately) beforehand.
- There are often multiple moving objects in a scene.

Visual guidance demands a real-time recursive SFM algorithm. Chiuso et al. [42] have impressive demonstrations of a recursive filter SFM system that works at a video frame rate of 30 Hz. However, once we start using Kalman filters to update estimates of vehicle (camera) state and feature location, some would argue that we enter the already very active realm of simultaneous localization and mapping (SLAM). The truth is that there are differences between SLAM and SFM and both have roles in visual guidance. Davison [43] has been very successful in using vision in a SLAM framework and Bosse [9] has published some promising work in indoor and outdoor navigation. The key to both of these is that they tackle a fundamental problem of using vision in SLAM: the relatively narrow FOV and recognizing features when revisiting a location. Davison used active vision in Reference 4 and wide-angle lenses in Reference 43 to fixate on a sparse set of dominant features whereas Bosse used a catadioptric camera and exploited vanishing points. SLAM often works well with 2D ladar by collecting and maintaining estimates of a sparse set of features with reference to world coordinate system. A problem with SFM occurs when features used for reference pass out of the FOV: in recursive mode, there is no guarantee at initiation that features will persist. Errors (drift) are introduced when the reference features are changed and the consequence is that a robot will have difficulty in returning home or knowing that it is revisiting a location. Chiuso has a scheme to reduce this problem but drift is still inevitable. On the other hand, SLAM has to rely on sparse data because it needs to maintain a full covariance matrix which will soon become computationally expensive if the number of data points is not restricted. It can be difficult to associate outdoor data when it is sparse.

The two techniques offer different benefits and a possible complementary role. SLAM is able to maintain a sparse map on a large scale for navigation but locally does not help much with terrain classification. SFM is useful for building a dense model of the immediate surroundings, useful for obstacle avoidance, path planning, and situation awareness. The availability of a 3D model (with texture and color) created by SFM will be beneficial for validation of the sensory data used in a SLAM framework: for example, associating an object type with range data; providing color (hue) as an additional state; and so on.

1.5 CONCLUSION

We have presented the essentials of a practical VGS and provided details on its sensors and capabilities such as road following, obstacle detection, and sensor fusion. Worldwide, there have been many impressive demonstrations of visual guidance and certain technologies are so mature that they are available commercially.

This chapter started with a road map for UGVs and we have shown that the research community is still struggling to achieve A-to-B mobility in tasks within large-scale environments. This is because navigating through open terrain is a highly complex problem with many unknowns. Information from the immediate surroundings is required to determine traversable surfaces among the many potential hazards. Vision has a role in the creation of terrain maps but we have shown that practically this is still difficult due to the physical limitations of available sensor technology. We anticipate technological advances that will enable the acquisition of high-resolution 3D data at fast frame rates.

Acquiring large amounts of data is not a complete solution. We argue that we do not make proper use of the information already available in 2D images, and that there is potential for exploiting algorithms such as SFM and vision-based SLAM. Another problem is finding alternative ways of representing the environment that are more natural for navigation; or how to extract knowledge from images and use this (state) information within algorithms.

We have made efforts to highlight problems and limitations. The task is complex and practical understanding is essential. The only way to make real progress along the road map is through testing sensors, systems, and algorithms in the field; and then seeing what can survive the challenges presented.

ACKNOWLEDGMENTS

The authors would like to acknowledge the support of A* STAR and DSTA (Singapore) in funding project activities that have contributed to the findings presented in this chapter.

REFERENCES

1. Committee on Army Unmanned Ground Vehicle Technology National Research Council. *Technology Development for Army Unmanned Ground Vehicles*. National Academy Press, Washington, 2002.
2. J. Kuamgai. Sand trap. *IEEE Spectrum*, 41: 34–40, 2004.
3. T. W. Fong, C. Thorpe, and C. Baur. Advanced interfaces for vehicle teleoperation: collaborative control, sensor fusion displays, and remote driving tools. *Autonomous Robots*, 11: 77–85, 2001.
4. A. J. Davison. *Mobile Robot Navigation Using Active Vision*. PhD thesis, Department of Engineering Science, University of Oxford, Oxford, UK, February 1998.
5. R. C. Gonzalez and R. E. Woods. *Digital Image Processing*. Addison-Wesley, Reading, MA, 2nd edition, 1992.
6. N. Zeng and J. D. Crisman. Categorical color projection for robot road following. In *Proceedings of 12th International Conference on Robotics and Automation*, pp. 1080–1085, 1995.
7. J. D. Crisman and C. E. Thorpe. Scarf: a color vision system that tracks roads and intersections. *IEEE Transactions on Robotics and Automations*, 9: 49–58, 1993.
8. C. Geyer and K. Daniilidis. A unifying theory for central panoramic systems and practical applications. In *ECCV (2)*, pp. 445–461. Springer-Verlag, Heidelberg, 2000.
9. M. Bosse, R. J. Rikoski, J. J. Leonard, and S. J. Teller. Vanishing points and three-dimensional lines from omni-directional video. *The Visual Computer*, 19: 417–430, 2003.
10. F. Tang, M. D. Adams, J. Ibanez-Guzman, and W. S. Wijesoma. Pose invariant, robust feature extraction from range data with a modified scale space approach. In *Proceedings of IEEE International Conference on Robotics and Automation*, New Orleans, LA, USA, April 2004.
11. T. C. Ng and J. C. Tan. Development of a 3D ladar system for autonomous navigation. In *Proceedings of IEEE Conference on Robotics, Automation and Mechatronics (RAM 04)*, Singapore, pp. 792–797, 1–3 December 2004.
12. W. C. Stone, M. Juberts, N. Dagalakis, J. Stone, and J. Gorman. Performance analysis of next generataion ladar for manufacturing, construction and mobility.

Technical Report NISTIR 7117, NIST, Building and Fire Research Laboratory and Manufacturing Engineering Laboratory, Maryland, USA, May 2004.

13. R. I. Hartley and A. Zisserman. *Multiple View Geometry in Computer Vision*. Cambridge University Press, London, New York, ISBN: 0521623049, 2000.

14. O. Faugeras and Q.-T. Luong. *The Geometry of Multiple Images*. MIT, Cambridge, MA, 1999.

15. Y. Ma, S. Soatto, J. Kosecka, and S. Sastry. *An Invitation to 3D Vision: From Images to Geometric Models*. Springer-Verlag, Heidelberg, 2003.

16. C. C. Slama. *Manual of Photogrammetry*. 4th edition, American Society of Photogrammetry, Falls Church, VA, 1980.

17. J. Salvi, X. Armangu, and J. Batlle. A comparative review of camera calibrating methods with accuracy evaluation. *Pattern Recognition*, 35: 1617–1635, 2002.

18. R. Y. Tsai. An efficient and accurate camera calibration technique for 3D machine vision. In *Proceedings of IEEE Conference on Computer Vision and Pattern Recognition*, Miami Beach, Florida, pp. 364–374, 1986.

19. J. Weng, P. Cohen, and M. Herniou. Camera calibration with distortion models and accuracy evaluation. *IEEE Transactions on Pattern Analysis and Machine Intelligence*, 14: 965–980, 1992.

20. J. Heikkila and O. Silven. A four-step camera calibration procedure with implicit image correction. In *Proceedings of the 1997 Conference on Computer Vision and Pattern Recognition (CVPR '97)*, San Juan, Puerto Rica, pp. 1106–1112, IEEE Computer Society, Washington, June 1997.

21. J.-Y. Bouguet. A camera calibration toolbox for matlab.
http://www.vision.caltech.edu/bouguetj/calib_doc/.

22. Intel. The open cv library.
http://sourceforge.net/projects/opencvlibrary/.

23. H. Tsai, S. B. Balakirsky, E. Messina, T. Chang, and M. Schneier. A hierarchical world model for an autonomous scout vehicle. *Proceedings of SPIE*, 4715: 343–354, July 2002.

24. M. Hebert, C. Thorpe, and A. Stentz. *Intelligent Unmanned Ground Vehicles: Autonomous Navigation Research at Carnegie Mellon*. Kluwer Academic Publishers, Dordrecht, 1997.

25. M. Hebert, T. Kanade, and I. Kweon. 3D vision techniques for autonomous vehicles CMU-RI-TR-88-12. Technical Report, Robotics Institute, Carnegie Mellon University, August 1988.

26. E. D. Dickmanns. An advanced vision system for ground vehicles. In *First International Workshop on In-Vehicle Cognitive Computer Vision Systems (IVCCVS)*, Graz, Austria, pp. 1–12, 3 April 2003.

27. R. Gregor, M. Lutzeler, M. Pellkofer, K.-H. Siedersberger, and E. D. Dickmanns. EMS-vision: a perceptual system for autonomous vehicles. *IEEE Transactions on Intelligent Transportation Systems*, 3: 48–59, 2002.

28. E. D. Dickmanns. The development of machine vision for road vehicles in the last decade. In *Procceedings of IEEE Intelligent Vehicle Symposium, IV2002*, Verailles, France, vol. 1, pp. 268–281, 17–21 June 2002.

29. A. Broggi, M. Bertozzi, A. Fascioli, and G. Conte. *Automatic Vehicle Guidance: The Experience of the ARGO Autonomous Vehicle.* World Scientific, 1999.
30. Chauffeur II Final Report. Technical Report IST-1999-10048 D24, The Promote-Chauffeur II Consortium, July 2003.
31. T. Chang, H. Tsai, S. Legowik, and M. N. Abrams. Concealment and obstacle detection for autonomous driving. In M. H. Hamza (ed.), *Proceedings of IASTED Conference, Robotics and Applications 99*, pp. 147–152. ACTA Press, Santa Barbara, CA, 28–30 October 1999.
32. T. Okoshi. *Three-Dimensional Imaging Techniques.* Academic Press, New York, 1976.
33. N. Ayache. *Artificial Vision for Mobile Robots: Stereo Vision and Multisensory Perception.* The MIT Press, Cambridge, MA, 1991.
34. Y. Nakamura, K. Satoh, T. Matsuura, and Y. Ohta. Occlusion detectable stereo-occlusion patterns in camera matrix. In *Proceedings of IEEE International Conference on Computer Vision and Pattern Recognition*, San Francisco, CA, pp. 371–378, 18–20 June 1996.
35. C. L. Zitnick and J. A. Webb. Multi-baseline stereo using surface extraction. Technical Report, School of Computer Science, Carnegie Melllon University, November 1996. CMU-CS-96-196.
36. M. Okutomi and T. Kanade. A multiple-baseline stereo. *IEEE Transactions on Pattern Analysis and Machine Intelligence*, 15: 353–363, 1993.
37. J. Xu, H. Wang, J. Ibanez-Guzman, T. C. Ng, J. Shen, and C. W. Chan. Isodisparity profile processing for real-time 3D obstacle identification. In *Proceedings of IEEE International Conference on Intelligent Transportation Systems (ITS)* Shanghai, China, pp. 288–292, October 2003.
38. C. Robl and G. Faerber. System architecture for synchronizing, signal level fusing, simulating and implementing sensors. In *Proceedings of the 2000 IEEE International Conference on Robotics and Automation*, San Francisco, CA, pp. 1639–1644, April 2000.
39. J. Huang, A. B. Lee, and D. Mumford. Statistics of range images. In *Proceedings of IEEE Conference on Computer Vision and Pattern Recognition*, vol. 1, pp. 324–331, 2000.
40. M. Hebert, N. Vandapel, S. Keller, and R. R. Donamukkala. Evaluation and comparison of terrain classification techniques from ladar data for autonomous navigation. In *Twentythird Army Science Conference*, Orlando, FL, USA, December 2002.
41. N. Vandapel, D. F. Huber, A. Kapuria, and M. Hebert. Natural terrain classification using 3D ladar data. In *Proceedings of IEEE International Conference on Robotics and Automation*, New Orleans, LA, USA, Vol. 5, pp. 5117–5122, 26 April–1 May 2004.
42. A. Chiuso, P. Favaro, H. L. Jin, and S. Soatto. Structure from motion causally integrated over time. *IEEE Transactions on Pattern Analysis and Machine Intelligence*, 24: 523–535, 2002.
43. A. J. Davison, Y. González Cid, and N. Kita. Real-time 3D SLAM with wide-angle vision. In *5th IFAC/EURON Symposium on Intelligent Autonomous Vehicles*, Lisboa, Portugal. IFAC, Elsevier Science, 5–7 July 2004.

BIOGRAPHIES

Andrew Shacklock has 20 years of experience in mechatronics and sensor guided robotic systems. He graduated with a B.Sc. from the University of Newcastle Upon Tyne in 1985 and a Ph.D. from the University of Bristol in 1994. He is now a research scientist at the Singapore Institute of Manufacturing Technology. His main research interest is in machine perception and sensor fusion, in particular for visual navigation.

Jian Xu received the bachelor of engineering degree and master engineering degree in electrical engineering from Shanghai Jiao Tong University, China in 1982 and 1984, respectively. He received his Doctor of Engineering from Erlangen-Nuremberg University, Germany in 1992. He is currently a research scientist at the Singapore Institute of Manufacturing Technology, Singapore. His research interests include 3D machine vision using photogrammetry and stereo vision, camera calibration, sensor fusion, subpixeling image processing, and visual guidance system for autonomous vehicle.

Han Wang is currently an associate professor at Nanyang Technological University, and senior member of IEEE. His research interests include computer vision and AGV navigation. He received his bachelor of engineering degree from Northeast Heavy Machinery Institute in 1982 and Ph.D. from Leeds University in 1989. He has been a research scientist at CMU and research officer at Oxford University. He spent his sabbatical in 1999 in Melbourne University.

2 Millimeter Wave RADAR Power-Range Spectra Interpretation for Multiple Feature Detection

Martin Adams and Ebi Jose

CONTENTS

2.1 INTRODUCTION

Current research in autonomous robot navigation [1,2] focuses on mining, planetary-exploration, fire emergencies, battlefield operations, as well as on agricultural applications. Millimeter wave (MMW) RADAR provides consistent and accurate range measurements for the environmental imaging required to navigate in dusty, foggy, and poorly illuminated environments [3]. MMW RADAR signals can provide information of certain distributed targets that appear in a single line-of-sight observation. This work is conducted with a 77-GHz frequency modulated continuous wave (FMCW) RADAR which operates in the MMW region of the electromagnetic spectrum [4,5].

For localization and map building, it is necessary to predict the target locations accurately given a prediction of the vehicle/RADAR location [6,7]. Therefore, the first contribution of this chapter offers a method for predicting the power–range spectra (or range bins) using the RADAR range equation and knowledge of the noise distributions in the RADAR. The predicted range bins are to be used ultimately as predicted observations within a mobile robot RADAR-based navigation formulation. The actual observations take the form of received power/range readings from the RADAR.

The second contribution of this chapter is an algorithm which makes optimal estimates of the range to multiple targets down-range, for each range spectra based on received signal-to-noise power. We refer to this as feature detection based on target presence probability. Results are shown which compare probability-based feature detection with other feature extraction techniques such as constant threshold [9] on raw data and constant false alarm rate (CFAR) techniques [24]. The results show the merit of the proposed algorithm which can detect features in typically cluttered outdoor environments with a higher success rate compared to other feature detection techniques. This work is a step toward robust outdoor robot navigation with MMW-RADAR-based continuous power spectra.

Millimeter wave RADAR can penetrate certain nonmetallic objects, meaning that multiple line-of-sight objects can sometimes be detected, a property which can be exploited in mobile robot navigation in outdoor unstructured environments. This chapter describes a new approach in predicting RADAR range bins which is essential for simultaneous localization and map building (SLAM) with MMW RADAR.

The third contribution of this chapter is a SLAM formulation using an augmented state vector which includes the normalized RADAR cross sections (RCS) and absorption cross sections of features as well as the usual feature Cartesian coordinates. The term "normalized" is used as the actual RCS is incorporated into a reflectivity parameter. Normalization results, as it is assumed that the sum of this reflectivity parameter and the absorption and transmittance parameters is unity. This is carried out to provide feature-rich representations of the environment to significantly aid the data association process in SLAM.

The final contribution is a predictive model of the range bins, from differing vehicle locations, for multiple line-of-sight targets. This forms a predicted power–range observation, based on estimates of the augmented SLAM state. The formulation of power returns from multiple objects down-range is derived and predicted RADAR range spectra are compared with real spectra, recorded outdoors. This prediction of power–range spectra is a step toward a full, RADAR-based SLAM framework.

Section 2.2 summarizes related work, while Section 2.3 describes FMCW RADAR operation and the noise affecting the range spectra, in order to understand the noise distributions in both range and power. Section 2.4 describes how power–range spectra can be predicted (predicted observations). This utilizes the RADAR range equation and an experimental noise analysis. Section 2.5 analyzes a feature detector based on the CFAR detection method. The study also shows ways to compensate for the inaccuracies of the power–range compensating high-pass filter, contained in FMCW RADARs, and thereby improve the feature detection process. A method for estimating the true range to objects from power–range spectra is given in Section 2.6 in the form of a new robust feature detection technique based on target presence probability. Section 2.6.1 shows the merits of the target presence probability-based algorithm which can detect ground level features with greater reliability than other feature detection techniques such as constant threshold on raw RADAR data and CFAR techniques. An augmented state vector is introduced in Section 2.8 where, along with the vehicle and feature positions, normalized RCS and absorption cross sections of features are added together with the RADAR losses. Finally, Section 2.9 shows full predicted range spectra and the results are compared with the measured range bins in the initial stages of a simple SLAM formulation.

2.2 RELATED WORK

In recent years RADAR, for automotive purposes, has attracted a great deal of interest in shorter range (<200 m) applications. Most of the work in short-range RADAR has focused on millimeter waves as this allows narrow beam shaping, which is necessary for higher angular resolution [5]. Some of the work to date in autonomous navigation using MMW RADAR is summarized here.

Boehmke et al. [8] succeeded in producing three-dimensional (3D) terrain maps using a pulsed RADAR with a narrow beam of 1° and high sampling rate. The 1° RADAR beam width has a large antenna sweep volume and its physical size is large for robotic applications. The efforts by Boehmke et al. show the compromise between a narrow beam and antenna size, where a narrow beam provides better angular resolution.

Steve Clark [9] presented a method for fusing RADAR readings from different vehicle locations into a two-dimensional (2D) representation. The method selects one range point per RADAR observation at a particular bearing angle based on a certain received signal power threshold level. This method takes only one range reading per bin which is the nearest power return to exceed that threshold to the RADAR, discarding all others. Clark [10] shows a MMW-RADAR-based navigation system which utilizes artificial beacons for localization and an extended Kalman filter for fusing multiple observations. The fixed threshold can be used when the environment is known with no clutter.[1] However, in a realistic environment (containing features having various RCS) fixed thresholding on raw data will cause an exorbitant number of false alarms if the threshold is low or missed detections if the threshold is too high. Manual assistance is required in adjusting the threshold as the returned signal power depends on various objects' RCS. This method of feature detection is environment-dependent.

Foessel [11] shows the usefulness of evidence grids for integrating uncertain and noisy sensor information. Foessel et al. [12] show the development of a RADAR sensor model for certainty grids and also demonstrate the integration of RADAR observations for building 3D outdoor maps. Certainty grids divide the area of interest into cells, where each cell stores a probabilistic estimate of its state [13,14]. The proposed 3D model by Foessel et al. has shortcomings such as the necessity of rigorous probabilistic formulation and difficulties

---

[1] Clutter in this research is assumed to be the backscatter from land and is difficult to model. Land clutter is dependent on the type of terrain, its roughness, and dielectric properties.

in representing dependencies due to occlusion. Jose and Adams [15] show a method of feature detection from MMW RADAR noisy data.

2.3 FMCW RADAR Operation and Range Noise

This section gives a brief introduction to the RADAR sensor used in this work and the FMCW technique for obtaining target range. This is necessary for RADAR signal interpretation and for understanding and quantifying the noise in the range/power estimates. This is ultimately used in predicting range bin observations given the predicted vehicle state, in a mobile robot navigation framework — which is one of the goals of this chapter. By analyzing the FMCW technique it will be shown which noise sources affect both the range and received power estimates, and how each of these is affected.

The RADAR unit (from Navtech Electronics) is a 77-GHz FMCW system. The transmitted power is 15 dBm and the swept bandwidth is 600 MHz [16]. The RADAR is shown in Figure 2.1, mounted on a four-wheel steerable vehicle. Figure 2.2 shows a schematic block diagram of an FMCW RADAR transceiver. In Figure 2.2, the input voltage to the voltage control oscillator (VCO) is

FIGURE 2.1   A 360° scanning MMW RADAR mounted on a vehicle test bed for SLAM experiments within the NTU campus.

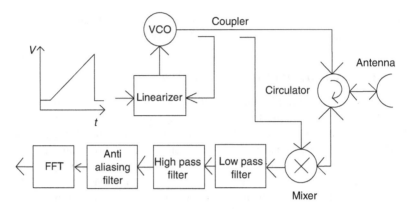

FIGURE 2.2   Schematic block diagram of a MMW RADAR transceiver.

a ramp signal. The VCO generates a signal of linearly increasing frequency δf in the frequency sweep period T_d. This linearly increasing chirp signal is transmitted via the antenna. An FMCW RADAR measures the distance to an object by mixing the received signal with a portion of the transmitted signal [17].

Let the transmitted signal $v_T(t)$ as a function of time, t, be represented as

$$v_T(t) = [A_T + a_T(t)] \cos \left[\omega_c t + A_b \int_0^t t \, dt + \phi(t) \right]$$

$$= [A_T + a_T(t)] \cos \left[\omega_c t + \frac{A_b}{2} t^2 + \phi(t) \right] \tag{2.1}$$

where A_T is the amplitude of the carrier signal, A_b is the amplitude of the modulating signal, ω_c is the carrier frequency (i.e., $2\pi \times 77$ GHz), $a_T(t)$ is the amplitude noise, and $\phi(t)$ is the phase noise present in the signal which occurs inside the transmitting electronic sections.

At any instant of time, the received echo signal, v_R is shifted in time from the transmitted signal by a round trip time, τ. The received signal is

$$v_R(t - \tau) = [A_R + a_R(t - \tau)] \cos \left[\omega_c(t - \tau) + \frac{A_b}{2}(t - \tau)^2 + \phi(t - \tau) \right] \tag{2.2}$$

where A_R is the received signal amplitude, $a_R(t - \tau)$ is the amplitude noise, and $\phi(t - \tau)$ is the phase noise. The sources of noise affecting the signal's amplitude consist of external interference to the RADAR system (e.g., atmospheric noise, man-made interference signals) and internally produced noise at the receiver antenna and amplifiers in the system.

In the mixer, the received signal is mixed with a portion of the transmitted signal with an analog multiplier.

$$v_T(t)v_R(t - \tau) = [A_T + a_T(t)][A_R + a_R(t - \tau)]$$

$$\times \left\{ \cos \left[\omega_c t + \frac{A_b}{2}t^2 + \phi(t) \right] \right\}$$

$$\times \left\{ \cos \left[\omega_c(t - \tau) + \frac{A_b}{2}(t - \tau)^2 + \phi(t - \tau) \right] \right\} \quad (2.3)$$

The output of the mixer, $v_{out}(t)$ is (using the trigonometric identity for the product of two sine waves $\cos A \cos B = 0.5[\cos(A + B) + \cos(A - B)]$)

$$v_{out}(t - \tau) = \frac{[A_T + a_T(t)][A_R + a_R(t - \tau)]}{2}[B_1 + B_2] \quad (2.4)$$

where $B_1 = \cos[(2t - \tau)(\omega_c - A_b\tau/2) + A_b t^2 + \phi(t) + \phi(t - \tau)]$ and $B_2 = \cos[(\omega_c - A_b(\tau/2 - t))\tau + \phi(t) - \phi(t - \tau)]$.

The second cosine term, B_2, is the signal containing the beat frequency. The output of the low pass filter consists of the beat frequency component, B_2 and noise components with similar frequencies to the beat frequency, while other components are filtered out. The beat frequency, f_b, is directly proportional to the delay time, τ which is directly proportional to the round trip time to the target. The relationship between beat frequency and target distance is

$$R = \frac{cT_s}{2}\frac{1}{f_s}f_b \quad (2.5)$$

where R is the range of the object, c is the velocity of the electromagnetic wave, T_s is the frequency sweep period, and f_s is the swept frequency bandwidth [18].

2.3.1 Noise in FMCW Receivers and Its Effect on Range Detection

As described above, the low pass filter output at the RADAR receiver can be represented by

$$v_{beat}(t, \tau) = \frac{A'}{2}\cos\left\{ \left[\omega_c - A_b\left(\frac{\tau}{2} - t\right)\right]\tau + \Delta\phi(t, \tau) \right\} \quad (2.6)$$

where $A' = [A_T + a_T(t)][A_R + a_R(t - \tau)]$ is the signal amplitude along with the noise affecting the amplitude. $\Delta\phi(t, \tau) = \phi(t) - \phi(t - \tau)$ is called the differential phase noise which occurs due to the nonlinear frequency chirp from imperfect VCO operation [19]. This phase noise affects the range accuracy [20]. The amplitude and phase noise will affect the beat frequency signal in two ways:

1. The amplitude noise will contain a signal frequency component which is the same as the beat frequency. This noise component will affect the amplitude of the beat frequency signal. This noise will introduce uncertainty into the *returned power*.
2. The noise components with frequencies lying close to the beat frequency (i.e., phase noise) distort the signal along the frequency axis. This introduces noise into the beat frequency value and hence into the range value. This will broaden the receiver power peaks and therefore introduce noise into the *range estimate*.

2.4 RADAR RANGE SPECTRA INTERPRETATION

Figure 2.3 shows a real single RADAR range spectra, which is the received power vs. range at a constant RADAR bearing angle. The RADAR can provide multiple returns in a single range bin. An entire range spectra at any particular bearing can be obtained. The range bin, is obtained by keeping the RADAR pointed toward a RADAR corner reflector of RCS 10 m^2 kept arbitrarily at 7.8 m and the second dominant reflection occurs from a concrete wall which is 23.7 m from the RADAR. That is, the RADAR waves are reflected from the corner reflector as well as from the wall. This is possible due to the RADAR's beam width. The corner reflector is of known RCS and can give good reflections (high signal power) back to the RADAR. The spectrum has two main features. First the signal return from the targets and second, noise. As shown in Figure 2.3, for the particular RADAR used here, these signals are riding over a low frequency signal which increases its amplitude up to a certain range ($\sim$150 m) and decreases toward the maximum range (200 m). This is due to the effect of the signal conditioning sections (filter roll-off) in the RADAR receiver. To compensate the reduction in received power as range increases (as will be shown in Section 2.4.1), a high pass filter^2 is usually used. The power

---

^{2} Assuming the RADAR range equation to be correct, a high pass filter with a gain of 40 dB/decade should produce a flat power response for particular targets at various ranges. Figure 2.3 shows a power–range spectrum recorded from the RADAR, which is fitted with a range compensating high pass filter. It can be seen from Figure 2.3, that the power range response is not flat. For this particular RADAR it makes sense to either determine the bias in the power–range spectra or model the high pass filter as having a gain of 60 dB/decade, which would better approximate the power–range relationship actually produced in Figure 2.3.

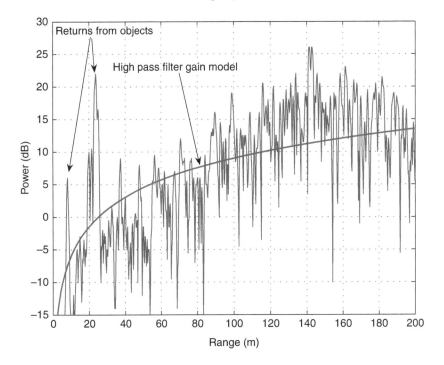

FIGURE 2.3   Range spectrum from a MMW RADAR. The X axis is the range (in meters) and the Y axis is the returned power (in decibel). The first reflection is from a corner reflector and the second one is from a concrete wall. Multiple reflections are obtained due to the beam width of the RADAR. The gain model of the high pass filter is also shown in the figure.

return of the RADAR spectra decreases near the maximum range (200 m) due to the low pass filter roll-off, which occurs before the high pass filter stage (Figure 2.2).

To understand the MMW RADAR range spectrum and to predict it accurately, it is necessary to use the RADAR range equation and knowledge of the noise distributions in the RADAR spectrum. A method for predicting the RADAR range spectra is now presented. An introduction is given explaining the relationship between RADAR signal returned power and range. Then, a method for establishing the relationship between the RCS and the range of objects in outdoor environments is shown. A noise analysis during signal absence and presence is then shown. This is necessary for predicting the range bins accurately during target presence and target absence. RADAR range bins are then predicted and it will be shown that the results compare reasonably well with actual (recorded) range bins recorded at various robot poses.

2.4.1 RADAR Range Equation

According to the simple RADAR equation, the returned power P_r is proportional to the RCS of the object, σ and inversely proportional to the fourth power of range, R [21]. The simple RADAR range equation is formally written as

$$P_r = \frac{P_t G^2 \lambda^2 \sigma}{(4\pi)^3 R^4 L} \tag{2.7}$$

where P_t is the RADAR's transmitted power, G is the antenna gain, λ is the wavelength (i.e., 3.89 mm in this case), and L the RADAR system losses. A high pass filter (shown in Figure 2.2) is used to compensate for the R^4 drop in received signal power. In an FMCW RADAR, closer objects produce signals with low beat frequencies and vice-versa (Equation [2.5]). Therefore by attenuating low frequencies and amplifying high frequencies, it is possible to correct the range-based signal attenuation [18]. To compensate the returned power loss due to increased range, the high pass filter is modeled in two ways:

1. The bias in the received power spectra is estimated.
2. By modeling the high pass filter with a gain of 60 dB/decade, instead of the usual 40 dB/decade, to comply with the characteristics of the particular RADAR used here.

The aim of this is to give a constant received signal power with range. The actual compensation which results in our system was shown in Figure 2.2 where it can be seen that the ideal flat response is not achieved by the internal high pass filter.

2.4.2 Interpretation of RADAR Noise

This section analyzes the sources of noise in MMW RADARs and quantifies the noise power in the received range spectra (seen in Figure 2.3). In most robot navigation formulations, observations must be predicted, and for the estimation algorithms to run correctly, the actual observations are assumed to equal the predictions, except that they are corrupted with Gaussian noise. It is therefore the aim of this section to determine the type of noise distributions in the actual received power and range values to determine how the predicted power–range spectra can be used correctly in a robot navigation formulation.

RADAR noise is the unwanted power that impedes the performance of the RADAR. For the accurate prediction of range bins, the characterization of noise is important. The two main components are thermal and phase noise. Thermal noise affects the power reading while phase noise affects the range estimate.

2.4.2.1 Thermal noise

Thermal noise is generated in the RADAR receiver electronics. The noise power is given by P_N (in Watts), where

$$P_N = kT_0\beta \qquad (2.8)$$

where k is the Boltzmann constant, T_0 is the temperature, and β is the receiver bandwidth [22]. As shown in Section 2.3, the power in the beat frequency signal (found from the FFT of this signal) is affected by the thermal noise power $a_R(t-\tau)$, which contributes to A' in Equation (2.6). It can be shown by analyzing the transition of this thermal (Gaussian) noise through the entire FMCW range detection process that when a target is present (strong received signal) the noise in the power–range spectrum follows a Gaussian distribution. When no target is present (weak or no reflected signal) it will be demonstrated from the results that the noise power follows a Weibull distribution. Therefore measurements with target presence/absence were made to verify these distributions and to quantify the power variance during target absence/presence.

2.4.2.2 Phase noise

Another source of noise which affects the range spectra is the phase noise. The phase noise is generated by the frequency instability of the oscillator due to the thermal noise. Ideally for a particular input voltage to the VCO, the output has a single spectral component. In reality, the VCO generates a spectrum of frequencies with finite bandwidth which constitutes phase noise. This is shown in Equation (2.6), where a band of noise frequencies with different phase components, $\Delta\phi(t,\tau)$ affects the desired signal frequency, which corresponds to range. The phase noise broadens the received power peaks and reduces the sensitivity of range detection [11] as shown in Figure 2.4.[3] This introduces noise into the range estimate itself. Experimental data provides insight into the phase noise distribution. For predicting the RADAR range spectra, the peaks at predicted targets are broadened by a small constant amount. This broadening is based on real measurements, which have shown the peaks[4] to have widths ranging from 2.5 to 3.5 m. This has been observed from targets, of different RCS, placed at different distances from the RADAR.

Figure 2.5 shows 1000 superimposed range bins obtained for the same RADAR swash plate bearing angle. Figure 2.5a shows the entire range bins over the full 200 m range, while Figure 2.5b shows a zoomed view of the spectra obtained from the feature at 10.25 m. From the figures, it is evident that

---

[3] The peaks and skirts shown in Figure 2.4 occur due to the leakage of signals from the transmitter into the mixer through the circulator and also due to the antenna impedance mismatch [11].

[4] At their intersections with the high pass filter gain curve shown in Figure 2.3.

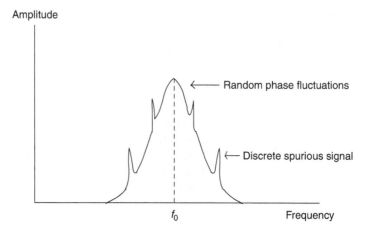

FIGURE 2.4 Phase noise in the FMCW transceiver occurs due to the instability of the VCO.

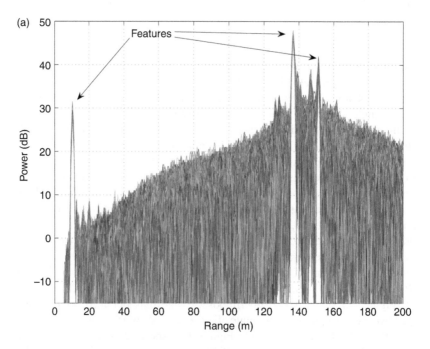

FIGURE 2.5 Thousand range bins plotted together for the same azimuth. (a) It shows the full range bin (200 m). (b) It shows the power returns from the feature at a distance of 10.25 m. The power noise affecting the returned power peaks is less than that during target absence within the range bin.

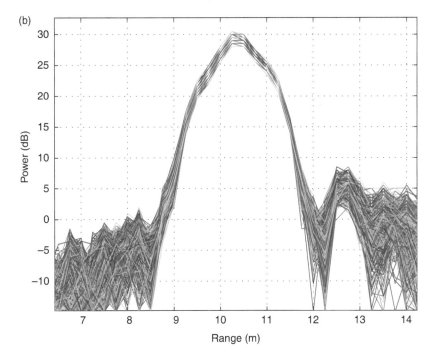

FIGURE 2.5 Continued.

the power variance of the noise at the peaks is less than that in the rest of the signal.

2.4.3   Noise Analysis during Target Absence and Presence

As indicated in Figure 2.5b, the power noise variance is different at the power peaks (target presence) and nonpeaks (target absence) sections of the power range spectra. Therefore, the noise statistics at the RADAR receiver outputs during target absence and presence will now be derived. Knowledge of the noise distributions is necessary for accurately predicting the RADAR spectra for prediction of feature location in robot navigation.

2.4.3.1   Power-noise estimation in target absence

The noise in the voltage signal entering the mixer stage is assumed to be Gaussian distributed with zero mean. A theoretical analysis to determine the power-noise distribution, after this signal has passed through the low pass and high pass filter stages, and the FFT process has been given in Reference 15. However, due to the unknown nature of the exact internal components within the RADAR used in this work, an experimental determination of the power noise distributions is used here.

To determine the power bias and variance of the range bins with no targets present, range bins were recorded at a fixed RADAR bearing angle, with no targets present. These were recorded by pointing the RADAR toward the open sky. The mean power and standard deviation of the noisy power–range spectra across the complete range of the RADAR is shown in Figure 2.6. The standard deviation of the noise is noticeably less at shorter ranges (<45 m), as the particular RADAR used can only output a minimum received power value of −15 dB, and any received power value less than this, will simply be output as −15 dB. The noise power values significantly increase above the minimum −15 dB at higher ranges due to the higher gain of the high pass (range compensation) filter at higher ranges.

Examination of the power distributions obtained at different ranges during target absence, suggests that a suitable approximation to the distributions is

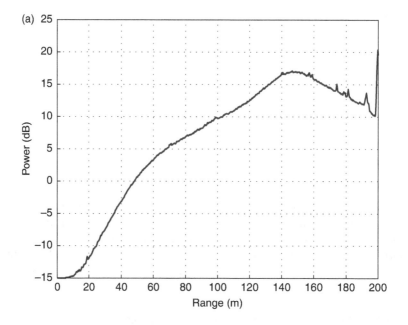

FIGURE 2.6 Mean and standard deviation of the noise during target absence over the complete range of the RADAR. The figures are obtained from noise only range bins by pointing the RADAR toward the sky. (a) Mean power bias as a function of RADAR range. (b) The standard deviation in power as a function of RADAR range. The standard deviation is less at shorter distances due to the lower amplification of the high pass filter at those ranges.

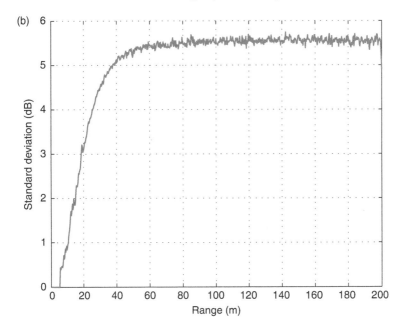

FIGURE 2.6 Continued.

the Weibull distribution [23]. This can be seen in Figure 2.7, where power distributions at arbitrary ranges of 10 and 100 m are shown.

The Weibull probability distribution function can be written as

$$f(x) = \frac{\xi}{\psi} \left(\frac{x}{\psi}\right)^{\xi-1} e^{-(x/\psi)^{\xi}}, \qquad \forall\, x > 0 \qquad (2.9)$$

where x is the random variable, with scale parameter $\psi > 0$ and shape parameter $\xi > 0$. The mean of x is $\mu = \psi\Gamma(1 + (1/\xi)) - 15$ and variance, $\sigma^2 = \psi^2\Gamma(1+(2/\xi)-\psi^2[\Gamma(1+(1/\xi))]^2)$, where $\Gamma(\cdots)$ is the Gamma function [23].

For scaling purposes, in this case the random variable x equals the received power $P_r + 15$ dB, in order to fit Equation (2.9).

For a range of 10 m (Figure 2.7a), suitable parameters for an equivalent Weibull distribution, ψ and ξ are 0.0662 and 0.4146, respectively.[5] At low ranges, this distribution is approximately equivalent to an exponential distribution, with mean, $\mu = -14.8$ dB and variance $\sigma^2 = 0.3501$ dB^2.

For a range of 100 m (Figure 2.7b), suitable Weibull parameters have been obtained as $\psi = 26.706$ and $\xi = 5.333$. The distribution has a mean,

---

[5] These values are obtained using MATLAB to fit Equation (2.9) to the experimentally obtained distribution of Figure 2.7a.

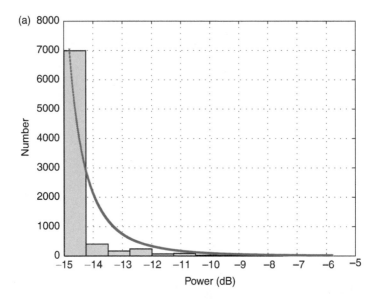

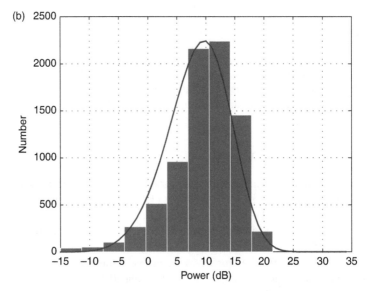

FIGURE 2.7 Experimental estimation of power noise distributions with no targets in the environment. (a) Experimental estimation of the noise distribution obtained from a 10 m distance. The distance has been chosen arbitrarily. (b) Experimental estimation of the noise distribution obtained from a 100 m distance. The distance has been chosen arbitrarily.

$\mu = 9.612$ dB and variance, $\sigma^2 = 28.239$ dB^2. These ranges have been selected arbitrarily to show the noise distributions for shorter (<45 m) and longer ranges ($45 <$ range < 200).

Therefore, to predict the power noise in the predicted power–range spectra, for ranges above approximately 45 m, Equation (2.9) can be used with the constant Weibull parameters determined at a range of 100 m. For ranges below this value, an exponential distribution is assumed, which uses a standard deviation value which is related to range as in Figure 2.6b.

2.4.3.2 Power-noise estimation in target presence

The receiver noise will also affect the signal when there is a target present. The resultant distribution is the convolution of both the signal and noise and is distributed normally [11]. The histogram in Figure 2.8a shows an approximately normal distribution obtained experimentally for 5000 observations of a RADAR retro-reflector at 10.25 m (the distance and the number of observations were selected arbitrarily). The experiment has been repeated for obtaining the distribution from a wall at a distance of 150 m approximately. This is shown in Figure 2.8b. The two histograms are approximately normally distributed and have variances of 4.07 and 5.76 dB^2, respectively. It is evident from Figure 2.8a, b and from Figure 2.5a that the noise variances affecting the signal during target presence are similar.

For an FMCW RADAR, features close to the RADAR give beat frequency signals with lower frequency and distant features give high frequency signals. By attenuating lower frequencies and amplifying higher frequencies, it is possible to achieve a constant returned power for an object with a particular RCS at all ranges. The graph shown in Figure 2.9 shows the calculated received power from two objects with RCS values of 1000 and 0.001 m^2 for all range values without the high pass filter effect. These have been calculated from the simple RADAR equation, using the parameters of the particular RADAR used here. The typical inverse range to the fourth power is still obtained even as the RCS of the target reduces significantly. Hence in practice, even the small signal reflections from atmospheric particles combined with the noise generated inside the RADAR's internal electronics will produce power–range relations of this form (such as, e.g., Figure 2.10). Therefore, an ideal high pass filter will give an approximately constant power noise variance for all ranges, for both target presence and target absence [11]. From the noise variances under signal absence and presence conditions shown above, it is evident that the high pass filter is close to its ideal state. (The power noise variance during target absence and target presence are similar irrespective of ranges.) The estimation of the noise statistics is helpful in accurately interpreting the range spectra as well as predicting the RADAR spectra for feature location prediction in robot navigation.

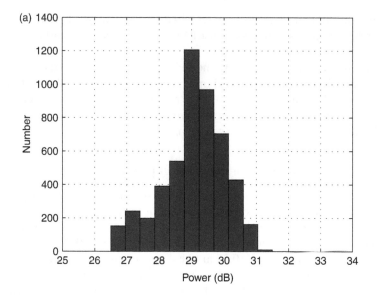

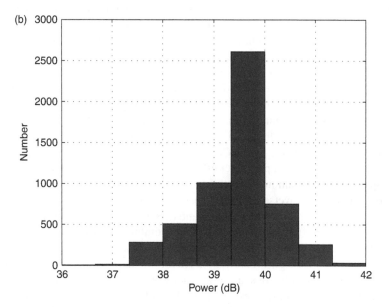

FIGURE 2.8 Experimental power distributions obtained from targets at differing ranges. (a) Experimental estimation of a noisy signal distribution. The distribution is obtained from a target (a RADAR corner reflector of RCS 10 m^2) at 10.25 m. (b) Experimental estimation of a noisy signal distribution obtained from a wall at approximately 150 m.

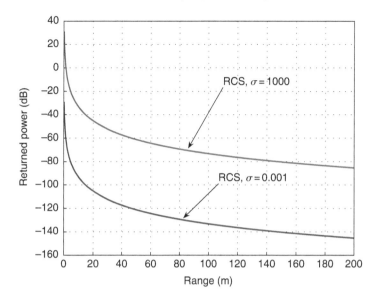

FIGURE 2.9   Expected curves of return power vs. distance for two objects with RCS values of 1000 and 0.001 m².

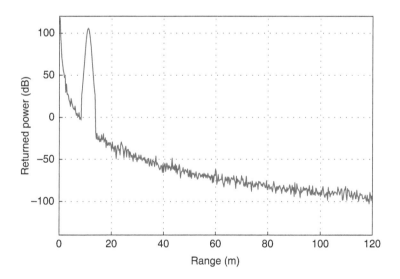

FIGURE 2.10   Range spectra prediction without range compensation.

2.4.4 Initial Range Spectra Prediction

The tools are now complete to simulate/predict RADAR spectra. In Figure 2.10, an object with a known RCS (10 m^2) is assumed to be at a distance of 10.25 m. A Monte Carlo method has been used for simulating the noise in the figure. A Gaussian noise distribution with a variance of 26.57 dB^2 is used when there is signal presence, and during signal absence Weibull distributions with parameters explained in the previous section have been used. The values are obtained from the experimental estimation of the noise distributions in target absence and presence (Figure 2.7 and Figure 2.4.3). The simulated result of applying the high pass 60 dB/decade filter is shown in Figure 2.11a. Analyzing the predicted (Figure 2.11a) and actual range bin (Figure 2.11b) shows a slight mismatch in the noise frequency with respect to range which is evident in the real spectra. This mismatch is due to the phase noise throughout the entire range bin. The phase noise, approximately quantified in Section 2.4.2, is taken into account only during the parts of the range bin which are predicted to have targets, as explained above. During sections of the range bin with no targets (i.e., beyond 11 m in Figure 2.11a) it is not modeled, since this part of the spectra is of little interest in target estimation.

A predicted and actual RADAR range spectra, obtained from an outdoor environment, is shown in Figure 2.12. Figure 2.13a and b show the results of a chi-squared test to determine any bias or inconsistency in the power–range bin predictions. The difference between the measured and the predicted range bins is plotted together with 99% confidence interval. The value of 99% bound, $= \pm 16.35 \text{ dB}$, has been found experimentally by recording several noisy power–range bins in target absence (RADAR pointing toward open space) as explained previously (3 × steady state standard deviation of Figure 2.6b) [15]. Close analysis of Figure 2.13a shows that the error has a negative bias. This is due to the approximate assumption of the high pass filter gain. For the RADAR used here, the gain of the high pass filter used in the predicted power–range bins was set to 60 dB/decade, as explained earlier. Figure 2.13b shows a chi-squared test on the difference between a measured bin and its predicted bin with the mean high pass filter bias of Figure 2.6a subtracted. Although the error in Figure 2.13b is less biased than Figure 2.13a, a gain of 60 dB/decade with the small bias (Figure 2.13a) is still acceptable as most of the error values are well within 99% confidence limit and also taking the high pass filter effect role into consideration.

A method for predicting the RADAR range spectra has been shown here which can be used for predicting observations, based on an estimate of a targets range and RCS value. Clearly a restriction of this method is that as a mobile robot moves with respect to objects within the environment, range bins can only be predicted assuming that the RCS does not change as the RADAR to target angle of incidence changes. In general this is clearly not a valid assumption, but

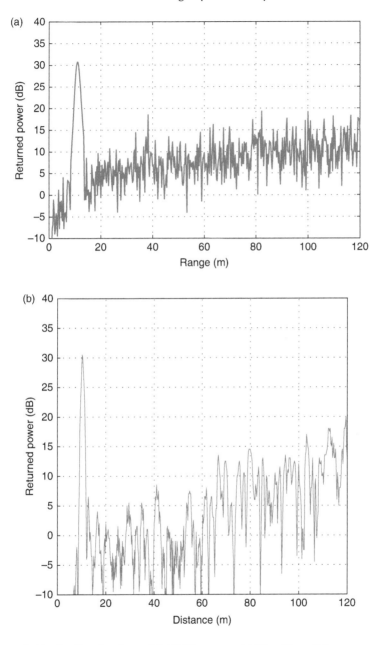

FIGURE 2.11 Predicted and actual RADAR spectra. (a) The effect of the range compensation (high pass) filter of 60 dB/decade. (b) Power vs. range of a single range bin obtained from an actual RADAR scan. A reflection is received from a target of RCS 10 m^2 at 10.25 m.

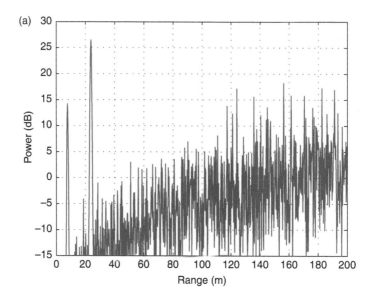

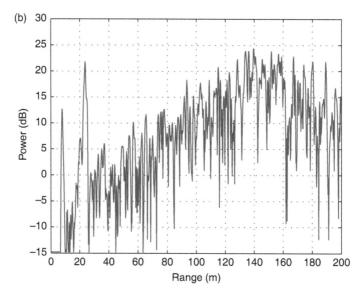

FIGURE 2.12 Predicted and actual range bins for multiple targets down-range. (a) Predicted power vs. range of a single range bin with two features down-range. (b) Power vs. range of a single range bin obtained from a RADAR scan with two features down-range.

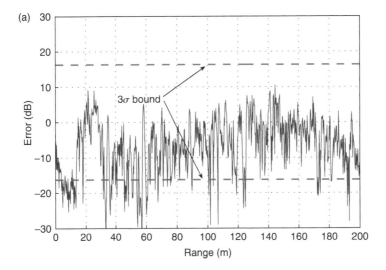

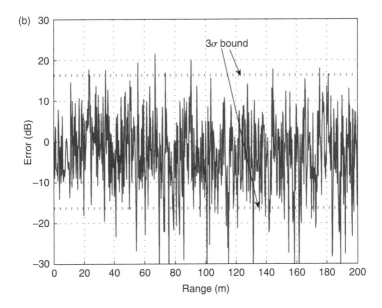

FIGURE 2.13 The difference between predicted and measured range bins, using two different approximations for the power bias. (a) The difference between predicted (using the 60 dB approximation for the high pass filter) and measured range bins containing two features down-range. This error is shown with the 3σ bounds. (b) The difference between predicted and measured range bins containing two features down-range. This error is shown with the 3σ bounds. The average error lies close to zero, as the gain of the high pass filter is obtained from the real measurements.

becomes acceptable for objects that are small and cylindrical in shape, making their RCS approximately view-point invariant, such as lamp posts, trees, etc., which can be used for outdoor navigation.

2.5 CONSTANT FALSE ALARM RATE PROCESSOR FOR TRUE TARGET RANGE DETECTION

To extract the true range values, previous methods have used a power threshold on the range bins (the closest power value to exceed some threshold gives the closest object) [9] or constant false alarm rate (CFAR) techniques [21,24]. The problem with thresholding is, it requires manual adjustment of the threshold as the RCS of objects in an outdoor natural environment will vary. The function of CFAR processors is to maintain a constant and low rate of false alarms in detecting true range values [25].

A cell averaging (CA) detector is useful for maintaining a CFAR where the power noise-plus-clutter observations $x = x_1, \ldots, x_i, \ldots, x_n$ follow a Weibull random distribution shown in Equation (2.9). The structure of the applied CA-CFAR is shown in Figure 2.14. This figure shows $M/2$ reference cells (where $M = 70$) on each side of the cell, Y, under investigation. Guard cells are present to account for the broadened target reflection [26]. A moving window of width $M = 70$ range points is then used to sum the local noisy power values in the

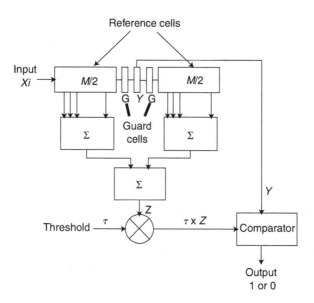

FIGURE 2.14   The structure of the applied CA-CFAR detector.

range bin as shown in Equation (2.10) [27].

$$Z = \sum_{i=1}^{M} x_i \qquad (2.10)$$

This sum is multiplied by a threshold, τ (in this case $\tau = 0.033$), for later comparison with a test sample power value. The value for τ is chosen for achieving the desired value of P_{fa}, the design false alarm probability, in the absence of targets [28]. The scalar τ is a function of the number of reference cells M (here $M = 70$) and P_{fa} is (1×10^{-6}) for the RADAR used here [10]. The test sample Y is either a noise-plus-clutter observation or a target return. The variable threshold τZ is compared with Y. A target is declared to be present if

$$Y > \tau Z \qquad (2.11)$$

The range bin in Figure 2.15 was obtained from an environment containing a concrete wall at approximately 18 m. The detected features are indicated along with the adaptive threshold. The moving average will set the threshold above which targets are considered detected. Due to the phase noise, the power returned from the target is widened along the range axis, resulting in more feature detections at approximately 18 m. In Figure 2.15a and b, CFAR "picks out" features which lie at closest range. Features at a longer range, however, will not be detected as the noise power variance estimate by the CFAR processor becomes incorrect due to the range bin distortion caused by the high pass filter.

2.5.1 The Effect of the High Pass Filter on CFAR

In general, since the gain of the high pass filter is not linear (Figure 2.6a) the sum of the noisy received power values in Equation (2.10) is inaccurate at higher ranges, which ultimately results in the missed detection of targets at these range values. This is evident from Figure 2.15b where CFAR detects a feature (corner reflector) at 10.25 m while it misses a feature (building) at 138 m. The second reflection is due to the beam-width of the RADAR, as part of the transmitted signal passes the corner reflector. It would therefore be useful to reduce the power–range bias before applying the CFAR method. Therefore, to correctly implement the CA-CFAR method here, first, the average of two noise only range bins can be obtained,[6] the result of which should be subtracted from the range bin under consideration. This is carried out to obtain a range independent, high pass filter gain for the resultant bin.

The CFAR method has been applied to the range bin of Figure 2.11b, the full 200 m bin of which is shown in Figure 2.16a, after subtracting the high

---

[6] The noise only range bins are obtained by pointing the RADAR toward open space.

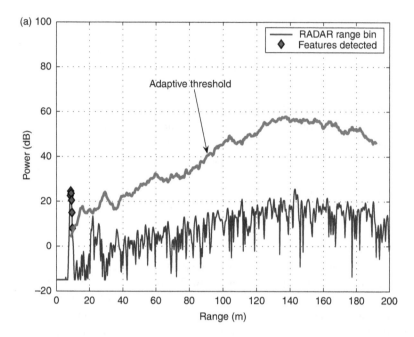

FIGURE 2.15   CFAR target detection. (a) The detection of a target (concrete wall approx-imately at 18 m) using a CA-CFAR detector. A series of targets around the 18 m mark are obtained due to the phase noise in the returned peak. (b) The missed detection of a feature (a building at 138 m) by a CA-CFAR detector. Due to the gain of the high pass filter, the noise estimation is inaccurate at higher ranges resulting in missed detection of features.

pass filter bias of Figure 2.6a. This figure shows the result from an environment, containing a corner reflector at 10.25 m and a building at approximately 138 m. By reducing the high pass filter effect (range independent gain for all the ranges), the CFAR detection technique finds features regardless of range as shown in Figure 2.16a. It is clearly necessary to compensate for any nonideal high pass filter characteristics, in the form of power–range bias, before CA-CFAR can be applied correctly.

Problems still arise however, as CFAR can misclassify targets as noise (missed-detection) and noise as targets (false-alarm). Both of these are evident and labeled in the CFAR results of Figure 2.16a.

2.5.1.1   Missed detections with CFAR

In a typical autonomous vehicle environment the clutter level changes. As the RADAR beam width increases with range, the returned range bin may have multiple peaks from features.

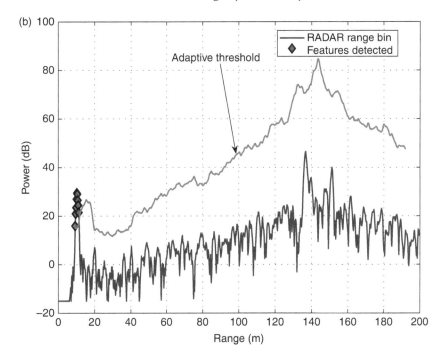

FIGURE 2.15 Continued.

From Figure 2.14, it can be seen that if two or more targets are separated by less than the window width M, the local power sum in Equation (2.10) will become large, causing the adaptive threshold to increase, resulting in a missed detection [29]. This is also shown in Figure 2.16b where a return from an object, which lies within M range samples of the first feature is completely missed by the CFAR detector.

2.5.1.2 False alarms with CFAR

Due to the filtering elements within the RADAR, the power noise in the RADAR range bins is correlated. Therefore, if the window size is too small, all of its power–range samples will be highly correlated. This means that the sum of the power values, calculated in Equation (2.10), will misrepresent the true sum which would be obtained from a set of uncorrelated values. This can ultimately result in the adaptive threshold being set too low, meaning that even noise only power values can exceed it. This gives false alarms. This can be overcome by increasing the window width. However, as explained above, a larger window width can result in the missed detection of features. The occurrence of false alarms is shown in Figure 2.16a and b.

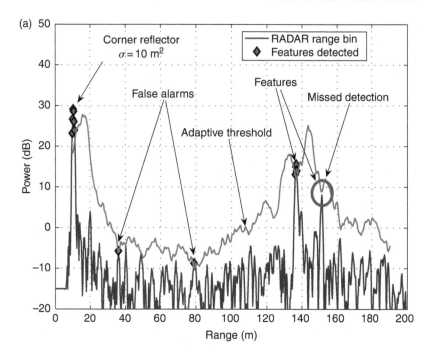

FIGURE 2.16 Target estimation with CFAR. (a) The graph shows target detection using a CFAR detector. The effect of the high pass filter is removed from the range bin. (b) The figure shows a missed detection of a feature (at 38 m) by the CA-CFAR processor. The first feature is at 22 m and the second feature is at 38 m approximately. The effect of the high pass filter is removed from the range bin.

In general, the CFAR method tends to work well with aircraft in the air having relatively large RCS, while surrounded by air (with extremely low RCS). At ground level, however, the RCS of objects is comparatively low and also there will be clutter (objects which cannot be reliably extracted). Further, as the CFAR method is a binary detection technique, the output is either a one or a zero (Equation [2.11]), that is, no probabilistic measures are given for target presence or absence.

2.6 TARGET PRESENCE PROBABILITY ESTIMATION FOR TRUE TARGET RANGE DETECTION

For typical outdoor environments, the RCS of objects may be small. The smaller returned power from these objects can be buried in noise. For reducing the

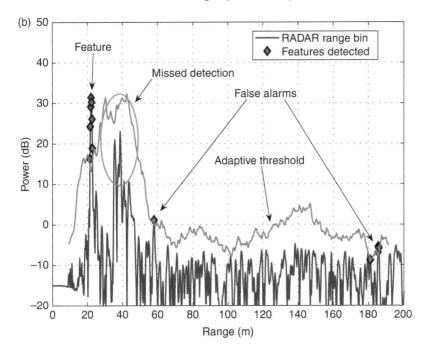

FIGURE 2.16   Continued.

noise and extracting smaller signal returns along with the higher power returns, a method is now introduced which uses the probability of target presence [30] for feature detection [15]. This method is appealing compared to CFAR and constant threshold methods at ground level, as a threshold can be applied on the target presence probability. By setting a threshold value to be dependent on target presence probability and independent of the returned power in the signal, a higher probability threshold value is more useful for target detection. The proposed method does not require manual assistance. The merits of the proposed algorithm will be demonstrated in the results in Section 2.6.1. The detection problem described here can be stated formally as a binary hypothesis testing problem [31]. Feature detection can be achieved by estimating the noise power contained in the range spectra. The noise is estimated by averaging past spectral power values and using a smoothing parameter. This smoothing parameter is adjusted by the target presence probability in the range bins. The target presence probability is obtained by taking the ratio between the local power of range spectra containing noise and its minimum value. The noise power thus estimated is then subtracted from the range bins to give a reduced noise range spectra.

Let the power of the noisy range spectra be smoothed by a w-point window function $b(i)$ whose length is $2w + 1$

$$\check{P}(k, l) = \sum_{i=-w}^{w} b(i) \check{P}(k - i, l) \qquad (2.12)$$

where $\check{P}(k, l)$ is the kth power value of lth range spectra.

Smoothing is then performed by a first order recursive averaging technique:

$$\check{P}(k, l) = \alpha_s \check{P}(k, l - 1) + (1 - \alpha_s) \check{P}(k, l) \qquad (2.13)$$

where α_s is a weighting parameter ($0 \le \alpha_s \le 1$). First the minimum and temporary values of the local power are initialized to $P_{\min}(k, 0) = P_{tmp}(k, 0) = \check{P}(k, 0)$. Then a range bin-wise comparison is performed with the present bin l and the previous bin $l - 1$.

$$P_{\min}(k, l) = \min\{P_{\min}(k, l - 1), \check{P}(k, l)\} \qquad (2.14)$$

$$P_{tmp}(k, l) = \min\{P_{tmp}(k, l - 1), \check{P}(k, l)\} \qquad (2.15)$$

When a predefined number of range bins have been recorded at the same vehicle location, and the same sensor azimuth, the temporary variable, P_{tmp} is reinitialized as

$$P_{\min}(k, l) = \min\{P_{tmp}(k, l - 1), \ \check{P}(k, l)\} \qquad (2.16)$$

$$P_{tmp}(k, l) = \check{P}(k, l) \qquad (2.17)$$

Let the signal-to-noise power (SNP), $P_{SNP}(k, l) = \check{P}(k, l)/P_{\min}(k, l)$ be the ratio between the local noisy power value and its derived minimum.

In the Neyman–Pearson test [32], the optimal decision (i.e., whether target is present or absent) is made by minimizing the probability of the type II error (see Appendix), subject to a maximum probability of type I error as follows.

The test, based on the *likelihood ratio*, is

$$\frac{p(P_{SNP}|H_1)}{p(P_{SNP}|H_0)} \overset{H_1}{\underset{H_0}{\gtrless}} \delta \qquad (2.18)$$

where δ is a threshold,[7] H_0 and H_1 designate hypothetical target absence and presence respectively. $p(P_{SNP}|H_0)$ and $p(P_{SNP}|H_1)$ are the conditional probability density functions. The decision rule of Equation (2.18) can be expressed as

$$P_{SNP}(k,l) \overset{H_1}{\underset{H_0}{\gtrless}} \delta \qquad (2.19)$$

An indicator function, $I(k,l)$ is defined where, $I(k,l) = 1$ for $P_{SNP} > \delta$ and $I(k,l) = 0$ otherwise.

The estimate of the conditional target presence probability,[8] $\hat{p}'(k,l)$ is

$$\hat{p}'(k,l) = \alpha_p \hat{p}'(k,l-1) + (1 - \alpha_p)I(k,l) \qquad (2.20)$$

This target presence probability can be used as a target likelihood within mobile robot navigation formulations. α_p is a smoothing parameter $(0 \le \alpha_p \le 1)$. The value of α_p is chosen in such a way that the probability of target presence in the previous range bin has very small correlation with the next range bin (in this case $\alpha_p = 0.1$).

It is of interest to note that, as a consequence of the above analysis, the noise power, $\hat{\lambda}_d(k,l)$ in kth range bin is given by

$$\hat{\lambda}_d(k,l) = \tilde{\alpha}_d(k,l)\hat{\lambda}_d(k,l-1) + [(1 - \tilde{\alpha}_d(k,l))] \check{P}(k,l) \qquad (2.21)$$

where

$$\tilde{\alpha}_d(k,l) = \alpha_d + (1 - \alpha_d)p'(k,l) \qquad (2.22)$$

and α_d is a smoothing parameter $(0 \le \alpha_d \le 1)$. This can be used to obtain a noise reduced bin, $\hat{P}_{NR}(k,l)$ using the method of power spectral subtraction [34]. In the basic spectral subtraction algorithm, the average noise power, $\hat{\lambda}_d(k,l)$ is subtracted from the noisy range bin. To overcome the inaccuracies in the noise power estimate, and also the occasional occurrence of negative power estimates, the following method can be used [35]

$$\hat{P}_{NR}(k,l) = \begin{cases} \check{P}(k,l) - c \times \hat{\lambda}_d(k,l) & \text{if } \check{P}(k,l) > c \times \hat{\lambda}_d(k,l) \\ d \times \hat{\lambda}_d(k,l) & \text{otherwise} \end{cases}$$

---

[7] This threshold can be chosen based upon the received SNP, at which the signal can be trusted not to be noise. Note that this does not have to be changed for differing environments, or types of targets.
[8] Conditioned on the indicator function $I(k,l)$ [33].

where c is an over-subtraction factor ($c \geq 1$) and d is spectral floor parameter ($0 < d < 1$). The values of c and d are empirically determined for obtaining an optimal noise subtraction level at all ranges and set to be 4 and 0.001.

Although a reduced noise range bin can be useful in other detection methods, the target presence probability estimate (Equation [2.20]), will be demonstrated further in the results. This method shows improved performance over CFAR methods as the threshold can be applied on the target presence probability instead of SNP. Setting an arbitrary threshold value on the probability of target presence (≥ 0.8) is sufficient for target detection. Based on the results, this is a robust method and requires no adjustments when used in different environments.

2.6.1 Target Presence Probability Results

The results of the proposed target detection algorithm are shown in Figure 2.17 where a noisy RADAR range bin (Figure 2.17a), the corresponding estimated target presence probability (Figure 2.17b) from Equation (2.20) and the reduced noise range spectra (Figure 2.17c) have been plotted. In Figure 2.17a, the range bin contains three distinct peaks of differing power values, whereas the target presence probability plot shows the three peaks with a more uniform range width and similar probabilistic values. This result shows that although the return power values varies from different objects, the corresponding target presence probability values will be similar.

The target presence probability-based feature detector is easier to interpret as shown in Figure 2.18 and Figure 2.19 where the target presence probability plot is shown along with the corresponding raw RADAR data. Figure 2.18a and Figure 2.19a show the raw RADAR data obtained in an indoor sports hall and outdoor sports field, respectively. The corresponding target presence probabilities are shown in Figure 2.18b and Figure 2.19b, respectively. Figure 2.18b shows the target presence probability plot of an indoor stadium. The four walls of the stadium are clearly obtained by the proposed algorithm. The other probability values at higher ranges arise due to the multipath effects in the RADAR range spectrum. Figure 2.19b is obtained from an outdoor field. The detected features are marked in the figure. The clutter shown in Figure 2.19b is obtained when the RADAR beam hits the ground due the unevenness of the field surface.

The merit of the proposed algorithm is shown in Figure 2.20 where plots obtained using different power thresholds applied to raw RADAR range spectra are shown and compared with the threshold (0.8) applied to the probability plot. Figure 2.20a shows the comparison of 2D plots obtained by choosing a constant threshold of 25 dB applied to the raw RADAR data and the target presence probability plot. Figure 2.20b shows the comparison of plots obtained by constant

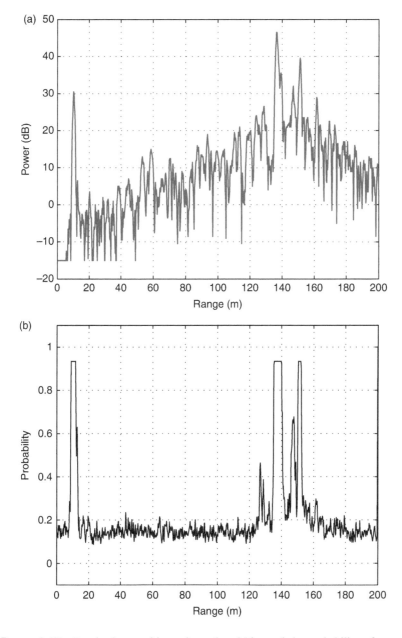

FIGURE 2.17   Received range bin, noise reduced bin, and the probability of target presence vs. range plot. (a) Received noisy RADAR range bin. (b) Target presence probability of the corresponding range bin. (c) Noise reduced RADAR range bin.

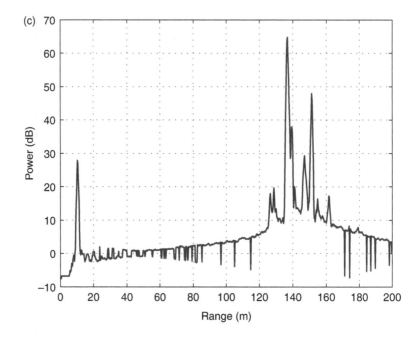

FIGURE 2.17   Continued.

threshold of 40 dB applied against the raw RADAR data and the target presence probability. Further results conducted show the target presence probability of objects will be the same and is found to be more than 0.8. Feature detection using the target presence probability is then carried out by keeping the threshold at 0.8. The results shown in Figures 2.18 to 2.20 clearly show that the target presence probability-based feature detection is easier to interpret and has lower false alarms compared to constant threshold-based feature detection in the typical indoor and outdoor environments tested [36].

2.6.2   Merits of the Proposed Algorithm over Other Feature Extraction Techniques

The constant threshold applied to raw RADAR data requires manual intervention for adjusting the threshold depending on the environment. In CA-CFAR, the averaging of power values in the cells provides an automatic, local estimate of the noise level. This locally estimated noise power is used to define the adaptive threshold (see e.g., Figure 2.16a). The test window compares the threshold with the power of the signal and classifies the cell content as signal or noise.

(a)

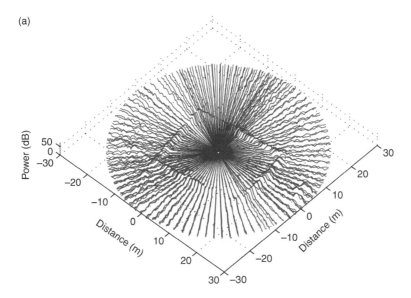

FIGURE 2.18   Raw RADAR data and corresponding target presence probability plots obtained from an indoor sports hall. (a) Power vs. range of a 2D RADAR scan from an indoor environment. (b) Target presence probability vs. range of a 2D RADAR scan in indoor environment. The probability of the targets detected (i.e., walls) are shown in the figure.

When the signal and noise distributions are distinctly separated in range, CFAR works well. But when the signal and noise distributions lie close together, which is often the case at ground level (as shown in Figure 2.21), the method misclassifies noise as signal and vice versa. This is the reason for the poor performance of the CFAR technique with noisy RADAR data. Figure 2.22 shows features obtained by target presence probability and the CA-CFAR technique. The dots are the features obtained by target presence probability while the "+" signs are the features obtained from the CFAR-based target detector. From the figures it can be seen that the target presence-based feature detection has a superior performance to CA-CFAR detector in the environment tested. Figure 2.23 shows the difference between the ground truth and the range observation obtained from the target presence probability. The ground truth has been obtained by manually measuring the distance of the walls from the RADAR location. The peaks in Figure 2.23 are to some extent due to inaccurate ground truth estimates, but mainly due to multi-path reflections.

The proposed algorithm for feature extraction appears to outperform the CFAR method because the CFAR method finds the noise locally, while the target presence probability-based feature detection algorithm estimates

(b)

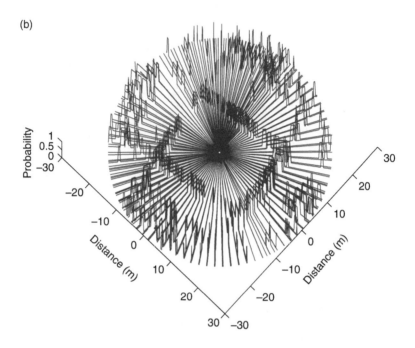

FIGURE 2.18   Continued.

the noise power by considering more than one range bin (Equation [2.16]). The target presence probability-based feature extraction, unlike the CFAR detector, is not a binary detection process as is shown in Figure 2.17c. This method of feature detection is useful in data fusion as the feature representation is probabilistic.

2.7   MULTIPLE LINE-OF-SIGHT TARGETS — RADAR PENETRATION

At 77 GHz, millimeter waves can penetrate certain nonmetallic objects, which sometimes explains the multiple line-of-sight objects within a range bin.[9] This limited penetration property can be exploited in mobile robot navigation in outdoor unstructured environments, and is explored further here.

For validating the target penetration capability of the RADAR, tests were carried out with two different objects. In the section of the RADAR scan, shown in Figure 2.24a, a RADAR reflector of RCS 177 m² and a sheet of

---

[9] Although it should be noted that these can be the results of specular and multiple path reflections also.

(a)

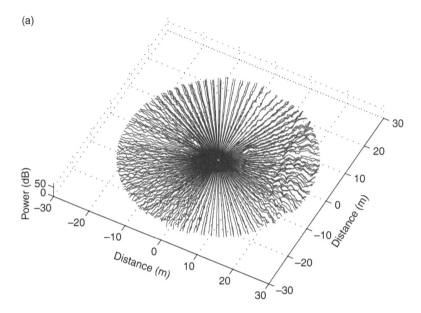

FIGURE 2.19 Raw RADAR data and corresponding target presence probability obtained from an outdoor environment. (a) Power vs. range of a 2D RADAR scan from an outdoor environment. (b) Target presence probability vs. range of a 2D RADAR scan in outdoor environment. The probability of the targets detected (i.e., RADAR reflectors, wall, and tree) are shown in the figure.

wood of thickness 0.8 cm were placed at ranges of 14 and 8.5 m respectively, to visually occlude the reflector from the RADAR. This ensured that no part of the RADAR reflector fell directly within the beam width of the RADAR, so that if it was detected, it must be due to the radio waves penetrating the wood. Figure 2.24a shows the detection of the two features down-range even though, visually, one occludes the other. The experiment was also repeated for a perspex sheet of thickness 0.5 cm (Figure 2.24b). The results of object penetration by RADAR waves motivates further development of power spectra prediction with multiple line-of-sight features which is one of the contributions of this chapter. For feature-based SLAM, it is necessary to predict the target/feature locations reliably, given a prediction of the vehicle/RADAR location. As RADAR can penetrate certain nonmetallic objects it can give multiple range information. A method for predicting the power–range spectra (or range bins) using the RADAR range equation and knowledge of various noise distributions in the RADAR has already been explained in this chapter.

For SLAM, the measurements taken from the RADAR used here are the range, R, bearing, θ, and the received power, P_R, from the target at range R.

(b)

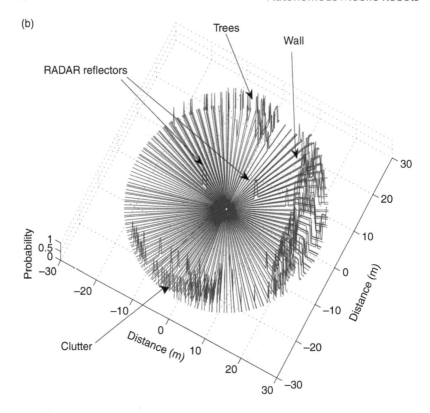

<small>**Figure 2.19** Continued.</small>

One of the contributions of this chapter is to predict range bins from new robot positions given an estimate of the vehicle and target states. A new augmented state vector is introduced here which, along with the usual feature coordinates x and y, contains that feature's normalized RCS, Υ_R, and absorption RCS, Υ_a, and the RADAR losses, L.

To illustrate this, Figure 2.25 shows a 360° RADAR scan obtained from an outdoor field. Objects in the environment consist of lamp-posts, trees, fences, and concrete steps. The RADAR penetrates some of the nonmetallic objects,[10] and can observe multiple targets down line. This is shown in Figure 2.26, which is the received power vs. range for the particular bearing of 231° marked in Figure 2.25. Multiple targets down range can occur due to either the beam width of the transmitted wave intersecting two or more objects at differing ranges or due to penetration of the waves through certain objects. The RADAR used here

---

[10] At 77 GHz the attenuation through paper, fiberglass, plastic, wood, glass, foliage, etc., are relatively low while attenuation through brick and concrete is high [37].

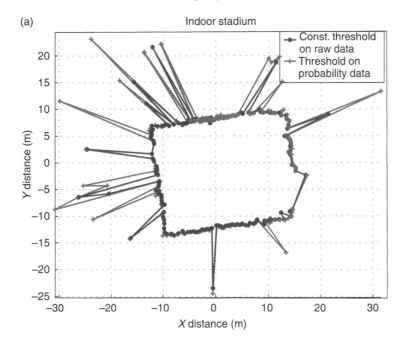

FIGURE 2.20 Target presence probability vs. range spectra and the corresponding power vs. range taken from a 2D RADAR scan in an indoor environment. The figures shows a comparison of the proposed feature detection algorithm with the constant threshold method. (a) A constant power threshold of 25 dB is chosen and compared with the threshold (0.8) applied on probability-range spectra. (b) A constant power threshold of 40 dB is chosen and compared with the threshold applied to the probability–range spectra.

is a pencil beam device, with a beam width of 1.8°. This means that multiple returns within the range spectra occur mostly due to penetration. Therefore a model for predicting entire range spectra, based on target penetration is now given.

2.8 RADAR-BASED AUGMENTED STATE VECTOR

The state vector consists of the normalized RADAR cross section, Υ_R, absorption cross section, Υ_a, and the RADAR loss constants, L, along with the vehicle state and feature locations. The variables, Υ_R, Υ_a, and L are assumed unique to a particular feature/RADAR. Hence, this SLAM formulation makes the (very) simplified assumption that all features are stationary and that the changes in the normalized values of RCS and absorption cross sections of features when sensed from different angles, can be modeled using Gaussian random variables v_{Υ_i}.

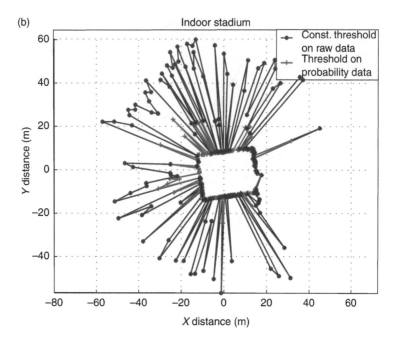

(b)

FIGURE 2.20 Continued.

This is a reasonable assumption only for small circular cross sectioned objects such as trees, lamp posts, and pillars, however, as will be shown the method produces good results in semi-structured environments even for the targets which do not conform to these assumptions. The SLAM formulation here can handle *multiple line-of-sight targets*.

2.8.1 Process Model

A simple vehicle predictive state model is assumed with stationary features surrounding it. The vehicle state, $\mathbf{x}_v(k)$ is given by $\mathbf{x}_v(k) = [x(k), y(k), \theta_R(k)]^T$ where $x(k)$, $y(k)$, and $\theta_R(k)$ are the local position and orientation of the vehicle at time k. The vehicle state, $\mathbf{x}_v(k)$ is propagated to time $(k+1)$ through a simple steering process model [38].

The model, with control inputs, $\mathbf{u}(k)$ predicts the vehicle state at time $(k+1)$ together with the uncertainty in vehicle location represented in the covariance matrix $\mathbf{P}(k+1)$ [39].

$$\mathbf{x}_v(k+1) = \mathbf{f}(\mathbf{x}_v(k), \mathbf{u}(k)) + \mathbf{v}(k) \qquad (2.23)$$

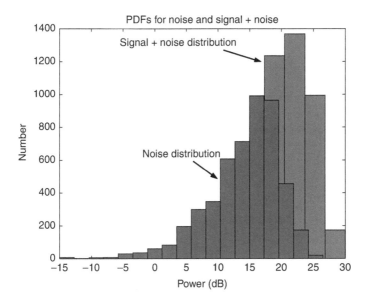

FIGURE 2.21 Experimental estimation of signal and noise distributions. In the CFAR method, the local noise-plus-clutter power (Equation [2.10]) in the window is used to set the detection threshold. The method compares the signal in the test window and the detection threshold. The method fails when there are multiple detections within a range-bin and in cluttered environments.

$\mathbf{u}(k) = [v(k), \alpha(k)]$. $v(k)$ is the velocity of the vehicle at time k and $\alpha(k)$ is the steering angle. In full, the predicted state at time, $(k + 1)$ becomes

$$
\begin{bmatrix}
\hat{x}(k+1|k) \\
\hat{y}(k+1|k) \\
\hat{\theta}_R(k+1|k) \\
x_{p_1}(k+1|k) \\
y_{p_1}(k+1|k) \\
\Upsilon_{R_1}(k+1|k) \\
\Upsilon_{a_1}(k+1|k) \\
\vdots \\
x_{p_N}(k+1|k) \\
y_{p_N}(k+1|k) \\
\Upsilon_{R_N}(k+1|k) \\
\Upsilon_{a_N}(k+1|k) \\
L(k+1|k)
\end{bmatrix}
=
\begin{bmatrix}
\hat{x}(k|k) \\
\hat{y}(k|k) \\
\hat{\theta}_R(k|k) \\
x_{p_1}(k|k) \\
y_{p_1}(k|k) \\
\Upsilon_{R_1}(k|k) \\
\Upsilon_{a_1}(k|k) \\
\vdots \\
x_{p_N}(k|k) \\
y_{p_N}(k|k) \\
\Upsilon_{R_N}(k|k) \\
\Upsilon_{a_N}(k|k) \\
L(k|k)
\end{bmatrix}
+
\begin{bmatrix}
\Delta x(k) \\
\Delta y(k) \\
\alpha(k) \\
0_{p_1} \\
0_{p_1} \\
0_{p_1} \\
0_{p_1} \\
\vdots \\
0_{p_N} \\
0_{p_N} \\
0_{p_N} \\
0_{p_N} \\
0
\end{bmatrix}
\tag{2.24}
$$

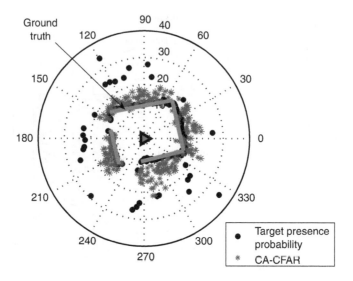

FIGURE 2.22 Comparison of CA-CFAR detector-based feature extraction and feature detection based on target presence probability.

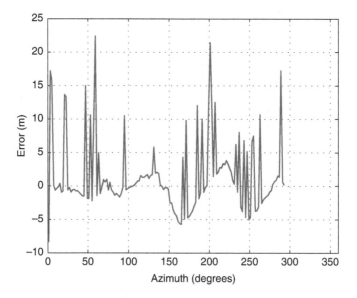

FIGURE 2.23 The difference between the ground truth range values and the range estimates from the target presence probability.

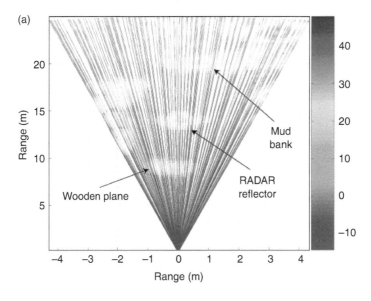

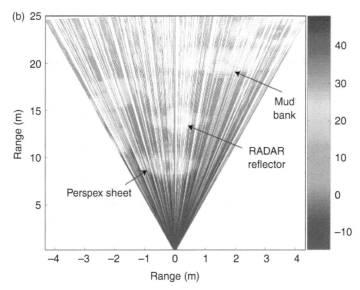

FIGURE 2.24   Initial test results carried out to show the RADAR wave penetration with different objects. (a) A scan of a RADAR reflector of RCS 177 m², 14 m from the RADAR, and a wooden sheet of thickness 0.8 cm visually occluding the reflector from the RADAR. The wooden sheet is 8.5 m from the RADAR. (b) A RADAR reflector of RCS 177 m², 14 m from the RADAR, and a perspex sheet of thickness 0.5 cm, 8.5 m from the RADAR. Again, the reflector is visually occluded from the RADAR.

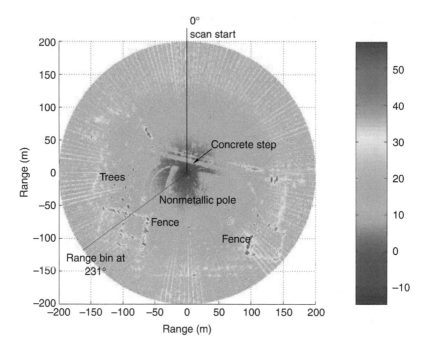

FIGURE 2.25 A 360° RADAR range spectra obtained from an outdoor field, containing trees, nonmetallic poles, fences, and concrete walls. The received power value is represented in color space, as shown by the right hand color bar, with power units in decibel.

where $\Delta x(k) = v(k)\Delta t \cos(\hat{\theta}_R(k|k) + \alpha(k))$, $\Delta y(k) = v(k)\Delta t \sin(\hat{\theta}_R(k|k) + \alpha(k))$ and Δt is the sampling time.

The augmented state vector is then $\mathbf{x}(k) = [x_v, \{F_1, \Upsilon_{R_1}, \Upsilon_{a_1}\}, \dots, \{F_i, \Upsilon_{R_i}, \Upsilon_{a_i}\}, \dots, \{F_N, \Upsilon_{R_N}, \Upsilon_{a_N}\}, L]^T$ where x_v is the vehicle's pose, $F_i = [x_{p_i}, y_{p_i}]^T$ is the ith feature's location, where $1 \leq i \leq N$. Υ_{R_i} is the normalized RCS of the ith feature, Υ_{a_i} is its normalized absorption cross section, L represents the RADAR loss, and $\mathbf{v}(k) = [\mathbf{v}_v(k), 0_{p_1}, 0_{p_1}, v\Upsilon_{R_1}, v\Upsilon_{a_1}, \dots, 0_{p_i}, 0_{p_i}, v\Upsilon_{R_i}, v\Upsilon_{a_i}, \dots, 0_{p_N}, 0_{p_N}, v\Upsilon_{R_N}, v\Upsilon_{a_N}, 0]^T$.

2.8.2 Observation (Measurement) Model

Another contribution of this chapter is the formulation of the observation model. The RADAR observation is used to estimate the vehicle's state once the vehicle's pose is predicted. During filter update, the prediction and estimation are fused. For each of the features in the map, the predicted range, $\hat{R}_i(k + 1|k)$, the RADAR bearing angle, $\hat{\beta}_i(k + 1|k)$, and the power, $\hat{P}_i(k + 1|k)$ are to be

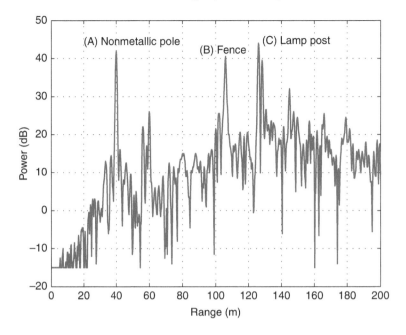

FIGURE 2.26   A single RADAR range bin, recorded at the bearing angle 231° shown in Figure 2.25, obtained from the outdoor field with multiple features down-range.

predicted from the predicted state in Equation (2.24). The predicted range and bearing observations are similar to the ordinary SLAM formulation, that is,

$$\hat{R}_i(k+1|k) = \sqrt{[\hat{x}_{p_i}(k+1|k) - \hat{x}_R(k+1|k)]^2 + [\hat{y}_{p_i}(k+1|k) - \hat{y}_R(k+1|k)]^2}$$

(2.25)

$$\hat{\beta}_i(k+1|k) = \hat{\theta}_R(k+1|k) - \tan^{-1}\left[\frac{\hat{y}_{p_i}(k+1|k) - \hat{y}_R(k+1|k)}{\hat{x}_{p_i}(k+1|k) - \hat{x}_R(k+1|k)}\right]$$

(2.26)

The predicted power for all targets, such as those in Figure 2.26, is the fundamental difference offered in this chapter.

2.8.2.1 Predicted power observation formulation

The assumptions made in the predicted power model are as follows:

- The environmental features of interest are assumed to have small circular cross-sections, so that the estimated normalized RCS sections and absorption coefficients are approximately the same in all directions with respect to that feature.

- The measured returned power should be independent of range (due to the built-in range compensation filter). This filter must first be removed or post-filtered to remove its effect, to produce range dependent power returns from all objects [15].
- The beam-width of the RADAR wave does not increase considerably with range.

A target is assumed to affect the incident electromagnetic radiation in three possible ways:

1. A portion of the incident energy Υ_R, $0 \leq \Upsilon_R \leq 1$, is reflected and scattered
2. A portion of the incident energy Υ_a, $0 \leq \Upsilon_a \leq 1$, is absorbed by the target
3. A portion of the incident energy $1 - (\Upsilon_R + \Upsilon_a)$ is further transmitted through the target

Υ_R is thus referred to as the "normalized" RCS section. Figure 2.27 shows a MMW RADAR in an environment with i-features down-range at a particular bearing. The following terms are used in formulating the predicted power observation:

- $P_{\text{INC}i}$ = Power incident on the ith feature
- $P_{\text{REF}i}$ = Power reflected from the ith feature
- $P_{\text{TRAN}i}$ = Power transmitted through the ith feature
- $P_{\text{INC}i1}$ = Power incident on the first feature which is reflected from the ith feature
- $P_{\text{REF}i1}$ = Power reflected back toward the ith feature from the first feature. This component will not reach the RADAR receiver directly and is not considered in this formulation
- $P_{\text{TRAN}i1}$ = Power transmitted through the first feature which is the reflection from the ith feature

The power incident at the first feature is given by

$$P_{\text{INC}1} = \frac{P_t G A_I}{4\pi R_1{}^2} \tag{2.27}$$

where P_t is the power transmitted by the RADAR, G is the antenna gain, and R_1 is the distance between RADAR and the first feature and A_I is the area of the object illuminated by the RADAR wave. Let Υ_{R_1} be the normalized

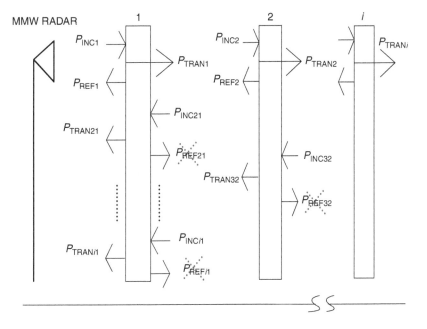

FIGURE 2.27 Power definitions for reflections, absorptions, and transmissions for i multiple line-of-sight features.

RCS and Υ_{a_1} be the normalized absorption cross section of the first feature. The power received by the RADAR receiver from the first feature is given by $\hat{P}'_{REF1} = P_{REF1}A_e/(4\pi R_1^2)$

$$\hat{P}'_{REF1} = \frac{P_t G \hat{\Upsilon}_{R_1} A_I}{(4\pi)^2 \hat{R}_1^4} A_e \qquad (2.28)$$

where A_e is the antenna aperture. It is shown in the RADAR literature that $A_e = G\lambda^2/4\pi$ [21]. Substituting for A_e in Equation (2.28), the power return from the first feature is

$$\hat{P}'_{REF1} = \frac{P_t G^2 \lambda^2 \hat{\Upsilon}_{R_1} A_I}{(4\pi)^3 \hat{R}_1^4} \qquad (2.29)$$

The power P_{TRAN1} that passes through the first feature is given by

$$P_{TRAN1} = \frac{P_t G A_I (1 - [\hat{\Upsilon}_{R_1} + \hat{\Upsilon}_{a_1}])}{(4\pi)\hat{R}_1^2} \qquad (2.30)$$

The power reflected from the second feature, P_{REF2} is given by

$$P_{REF2} = \frac{P_t GA_I{}^2 \hat{\Upsilon}_{R_2}(1 - [\hat{\Upsilon}_{R_1} + \hat{\Upsilon}_{a_1}])}{(4\pi)^2 \hat{R}_1^2 (\hat{R}_2 - \hat{R}_1)^2} \qquad (2.31)$$

The power then transmitted back to the first feature from the second feature is given by

$$P_{INC21} = \frac{P_t GA_I{}^3 \hat{\Upsilon}_{R_2}(1 - [\hat{\Upsilon}_{R_1} + \hat{\Upsilon}_{a_1}])}{(4\pi)^3 \hat{R}_1^2 (\hat{R}_2 - \hat{R}_1)^4} \qquad (2.32)$$

The power, P_{INC21} then passes through feature 1 and is given by

$$P_{TRAN21} = P_{INC21}(1 - [\hat{\Upsilon}_{R_1} + \hat{\Upsilon}_{a_1}]) \qquad (2.33)$$

The power returned from the second feature is then $\hat{P}'_{TRAN21} = P_{TRAN21} A_e / (4\pi \hat{R}_1^2)$

$$\hat{P}'_{TRAN21} = \frac{P_t GA_I{}^3 A_e \hat{\Upsilon}_{R_2}(1 - [\hat{\Upsilon}_{R_1} + \hat{\Upsilon}_{a_1}])^2}{(4\pi)^4 \hat{R}_1^4 (\hat{R}_2 - \hat{R}_1)^4} \qquad (2.34)$$

In general, the predicted power from the ith feature can be written as

$$\hat{P}'_{TRANil}(k+1|k) = \frac{KA_I{}^{(2i-1)} \hat{\Upsilon}_{R_i}(k+1|k)}{(4\pi)^{2i}}$$

$$\times \frac{\prod_{j=0}^{i-1}[1 - (\hat{\Upsilon}_{R_j}(k+1|k) + \hat{\Upsilon}_{a_j}(k+1|k))]^2}{\prod_{j=0}^{i-1}(\hat{R}_{j+1}(k+1|k) - \hat{R}_j(k+1|k))^4} \qquad (2.35)$$

where $K = P_t GA_e$, $A_e = G\lambda^2/4\pi$, $\hat{\Upsilon}_{R_0} = \hat{\Upsilon}_{a_0} = \hat{R}_0 = 0$ and, for the ith feature, $\hat{R}_i$ is related to the augmented state by Equation (2.25).

Equation (2.25), Equation (2.26), and Equation (2.35) between them comprise the observation. In order to generate realistic predictions of the range bins, knowledge of the power and range noise distributions is necessary. This has been studied extensively in previous work, and can be found in Reference 15.

The range and power noise are experimentally obtained [15]. The noise in range is the phase noise, which is obtained by observing the range bins containing reflections from objects with different RCSs at different locations. The noise statistics in power is obtained during both target presence and absence.

The angular standard deviation is assumed to be $1°$ as the RADAR wave is a pencil beam. The observation model is then given by

$$\mathbf{z}_i(k + 1) = [R_i(k + 1), \beta_i(k + 1), P_i(k + 1)]^T + \mathbf{w}_i(k + 1)$$
$$= \mathbf{h}(\mathbf{x}(k + 1)) + \mathbf{w}_i(k + 1) \tag{2.36}$$

where $\mathbf{z}_i(k + 1)$ is the observation, and $\mathbf{w}_i(k + 1)$ is the additive observation noise given by

$$\mathbf{w}_i(k + 1) = [v_R(k + 1)v_\beta(k + 1)v_p(k + 1)]^T \tag{2.37}$$

and $\mathbf{h}$ is the nonlinear observation function defined by Equation (2.25), Equation (2.26), and Equation (2.35).

2.9 MULTI-TARGET RANGE BIN PREDICTION — RESULTS

To validate the formulation for realistically predicting multiple line-of-site target range bins, tests using a RADAR unit from Navtech Electronics were carried out. Initially the vehicle was positioned at pose $\mathbf{x}_v(k)$ as demonstrated in Figure 2.28. The full $360°$ RADAR scan obtained from this vehicle location is shown in Figure 2.25. Range bins obtained from the initial vehicle location at two different bearing angles are shown in Figure 2.26 and Figure 2.29a. Figure 2.26 is obtained at azimuth $231°$ and is indicated by the black line in Figure 2.25. Features in the environment are marked in the figures. The next predicted vehicle location is calculated using the vehicle model and system inputs (Equation [2.24]). This corresponds to the new predicted vehicle pose $\hat{\mathbf{x}}_v(k + 1 \mid k)$ in Figure 2.28. The range spectra in all directions are then predicted from the new predicted vehicle location. For example, in the range bin predicted at angle $\hat{\beta}(k + 1 \mid k)$ in Figure 2.28, the predicted values for the range, bearing and received power of features A and D are calculated according to Equation (2.25), Equation (2.26), and Equation (2.35).

A single range prediction obtained from the predicted vehicle location $\mathbf{x}_v(k + 1 \mid k)$ is shown in Figure 2.29b having two features down-range. Equation (2.35) can be used to predict the received power as long as the power bias as a function of range incorporated into the RADAR electronics is taken into account. This simply requires knowledge of the RADAR's high pass filter circuitry which in an FMCW RADAR compensates for the fourth power of range loss, expected according to the simple RADAR Equation [15, 21].

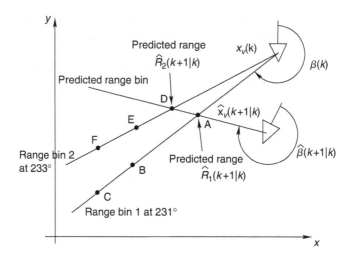

Figure 2.28 Vehicle motion and the features observed/predicted. Features observed/predicted down-range at different bearings are marked.

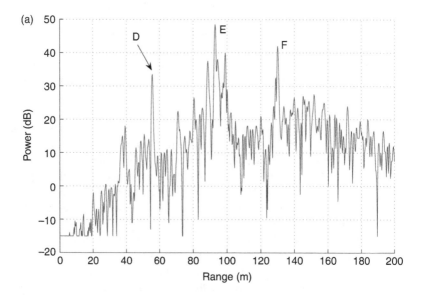

Figure 2.29 Observed and one step ahead predicted range spectra. (a) RADAR range spectra (233° azimuth) obtained at the starting robot location. Two features observed down-range are marked. (b) Predicted RADAR range spectra (at 234° bearing) obtained from the predicted vehicle location.

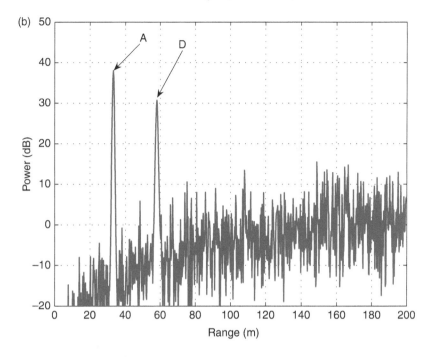

FIGURE 2.29   Continued.

The actual observation is obtained from the next vehicle location and is shown in Figure 2.30a which shows power peaks in close proximity to those predicted in Figure 2.29b. The predicted and actual received powers from the target at A are in close agreement in both figures whereas, the predicted value for the received power (30 dB) of the target at 58 m (feature D in Figure 2.29b) is slightly less than the actual received power (38 dB) in Figure 2.30a. The discrepancy for feature D can be due to violation of some of the assumptions made in the formulation — in particular that the normalized reflection and absorption cross-sections remain constant, independent of the RADAR to target angle of incidence.

Figure 2.30b shows the results of a chi-squared test to determine any bias or inconsistency in the power–range bin predictions. The difference between the measured and the predicted range bins is plotted together with 99% confidence interval. The value of 99% bound, $= \pm 16.35$ dB, has been found experimentally by recording several noisy power–range bins in target absence (RADAR pointing toward open space) [15]. Close analysis of Figure 2.30b shows that the error has a negative bias. This is due to the approximate assumption of the high pass filter gain. For the RADAR used here, the gain of the high pass filter

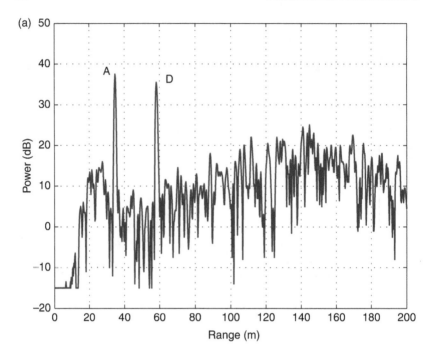

FIGURE 2.30 An actual range bin and the error between the predicted and observed spectra. (a) Actual RADAR range spectra (at 234° bearing) obtained at the next robot location. Features observed down-range are marked. (b) The difference between predicted and measured range bins containing two features down-range is shown. This error is compared against 3σ noise power bounds.

used in the predicted power–range bins was set to 60 dB/decade.[11] The result shows that this approximation for the high pass filter gain is acceptable, as a large portion of the error plot lies within the 3σ limits.

This formulation and analysis shows the initial stages necessary in implementing an augmented state, feature rich SLAM formulation with MMW RADAR. Future work will address the ease with which data association can be carried out using the multidimensional feature state estimates, and a full SLAM implementation in outdoor environments, will be tested.

---

[11] Assuming the RADAR range equation to be correct, a high pass filter with a gain of 40 dB/decade should produce a flat power response for particular targets at various ranges. Figure 2.26 shows a power–range spectrum recorded from the RADAR. It can be seen from Figure 2.26, that the power range response is not flat. For this particular RADAR it makes sense to either determine the bias in the power–range spectra or, model the high pass filter as having a gain of 60 dB/decade, which would better approximate the power–range relationship actually produced in Figure 2.26.

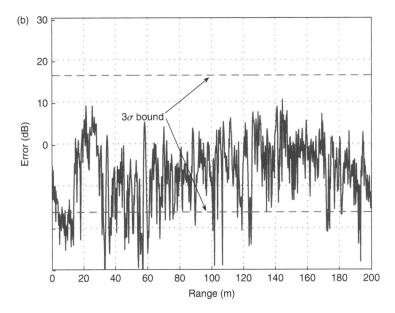

FIGURE 2.30   Continued.

2.10   CONCLUSIONS

This chapter describes a new approach in predicting RADAR range bins which is essential for SLAM with MMW RADAR.

A noise analysis during signal absence and presence was carried out. This is to understand the MMW RADAR range spectrum and to predict it accurately as it is necessary to know the power and range noise distributions in the RADAR power–range spectra. RADAR range bins are then simulated using the RADAR range equation and the noise statistics, which are then compared with real results in controlled environments. In this chapter, it is demonstrated that it is possible to provide realistic predicted RADAR power/range spectra, for multiple targets down-range.

Feature detection based on target presence probability was also introduced. Results are shown which compare probability-based feature detection with other feature extraction techniques such as constant threshold on raw data and CFAR techniques. A difficult compromise in the CA-CFAR method is the choice of the window size which results in a play-off between false alarms and missed detections. Variants of the CFAR method exist, which can be tuned to overcome the problem of missed detections, but the problem of false alarms remains inherent to these methods.

The target presence probability algorithm presented here does not rely on adaptive threshold techniques, but estimates the probability of target presence based on local signal-to-noise power estimates, found from several range bins. The results show that the algorithm can detect features in the typically cluttered outdoor environments tested, with a higher success rate compared to the constant threshold and CFAR feature detection techniques.

A SLAM formulation using an augmented state vector which includes the normalized RCS and absorption cross-sections of features, as well as the usual feature Cartesian coordinates, was introduced. This is intended to aid the data association process, so that features need not just be associated based on their Cartesian coordinates, but account can be taken of their estimated normalized reflection and absorption cross-sections also.

The final contribution is a predictive model of the form and magnitudes of the power–range spectra from differing vehicle locations, for multiple line-of-sight targets. This forms a predicted power–range observation, based on estimates of the augmented SLAM state. The formulation of power returns from multiple objects down-range is explained and predicted RADAR range spectra are compared with real spectra, recorded outdoors.

This work is a step toward building reliable maps and localizing a vehicle to be used in mobile robot navigation. Further methods of including the target presence probability of feature estimates into SLAM are being investigated.

APPENDIX

The binary hypothesis testing problem is a special case of decision problems. The decision space consists of target presence and target absence represented by δ_0 and δ_1, respectively. There is a hypothesis corresponding to each decision. H_0 is called null hypothesis (hypothesis accepted by choosing decision δ_0) and H_1 is called the alternative hypothesis. The binary hypothesis problem has four possible outcomes:

- H_0 was true, δ_0 is chosen : correct decision.
- H_1 was true, δ_1 is chosen : correct decision.
- H_0 was true, δ_1 is chosen : False alarm, also known as a type I error.
- H_1 was true, δ_0 is chosen : missed detection also known as a type II error.

ACKNOWLEDGMENTS

This work was funded under the first author's AcRF Grant, RG 10/01, Singapore. We gratefully acknowledge John Mullane for providing some of the outdoor RADAR scans and Javier Ibanez-Guzman, SIMTech Institute of Manufacturing Technology, Singapore, for use of the utility vehicle. We further acknowledge the valuable advice from Graham Brooker (Australian Centre for Field Robotics) and Steve Clark (Navtech Electronics, UK).

REFERENCES

1. J. J. Leonard and Hugh F. Durrant-Whyte. Dynamic map building for an autonomous mobile robot. In *IEEE International Workshop on Intelligent Robots and Systems*, pp. 89–96, Ibaraki, Japan, July 1990.

2. Fan Tang, Martin Adams, Javier Ibanez-Guzman, and Sardha Wijesoma. Pose invariant, robust feature extraction from range data with a modified scale space approach. In *IEEE International Conference on Robotics and Automation (ICRA)*, New Orleans, USA, April 2004.

3. Alex Foessel, Sachin Chheda, and Dimitrios Apostolopoulos. Short-range millimeter-wave radar perception in a polar environment. In *Proceedings of the Field and Service Robotics Conference*, pp. 133–138, Pittsburgh, PA, USA, August 1999.

4. Steve Clark and Hugh F. Durrant-Whyte. Autonomous land vehicle navigation using millimeter wave radar. In *International Conference on Robotics and Automation (ICRA)*, pp. 3697–3702, Leuven, Belgium, 1998.

5. Graham Brooker, Mark Bishop, and Steve Scheding. Millimetre waves for robotics. In *Australian Conference for Robotics and Automation*, Sydney, Australia, November 2001.

6. J. Leonard and Hugh F. Durrant-Whyte. *Directed Sonar Sensing for Mobile Robot Navigation.*, Kluwer Academic Publishers, Dordrecht, 1992.

7. Somajyoti Majumder. *Sensor Fusion and Feature Based Navigation for Subsea Robots*, The University of Sydney, August 2001.

8. Scott Boehmke, John Bares, Edward Mutschler, and Keith Lay. A high speed 3D radar scanner for automation. In *Proceedings of ICRA '98*, Vol. 4, pp. 2777–2782, May 1998.

9. Steve Clark and G. Dissanayake. Simultaneous localisation and map building using millimetre wave radar to extract natural features. In *IEEE International Conference on Robotics and Automation (ICRA)*, pp. 1316–1321, Detroit, Michigan, May 1999.

10. Steve Clark. *Autonomous Land Vehicle Navigation Using Millimetre Wave Radar*. PhD thesis, Australian Centre for Field Robotics, University of Sydney, 1999.

11. Alex Foessel. *Scene Modeling from Motion-Free Radar Sensing*. PhD thesis, Robotics Institute, Carnegie Mellon University, Pittsburgh, PA, January 2002.

12. Alex Foessel, John Bares, and William Red L. Whittaker. Three-dimensional map building with MMW radar. In *Proceedings of the 3rd International Conference on Field and Service Robotics*, Helsinki, Finland, June 2001. Yleisjljenns, Painnoprssi.

13. Hans Moravec and A. E. Elfes. High resolution maps from wide angle sonar. In *Proceedings of the 1985 IEEE International Conference on Robotics and Automation*, pp. 116–121, March 1985.

14. Sebastian Thrun. Learning occupancy grids with forward models. In *Proceedings of the Conference on Intelligent Robots and Systems*, Hawaii, 2001.

15. Ebi Jose and Martin D. Adams. Millimetre wave radar spectra simulation and interpretation for outdoor slam. In *International Conference on Robotics and Automation (ICRA)*, New Orleans, USA, April 2004.

16. Steve Clark and Hugh F. Durrant-Whyte. The design of a high performance MMW radar system for autonomous land vehicle navigation. In *FSR'97 Proceedings of the International Conference on Field and Service Robotics*, A. Zelinsky, Ed. Australian Robotic Association Inc, Sydney, NSW, Australia, pp. 292–299, 1997.

17. S. Scheding, G. Brooker, M. Bishop, R. Hennessy, and A. Maclean. Terrain imaging millimetre wave radar. In *International Conference on Control, Automation, Robotics and Vision*, Singapore, November 2002.

18. Dirk Langer. *An Integrated MMW Radar System for Outdoor Navigation*. PhD thesis, Robotics Institute, Carnegie Mellon University, Pittsburgh, PA, January 1997.

19. M. E. Adamski, K. S. Kulpa, M. Nalecz, and A. Wojtkiewicz. Phase noise in two-dimensional spectrum of video signal in FMCW homodyne radar. In *13th International Conference on Microwaves, Radar and Wireless Communications, (MIKON-2000)*, Vol. 2, pp. 645–648, 2000.

20. K. Nakamura, T. Hara, M. Yoshida, T. Miyahara, and H. Ito, Optical frequency domain ranging by a frequency-shifted feedback laser. *IEEE Journal of Quantum Electronics*, 36, 305–316, 2000.

21. M. I. Scolnik. *Introduction to Radar Systems*. McGraw Hill, New York, 1982.

22. F. R. Connor. *Noise. Introductory Topics in Electronics and Telecommunications*, 2nd ed. Edward Arnold, 1982.

23. Douglas C. Montgomery and George C. Runger. *Applied Statistics and Probability for Engineers*. John Wiley and Sons, Inc, 2nd ed., 1999.

24. N. C. Currie and C. E. Brown. *Principles and Applications of MMW Radar*. Artech House, Dedham, MA, 1987.

25. P. P. Gandhi and S. A. Kassam. Analysis of CFAR processors in nonhomogeneous background. *IEEE Transactions on AES*, 4, 427–445, 1988.

26. G. Davidson, H. D. Griffiths, and S. Ablett. Analysis of high-resolution land clutter. *IEE Proceedings — Visual Image Signal Processing*, 151, 86–91, 2004.

27. P. P. Gandhi and S. A. Kassam. Optimality of the cell averaging CFAR detector. *IEEE Transactions on Information Theory*, 40, 1226–1228, 1994.

28. S. Watts. The performance of cell-averaging CFAR systems in sea clutter. In *The Record of the IEEE 2000 International Radar Conference*, Alexandria, VA, USA, May 2000.

29. H. Rohling and R. Mende. Os CFAR performance in a 77 GHz radar sensor for car application. In *CIE International Conference of Radar*, pp. 109–114, October 1996.

30. Israel Cohen and Baruch Berdugo. Noise estimation by minima controlled recursive averaging for robust speech enhancement. *IEEE Signal Processing Letters*, 9, 12–15, 2002.

31. H. L. Van Trees. *Detection, Estimation and Modulation Theory — Part I*. Wiley, New York, 1968.

32. T. Kirubarajan and Y. Bar-Shalom. *Multisensor-Multitarget Statistics in Data Fusion Handbook*. CRC Press, Boca Raton, FL, 2001.

33. Robert M. Gray and Lee D. Davisson. *Introduction to Statistical Signal Processing*. Cambridge University Press, London, New York, December 2004.

34. S. F. Boll. Supression of acoustic noise in speech using spectral subtraction. *IEEE Transactions on Acoustic, Speech and Signal Processing*, ASSP-27, 113–120, 1979.
35. M. Berouti, R. Schwartz, and J. Makhoul. Enhancement of speech corrupted by acoustic noise. In *Proceedings of the IEEE ICASSP'79*, pp. 208–211, Washington, USA, 1979.
36. Ebi Jose and Martin D. Adams. Relative radar cross section based feature identification with millimetre wave radar for outdoor slam. In *International Conference on Intelligent Robots and Systems (IROS)*, Sendai, Japan, September 2004.
37. D. D. Ferris, Jr. and N. C. Currie. Microwave and millimeter-wave systems for wall penetration. In *Proceedings of the SPIE — The International Society for Optical Engineering*, Vol. 3375, pp. 269–279, Orlando, FL, USA, 1998.
38. J. Guivant, E. Nebot, and S. Baiker. Autonomous navigation and map building using laser range sensors in outdoor applications. *Journal of Robotic Systems*, 17, 565–583, 2000.
39. Y. Bar-Shalom and T. E. Fortmann. *Tracking and Data Association*. Academic Press, New York, 1988.

Biographies

Martin Adams is an associate professor in the School of Electrical and Electronic Engineering, Nanyang Technological University (NTU), Singapore. He obtained his first degree in engineering science at the University of Oxford, U.K. in 1988 and continued to study for a D.Phil. at the Robotics Research Group, University of Oxford, which he received in 1992. He continued his research in autonomous robot navigation as a project leader and part time lecturer at the Institute of Robotics, Swiss Federal Institute of Technology (ETH), Zurich, Switzerland. He was employed as a guest professor and taught control theory in St. Gallen (Switzerland) from 1994 to 1995. From 1996 to 2000, he served as a senior research scientist in robotics and control, in the field of semiconductor assembly automation, at the European Semiconductor Equipment Centre (ESEC), Switzerland.

He is currently the principal investigator of two robotics projects at NTU, both of which involve sensor data interpretation for SLAM and other mobile robot applications. His other research interests include active LADAR design, range data processing and data fusion with inertial navigation systems and other aiding devices.

Ebi Jose is a Ph.D. student at the School of Electrical and Electronic Engineering, Nanyang Technological University (NTU), Singapore. He obtained his B.Tech. degree in instrumentation from Cochin University of Science and Technology, Kerala, India in 2002. During his degree he worked in the semi-conductor testing industry where he designed and developed equipment for the nondestructive testing of semiconductor devices.

His current research interests include MMW RADAR sensor perception, RADAR signal processing, online feature detection, and autonomous navigation of land vehicles.

3 Data Fusion via Kalman Filter: GPS and INS

Jingrong Cheng, Yu Lu, Elmer R. Thomas, and Jay A. Farrell

CONTENTS

3.1 INTRODUCTION

Data fusion is the process of combining sensory information from different sources into one representational data format. The source of information may come from different sensors that provide information about completely different aspects of the system and its environment; or that provide information about the same aspect of the system and its environment, but with different signal quality or frequency. A group of sensors may provide redundant information, in this case, the fusion or integration of the data from different sensors enables the system to reduce sensor noise, to infer information that is observable but not directly sensed, and to recognize and possibly recover from sensor failure. If a group of sensors provides complementary information, data fusion makes it possible for the system to perform functions that none of the sensors could accomplish independently. In some cases data fusion makes it possible for the system to use lower cost sensors while still achieving the performance specification.

Data fusion is a large research area with various applications and methods [1–3]. In addition to having a thorough understanding of various data fusion methods, it is useful to understand which methods most appropriately fit the corresponding aspects of a particular application. In many autonomous vehicle applications it is useful to dichotomize the overall set of application information into (internal) vehicle information and (external) environmental information. One portion of the vehicle information is the vehicle state vector. Accurate estimation of the vehicle state is a small portion of the data fusion problem that must be solved onboard an autonomous vehicle to enable complex missions; however, accurate estimation of the vehicle state is critical to successful planning, guidance, and control. When it is possible to analytically model the vehicle state dynamics and the relation between the vehicle state and the sensor measurements, the Kalman filter (KF) and the extended Kalman filter (EKF) are often useful tools for accurately fusing the sensor data into an accurate state estimate. In fact, when certain assumptions are satisfied, the KF is the optimal state estimation algorithm. The KF and its properties are reviewed in Section 3.2.

This chapter has two goals. The first is to review the theory and application of the KF as a method to solve data fusion problems. The second is to discuss the use of the EKF for fusing information from the global positioning system (GPS) with inertial measurements to solve navigation problems for autonomous vehicles. Various fusion paradigms have been suggested in the literature — GPS, inertial navigation system (INS) only, INS with GPS resetting, INS with GPS position aiding (i.e., loose coupling), and INS with GPS range aiding (i.e., tight coupling). This chapter presents each approach and discusses the issues that are expected to affect performance. Discussion of latency, asynchronous, and nonlinear measurements are also included.

3.1.1 Data Fusion — GPS and INS

For planning, guidance, and control applications it is critical that the state of the vehicle be accurately estimated. For these applications, the state vector of the vehicle includes the three-dimensional (3D) position, velocity, and attitude. Often, it is also possible to estimate the acceleration and angular rate. Various sensor suites and data fusion methods have been considered in the literature [4–8]. This chapter focuses on one of the most common sensor suites [9–11] — fusion of data from an inertial measurement unit (IMU) and a GPS receiver. The chapter considers the positive and negative aspects of various methods that have been proposed for developing an integrated system.

An IMU provides high sample rate measurements of the vehicle acceleration and angular rate. An INS integrates the IMU measurements to produce position, velocity, and attitude estimates. INSs are self-contained and are not sensitive to external signals. Since an INS is an integrative process, measurement errors within the IMU can result in navigation errors that will grow without bound. The rate of growth of the INS errors can be decreased through the use of higher fidelity sensors or through sensor calibration. In addition, the INS errors (and calibrations) can be corrected through the use of external sensors. With a well-designed data fusion procedure, even an inexpensive INS can provide high frequency precise navigation information [12]. The rate of growth of the INS error will depend on the IMU characteristics and data fusion approach.

A GPS receiver measures information that can be processed to directly estimate the position and velocity of the receiver antenna. More advanced multi-antenna GPS approaches can also estimate the vehicle attitude [13–15]. The accuracy of the vehicle state estimate attained by GPS methods depends on the receiver technology and the processing method. Civilian nondifferential GPS users can attain position estimates accurate to tens of meters. Differential GPS users can attain position estimates accurate to a few meters. Differential GPS users capable of resolving carrier phase integer ambiguities can attain position estimates accurate to a few centimeters. The main disadvantage of state estimates determined using GPS is that the estimates are dependent on reception of at least four GPS satellite signals by the receiver. Satellite signal reception requires direct line of sight between the receiver and the satellite. While this line of sight is obstructed for a sufficiently large number of satellites, the GPS solution will not be available.

GPS and INS have complementary properties which have motivated various researchers to study methods to fuse the data from the two systems. The objective is to attain high performance for a higher percentage of the time than either approach could attain independently. This chapter uses GPS and INS to illustrate the use of the KF for data fusion. Section 3.2 reviews the KF, EKF, and various properties and application issues. Section 3.3 reviews the various issues related to the GPS system. Of particular interest will be various assumed dynamic models and issues affecting state estimation accuracy. Section 3.4 provides a brief review of INS and their error models. The main objective is to present the model information necessary to analyze alternative methods for combining GPS and INS information, which is done in Section 3.5.

3.2 KALMAN FILTER

Since R. E. Kalman published his idea in the early 1960s [16,17], the KF has been the subject of extensive research. It has been applied successfully to solve many practical problems in different fields, particularly in the area of autonomous navigation. The KF involves two basic steps: use of the system dynamic model to predict the evolution of the state between the times of the measurements and use of the system measurement model and the measurements to optimally correct the estimated state at the time of the measurements. It is well known that the KF is recursive, computationally efficient, and optimal in the sense of the minimum mean of the squared errors [18].

This section contains three subsections. Section 3.2.1 reviews the linear dynamic system models that are required for the prediction and measurement update steps of the KF. Section 3.2.2 reviews the KF algorithm, a few of its properties, and methods to address various implementation issues. Section 3.2.3 reviews the EKF algorithm which is applicable when either the dynamic or measurement model of the system is not linear. The EKF is needed in GPS–INS data fusion applications since the INS dynamic model is nonlinear and the GPS measurement model may be nonlinear.

3.2.1 Stochastic Process Models

Because the state of most physical systems evolve in continuous time, continuous-time dynamic models are of interest. The dynamic behavior of a linear continuous-time system driven by a random process $\omega(t)$ may be described mathematically by a set of ordinary differential equations:

$$\dot{\mathbf{x}}(t) = \mathbf{F}(t)\mathbf{x}(t) + \mathbf{G}(t)\omega(t) \tag{3.1}$$

$$\mathbf{y}(t) = \mathbf{H}(t)\mathbf{x}(t) + \mathbf{v}(t) \tag{3.2}$$

where $\mathbf{x}(t)$ is the n-element state vector of the system, $\mathbf{F}(t)$ is the system matrix, $G(t)$ is the input distribution matrix, $\mathbf{y}(t)$ is the measurement vector, $\mathbf{H}(t)$ is the measurement matrix, and $\mathbf{v}(t)$ is the measurement noise vector.

The vectors $\omega(t)$ and $\mathbf{v}(t)$ are assumed to be white and Gaussian with

$$E[\omega(t)] = \mathbf{0}, \quad E[\omega(t)\omega^{\mathrm{T}}(t+\tau)] = \mathbf{Q}_w(t)\delta(\tau) \tag{3.3}$$

$$E[\mathbf{v}(t)] = \mathbf{0}, \quad E[\mathbf{v}(t)\mathbf{v}^{\mathrm{T}}(t+\tau)] = \mathbf{R}(t)\delta(\tau) \tag{3.4}$$

where $\mathbf{Q}_w$ is the power spectral density (PSD) of the white noise $\omega(t)$ and $\mathbf{R}(t)$ is the covariance matrix of the measurement noise process $\mathbf{v}(t)$. If either $\omega(t)$ or $\mathbf{v}(t)$ is not white, then it may be possible to append linear dynamics to the model of Equation (3.1) and Equation (3.2) to still utilize the model of a linear system driven by white noise, see Reference 19. For the state estimation design discussions of this chapter, unless otherwise stated, assume that the system model has been manipulated into the form of Equation (3.1) and Equation (3.2) with white process and measurement noise.

In applications, such as those involving GPS, where the measurements occur at discrete instants of time, it is convenient to utilize a discrete-time formulation of the KF. If we denote the sequence of measurement times by $t_1, \ldots, t_k, t_{k+1}, \ldots$, then implementation of the discrete-time KF requires a model for propagating the state between measurement times and a model for the relation between the state and the measurement that is valid at the measurement time. Using linear system theory [20,21], the state transition valid between t_k and t_{k+1} is

$$\mathbf{x}(t_{k+1}) = \mathbf{\Phi}(t_{k+1}, t_k)\mathbf{x}(t_k) + \mathbf{w}(t_k) \tag{3.5}$$

where

$$\mathbf{w}(t_k) = \int_{t_k}^{t_{k+1}} \mathbf{\Phi}(t_{k+1}, \tau)G(\tau)\omega(\tau)\,\mathrm{d}\tau \tag{3.6}$$

and $\mathbf{\Phi}(t_{k+1}, t)$ is the state transition matrix from t to t_{k+1}. The measurement model valid at t_k is

$$\mathbf{y}(t_k) = \mathbf{H}(t_k)\mathbf{x}(t_k) + \mathbf{v}(t_k) \tag{3.7}$$

To simplify notation, these equations will be written as

$$\mathbf{x}_{k+1} = \mathbf{\Phi}_k\mathbf{x}_k + \mathbf{w}_k \tag{3.8}$$

$$\mathbf{y}_k = \mathbf{H}_k\mathbf{x}_k + \mathbf{v}_k \tag{3.9}$$

The process noise $\mathbf{w}_k$ and measurement noise $\mathbf{v}_k$ are assumed to be zero-mean, white noise with covariance properties as follows:

$$E[\mathbf{w}_k \mathbf{w}_j^T] = \begin{cases} \mathbf{Q}_k, & k = j \\ \mathbf{0}, & k \neq j \end{cases} \tag{3.10}$$

$$E[\mathbf{v}_k \mathbf{v}_j^T] = \begin{cases} \mathbf{R}_k, & k = j \\ \mathbf{0}, & k \neq j \end{cases} \tag{3.11}$$

$$E[\mathbf{w}_k \mathbf{v}_j^T] = \mathbf{0}, \quad \text{for all } k \text{ and } j \tag{3.12}$$

3.2.1.1 Computation of Φ and $\mathbf{Q}_k$

The covariance matrix associated with $\mathbf{w}(t_k)$ is:

$$\mathbf{Q}_k = \mathbf{Q}(t_{k+1}, t_k) = \int_{t_k}^{t_{k+1}} \Phi(t_{k+1}, \tau) \mathbf{G}(\tau) \mathbf{Q}_w \mathbf{G}^T(\tau) \Phi^T(t_{k+1}, \tau) \, d\tau \tag{3.13}$$

For systems where $\mathbf{F}(t)$, $\mathbf{G}(t)$, and $\mathbf{Q}_w(t)$ are accurately approximated as constant over the interval of integration, the transition matrix can be calculated by the inverse Laplace transform

$$\Phi(t_{k+1}, t_k) = \ell^{-1}\{[s\mathbf{I} - \mathbf{F}]^{-1}\}_{t=t_{k+1}-t_k} \tag{3.14}$$

Alternative methods to compute Φ_k and $\mathbf{Q}_k$ use matrix exponentials [22,23] or Taylor series expansions. A common method for computing Φ_k is the truncated power series:

$$\Phi(\Delta t) = e^{\mathbf{F}\Delta t} = \mathbf{I} + \mathbf{F}\Delta t + \frac{\mathbf{F}^2 \Delta t^2}{2!} + \frac{\mathbf{F}^3 \Delta t^3}{3!} + \cdots \tag{3.15}$$

where $\Delta t = t_{k+1} - t_k$ and the choice of the order of the power series depends on the system design requirements.

When $\mathbf{F}$ is time varying, it is necessary to subdivide Δt such that $\mathbf{F}$ can be considered as constant on the subintervals $\Delta \tau_i = \tau_i - \tau_{i-1}$ where $\tau_0 = t_k$, $\tau_N = t_{k+1}$, and $\tau_i = \tau_{i-1} + \Delta \tau_i$ for $i = 1, \ldots, N$. Let $\Delta t = \sum_{i=1}^{N} \Delta \tau_i$ then $\Phi_k \prod_{i=1}^{N} \Phi(\tau_i, \tau_{i-1})$. The matrix $\mathbf{Q}_k$ can be found by approximation techniques:

$$\mathbf{Q}_k = \mathbf{Q}(\tau_N, \tau_0) \tag{3.16}$$

where by subdividing (3.13) into subintegrals and using $\Phi(\tau_{i+1}, \tau_0) = \Phi(\tau_{i+1}, \tau_i)\Phi(\tau_i, \tau_0)$ we obtain

$$\mathbf{Q}(\tau_i, \tau_0) = \Phi(\tau_i, \tau_{i-1})[\mathbf{GQ}_w\mathbf{G}^\mathsf{T}\Delta\tau_i + \mathbf{Q}(\tau_{i-1}, \tau_0)]\Phi^\mathsf{T}(\tau_i, \tau_{i-1}) \qquad (3.17)$$

for $i = 1, \ldots, N$ with $\mathbf{Q}(\tau_0, \tau_0) = 0$.

Example 3.1   Since a common GPS measurement epoch uses $t_k = k$, this example considers computation of Φ_k and $\mathbf{Q}_k$ over the unit interval $t \in [k, k+1)$ where k is an integer. First, the unit interval is subdivided into N subintervals of length $d\tau = 1/N$ sec. Each subinterval is $[\tau_i, \tau_{i+1})$ where $\tau_i = k + id\tau$ for $i = 0, \ldots, N$. For small $d\tau$,

$$\Phi(\tau_{i+1}, \tau_i) = (\mathbf{I} + \mathbf{F}(\tau_i)d\tau)$$

and

$$\Phi(\tau_{i+1}, \tau_0) = \Phi(\tau_{i+1}, \tau_i)\Phi(\tau_i, \tau_0) \qquad (3.18)$$

therefore,

$$\Phi(\tau_{i+1}, \tau_0) = \Phi(\tau_i, \tau_0) + \mathbf{F}(\tau_i)\Phi(\tau_i, \tau_0)\,d\tau$$

Similarly, over each 1 sec interval, $\mathbf{Q}_k = \mathbf{Q}_k(\tau_N, \tau_0)$ can be integrated as follows:

$$\begin{aligned}
\mathbf{Q}_k(\tau_1, k) &= \Phi(\tau_1, \tau_0)\mathbf{GQ}_w\mathbf{G}^\mathsf{T}\Phi^\mathsf{T}(\tau_1, \tau_0)\,d\tau \\
\mathbf{Q}_k(\tau_2, k) &= \Phi(\tau_2, \tau_1)[\mathbf{GQ}_w\mathbf{G}^\mathsf{T}\,d\tau + \mathbf{Q}_k(\tau_1, k)]\Phi^\mathsf{T}(\tau_2, \tau_1) \\
&\ \ \vdots \\
\mathbf{Q}_k(\tau_N, k) &= \Phi(\tau_N, \tau_{N-1})[\mathbf{GQ}_w\mathbf{G}^\mathsf{T}\,d\tau + \mathbf{Q}_k(\tau_{N-1}, k)]\Phi^\mathsf{T}(\tau_N, \tau_{N-1})
\end{aligned} \qquad (3.19)$$

3.2.2 Basic KF

Since there are numerous books devoted to the derivation of the KF, such as References 19, 20, and 24, the derivation is not included herein. Instead, the KF algorithm and its properties are reviewed.

The KF estimates the state of a stochastic system. To determine optimal gains for the filter at time t_k, the KF compares the covariance of the state estimate at t_k with the covariance of the measurement at t_k. To enable this comparison, the KF algorithm will propagate the covariance of the state estimate as well as the state estimate. Prior to discussing the KF algorithm, it will be useful to summarize the new notation that will be used. The KF gain valid at time t_k

is $\mathbf{K}_k$. The state estimate at time t_j using all measurements up to time t_i will be denoted by $\hat{\mathbf{x}}_{j|i}$. Therefore, $\hat{\mathbf{x}}_{k|k}$ is the estimate of the state at time t_k using all measurements up to and including $\mathbf{y}_k$, while $\hat{\mathbf{x}}_{k|k-1}$ is the estimate of the state at time t_k using all measurements up to and including $\mathbf{y}_{k-1}$. Similarly, $\mathbf{P}_{k|k}$ denotes the covariance of the state estimation error at time t_k after using all measurements available up to and including $\mathbf{y}_k$, and $\mathbf{P}_{k|k-1}$ denotes the covariance of the state estimation error at time t_k after using all measurements available up to and including $\mathbf{y}_{k-1}$.

The KF algorithm is a recursive process. As such, it requires initialization prior to starting the recursion. Assume that the first measurement will occur at t_1 and denote the initialized state estimate and its associated error covariance matrix as $\hat{\mathbf{x}}_{0|0}$ and $\mathbf{P}_{0|0}$. These initial values should be

$$\hat{\mathbf{x}}_{0|0} = E(\mathbf{x}_0), \quad \mathbf{P}_{0|0} = \text{cov}(\mathbf{x}_0) \tag{3.20}$$

and $k - 0$. Often, it will be the case that $\mathbf{P}_{0|0}$ is diagonal with each element being large. The KF is implemented as follows:

1. Predict the state vector and error covariance matrix for the next measurement time:

$$\hat{\mathbf{x}}_{k+1|k} = \mathbf{\Phi}_k \hat{\mathbf{x}}_{k|k} \tag{3.21}$$

$$\mathbf{P}_{k+1|k} = \mathbf{\Phi}_k \mathbf{P}_{k|k} \mathbf{\Phi}_k^{\mathsf{T}} + \mathbf{Q}_k \tag{3.22}$$

    Then, increment $k = k + 1$.
2. Calculate the KF gain matrix for incorporation of $\mathbf{y}_k$:

$$\mathbf{K}_k = \mathbf{P}_{k|k-1} \mathbf{H}_k^{\mathsf{T}} [\mathbf{H}_k \mathbf{P}_{k|k-1} \mathbf{H}_k^{\mathsf{T}} + \mathbf{R}_k]^{-1} \tag{3.23}$$

3. Use $\mathbf{y}_k$ to correct $\hat{\mathbf{x}}_{k|k-1}$:

$$\hat{\mathbf{x}}_{k|k} = \hat{\mathbf{x}}_{k|k-1} + \mathbf{K}_k [\mathbf{y}_k - \mathbf{H}_k \hat{\mathbf{x}}_{k|k-1}] \tag{3.24}$$

4. Compute the error covariance of the state estimate after incorporating $\mathbf{y}_k$:

$$\mathbf{P}_{k|k} = [\mathbf{I} - \mathbf{K}_k \mathbf{H}_k] \mathbf{P}_{k|k-1} \tag{3.25}$$

where $\mathbf{I}$ is an n-dimensional identity matrix.

Steps 1–4 are iterated for each new measurement. This iteration can continue *ad infinitum*. A few facts are important to point out. First, the discrete measurements have not been assumed to be equally spaced in time. The only assumption is that we have a model available, of the form of Equation (3.8) and Equation (3.9), suitable for accurate propagation of the state estimate and state estimate error covariance matrix between measurement instants. Second, Step 3 is the only step that requires the measurement; therefore, when the next measurement time can be accurately predicted, then Steps 1 and 2 are often computed prior to the arrival of the next measurement. The purpose of doing this is to minimize the computational delay between the arrival of y_k and availability of $\hat{x}_{k|k}$ to the other online processes that need it (e.g., planning, guidance, or control). Third, the portions of the KF algorithm that require the majority of the computations are Equation (3.22), Equation (3.23), and Equation (3.25), which are related to maintaining the error covariance matrix and the Kalman gain.

3.2.2.1 Implementation issues

The performance of the KF depends on the accuracy of the process model and the measurement model. The implementation approach also affects the performance and computational load of the KF. This section discusses some of the important implementation issues related to the KF.

Sequential processing of independent measurements. When the system has m simultaneous, but independent measurements, the noise covariance matrix is diagonal:

$$\mathbf{R}_k = \begin{bmatrix} r_1 & 0 & 0 \\ 0 & \ddots & 0 \\ 0 & 0 & r_m \end{bmatrix} \tag{3.26}$$

In this case, it is computationally efficient to treat the measurements as sequential measurements. This replaces an m-dimensional matrix inversion with m scalar divisions. At time t_k, we introduce an auxiliary vector $\hat{x}_0$ and matrix p_0 which are initialized as

$$\mathbf{p}_0 = \mathbf{P}_{k|k-1} \quad \text{and} \quad \hat{x}_0 = \hat{x}_{k|k-1} \tag{3.27}$$

The following recursion is performed for $i = 1$ to m:

$$\mathbf{K}_i = \frac{\mathbf{p}_{i-1}\mathbf{h}_i^{\mathrm{T}}}{r_i + \mathbf{h}_i\mathbf{p}_{i-1}\mathbf{h}_i^{\mathrm{T}}}$$

$$\hat{x}_i = \hat{x}_{i-1} + \mathbf{K}_i[y_i - \mathbf{h}_i\hat{x}_{i-1}] \tag{3.28}$$

$$\mathbf{p}_i = [\mathbf{I} - \mathbf{K}_i\mathbf{h}_i]\mathbf{p}_{i-1}$$

where $\mathbf{h}_i$ is the ith row of the measurement matrix $\mathbf{H}_k$ and y_i is the ith element of the vector $\mathbf{y}$. After the mth step of the recursion, the state and error covariance are

$$\hat{\mathbf{x}}_{k|k} = \hat{\mathbf{x}}_m$$
$$\mathbf{P}_{k|k} = \mathbf{p}_m \tag{3.29}$$

Note that the state estimate $\hat{\mathbf{x}}_{k|k}$ and error covariance $\mathbf{P}_{k|k}$ that result from this scalar processing will exactly match (within numerical error) the results that would have been obtained via the vector processing implementation. The gain matrices $\mathbf{K}$ that result from the vector and scalar processing algorithms will be distinct, due to the different order in which each implementation introduces the measurements.

Rejection of bad measurements. In engineering applications, data does not always match theoretical expectations. Therefore, it is also necessary to set up some criteria to reject some measurements.

For example, if for a scalar measurement y_i the absolute value of the measurement residual $\mathrm{res}_i = y_i - \mathbf{h}_i \hat{\mathbf{x}}_{i-1}$ at time k is sufficiently larger than its standard deviation $\sqrt{\mathbf{h}_i \mathbf{P}_{k|k-1} \mathbf{h}_i^{\mathrm{T}} + r}$, then the measurement could be ignored. In this case, this kth measurement would be missed. Such situations are discussed below.

Missed measurements. Sometimes an expected measurement may be missing. One circumstance under which this could occur was discussed earlier. When a measurement is missing, the "measurement" contains no information; therefore, the uncertainty of the measurement is infinite (i.e., $\mathbf{R} = \alpha \mathbf{I}$ with $\alpha = \infty$). In this case, by Equation (3.23), $\mathbf{K}_k = \mathbf{0}$. Using this fact, in Steps 3 and 4 of the KF, yields

$$\hat{\mathbf{x}}_{k|k} = \hat{\mathbf{x}}_{k|k-1} \tag{3.30}$$

$$\mathbf{P}_{k|k} = \mathbf{P}_{k|k-1} \tag{3.31}$$

The missed measurement has no effect on the estimated state or its state error covariance matrix.

Divergence of the KF. The KF is the optimal state estimator for the modeled system. The KF is stable if certain technical assumptions, including observability and controllability from the process noise vector are met [19–21]. Lack of observability, absence of controllability from the process noise vector, or modeling error can cause the KF state estimate to diverge from the true state. These are issues that must be studied and addressed at the design stage.

Tuning. Ideally, the KF is applied to a well-modeled dynamic system with stochastic process noise and measurement noise satisfying the required assumptions. In such cases, the $\mathbf{Q}$ and $\mathbf{R}$ matrices can be computed correctly as a portion of the stochastic model. In some other applications, examples of which will occur in Section 3.3.3, the vector ω represents unknown factors that may not be truly random. In such applications, $\mathbf{Q}$ and $\mathbf{R}$ are often used as performance tuning parameters. As $\mathbf{Q}$ is decreased relative to $\mathbf{R}$, the KF trusts the dynamic model of the system more than the measurements; therefore, the states of the system converge more slowly since new information is weighted less. If $\mathbf{Q}$ is increased relative to $\mathbf{R}$, the measurements will be weighted more and the states will converge faster; however, the measurement noise will have a larger effect on the accuracy of the filtered solution. Note that in applications where $\mathbf{Q}$ and $\mathbf{R}$ are used as performance tuning parameters, all stochastic interpretations of $\mathbf{P}_{k|k}$ are lost. Instead, the KF is being used as an algorithm to estimate the state, but the KF optimality properties are not applicable.

Maintaining symmetry. The equation

$$\mathbf{P}_{k|k} = [\mathbf{I} - \mathbf{K}_k\mathbf{H}_k]\mathbf{P}_{k|k-1} \qquad (3.32)$$

is a simplified version of

$$\mathbf{P}_{k|k} = [\mathbf{I} - \mathbf{K}_k\mathbf{H}_k]\mathbf{P}_{k|k-1}[\mathbf{I} - \mathbf{K}_k\mathbf{H}_k]^{\mathrm{T}} + \mathbf{K}_k\mathbf{R}_k\mathbf{K}_k^{\mathrm{T}} \qquad (3.33)$$

Equation (3.32) is valid only when $\mathbf{K}_k$ is the optimal Kalman gain matrix. When $\mathbf{K}_k$ is defined by an equation other than Equation (3.23) and is not the KF optimal gain matrix, then Equation (3.33) should be used. Since $\mathbf{P}_{k|k}$ is the error covariance matrix, it should be symmetric and positive semidefinite. Although Equation (3.33) requires more computational operations than Equation (3.32) does, Equation (3.33) is a symmetric equation. However, the symmetry of either result can be guaranteed and the computational requirements are decreased by only computing the lower diagonal half of $\mathbf{P}_{k|k}$.

Maintaining definiteness. Neither Equation (3.32) nor Equation (3.33) guarantees that $\mathbf{P}_{k|k}$ is symmetric or positive semidefinite in the presence of numeric errors. One possible solution is to factorize $\mathbf{P}_{k|k}$ (e.g., $\mathbf{P} = \mathbf{U}\mathbf{D}\mathbf{U}^{\mathrm{T}}$ or $\mathbf{P} = \mathbf{QR}$) and derive algorithms that propagate the factors directly. Such factorized algorithms [20,21] have better numeric stability properties, especially in applications where computational error is an issue.

3.2.3 Extended KF

The previous sections have discussed only linear systems with zero-mean, white Gaussian process, and measurement noise. The optimality properties of the KF under these assumptions were briefly discussed in the previous sections. In many applications either the measurement model, the system dynamics, or both are nonlinear. In these cases the KF may not be the optimal estimator. Nonlinear estimation is a difficult problem without a general solution. Nonlinear estimation methods are discussed, for example, in References 25 and 26. For the navigation systems which are the main focus of this chapter, the EKF has proved very useful because the linearized dynamic and measurement models are accurate for the short periods of time between measurements. Due to its utility in the applications of interest, the EKF is reviewed in this section.

Such navigation systems can be described by the nonlinear stochastic differential equation

$$\dot{\mathbf{x}}(t) = \mathbf{f}(\mathbf{x}, \mathbf{u}, t) + \mathbf{g}(\mathbf{x}, t)\mathbf{w}'(t) \tag{3.34}$$

with a measurement model of the form

$$\mathbf{y}(t) = \mathbf{h}(\mathbf{x}, t) + \mathbf{v}'(t) \tag{3.35}$$

where $\mathbf{f}$ is a known nonlinear function of the state $\mathbf{x}$, the signal $\mathbf{u}$, and time; $\mathbf{g}$ is a known nonlinear function of the state and time; and $\mathbf{w}'$ and $\mathbf{v}'$ are continuous-time white noise processes.

For its covariance propagation and measurement updates, the EKF will use a linearization of Equation (3.34) and Equation (3.35). The linearization is performed relative to a reference trajectory $\mathbf{x}^*(t)$ which is a solution of

$$\dot{\mathbf{x}}^*(t) = \mathbf{f}(\mathbf{x}^*, \mathbf{u}, t)$$

between the measurement time instants. The corresponding reference measurement is

$$\mathbf{y}^*(t) = \mathbf{h}(\mathbf{x}^*, t)$$

The error state vector is defined as

$$\delta\mathbf{x} = \mathbf{x} - \mathbf{x}^*$$

and the measurement residual vector as

$$\delta\mathbf{y}(t) = \mathbf{y}(t) - \mathbf{y}^*(t)$$

The linearized dynamics of the error state vector are

$$\delta\dot{\mathbf{x}}(t) = \mathbf{F}(t)\delta\mathbf{x}(t) + G(t)\mathbf{w}'(t) + \mathbf{e}_x(t) \tag{3.36}$$

and the linearized (residual) measurement model is

$$\delta \mathbf{y}(t) = \mathbf{H}(t)\delta \mathbf{x}(t) + \mathbf{v}'(t) + e_y(t) \tag{3.37}$$

where

$$\mathbf{F}(t) = \left.\frac{\partial f}{\partial \mathbf{x}}\right|_{\mathbf{x}=\mathbf{x}^*}, \quad \mathbf{H}(t) = \left.\frac{\partial \mathbf{h}}{\partial \mathbf{x}}\right|_{\mathbf{x}=\mathbf{x}^*}, \quad \mathbf{G}(t) = g(\mathbf{x}^*, t)$$

and $e_x(t)$, $e_y(t)$ are linearization error terms.

From Equation (3.36) and Equation (3.37), the equivalent model for discrete measurements is

$$\delta \mathbf{x}_{k+1} = \mathbf{\Phi}_k \delta \mathbf{x}_k + \mathbf{w}_k$$

$$\delta \mathbf{y}_k = \mathbf{H}_k \delta \mathbf{x}_k + \mathbf{v}_k$$

where the state transition matrix and process noise covariance matrix can be calculated by the methods given in Section 3.2.1. The approximation will hold only for a short period of time and only if the reference trajectory is near the actual trajectory. For the systems that are the focus of this chapter, the linearization will occur around the computed navigation state. Time propagation occurs between GPS measurement epochs, which are typically separated by only a few seconds. Measurements at a given epoch are assumed to occur simultaneously. The purpose of the GPS corrections is to keep the navigation state near the state of the true system.

Implementation of the EKF algorithm is very similar to that of the KF, in fact, only the state propagation and measurement prediction steps will change. In addition, the $\mathbf{P}$ matrices computed in the algorithm are no longer true covariance matrices; although, we will still use that name in the following text.

To initialize the EKF algorithm, assume that the first measurement will occur at t_1 and denote the initialized state estimate, residual state estimate, and its associated error covariance matrix as $\hat{\mathbf{x}}_{0|0}$, $\delta\hat{\mathbf{x}}_{0|0}$, and $\mathbf{P}_{0|0}$, respectively. These initial values should be

$$\hat{\mathbf{x}}_{0|0} = E(\mathbf{x}_0), \quad \delta\hat{\mathbf{x}}_{0|0} = 0, \quad \mathbf{P}_{0|0} = \text{cov}(\mathbf{x}_0)$$

Since this is a nonlinear estimation process, it is important that $\mathbf{x}(0) - \hat{\mathbf{x}}_{0|0}$ be small. The equations and procedures for the EKF are summarized as follows:

1. Propagate the state estimate to the next measurement time t_{k+1} by integrating

$$\dot{\mathbf{x}}^*(t) = \mathbf{f}(\mathbf{x}^*, \mathbf{u}, t) \tag{3.38}$$

over the time interval $t \in [t_k, t_{k+1}]$ with initial condition $\mathbf{x}^*(t_k) = \hat{\mathbf{x}}_{k|k}$. At the completion of the integration, let $\hat{\mathbf{x}}_{k+1|k} = \mathbf{x}^*(t_{k+1})$. Along the solution $\mathbf{x}^*(t)$, compute

$$\mathbf{F}(t) = \left.\frac{\partial f(x)}{\partial x}\right|_{x=x^*} \quad \text{and} \quad \mathbf{G}(t) = \mathbf{g}(\mathbf{x}^*, t), \quad \text{for } t \in [t_k, t_{k+1}]$$

2. Compute the state transition matrix $\mathbf{\Phi}_k$ and compute the process noise covariance matrix $\mathbf{Q}_k$. Predict the error state vector and error covariance matrix:

$$\delta\hat{\mathbf{x}}_{k+1|k} = \mathbf{\Phi}_k \delta\hat{\mathbf{x}}_{k|k} = \mathbf{\Phi}_k 0 = 0 \tag{3.39}$$

$$\mathbf{P}_{k+1|k} = \mathbf{\Phi}_k \mathbf{P}_{k|k} \mathbf{\Phi}_k^T + \mathbf{Q}_k \tag{3.40}$$

The reason that $\delta\hat{\mathbf{x}}_{k|k}$ is set to $\mathbf{0}$ in (3.39) is clarified in the discussion following (3.43).

3. Increment $k = k + 1$.
4. Linearize the measurement matrix at $\mathbf{x}^*(t_k)$ and calculate the EKF gain matrix:

$$\mathbf{H}_k = \mathbf{H}(t_k) = \left.\frac{\partial \mathbf{h}}{\partial \mathbf{x}}\right|_{\mathbf{x}=\hat{\mathbf{x}}_{k|k-1}} \tag{3.41}$$

$$\mathbf{K}_k = \mathbf{P}_{k|k-1}\mathbf{H}_k^T[\mathbf{H}_k\mathbf{P}_{k|k-1}\mathbf{H}_k^T + \mathbf{R}_k]^{-1}$$

5. Compute the error states using the residual measurements:

$$\delta\hat{\mathbf{x}}_{k|k} = \delta\hat{\mathbf{x}}_{k|k-1} + \mathbf{K}_k[\delta\mathbf{y}_k - \mathbf{H}_k\delta\hat{\mathbf{x}}_{k|k-1}]$$
$$= \mathbf{K}_k\delta\mathbf{y}_k \tag{3.42}$$

where $\delta\hat{\mathbf{x}}_{k|k-1}$ is the error state vector estimated prior to the new measurements, which by Equation (3.39) is zero.

6. Update the estimated states $\hat{\mathbf{x}}_{k|k}$:

$$\hat{\mathbf{x}}_{k|k} = \hat{\mathbf{x}}_{k|k-1} + \delta\hat{\mathbf{x}}_{k|k} \tag{3.43}$$

Since the error state has been included in the state estimate, the error has been corrected; therefore, the new best estimate of the error state is zero. Therefore, $\delta\hat{\mathbf{x}}_{k|k}$ must be set to zero: $\delta\hat{\mathbf{x}}_{k|k} = 0$.

7. Update the posterior error covariance matrix:

$$\mathbf{P}_{k|k} = [\mathbf{I} - \mathbf{K}_k\mathbf{H}_k]\mathbf{P}_{k|k-1}$$

The EKF iterates Steps 1 to 7 for the duration of the application. Steps 1 and 2 perform computations related to time propagation of the state and the matrix **P**. Steps 4 to 7 perform computations related to the measurement update.

In the EKF algorithm, the computation, use, resetting, and time propagation of $\delta\hat{\mathbf{x}}_{k|k}$ often causes confusion. The above algorithm is a *total state* implementation. In an alternative *error state* implementation of the algorithm, Step 6 could be removed. Without Step 6, Equation (3.39) of Step 2 would have to be implemented to time propagate the error state and the simplification to Equation (3.42) of Step 5 would not be possible. In this alternative implementation, it is possible that, over time, $\delta\hat{\mathbf{x}}_{k|k}$ could become large. In this case, x* is not near the actual state. In this case, the linearized equations may not be accurate. The EKF algorithm as presented (using Step 6) includes $\delta\hat{\mathbf{x}}_{k|k}$ in **x*** resulting in a more accurate linearization. The total and error state implementations are discussed in greater detail in References 20 and 27.

3.3 GPS NAVIGATION SYSTEM

The purpose of this section is to discuss the advantages and disadvantages of various EKF approaches to state estimation using GPS measurements. Section 3.3.1 presents background information about GPS that is necessary for the subsequent discussions. Section 3.3.2 discusses position estimation based on GPS measurements. The EKF approaches to solving the GPS equations are compared in Section 3.3.3.

3.3.1 GPS Measurements

The GPS is designed to provide position, velocity, and time estimates to users at all times, in all weather conditions, anywhere on the Earth. The existing GPS signal for each satellite consists of a spectrum spreading code and data bits modulated onto a carrier signal. By accurately measuring the transit time of the code signal, the receiver can form a measurement of the pseudorange between the satellite and the receiver antenna. This measurement is referred to as a pseudorange as it is also affected by receiver and satellite clock errors. By processing the data bits to determine the clock error model and ephemeris data, the receiver can compute the satellite position and clock errors as a function of time. Tracking the satellite signal requires that the receiver acquire either frequency or phase lock to the satellite carrier signal. Phase information from the tracking loop has utility as an additional range measurement and the change in the phase measurement over a known period of time (referred to in the GPS literature as a Doppler measurement) can be used to estimate the receiver velocity. The GPS satellites broadcast signals on two frequencies: L1 and L2. Users with "two frequency" receivers can obtain pseudorange, phase, and Doppler measurements for each of the two frequencies.

The L1 and L2 code and carrier phase measurements from a given satellite can be modeled as

$$\tilde{\rho}_{L1} = R + b_u + c\Delta t_{sv} + \frac{f_2}{f_1}I_a + E_{cm} + MP_1 + \eta_1$$

$$\tilde{\rho}_{L2} = R + b_u + c\Delta t_{sv} + \frac{f_1}{f_2}I_a + E_{cm} + MP_2 + \eta_2$$

$$\tilde{\phi}_{L1}\lambda_1 + N_1\lambda_1 = R + b_u + c\Delta t_{sv} - \frac{f_2}{f_1}I_a + E_{cm} + mp_1 + n_1$$

$$\tilde{\phi}_{L2}\lambda_2 + N_2\lambda_2 = R + b_u + c\Delta t_{sv} - \frac{f_1}{f_2}I_a + E_{cm} + mp_2 + n_2$$

where $R = \|X_{sv} - X_u\|$ is the geometric distance between the satellite position X_{sv} and receiver antenna position X_u, b_u is the receiver clock bias, and $c\Delta t_{sv}$ is the satellite clock bias. The satellite clock bias can be partially corrected by ephemeris data. E_{cm} represents common errors other than dispersive effects such as ionosphere and I_a represents ionospheric error. The symbols η and n represent receiver measurement noise. The symbols mp and MP represent errors due to multipath. Note that the receiver clock bias is identical across satellites for all simultaneous pseudorange and phase measurements. Since the receiver phase lock loops can only track changes in the signal phase and the initial number of carrier wavelengths at the time of signal lock is not known, each phase signal is biased by an unknown constant integer number of carrier cycles represented by N_1 and N_2. Use of the phase measurements as pseudorange signals for position estimation also requires estimation of these unknown integers [28–32]. Use of the change in the phase over a known period of time to estimate the receiver velocity does not require estimation of these integers, since the integers are canceled in the differencing operation [33,34]. The standard GPS texts [34,35] include entire sections or chapters devoted to the physical aspects of the various quantities that have been briefly defined in this section.

Note that only R and b_u contain the position and receiver clock information that we wish to estimate. The symbols $c\Delta t_{sv}$, I_a, E_{cm}, MP, mp, η, and n all represent errors that decrease the accuracy of the estimated quantities. There are many techniques to reduce these measurement errors prior to the navigation solution. Dual frequency receivers can take advantage of the code measurements from $L1$ and $L2$ to estimate the ionospheric delay error I_a as

$$I_a = \frac{f_1 f_2}{f_2^2 - f_1^2}(\tilde{\rho}_{L1} - \tilde{\rho}_{L2})$$

Due to the differencing of measurements, this estimate is noisy; since I_a changes very slowly, it can be low-pass filtered to remove the noise. For measurements from single frequency receivers, it is possible to compensate part of the ionospheric delay errors by an ionospheric delay model [36]. Alternatively, differential operation using at least two receivers can effectively remove all common mode errors (i.e., $c\Delta t_{sv}$, I_a, E_{cm}).

The methods discussed in the subsequent sections can be used for the pseudorange or integer-resolved carrier phase measurements. We will not discuss Doppler measurements. To avoid redundant text for the code and integer-resolved carrier measurements, we will adopt the following general model for the range measurement to the ith satellite

$$\tilde{\rho}_i = \sqrt{(X_i - x)^2 + (Y_i - y)^2 + (Z_i - z)^2} + b_u + \varepsilon_i \qquad (3.44)$$

where $\tilde{\rho}$ could represent the code pseudorange measurements or integer-resolved carrier phase measurements. The variable b_u represents the receiver clock bias. The symbol ε represents the error terms appropriate for the different measurements. When a GPS receiver has collected range measurements from four or more satellites, it can calculate a navigation solution.

3.3.2 Single-Point GPS Navigation Solution

This section presents the standard GPS position solution method using nonlinear least squares. In the process, we will introduce notation needed for the subsequent sections. In this section, the state vector is defined as $\mathbf{x} = [x, y, z, b_u]^T$ where (x, y, z) is the receiver antenna position in earth centered earth fixed (ECEF) coordinates and b_u is the receiver clock bias.

Taylor series expansion of Equation (3.44) about the current state estimate $\hat{\mathbf{x}} = [\hat{x}, \hat{y}, \hat{z}, \hat{b}_u]$ yields

$$\tilde{\rho}_i(\mathbf{x}) = \hat{\rho}_i(\hat{\mathbf{x}}) + [\mathbf{h}_i, 1]\delta\mathbf{x} + \text{h.o.t.s} + \varepsilon_i$$

where

$$\delta\mathbf{x} = \mathbf{x} - \hat{\mathbf{x}} = [x - \hat{x},\ y - \hat{y},\ z - \hat{z},\ b_u - \hat{b}_u]^T$$

$$\hat{\rho}_i(\hat{\mathbf{x}}) = \sqrt{(X_i - \hat{x})^2 + (Y_i - \hat{y})^2 + (Z_i - \hat{z})^2} + \hat{b}_u \qquad (3.45)$$

$$\mathbf{h}_i = \left[\frac{\partial \rho_i}{\partial x},\ \frac{\partial \rho_i}{\partial y},\ \frac{\partial \rho_i}{\partial z} \right]\Big|_{(\hat{x}, \hat{y}, \hat{z})}$$

and

$$\frac{\partial \rho_i}{\partial x} = \frac{-(X_i - x)}{\sqrt{(X_i - x)^2 + (Y_i - y)^2 + (Z_i - z)^2}}$$

$$\frac{\partial \rho_i}{\partial y} = \frac{-(Y_i - y)}{\sqrt{(X_i - x)^2 + (Y_i - y)^2 + (Z_i - z)^2}}$$

$$\frac{\partial \rho_i}{\partial z} = \frac{-(Z_i - z)}{\sqrt{(X_i - x)^2 + (Y_i - y)^2 + (Z_i - z)^2}}$$

Given m simultaneous range measurements, all the measurements can be put in the matrix form

$$\delta\rho = \mathbf{H}\delta\mathbf{x} + \mathbf{v} \tag{3.46}$$

by making the definitions

$$\delta\rho = \begin{bmatrix} \Delta\rho_1 \\ \Delta\rho_2 \\ \vdots \\ \Delta\rho_m \end{bmatrix} \quad \text{and} \quad \mathbf{H} = \begin{bmatrix} \mathbf{h}_1, 1 \\ \mathbf{h}_2, 1 \\ \vdots \\ \mathbf{h}_m, 1 \end{bmatrix} \tag{3.47}$$

where the residual measurement is

$$\Delta\rho_i = \tilde{\rho}_i(\mathbf{x}) - \hat{\rho}_i(\hat{\mathbf{x}})$$

and $\mathbf{v}$ represents the high order terms (h.o.t.s) of the linearization plus the measurement noise.

To determine the state vector, a minimum of four simultaneous range measurements are required. The weighted least squares solution to Equation (3.46) is

$$\delta\mathbf{x} = [\mathbf{H}^{\mathrm{T}}\mathbf{R}^{-1}\mathbf{H}]^{-1}\mathbf{H}^{\mathrm{T}}\mathbf{R}^{-1}\delta\rho \tag{3.48}$$

The corrected position estimate is then

$$\hat{\mathbf{x}}^+ = \hat{\mathbf{x}} + \delta\mathbf{x} \tag{3.49}$$

To reduce the effects of the linearization error terms, the above process can be repeated using the same measurement data and the corrected position at the end of the current iteration as the starting point of the next iteration (i.e., $\hat{\mathbf{x}} = \hat{\mathbf{x}}^+$). The iteration is stopped when the error state vector $\delta\mathbf{x}$ converges to a sufficiently small value. Even after the convergence of $\delta\mathbf{x}$ has been achieved, there may still be significant error $\eta_x = \mathbf{x} - \hat{\mathbf{x}}^+$ between the actual state and the best estimate after incorporating the measurements. The measurement noise covariance matrix is

$$\mathbf{R}_\rho = \text{cov}(\mathbf{v}) = \begin{bmatrix} \sigma_1^2 & 0 & \cdots & 0 \\ \vdots & & & \vdots \\ 0 & \cdots & 0 & \sigma_m^2 \end{bmatrix} \tag{3.50}$$

The value of σ_i^2 for the ith satellite could be determined based on time-series analysis of measurement data, the S/N ratio determined in the tracking loop for that channel, or computed based on satellite elevation. The covariance of η_x is

$$\mathbf{R_x} = \text{cov}(\eta_{\mathbf{x}}) = [\mathbf{H}^T \mathbf{R}^{-1} \mathbf{H}]^{-1} \tag{3.51}$$

It is important to note that this matrix is not diagonal. Therefore, errors in the GPS position estimates at a given epoch are correlated.

The above solution approach can be repeated (independently) for each epoch of measurements. This calculation of the position described so far, at each epoch, results in a series of single-point solutions. At each epoch, at least four simultaneous measurements are required and the solution is sensitive to the current measurement noise. There is no information sharing between epochs. Such information sharing between epochs could decrease noise sensitivity and decrease the number of satellites required per epoch; however, information sharing across epochs will require use of a dynamic model. Section 3.3.3 discusses advantages and disadvantages of alternative models and EKF solutions for GPS-only solutions. Section 3.4 discusses methods for combining GPS and IMU data.

Example 3.2 Throughout the remainder of this chapter we will extend the example that begins here. The example will be analyzed in $\Re^2$. By this we mean that we are analyzing a 2D world, not a 2D solution in a 3D world. We restrict the analysis to a 2D world for a few reasons (1) the analysis will conveniently fit within the page constraints of this chapter; (2) graphical illustrations are convenient; and (3) several important theoretical issues can be conveniently illustrated within the 2D example. The main conclusions from the 2D example have exact analogs in the 3D world (discussed in Example 3.6).

In a 2D world, $\mathbf{p}(t), v(t) \in \Re^2$ and there is a single angular rotation angle $\psi(t) \in \Re$ with $\omega(t) = \dot{\psi}(t) \in \Re$. All positions and ranges will be in meters. All angles are measured in degrees. The quantities ψ and ω are not used in this example, but are defined here for completeness as they are used in Example 3.6.

To find the position corresponding to range measurements in the 2D example, we define the position and clock bias error vector as

$$\delta \mathbf{p} = \mathbf{x} - \hat{\mathbf{x}} = [x - \hat{x}, y - \hat{y}, b\hat{u} - \hat{b}_u]^T$$

The range is computed as

$$\hat{\rho}_i(\hat{\mathbf{x}}) = \sqrt{(X_i - \hat{x})^2 + (Y_i - \hat{y})^2} + \hat{b}_u$$

The line-of-sight vector (from satellite to user) is

$$\mathbf{h}_i = \left[\frac{-(X_i - \hat{x})}{\sqrt{(X_i - \hat{x})^2 + (Y_i - \hat{y})^2}}, \frac{-(Y_i - \hat{y})}{\sqrt{(X_i - \hat{x})^2 + (Y_i - \hat{y})^2}} \right] \qquad (3.52)$$

Because there are three unknowns, measurements from at least three satellites will be required. Let us assume that there are satellites at locations $\mathbf{P}_i = 10 \times 10^6 \left[\begin{smallmatrix} \sin \theta_i \\ \cos \theta_i \end{smallmatrix} \right] m$ for $\theta_1 = 90°$, $\theta_2 = 85°$, $\theta_3 = 20°$, and $\theta_4 = -85°$ with corresponding range measurements of $\rho_1 = 9.513151e6$, $\rho_2 = 9.469241e6$, $\rho_3 = 9.363915e6$, and $\rho_4 = 10.468545e6$. Then, if the initial position estimate is $\hat{\mathbf{x}} = [0.00, 0.00, 0.00]^T$, the sequence of positions and position corrections computed by iterating Equation (3.48) and Equation (3.49) with $\mathbf{R} = \mathbf{I}$, is shown in Table 3.1. Note that if the initial estimate, possibly obtained by propagation of the estimate from a previous epoch, was accurate to approximately 10 m, then one or possibly two iterations would provide convergence of a new estimate consistent with the measurements of the current epoch to better than millimeter accuracy. Also, even after the estimate of $\mathbf{x}$ has converged to micrometer accuracy, the error in the estimated measurement is still 0.44 m. This is the least squared error that can be achieved by adjusting the three elements of $\mathbf{x}$ to fit the four measurements of ρ.

TABLE 3.1
Results of Computations for Example 3.2

Iteration	$\delta \mathbf{x}$	$\|\delta \mathbf{x}\|$	$\hat{\mathbf{x}}$	$\|\rho - \hat{\rho}\|$
0	NA	NA	$[0, 0, 0]$	
1	$[5.01, 5.09, 0.14]e5$	7.1e5	$[5.011961, 5.090871, 1.364810]e5$	23368.75
2	$[-0.12, -0.91, -1.36]e4$	1.6e4	$[5.000000, 5.000046, 0.000062]e5$	7.33
3	$[0.01, -4.29, -4.21]e0$	6.0e0	$[5.000000, 5.000000, 0.000002]e5$	0.44
4	$[-0.26, -8.53, -9.29]e-7$	1.3e-6	$[5.000000, 5.000000, 0.000002]e5$	0.44

The error covariance matrix for the estimated position vector is

$$\mathbf{R_x} = \begin{bmatrix} 0.377 & 0.040 & 0.133 \\ 0.040 & 1.492 & 0.376 \\ 0.133 & 0.376 & 0.384 \end{bmatrix}$$

The variance of x is smaller than the variance of y due to the geometric alignment of the satellites. Also, the estimates of x and y are correlated. Note that receivers do not typically provide this covariance matrix as an output.

If only the first three measurements are used to estimate $\mathbf{x}$, then the measurements can be perfectly fit, but the error in the estimated position and the error covariance matrix will increase.

3.3.3 KF for Stand-Alone GPS Solutions

This section discusses methods that have been proposed in the literature to achieve improved performance by using the EKF to share information across measurement epochs. Higher performance can be represented by increased position accuracy or decreased requirements on the number of required satellites per epoch.

Sharing information across measurement epochs requires models for the dynamics of the user receiver and the receiver clock error. The receiver clock dynamic model is discussed briefly in Section 3.3.3.1. Various possible dynamic models for the receiver antenna position are discussed in Section 3.3.3.2 to Section 3.3.3.4. Each of these dynamic models will be linear; however, since the GPS measurement model is nonlinear, the solution still requires an EKF.

The use of the EKF algorithm of Section 3.2.3 to solve the GPS navigation problem is illustrated as a block diagram in Figure 3.1. The dynamic motion equations are integrated between measurement times to predict the receiver antenna position at subsequent measurement times. The measurement prediction equations use the predicted antenna position and the computed satellite position to predict range measurement for each satellite. The residuals between the GPS measurements and the predicted measurements drive the EKF which outputs the error state estimates. The error state estimates are fed back to correct the predicted states, which are used to initialize the prediction step for the next epoch.

Section 3.3.3.2 to Section 3.3.3.4 discuss the advantages and disadvantages of three possible receiver state estimation algorithms. The EKF algorithm and Figure 3.1 are applicable to all three approaches. The main distinctions between the approaches are the definitions of the state vector and the dynamic model for the state vector.

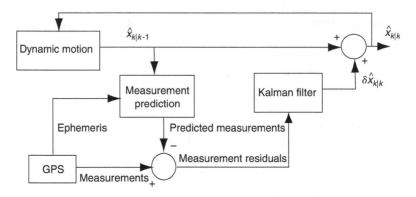

FIGURE 3.1 Block diagram representation for a GPS-only navigation system solved via KF. The dynamic motion prediction by either Equation (3.21) or Equation (3.38) extrapolates from the present state estimate using the assumed dynamic model.

3.3.3.1 Clock model

Global positioning system receivers use oscillators with very stable frequencies. Integration of this frequency provides the basis for the receiver clock time signal. The error between the oscillator frequency and its specified frequency represents the receiver clock drift rate. It is common to model the clock drift rate as a random walk process. We scale these quantities by the speed of light to represent the clock bias b_u and drift rate f_u in meters and meters per second. The dynamic model for $\mathbf{x}_c = [b_u, f_u]^{\mathrm{T}}$ is

$$\dot{\mathbf{x}}_c = \mathbf{F}_c \mathbf{x}_c + \mathbf{w}_c \tag{3.53}$$

where

$$\mathbf{F}_c = \begin{bmatrix} 0 & 1 \\ 0 & 0 \end{bmatrix}, \quad \mathbf{w}_c = \begin{bmatrix} \omega_b \\ \omega_f \end{bmatrix}, \tag{3.54}$$

and the power spectral density S_b and S_f of the process noise ω_b and ω_f are determined by the characteristics of the receiver clock [20]. The corresponding state transition matrix and process noise covariance matrix for the discrete clock model are:

$$\Phi_k^c = \Phi^c(t_k, t_{k+1}) = \begin{bmatrix} 1 & \Delta t \\ 0 & 1 \end{bmatrix}, \quad \mathbf{Q}_k^c = \begin{bmatrix} S_b \Delta t + S_f \frac{\Delta t^3}{3} & S_f \frac{\Delta t^2}{2} \\ S_f \frac{\Delta t^2}{2} & S_f \Delta t \end{bmatrix} \tag{3.55}$$

where $\Delta t = t_{k+1} - t_k$. This clock model will be included as a portion of the model in each of the following sections.

3.3.3.2 Stationary user (P model)

If it is known that the receiver antenna is stationary, then the position vector $\mathbf{x}_p = [x, y, z]^T$ satisfies $\dot{\mathbf{x}}_p = \mathbf{0}$. Combining the receiver position model with the clock model, the dynamic model for a stationary user is

$$\begin{bmatrix} \dot{\mathbf{x}}_p \\ \dot{\mathbf{x}}_c \end{bmatrix} = \begin{bmatrix} \mathbf{0} & \mathbf{0} \\ 0 & \mathbf{F}_c \end{bmatrix} \begin{bmatrix} \mathbf{x}_p \\ \mathbf{x}_c \end{bmatrix} + \begin{bmatrix} \mathbf{w}_p \\ \mathbf{w}_c \end{bmatrix} \tag{3.56}$$

where $\mathbf{w}_p = [\omega_x, \omega_y, \omega_z]^T$ is the process noise for the position states. The state transition matrix and process noise covariance matrix for the equivalent discrete model are:

$$\Phi_k^s = \begin{bmatrix} \mathbf{I} & 0 \\ 0 & \Phi_k^c \end{bmatrix}, \quad \mathbf{Q}_k^s = \begin{bmatrix} \mathbf{Q}_k^p & 0 \\ 0 & \mathbf{Q}_k^c \end{bmatrix} \tag{3.57}$$

where $\mathbf{Q}_k^p$ is the process noise covariance matrix corresponding to $\mathbf{w}_p$ in the sense of Equation (3.6) and Equation (3.10), and $\mathbf{I}$ is a 3×3 identity matrix. The linearized measurement model is

$$\mathbf{y} = \delta\rho = \begin{bmatrix} \mathbf{h_1} & \mathbf{h}_c \\ \mathbf{h_2} & \mathbf{h}_c \\ \vdots & \\ \mathbf{h}_m & \mathbf{h}_c \end{bmatrix} \begin{bmatrix} \delta\mathbf{x}_p \\ \delta\mathbf{x}_c \end{bmatrix} + \mathbf{v} \tag{3.58}$$

where $\mathbf{h}_c = [1, 0]$ characterizes the effect of the clock state $\delta\mathbf{x}_c$ on the measurement, $\delta\rho$ is defined in Equation (3.47), $\mathbf{h}_i$ is defined in Equation (3.45), $\delta\mathbf{x}_p = [\delta x, \delta y, \delta z]^T$ is the position error vector, $\mathbf{R}_k = \mathbf{R}_\rho$ denotes the covariance of $\mathbf{v}$ as defined in (3.50), and $\delta\mathbf{x}_c = [\delta b_u, \delta f_u]^T$ is the clock state error vector. With the above specifications, the parameters $\Phi_k^s, \mathbf{H}_k, \mathbf{Q}_k^s$, and $\mathbf{R}_k$ required for the EKF implementation described in Section 3.2.3 have all been defined.

For a receiver that is in fact stationary, $\mathbf{w}_p = [\omega_x, \omega_y, \omega_z]^T = \mathbf{0}$. In this case, $\mathbf{Q}^p = 0\mathbf{I}$. If the EKF algorithm is designed using $\mathbf{Q}^p = 0\mathbf{I}$, then portions of the diagonal of the state error covariance matrix $\mathbf{P}$ and of the gain matrix $\mathbf{K}$ will asymptotically approach zero. This is desirable when the model is accurate and the antenna is stationary. If the receiver antenna position is not stationary or if the model is not accurate (e.g., time correlated multipath errors have been ignored), then this property is not desirable. An ad hoc approach is to treat the matrices $\mathbf{Q}^p$ and $\mathbf{R}$ as tuning parameters to adjust the convergence rate of the filter. However, a far better approach is to develop a more appropriate system model.

3.3.3.3 Low dynamic user (PV Model)

For a receiver in a "low dynamic environment" it may be reasonable to assume that the velocity vector is a random walk process. In this case, an eight state model is appropriate with the acceleration vector modeled as white noise. The state vector includes the position state $\mathbf{x}_p$, receiver clock state $\mathbf{x}_c$, and velocity state $\mathbf{x}_v = [v_x, v_y, v_z]^T$; therefore, the dynamic model is

$$\begin{bmatrix} \dot{\mathbf{x}}_p \\ \dot{\mathbf{x}}_v \\ \dot{\mathbf{x}}_c \end{bmatrix} = \begin{bmatrix} \mathbf{0} & \mathbf{I} & \mathbf{0} \\ \mathbf{0} & \mathbf{0} & \mathbf{0} \\ \mathbf{0} & \mathbf{0} & \mathbf{F}_c \end{bmatrix} \begin{bmatrix} \mathbf{x}_p \\ \mathbf{x}_v \\ \mathbf{x}_c \end{bmatrix} + \begin{bmatrix} \mathbf{0} \\ \mathbf{w}_v \\ \mathbf{w}_c \end{bmatrix} \tag{3.59}$$

where $\mathbf{w}_v = [\omega_{vx}, \omega_{vy}, \omega_{vz}]^T$ represents the process noise representation of the unknown acceleration. The measurement model is

$$\mathbf{y} = \delta\rho = \begin{bmatrix} \mathbf{h}_1 & \mathbf{0} & \mathbf{h}_c \\ \mathbf{h}_2 & \mathbf{0} & \mathbf{h}_c \\ & \vdots & \\ \mathbf{h}_m & \mathbf{0} & \mathbf{h}_c \end{bmatrix} \begin{bmatrix} \delta\mathbf{x}_p \\ \delta\mathbf{x}_v \\ \delta\mathbf{x}_c \end{bmatrix} + \mathbf{v} \tag{3.60}$$

with the measurement noise covariance defined in Equation (3.50).

We will not provide an in-depth discussion of this model here, as the majority of the comments about the model of the following section are also applicable to the model of this section. However, it is important to note that few applications involve white acceleration processes. In fact, the acceleration process is rarely even stationary. Therefore, with this assumed dynamic model, the matrix $\mathbf{Q_w}$ is best considered as a tuning parameter and proper stochastic interpretations of the various variables in the algorithm are no longer applicable.

3.3.3.4 High dynamic user (PVA model)

A GPS receiver may (and typically will) operate in applications where the acceleration vector is time varying. In such "high dynamic environments," it is necessary to augment the three acceleration states $\mathbf{x}_a = [a_x, a_y, a_z]^T$ to the system model. With the acceleration states modeled as first-order Markov processes, the process model for a high dynamic user is

$$
\begin{bmatrix} \dot{\mathbf{x}}_p \\ \dot{\mathbf{x}}_v \\ \dot{\mathbf{x}}_a \\ \dot{\mathbf{x}}_c \end{bmatrix} = \begin{bmatrix} \mathbf{0} & \mathbf{I} & \mathbf{0} & \mathbf{0} \\ \mathbf{0} & \mathbf{0} & \mathbf{I} & \mathbf{0} \\ \mathbf{0} & \mathbf{0} & \mathbf{D} & \mathbf{0} \\ \mathbf{0} & \mathbf{0} & \mathbf{0} & \mathbf{F}_c \end{bmatrix} \begin{bmatrix} \mathbf{x}_p \\ \mathbf{x}_v \\ \mathbf{x}_a \\ \mathbf{x}_c \end{bmatrix} + \begin{bmatrix} \mathbf{0} \\ \mathbf{0} \\ \mathbf{w}_a \\ \mathbf{w}_c \end{bmatrix} \qquad (3.61)
$$

where $\mathbf{D} = \mathrm{diag}\left[-\frac{1}{\tau_h}, -\frac{1}{\tau_h}, -\frac{1}{\tau_v}\right]$ is a design matrix containing the reciprocal of the acceleration correlation times and $\mathbf{w}_a = [\omega_{ax}, \omega_{ay}, \omega_{az}]^{\mathrm{T}}$ represents the acceleration state process driving noise. The measurement model is

$$
\mathbf{y} = \delta\rho = \begin{bmatrix} \mathbf{h}_1 & \mathbf{0} & \mathbf{0} & \mathbf{h}_c \\ \mathbf{h}_2 & \mathbf{0} & \mathbf{0} & \mathbf{h}_c \\ & & \vdots & \\ \mathbf{h}_m & \mathbf{0} & \mathbf{0} & \mathbf{h}_c \end{bmatrix} \begin{bmatrix} \delta\mathbf{x}_p \\ \delta\mathbf{x}_v \\ \delta\mathbf{x}_a \\ \delta\mathbf{x}_c \end{bmatrix} + \mathbf{v} \qquad (3.62)
$$

with the measurement noise covariance defined in Equation (3.50).

As with the P and PV models, there is no rigorous method to properly select the $\mathbf{D}$ and $\mathbf{Q}_\mathbf{w}$ parameters of the PVA model. The above model assumes different correlation times for the horizontal vs. vertical accelerations. This assumption obviously depends on whether the receiver is used in an aircraft, sea surface, or land vehicle application. Although the receiver user may specify an application class, correct values for $\mathbf{D}$ and $\mathbf{Q}_\mathbf{w}$ may not exist or be known for this design approach. Therefore, these quantities are used to tune the performance of the EKF algorithm. Even though there is no direct measurement of acceleration, the augmented states may enable the filter to improve the accuracy of the navigation solution by fusing sensor information across measurement epochs. Compared to single epoch solutions, improved accuracy would be obtained by the EKF methods if the vehicle were not accelerating during a period of time when an insufficient number of satellites were available; however, receiver acceleration would affect the estimation accuracy.

3.3.3.5 GPS KF examples

This section presents two examples of the use of the EKF in the solution of the GPS state estimation problem. In each example, we work in the 2D world introduced in Example 3.2 and include sufficient details to allow duplication of the results by interested readers.

Example 3.3 This example considers the situation where a stationary receiver is in operation with a PVA model. The state model is defined in Equation (3.61). Using a 1 sec measurement epoch with $R = 1\,\mathrm{m}^2, \mathrm{cov}(\mathbf{w}_a^{\mathrm{T}}\mathbf{w}_a) = Q\mathbf{I}_2$,

$S_b = 1.10 \times 10^{-10}$, $S_f = 0.65 \times 10^{-10}$, and $\tau_h = 1.0$, the resulting discrete-time state transition and process noise covariance matrices are

$$\Phi_k = \begin{bmatrix} \mathbf{I}_2 & \mathbf{I}_2 & 0.37\mathbf{I}_2 & \mathbf{0}_2 \\ \mathbf{0}_2 & \mathbf{I}_2 & 0.63\mathbf{I}_2 & \mathbf{0}_2 \\ \mathbf{0}_2 & \mathbf{0}_2 & 0.37\mathbf{I}_2 & \mathbf{0}_2 \\ \mathbf{0}_2 & \mathbf{0}_2 & \mathbf{0}_2 & \Phi_c \end{bmatrix}$$

and

$$\mathbf{Q}_k = \begin{bmatrix} 0.1Q \begin{bmatrix} 0.30\mathbf{I}_2 & 0.68\mathbf{I}_2 & 0.64\mathbf{I}_2 \\ 0.68\mathbf{I}_2 & 1.68\mathbf{I}_2 & 2.00\mathbf{I}_2 \\ 0.64\mathbf{I}_2 & 2.00\mathbf{I}_2 & 4.33\mathbf{I}_2 \end{bmatrix} \begin{bmatrix} \mathbf{0}_2 \\ \mathbf{0}_2 \\ \mathbf{0}_2 \end{bmatrix} \\ \mathbf{0}_2 \qquad \mathbf{0}_2 \qquad \mathbf{0}_2 \qquad \mathbf{Q}_c \end{bmatrix}$$

where $\mathbf{I}_2$ is a 2D identity matrix, $\mathbf{0}_2$ is a 2D null matrix, and

$$\Phi_c = \begin{bmatrix} 1 & 1 \\ 0 & 1 \end{bmatrix} \quad \text{and} \quad \mathbf{Q}_c = \text{cov}(\mathbf{w}_c^T \mathbf{w}_c) = \begin{bmatrix} 1.32 & 0.32 \\ 0.32 & 0.65 \end{bmatrix} \times 10^{-10}$$

The scalar parameter Q, which theoretically represents the spectral density of the "acceleration driving noise," is used to tune the size of the $\mathbf{Q}_k$ matrix. We generate noisy measurements using the following procedure: compute exact ranges between the user and each satellite, add the clock bias b_u, and add Gaussian random noise with unit variance. The clock bias in the simulation grows at a unit rate (i.e., $b_u = 1.0t$). The initial $\mathbf{P}$ matrix is defined by the diagonal [1e6, 1e6, 1e2, 1e2, .1, .1, 1e6, 1]. At this point, we have enough information to implement the discrete-time EKF.

The norm of the sequence of position estimation errors is shown in Figure 3.2a which is the left column of Figure 3.2. Each row of the figure shows the estimation error for the same sequence of measurements when only the value of Q is changed in the EKF design. When the design specifies a large acceleration driving noise (e.g., $Q = 10$), the estimation error is large with significant energy at high frequencies. This is due to the fact that the large value of Q causes the EKF computations to keep the Kalman gain relatively large, favoring current measurements over information from past measurements that is represented by the state estimate. When the design specifies a small acceleration driving noise (e.g., $Q = 0.001$), the estimation error is smaller in magnitude with significantly less energy at high frequencies. This is due to the fact that the small value of Q causes the EKF computations to decrease the Kalman gain over time causing the current measurements to make smaller corrections to the information from past measurements that is represented by the state estimate.

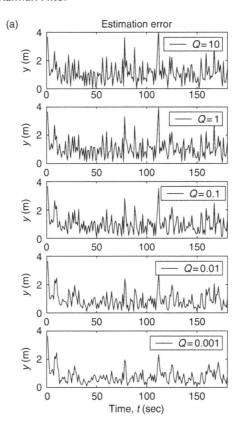

FIGURE 3.2   EKF-based GPS solutions for Examples 3.3 and 3.4. (a) Estimation error for a stationary receiver using the PVA model and EKF with different settings of the "covariance" parameter Q. (b) Actual (solid line) and estimated trajectory (dots) for a moving receiver using the PVA model, EKF estimation, and different settings of the "covariance" parameter Q.

This example shows the possible benefit of using the EKF to combine measurements over time to attain improved accuracy. The performance that is achieved will depend on the EKF parameter settings relative to the actual dynamic situation of the receiver. If the process noise covariance matrix $\mathbf{Q}_k$ is too large, then the past information encapsulated in the prior estimate of the state will be largely ignored in the computation by the EKF of the posterior state estimate. If the matrix $\mathbf{Q}_k$ is too small, then the estimated state may significantly lag the actual state. This is further illustrated in the next example.

Example 3.4   In this example, the receiver is attached to a moving platform. The platform trajectory is illustrated by the solid curve in each subgraph of

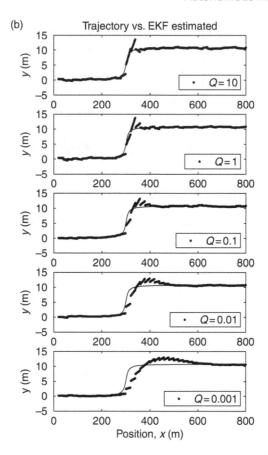

FIGURE 3.2   Continued.

Figure 3.2b. For the majority of the simulation, the motion is parallel to the x-axis at 20 m/sec, except for a short period of time near $t = 15$ sec when the platform performs a maneuver similar to an automobile lane change that involves a nonzero yaw rate and lateral acceleration. The discrete-time model, estimator design, and method of computing noisy measurements are the same as in Example 3.3.

Figure 3.2b shows the estimated positions on an x–y graph to allow straightforward comparison between the estimated and actual trajectory for various settings of the design parameter Q. The estimated positions are marked by a dot every 0.1 sec even though the GPS measurement epoch is still 1.0 sec to clearly indicate the estimated velocity (i.e., the slope of the estimated position curve between GPS epochs).

When the design uses a large Q, the variance of the estimated position is large, but the estimator rapidly adjusts the estimated state so that the estimate does not significantly lag the actual state following the vehicle maneuver. When the design uses a small value for Q, the variance of the estimated position is smaller; however, the estimated state significantly lags the actual state following the vehicle maneuver.

Figure 3.2a and b are intentionally placed side-by-side to emphasize the fact that there is no single optimal choice for the design parameter Q. The desirable setting of Q depends on the application and maneuvering conditions. Some receivers allow the user to effect the receiver estimation procedure (either the model structure or the value of Q) through the user interface. It is the responsibility of the user to understand the settings and their tradeoffs relative to the application. This is especially true when the state estimate is being used as the input to a control system.

Due to the structure of the $\mathbf{\Phi}_k$ matrix, if the GPS $\mathbf{H}$ matrix has a null direction $\mathbf{d}$ such that $\mathbf{Hd} = 0$, then position, velocity, and acceleration errors parallel to $\mathbf{d}$ will not be observable from the GPS measurements. Note that the rows of the $\mathbf{H}$ matrix contain the line-of-sight unit vectors between the receiver antenna and the satellite. Therefore, to accurately and rapidly track the platform motion during (and after) a maneuver, the receiver must be tracking at least one satellite located in a direction such that the line-of-sight unit vector has a significant component in the same direction as the acceleration unit vector; otherwise, the GPS measurements will be insensitive to the acceleration. In particular, if a receiver is operating in an urban canyon[1] type of environment and accelerates parallel to the direction in which the satellite signals are blocked then the position, velocity, and acceleration accuracy in that direction will deteriorate.

No amount of signal processing can help, unless additional sensors (e.g., inertial, wheel speed, vision, precision clock) are added.

Finally, it is critical to note that estimation errors, even restricted to the GPS measurement epochs, are correlated. They are not white discrete-time processes. This is clearly illustrated in Figure 3.2b for small values of Q, but is also true for larger values of Q. The fact that the position estimation errors are not white is critical to understanding one of the drawbacks of using the GPS position estimates to aid an INS (see Section 3.5.2.1).

3.3.3.6 Summary

The approaches discussed in the previous three sections have several aspects that should be pointed out. First, as discussed following Equation (3.50),

---

[1] This is a canyon created by the urban environment (e.g., a road between tall buildings) that may block satellite signals in specific directions [37–39].

at each epoch the components of the estimated state vector will be correlated. In addition, due to the fusion of measurements across epochs, even if the measurement noise **v** is white, the state estimation error will be a colored noise process. Second, the three preceding sections discussed estimation algorithms in the order of increasing complexity. The required number of computations to implement the EKF increases as the size of the state vector increases. Third, performance will suffer if the application conditions do not match the algorithm assumptions. If, for example, a P or PV algorithm or too small a value of Q is used in a "high-dynamic environment," then the estimated position may have significant lag relative to the actual position. Fourth, use of Doppler measurements can increase convergence rates, but opens up other modeling issues [40]. Fifth, if GPS measurements are unavailable for some period of time, the dynamic model is available to propagate the state estimates; however, acceleration of the system during this time period can have serious adverse effects on the accuracy of such predictions. This issue can be addressed well by, for example, properly incorporating an inertial measurement system. Sixth, a recurrent issue in the approaches of this section is that the stochastic model parameter Q could not be properly selected. Instead it was used as performance tuning parameter. Proper incorporation of IMU data into the approach will also allow proper selection and interpretation of the parameter in a stochastic sense. Addition of an IMU will increase the cost of the system, but offers the potential for higher performance (e.g., bandwidth, accuracy, coast time, and sample rate) and availability.

3.4 INERTIAL NAVIGATION SYSTEM

A strapdown INS incorporates an IMU that measures the acceleration and angular rate of the system and analytic routines on a computer that integrate the inertial measurements to provide estimates of the vehicle position, velocity, and attitude in a desired coordinate frame. This section reviews the strapdown INS mechanization equations and the dynamic error model of the INS system. Various methods for fusing the GPS and INS information are reviewed and discussed in Section 3.5. The example in a 2D world is continued to highlight various important issues related to GPS–INS integration.

This paragraph briefly defines the various coordinate frames that will be used in the subsequent discussion. All coordinate frames are defined by orthogonal axes in a right-handed sense. The body frame is attached to and moves with the vehicle. The inertial measurements are resolved along the axes of the platform frame. To simplify the discussion, we assume that the body and platform frames are identical. The navigation frame is attached to the earth at a convenient point of reference and determines the desired frame in which to resolve the vehicle position and velocity vectors. The ECEF frame is attached to the center of and rotates with the Earth. A local geodetic frame has its origin fixed on the surface of the earth and axes aligned with the directions of true north, east, and down (along the parallel to the ellipsoid normal vector to complete the right-handed coordinate frame).

3.4.1 Strapdown System Mechanizations

As illustrated in Figure 3.3, the accelerometers measure the specific force vector $\mathbf{f}^b$ in the body frame-of-reference and the gyros measure the angular rate of the vehicle with respect to an inertial frame-of-reference $\omega_{ib}^b = [p, q, r]^{\mathrm{T}}$ in the body frame-of-reference. The gyro measurements are integrated to compute the attitude of the vehicle frame with respect to the navigation frame. The attitude is used to compute the rotation matrix $\mathbf{C}_b^n$ required to transform vectors between the body and navigation frames. In particular, the specific force in the navigation frame is

$$\mathbf{f}^n = \mathbf{C}_b^n \mathbf{f}^b \tag{3.63}$$

This rotation matrix can be represented as a direction cosine matrix which is the solution of

$$\dot{\mathbf{C}}_b^n = \mathbf{C}_b^n \mathbf{\Omega}_{nb}^b \tag{3.64}$$

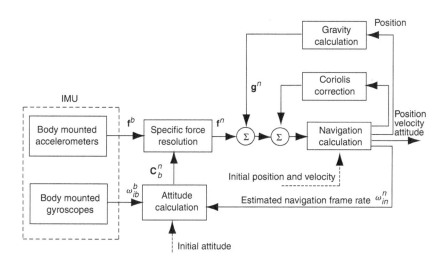

FIGURE 3.3 Block diagram representation (similar to figure 3.12 in Reference 41) of a strapdown INS.

where $\mathbf{\Omega}_{nb}^{b}$ is the skew matrix form of ω_{bn}^{b} and

$$\omega_{bn}^{b} = \omega_{ib}^{b} - \mathbf{C}_{n}^{b}(\omega_{ie}^{n} + \omega_{en}^{n}) \tag{3.65}$$

The symbols ω_{ie}^{n} and ω_{en}^{n} represent the rotation rate of the ECEF frame relative to an inertial frame and the rotation rate of the navigation frame relative to the earth frame (i.e., transport rate), respectively. Both vectors are resolved in the navigation frame. The second term on the right of (3.65) compensates the gyro measurements for the rotation rate of the navigation frame relative to an inertial frame.

The dynamic equations for an INS system have different forms for different navigation frames. Detailed derivations of the navigation equations with respect to different navigation frames can be found in various references, for example, in References 27, 34, and 41–43. The navigation equations for a terrestrial navigation system operating in the local geodetic frame are:

$$\dot{\mathbf{v}}_{e}^{n} = \mathbf{f}^{n} - (2\omega_{ie}^{n} + \omega_{en}^{n}) \times \mathbf{v}_{e}^{n} + \mathbf{g}_{l}^{n} \tag{3.66}$$

where $\mathbf{v}_{e}^{n} = [v_{n}, v_{e}, v_{d}]^{\mathrm{T}}$ is the velocity with respect to the Earth expressed in the local geodetic frame (i.e., navigation frame), $\mathbf{f}^{n} = [f_{n}, f_{e}, f_{d}]^{\mathrm{T}}$ is the specific force resolved to this navigation frame, and $\mathbf{g}_{l}^{n}$ is the local gravity vector expressed in the navigation frame. The local gravity vector is

$$\mathbf{g}_{l} = \mathbf{g} - \omega_{ie}^{i} \times [\omega_{ie}^{i} \times \mathbf{R}^{i}] \tag{3.67}$$

which accounts for the mass attraction of the earth $\mathbf{g}$ and the centripetal acceleration caused by the Earth's rotation. Note that $\mathbf{g}_{l}$ is the acceleration sensed by a stationary accelerometer located on the surface of the earth. Note also that $\mathbf{g}_{l}$ is a function of position.

Given an initial velocity, Equation (3.66) integrates acceleration to estimate the velocity as a function of time. Prior to integration, the measured specific force vector is corrected for Coriolis effects (second term) and gravity (third term). Given an initial position, integration of velocity provides an INS estimate of position. Given high rate IMU measurements (and a sufficiently fast computer), the INS can integrate the above equations to provide high rate estimates of position, velocity, attitude, angular rate, and acceleration. Since the INS is an integrative process, the INS attenuates the high frequency measurement noise from the IMU, but amplifies low frequency measurement errors such as biases. Calibration and removal of the INS state and IMU instrument errors can be accomplished through EKF data fusion, once the designer obtains a dynamic model for the INS and IMU error processes.

3.4.2 Error Model of INS System

The INS dynamics as represented in Equation (3.66) are nonlinear functions of the INS state variables. As discussed in Section 3.2.3, the EKF will utilize a linearized error-state model. The dynamic model for the error state of the INS is derived in several references, for example, in References 27 and 41–43. After minor simplification, it can be expressed as

$$
\begin{bmatrix} \delta \dot{\mathbf{p}} \\ \delta \dot{\mathbf{v}} \\ \delta \dot{\rho} \\ \dot{\mathbf{x}}_a \\ \dot{\mathbf{x}}_g \end{bmatrix} = \begin{bmatrix} 0 & \mathbf{F_{pv}} & 0 & 0 & 0 \\ \mathbf{F_{vp}} & \mathbf{F_{vv}} & \mathbf{F_{v\rho}} & \mathbf{F_{vx_a}} & 0 \\ 0 & 0 & \mathbf{F}_{\rho\rho} & 0 & \mathbf{F}_{\rho \mathbf{x_g}} \\ 0 & 0 & 0 & \mathbf{F_{x_a x_a}} & 0 \\ 0 & 0 & 0 & 0 & \mathbf{F_{x_g x_g}} \end{bmatrix} \begin{bmatrix} \delta \mathbf{p} \\ \delta \mathbf{v} \\ \delta \rho \\ \mathbf{x}_a \\ \mathbf{x}_g \end{bmatrix} + \begin{bmatrix} \omega_p \\ \omega_v + v_a \\ \omega_\rho + v_g \\ \omega_{\mathbf{a}} \\ \omega_{\mathbf{g}} \end{bmatrix} \quad (3.68)
$$

which is the required time-varying linear model in the form of Equation (3.1). The 15 state errors are defined as: tangent plane position error $\delta \mathbf{p} = [\delta n, \delta e, \delta d]^T$, tangent plane velocity error $\delta \mathbf{v} = [\delta v_N, \delta v_E, \delta v_D]^T$, attitude error $\delta \rho = [\epsilon_N, \epsilon_E, \epsilon_D]^T$, platform frame accelerometer bias error $\mathbf{x}_a$, and platform frame gyro bias error $\mathbf{x}_g$. Additional IMU error calibration states (e.g., gyro scale factors) could be considered for state augmentation. The various $\mathbf{F}$ matrices in Equation (3.68) are derived and defined, for example, in chapter 6 of Reference 27. The $\mathbf{F}$ matrix is time-varying, since it is a function of velocity, attitude, angular rate, and specific force. The $\mathbf{F}$ matrix contains unstable and neutrally stable components; therefore, initial condition errors and measurement errors can cause the INS error state to diverge. Fusion of the INS state with external sensors, such as GPS, using the EKF can estimate and compensate for these errors.

3.4.3 EKF Latency Compensation

During each INS integration step, the INS will first compensate the IMU measurements for the calibration factors (e.g., biases) estimated by the EKF. Next, the INS will integrate Equation (3.64) and Equation (3.66) (or similar equations depending on the choice of navigation frame and attitude representation). The equations are integrated for the duration of the application regardless of the availability of aiding measurements. With regard to aiding, two time instants should be distinguished. The *time-of-applicability* of an aiding measurement is the time at which the measurement is accurate. The *time-of-availability* of a measurement is the time at which the aiding measurement is available for use by the computer performing the data fusion operation. While the INS integration process is ongoing, the INS state must be saved at the time-of-applicability of the aiding measurements.

In particular, for GPS aiding, the time-of-availability of a GPS measurement is typically delayed from its time-of-applicability due to latency within the receiver and communication delay between the receiver and the EKF processor. A typical latency between the times of applicability and availability is on the order of a few hundred milliseconds (i.e., typically <0.25 sec). Fortunately, most receivers provide a one-pulse-per-second (1PPS) output signal that can be configured to align in time with the GPS second. In addition, assuming a one second GPS measurement epoch, the time-of-applicability of the GPS measurement can be aligned with the GPS second. When the EKF processor receives the 1PPS signal, it saves the INS state. By doing this, the EKF will have the INS state coincident with the GPS measurement even though the GPS measurement will not arrive until a significant fraction of a second later. At the time-of-availability of the EKF estimated correction, the EKF can use the state transition matrix to propagate the correction from its time-of-applicability to its time-of-availability.

Example 3.5 Let t denote an integer GPS second. At time t, the EKF processor detects the 1PPS signal and saves the INS state $\mathbf{x}(t)$. In addition, the GPS processor saves the receiver tracking data and computes the pseudoranges $\rho(t)$. The pseudorange measurements are sent to the EKF processor arriving at time $t_1 = t + \tau$ where $0 < \tau < 1$ sec. At time t_1 the EKF processes the pseudoranges to compute $\delta\mathbf{x}(t)$ which is available at some $t_2 > t_1$. At this point in time it is not correct to simply add the correction to the current INS state, since $\delta\mathbf{x}(t) \neq \delta\mathbf{x}(t_2)$ (i.e., $\mathbf{x}(t_2) + \delta\mathbf{x}(t)$ would not be correct). Note that the time t_2 is known to the EKF processor and that the processor is already propagating the state transition matrix $\mathbf{\Phi}$ by a method such as Equation (3.18), because $\mathbf{\Phi}$ is required to propagate the state estimation error covariance matrix. With these quantities being known and available, it is straightforward for the EKF processor to propagate the correction from its time-of-applicability to its time-of-availability t_2 as

$$\delta\mathbf{x}(t_2) = \mathbf{\Phi}(t_2, t)\delta\mathbf{x}(t)$$

Then, $\delta\mathbf{x}(t_2)$ can be added to the INS state $\mathbf{x}(t_2)$ to properly compensate the system.

Alternative latency compensation methods are described in the literature, see, for example, Reference 44.

3.5 INTEGRATION OF GPS AND INS

Due to their complementary characteristics, various methods have been suggested to implement a system to integrate GPS and INS with the goal of achieving performance that is superior to which either system could attain on its own. This section will discuss different approaches for GPS/INS integration. The main objective is to compare the relative advantages and disadvantages between the alternative approaches.

3.5.1 INS with GPS Resetting

In this approach, the INS is integrated to provide its state estimate between the GPS measurement epochs. At a GPS measurement epoch, the methods of Section 3.3 are used to compute the position and velocity based only on the GPS measurement data. The GPS position and velocity estimates are used as the initial conditions for the INS state during the next period of integration.

Often, the reason that this approach is proposed is its extreme simplicity. For example, GPS receivers directly output user position and velocity. In this approach, where the designer treats the position and velocity computed by the GPS receiver as measurements for the state estimation process, the designer of the integrated system need not solve the GPS system equations. In addition, the design of this approach does not involve a KF (outside of the receiver). However, the disadvantage of this simplicity is a low level of performance relative to the level that could be achieved by a more advanced approach. Note, for example, that the IMU errors are not estimated or compensated. Therefore, the rate of growth of the INS error state does not decrease over time. Also, additional sensors or multiple GPS antennae and additional processing are required to maintain the attitude accuracy.

Various ad hoc procedures can be defined to improve performance of the resetting approach, but performance analysis is typically not possible. The resetting approach is not a recommended approach. Note that this approach does not involve any advanced form of data fusion. The only point at which information is exchanged is after the GPS measurement, when the INS state is reset. Significantly better performance can be obtained by the methods described in the following section.

3.5.2 GPS Aided INS

The following two sections discuss the EKF as a tool to use GPS measurements to calibrate INS errors. In both approaches, the INS integrates the vehicle state based on IMU measurements. In Step 1 of the EKF algorithm of Section 3.2.3, the INS state is represented by $\mathbf{x}^*$ and the IMU input is represented by $\mathbf{u}$. The linearized $\mathbf{F}$ matrix is given by Equation (3.68). The matrix $\mathbf{Q}_k$ represents the covariance of the integrated accelerometer and gyro measurement noise processes. The matrix $\mathbf{Q}_k$ can be computed accurately (see Example 3.1) and is determined by the quality of the IMU. The only remaining quantities that

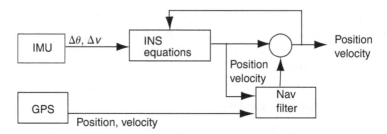

FIGURE 3.4   Block diagram of a loosely coupled GPS aided INS.

must be specified for implementation of the EKF are the matrices $\mathbf{H}$ and $\mathbf{R}$. These matrices are distinct for the two methods to be discussed and will be specified below.

3.5.2.1 Loosely coupled system

As illustrated in Figure 3.4, in a loosely coupled system, the EKF measurements are the GPS position (or velocity or both). Residual measurements are formed with the INS estimates of position (or velocity or both). The position measurement residual is

$$\mathbf{y} = \mathbf{p}_{\text{gps}} - \mathbf{p}_{\text{ins}} = \begin{bmatrix} 1 & 0 & 0 \\ 0 & 1 & 0 \\ 0 & 0 & 1 \end{bmatrix} \begin{bmatrix} \delta n \\ \delta e \\ \delta d \end{bmatrix} + \eta_x \qquad (3.69)$$

If the INS error state is ordered as $\delta \mathbf{x} = [\delta \mathbf{p}^{\mathrm{T}}, \delta \mathbf{v}^{\mathrm{T}}, \delta \rho^{\mathrm{T}}, \mathbf{x}_a^{\mathrm{T}}, \mathbf{x}_g^{\mathrm{T}}]^{\mathrm{T}}$ as in Equation (3.68); then, for Step 4 of the EKF algorithm, the linearized position measurement matrix is

$$\mathbf{H}_k = [\mathbf{I}, \mathbf{0}]$$

where $\mathbf{I}$ is a 3×3 identity matrix and $\mathbf{0}$ is a 3×12 matrix of zeros. In this approach, the receiver clock model and associated error states need not be included in the EKF model, as the receiver has already accounted for the receiver clock bias in the estimation of the receiver antenna position.

For the implemented system to attain performance near that predicted theoretically, it is critical for the designer to understand at least the following practical issues:

Correlated GPS position error vector: As discussed in Section 3.3.2 and demonstrated in Example 3.2, at any given epoch, the components of the vector of position errors are correlated. For the EKF estimation of the INS error, the position error correlation matrix $\mathbf{R}_\mathbf{x}(t)$ must be available and due to the cross-correlation scalar measurement processing cannot be used. Typically, GPS receivers will *not* provide $\mathbf{R}_\mathbf{x}(t)$ along with the estimated position.

Nonwhite measurement error processes: As shown in Examples 3.3 and 3.4 the GPS position error processes are not white, but may have significant time correlation. The time correlation may come from nonwhite GPS measurement errors such as multipath or from the GPS solution method. In particular, when the GPS receiver position solution incorporates a KF [45], then the time correlation of the GPS position errors is increased. The designer of a loosely coupled GPS aided INS approach should ensure that the GPS receiver is configured to determine epoch-wise position and velocity solutions.

Doppler: The GPS "Doppler" measurement is typically not a true Doppler measurement. Typically, the Doppler measurement is the change of the phase of the carrier signal over some interval of time [40]. The interval of time is often 1.0 sec. Because of this, the GPS velocity output computed from the Doppler measurement is not the instantaneous velocity at some specific time-of-applicability.

Lever arm: The INS computes the position of the IMU effective center location. The GPS computes the position of the antenna phase center. These two positions are not the same. The vector offset is referred to as the lever arm and should be compensated for the EKF data fusion procedure.

The main motivation for the use of a loosely coupled approach, instead of a tightly coupled approach, is that the former is simpler. A loosely coupled approach can be implemented with an off-the-shelf GPS receiver and an off-the-shelf INS. The designer need not work with clock models, GPS ephemeris data, ephemeris calculations, or GPS basic measurements. Note that this approach does attempt to estimate IMU calibration parameters (e.g., biases). As those errors are calibrated, the rate of growth of the INS errors will decrease. However, depending on the extent of the simplifications made in implementing the EKF and the extent to which the above issues are addressed, the INS errors may not be correctly estimated.

3.5.2.2 Tightly coupled system

As illustrated in Figure 3.5, in a tightly coupled system, the EKF measurements are the GPS range (or phase change) measurements. Residual measurements are formed with the INS estimates of range (or phase change). The INS estimates the range using Equation (3.44) with $\varepsilon_i = 0$. To utilize that equation, the satellite position is computed using ephemeris data downloaded through the GPS receiver. Similarly, the GPS pseudorange or carrier phase measurements are output by the GPS receiver.

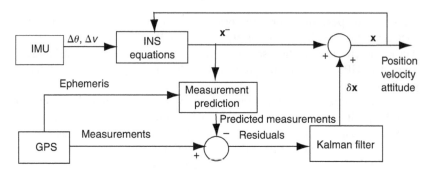

FIGURE 3.5   Block diagram of a tightly coupled GPS aided INS.

From Section 3.3, the range measurement residual is

$$\mathbf{y} = \delta\rho = \begin{bmatrix} \mathbf{h}_1 \mathbf{C}_n^e, & 1 \\ \mathbf{h}_2 \mathbf{C}_n^e, & 1 \\ \vdots & \\ \mathbf{h}_m \mathbf{C}_n^e, & 1 \end{bmatrix} \begin{bmatrix} \delta n \\ \delta e \\ \delta d \\ b_u \end{bmatrix} \tag{3.70}$$

where $\mathbf{C}_n^e$ is the rotation matrix for transforming the representation of vectors in navigation frame to the ECEF frame that is valid at the measurement epoch. When using this implementation approach, the designer is responsible for accommodating the receiver clock bias. As an alternative to including clock bias states in the error model, the clock bias can be addressed by subtracting the measurement of one satellite from the measurement of all other satellites, but the resulting differenced signals then have correlated measurement errors.

If the INS error state is ordered as $\delta\mathbf{x} = [\delta\mathbf{p}^T, b_u, \delta\mathbf{v}^T, \dot{b}_u, \delta\rho^T, \mathbf{x}_a^T, \mathbf{x}_g^T]^T$ with the INS error dynamics as in Equation (3.68) and the receiver clock dynamics as in Section 3.3.3.1; then, for Step 4 of the EKF algorithm, the linearized pseudorange measurement matrix is

$$\mathbf{H}_k = [\mathbf{H}\mathbf{C}_n^e, \mathbf{0}]$$

where $\mathbf{H}$ is defined in Equation (3.47), $\mathbf{0}$ is an m by 13 matrix of zeros, and m is the number of satellites available. Note that the components of the error in this vector of measurements are uncorrelated. Whether or not the measurement error can be considered white depends on which GPS error correction approaches are used and the time between measurement epochs. If significantly correlated measurement errors exist, then they should be addressed through state augmentation and possibly a Schmidt–Kalman filter implementation approach [19,20].

As opposed to a loosely coupled system, the designer of a tightly coupled system must implement ephemeris calculations, implement a receiver clock model, and be familiar with various receiver specific issues and peculiarities. The payoff for this increased level of understanding is potentially better performance. The higher performance is achievable because the various measurement errors and their covariance can be properly modeled and incorporated in the design approach. As in the loosely coupled approach, the tightly coupled approach does attempt to estimate IMU calibration parameters (e.g., biases). As the errors are calibrated, the rate of growth of the INS errors will decrease.

Example 3.6 This example uses the same hypothetical 2D world as in Example 3.4. Simulation results are shown in Figure 3.6. The vehicle trajectory is also similar to that in the previous example. In this example, using GPS measurement epochs that have 1 sec duration, at the $(k+1)$th measurement epoch (i.e., $t = k + 1$) the GPS range vector will be used as measurements in the EKF to estimate the INS error state. The GPS measurements are computed as the actual range plus a linearly increasing clock bias, and Gaussian random noise with unit variance. In addition to the GPS receiver, the vehicle is equipped with an IMU and a computer capable of integrating the INS equations.

In two dimensions, the INS integrates the equation

$$
\dot{\hat{\mathbf{x}}} = \begin{bmatrix} \dot{\hat{n}} \\ \dot{\hat{e}} \\ \dot{\hat{v}}_n \\ \dot{\hat{v}}_e \\ \dot{\hat{\psi}} \end{bmatrix} = \begin{bmatrix} 0 & 0 & 1 & 0 & 0 \\ 0 & 0 & 0 & 1 & 0 \\ 0 & 0 & 0 & 0 & 0 \\ 0 & 0 & 0 & 0 & 0 \\ 0 & 0 & 0 & 0 & 0 \end{bmatrix} \begin{bmatrix} \hat{n} \\ \hat{e} \\ \hat{v}_n \\ \hat{v}_e \\ \hat{\psi} \end{bmatrix} + \begin{bmatrix} 0 & 0 & 0 \\ 0 & 0 & 0 \\ \cos \hat{\psi} & -\sin \hat{\psi} & 0 \\ \sin \hat{\psi} & \cos \hat{\psi} & 0 \\ 0 & 0 & 1 \end{bmatrix} \begin{bmatrix} \tilde{a}_u \\ \tilde{a}_v \\ \tilde{\omega}_r \end{bmatrix}
$$

(3.71)

between GPS measurement epochs, that is, $t \in [k, k+1)$ sec. In these equations, for a generic variable z, $\hat{z}$ denotes the computed value of z and $\hat{z}$ denotes the measured value of z. Using this notation, $[\delta\hat{a}_u, \delta\hat{a}_v, \delta\hat{\omega}_r]$ are the estimated values of the IMU biases $[\delta a_u, \delta a_v, \delta \omega_r]$.

Let $(\tilde{a}_u, \tilde{a}_v)$ be the measured acceleration vector and $\tilde{\omega}_r$ be the measured yaw rate in body frame.

Considering bias errors, scale factor errors, and white measurement noise, the assumed relations between the IMU measurements $(\tilde{a}_u, \tilde{a}_v, \tilde{\omega}_r)$ and the actual

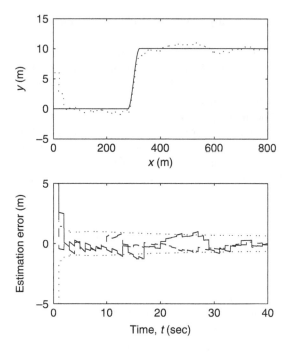

FIGURE 3.6   EKF based GPS solutions for Example 3.6. (a) Position estimation results for Example 3.6. Top — Estimated position trajectory (dotted) overlaid on the actual trajectory (solid). Bottom — Position estimation errors vs. time (solid and dashed curves), and EKF estimate of the position error standard deviation (dotted). (b) Estimation results for Example 3.6. Top — Velocity estimation errors vs. time (solid curves) and EKF estimate of the velocity error standard deviation (dotted). Middle — IMU bias estimation errors vs. time. Bottom — Yaw estimation error vs. time (solid) and EKF estimate of the yaw error standard deviation (dotted).

values (a_u, a_v, ω_r) are:

$$\tilde{a}_u = (1 + \delta k_u)a_u - \delta a_u + n_u \tag{3.72}$$

$$\tilde{a}_v = (1 + \delta k_v)a_v - \delta a_v + n_v \tag{3.73}$$

$$\tilde{\omega}_r = (1 + \delta k_r)\omega_r - \delta \omega_r + n_r \tag{3.74}$$

where $\delta a_u, \delta a_v, \delta \omega_r$ are bias errors, n_u, n_v, n_r represent white noise processes with variance of $(5.0 \times 10^{-4},\ 5.0 \times 10^{-4},\ 5.0 \times 10^{-6})$ respectively, and $(\delta k_u, \delta k_v, \delta k_r)$ represent sensor scale factor errors. We have included scale factor errors at this point due to their importance, but will assume that the scale factor errors are known to be identically zero in the following discussion.

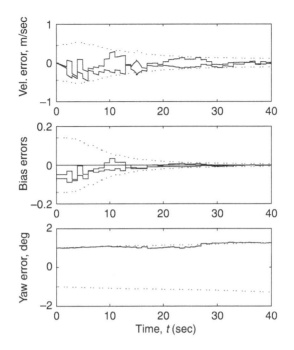

Figure 3.6 Continued.

To estimate the IMU bias vector, we append the bias error to the state vector

$$\delta \mathbf{x} = [\delta n, \delta e, \delta v_n, \delta v_e, \delta \psi, \delta a_u, \delta a_v, \delta \omega_r]$$

and specify a dynamic model for the appended states. By its design, the IMU performance is independent of vehicle maneuvering, as long as the IMU is used within its bandwidth and output range specifications. Therefore, specification of the IMU bias stochastic models can be based on data acquired in the lab. It is often sufficient to consider the IMU bias errors as random walk variables

$$\delta \dot{a}_u = n_{b_1}$$
$$\delta \dot{a}_v = n_{b_2}$$
$$\delta \dot{\omega}_r = n_{b_3}$$

In this simulation example, $(n_{b_1}, n_{b_2}, n_{b_3})$ have variance of $(1.0 \times 10^{-8}, 1.0 \times 10^{-8}, 5.0 \times 10^{-12})$ respectively. The augmented, linearized, dynamic model for the error state (used to implement the EKF) is

$$
\begin{bmatrix} \delta \dot{n} \\ \delta \dot{e} \\ \delta \dot{v}_n \\ \delta \dot{v}_e \\ \delta \dot{\psi} \\ \delta \dot{a}_u \\ \delta \dot{a}_v \\ \delta \dot{\omega}_r \end{bmatrix} =
\begin{bmatrix}
0 & 0 & 1 & 0 & 0 & 0 & 0 & 0 \\
0 & 0 & 0 & 1 & 0 & 0 & 0 & 0 \\
0 & 0 & 0 & 0 & -\hat{a}_e & \cos\psi & -\sin\psi & 0 \\
0 & 0 & 0 & 0 & \hat{a}_n & \sin\psi & \cos\psi & 0 \\
0 & 0 & 0 & 0 & 0 & 0 & 0 & 1 \\
0 & 0 & 0 & 0 & 0 & 0 & 0 & 0 \\
0 & 0 & 0 & 0 & 0 & 0 & 0 & 0 \\
0 & 0 & 0 & 0 & 0 & 0 & 0 & 0
\end{bmatrix}
\begin{bmatrix} \delta n \\ \delta e \\ \delta v_n \\ \delta v_e \\ \delta \psi \\ \delta a_u \\ \delta a_v \\ \delta \omega_r \end{bmatrix}
$$

$$
+ \begin{bmatrix}
0 & 0 & 0 & 0 & 0 & 0 \\
0 & 0 & 0 & 0 & 0 & 0 \\
\cos\psi & -\sin\psi & 0 & 0 & 0 & 0 \\
\sin\psi & \cos\psi & 0 & 0 & 0 & 0 \\
0 & 0 & 1 & 0 & 0 & 0 \\
0 & 0 & 0 & 1 & 0 & 0 \\
0 & 0 & 0 & 0 & 1 & 0 \\
0 & 0 & 0 & 0 & 0 & 1
\end{bmatrix}
\begin{bmatrix} n_u \\ n_v \\ n_r \\ n_{b1} \\ n_{b2} \\ n_{b3} \end{bmatrix}
\tag{3.75}
$$

where

$$
\begin{bmatrix} \hat{a}_n \\ \hat{a}_e \end{bmatrix} = \begin{bmatrix} \cos\hat{\psi} & -\sin\hat{\psi} \\ \sin\hat{\psi} & \cos\hat{\psi} \end{bmatrix} \begin{bmatrix} \tilde{a}_u \\ \tilde{a}_v \end{bmatrix} = \hat{\mathbf{C}}_b^n \begin{bmatrix} \hat{a}_u \\ \hat{a}_v \end{bmatrix}
\tag{3.76}
$$

The clock model and clock error states must also be appended. The resulting equation can be written as

$$
\delta \dot{\mathbf{x}}_{\text{ins}} = \mathbf{F}_{\text{ins}} \delta \mathbf{x}_{\text{ins}} + \Gamma \mathbf{w}_{\text{ins}}
\tag{3.77}
$$

With the variances specified above, the matrix $\mathbf{Q}$ is known. Note that in this approach the matrices $\mathbf{Q}$ and $\mathbf{R}$ are well defined based on the physics of the problem; they are not ad hoc tuning parameters as they were in Section 3.3.3.

Between GPS measurement epochs that are separated by 1 sec (i.e., $t \in [k, k+1)$ sec for the $(k+1)$th epoch) the INS propagates the state estimate using the IMU data. The INS also propagates the error covariance matrix $\mathbf{P}$ according to Equation (3.40). The error covariance propagation does depend on the IMU data because the $\mathbf{F}$ matrix includes a_n, a_e, and ψ. Due to the dependence of $\mathbf{F}$ on the IMU data, the matrices $\mathbf{\Phi}$ and $\mathbf{Q}_k$ must be computed during operation as discussed in Section 3.2.1.1.

At the GPS measurement epoch, the GPS pseudorange measurements are used in an EKF to estimate the INS error state. When the INS error state is available from the EKF, it is used to correct the INS state according to Equation (3.43). As time progressed the IMU errors are calibrated and the rate of growth of the INS errors decreases.

The top graph in Figure 3.6a shows both the estimated and the actual vehicle trajectories. The lateral maneuver occurs at approximately $t = 15$ sec. The bottom graph shows the position estimation error components as a function of time. In addition to the estimation errors, the graph shows $\pm\sqrt{P_{11} + P_{22}}$ which represents the EKF prediction of the standard deviation of the position estimation error. The variance of the position error decreases steadily over the period of the simulation due to the decay of the initial position error, the estimation of velocity, and the balancing of the acceleration biases with the yaw estimation error.

The top graph of Figure 3.6b shows velocity estimation error components and the EKF prediction of the standard deviation of the velocity estimation error as functions of time. After the initial transients, the velocity estimation error decreases steadily due to the decay of the initial velocity error and the balancing of the acceleration biases with the yaw estimation error. The middle graph shows the bias estimation error components as functions of time. The bottom graph shows the yaw estimation error and the EKF prediction of the standard deviation of the yaw estimation error as functions of time. Analysis of Equation (3.76) shows that the yaw angle and gyro bias errors are observable only when the acceleration vector $[a_n(t), a_e(t)]$ is nonzero. Therefore, the yaw error is not adjusted by the EKF except for a brief interval following the maneuver. Close inspection of Figure 3.6b shows that the yaw error standard deviation is slowly increasing due to the accumulation of gyro measurement noise during the attitude integration process. Note that the yaw estimation error does not approach zero; however, its net effect on the velocity and position does approach zero (in the absence of maneuvering). From Equation (3.71) we see that (neglecting noise)

$$\delta\dot{v}_n = \dot{v}_n - \dot{\hat{v}}\hat{n} = a_n - (\tilde{a}_u \cos\hat{\psi} - \tilde{a}_v \sin\hat{\psi})$$

$$\delta\dot{v}_n = a_n - ((a_u - \delta\hat{a}_u)\cos(\psi - \delta\psi) - (a_v - \delta\hat{a}_v)\sin(\psi - \delta\psi))$$

Even when the acceleration vector is zero, we have

$$\delta\dot{v}_n = (\cos(\psi)\cos(\delta\psi) + \sin(\psi)\sin(\delta\psi))\delta\hat{a}_u$$

$$- (\sin(\psi)\cos(\delta\psi) - \cos(\psi)\sin(\delta\psi))\delta\hat{a}_v \qquad (3.78)$$

Although the linearization of Equation (3.78) is used to formulate the third row of Equation (3.76), for fixed ψ the equation

$$(\cos(\psi)\cos(\delta\psi) + \sin(\psi)\sin(\delta\psi))\delta\hat{a}_u$$
$$- (\sin(\psi)\cos(\delta\psi) - \cos(\psi)\sin(\delta\psi))\delta\hat{a}_v = 0$$

defines a surface of $[\delta\hat{a}_u, \delta\hat{a}_v, \delta\psi]$ values such that $\delta\dot{v}_n = 0$. The δv_e dynamics provide a second such null surface. As long as the EKF drives the vector $[\delta\hat{a}_u, \delta\hat{a}_v, \delta\psi]$ to the intersection of these two surfaces, the net effect of these errors are balanced. For this 2D example, the intersection of the two surfaces is defined by

$$\begin{bmatrix} 0 \\ 0 \end{bmatrix} = \begin{bmatrix} \cos\psi & -\sin\psi \\ \sin\psi & \cos\psi \end{bmatrix} \begin{bmatrix} \cos\delta\psi & \sin\delta\psi \\ -\sin\delta\psi & \cos\delta\psi \end{bmatrix} \begin{bmatrix} \delta\hat{a}_u \\ \delta\hat{a}_v \end{bmatrix} \qquad (3.79)$$

In particular, the EKF causes the accelerometer bias estimation errors $\delta\hat{a}_u$ and $\delta\hat{a}_v$ to converge to zero, but $\delta\psi$ need not converge to zero. This is the practical result of the fact that, without acceleration, the yaw error is not observable.

In real 3D applications, the situation is more complex, since without maneuvering the errors in estimating pitch and roll have similar effects as accelerometer bias errors. Therefore, the linearized dynamics have unobservable subspaces. As the vehicle maneuvers, the null surfaces change. Over time, if the null surfaces change sufficiently, then the yaw and bias estimation errors will converge toward zero (until the maneuvering stops).

Note that if GPS measurements are unavailable, the integration of the IMU measurements by the INS is not interrupted. Therefore, this approach does increase the availability of the state estimate (higher frequency and no dropouts due to missing satellites). The bandwidth of the state estimate is also increased since it is determined by the bandwidth of the IMU not the bandwidth of the GPS receiver. The accuracy of the integrated solution will depend on the quality of the IMU and the GPS receiver. The length of time that the INS can maintain a specified level of accuracy after losing reception of GPS signals will predominantly depend on the quality of the IMU.

3.6 Chapter Summary

This chapter has reviewed the use of the KF and EKF as a tool for fusing information from various sensors that provide information about the state of a dynamic system. Preconditions necessary for the use of these methods are analytic models for the state dynamics and the relation between the state and the measurement. One prominent application of these tools that satisfies these preconditions is the integration of GPS and INS. We have presented an analytic overview of a few of the existing uses of the EKF in this application. Many other alternatives have been suggested in the literature. We have used a 2D example to work through various design issues and to illustrate various implementation issues.

While the theory of this chapter has reviewed GPS aided INS in standard vector form, four of the examples have utilized a fictional 2D world. Therefore, it is useful to briefly consider how the conclusions of those examples generalized to the 3D world in which an actual system must function. The objectives of Example 3.2 were to illustrate the standard method of solution of the GPS positioning problem and to demonstrate that the components of the position estimate error vector were correlated (i.e., $\mathbf{R}_x$ is not diagonal). The objectives of Examples 3.3 and 3.4 were to illustrate the use of the $\mathbf{Q}$ matrix as a tuning parameter, to reinforce the fact that such tuning removes the optimal stochastic properties from the KF, and to illustrate the fact that there are not optimal settings of the tuning parameters that apply in all user situations. In addition, that example demonstrates that the position estimate error vector is not white, but has significant time correlation. The objectives of Example 3.6 were to illustrate the error state modeling approach which allows a proper stochastic interpretation of KF implementations,[2] to illustrate the state augmentation process used for instrument calibration, to illustrate that in this approach the $\mathbf{Q}$ and $\mathbf{R}$ matrices are not tuning parameters but are physically determined, to illustrate that the observability of certain subspaces of the error state are dependent on the vehicle motion, and to illustrate that the state estimation error is uncorrelated with the vehicle motion due to the IMU and INS. All these issues were more convenient to illustrate in a 2D example, but are equally applicable to our 3D world.

Another implementation approach, referred to in the literature as *Deep* or *Ultratight* integration, feeds information from the INS back into the GPS receiver [46–48]. We have not discussed these methods in this chapter as their implementation requires access to GPS receiver source code, which is not available to most GPS users. The objective of these techniques is to use the INS estimates of the GPS receiver position and velocity to aid the receiver in acquiring and tracking the GPS satellite signals. This would be especially beneficial in low signal-to-noise ratio situations.

ACKNOWLEDGMENTS

The authors gratefully thank California PATH and CalTrans for their continued interest in and support of related research at the University of California Riverside. Specifically, this chapter was written with support from CalTrans

---

[2] It is only proper to first order for EKF implementations.

under Research Technical Agreement No. 65A0178. We are also grateful to Y. Zhao at the Department of Electrical Engineering of U.C. Riverside for her review and comments.

REFERENCES

1. A. Abidi and C. Gonzalez. *DATA FUSION in Robotics and Machine Intelligence*. Academic Press, San Diego, CA, 1992.
2. E. R. Dougherty and C. R. Giardina. *Mathematical Methods for Artificial Intelligence and Autonomous Systems*. Prentice Hall, Englewood Cliffs, NJ, 1988.
3. C. W. Therrien. *Decision, Estimation, and Classification: An Introduction to Pattern Recognition and Related Topics*. Wiley, New York, 1989.
4. J. Manyika and H. F. Durrant-Whyte. *Data Fusion and Sensor Management: A Decentralized Information Theoretic Approach*. Ellis Horwood, New York, 1994.
5. L. D. Stone, C. A. Barlow, and T. L. Corwin. *Bayesian Multiple Target Tracking*. Artech House, Boston and London, 1999.
6. K. Varshney. *Distributed Detection and Data Fusion*. Springer-Verlag, New York, 1996.
7. D. L. Hall. *Mathematical Techniques in Multisensor Data Fusion*. Artech House, Norwood, MA, 1992.
8. E. Waltz and J. Llinas. *Multisensor Data Fusion*. Artech House, Norwood, MA, 1990.
9. D. B. Cox. Integration of GPS with Inertial Navigation Systems. *Proceedings of ION GPS 1980*, 1980.
10. Y. Yang, J. A. Farrell, and M. Barth. High-Accuracy, High-Frequency Differential Carrier Phase GPS Aided Low-Cost INS. *Proceedings of IEEE PLANS*, March 13–16, 2000.
11. D. Eller. GPS/IMU Navigation in a High Dynamics Environment. *Proceedings of the First International Symposium on Precise Positioning with GPS*, April 15–19, 1985.
12. J. A. Farrell, T. Givargis, and M. Barth. Real-Time Differential Carrier Phase GPS Aided INS. *IEEE Transactions on Control Systems Technology*, 8: 709–721, 2000.
13. A. K. Brown and W. M. Bowles. Interferometric Attitude Determination Using GPS. *Proceedings of the Third Geodetic Symposium on Satellite Doppler Positioning*, February 1982.
14. J. L. Crassidis, E. G. Lightsey, and F. L. Markley. Efficient and Optimal Attitude Determination Using Recursive Global Positioning System Signal Operations. *Journal of Guidance, Control, and Dynamics*, 22: 193–201, 1999.
15. S. J. Fujikawa and D. F. Zimbelman. Spacecraft Attitude Determination by Kalman Filtering of Global Positioning System Signals. *Journal of Guidance, Control, and Dynamics*, 18: 1365–1371, 1995.
16. R. E. Kalman. A New Approach to Linear Filtering and Prediction Problems. *ASME Journal of Basic Engineering*, series D: 35–45, 1960.

17. R. E. Kalman. New Results in Linear Filtering and Prediction Theory. *ASME Journal of Basic Engineering*, series D: 95–108, 1961.
18. S. Haykin. *Adaptive Filter Theory*. Prentice Hall, Englewood Cliffs, NJ, 1996.
19. A. Gelb. *Applied Optimal Estimation*. MIT, Cambridge, MA, 1974.
20. R. G. Brown and Y. C. Hwang. *Introduction to Random Signals and Applied Kalman Filtering*. Wiley, New York, 1992.
21. M. S. Grewal and A. P. Andrews. *Kalman Filtering: Theory and Practice*. Prentice Hall, Englewood Cliffs, NJ, 1993.
22. C. Van Loan. Computing Integrals Involving the Matrix Exponential. *IEEE Transactions on Automatic Control*, AC. 23: 395–404, 1978.
23. C. Moler and C. Van Loan. Nineteen Dubious Ways to Compute the Exponential of a Matrix. *SIAM*, Rev. 20: 801–836, 1978.
24. B. D. O. Anderson and J. B. Moore. *Optimal Filtering*. Prentice Hall, Englewood Cliffs, NJ, 1979.
25. G. J. S. Ross. *Nonlinear Estimation*. Springer-Verlag, New York, 1990.
26. D. D. Denison. *Nonlinear Estimation and Classification*. Springer, New York, 2003.
27. J. A. Farrell and M. Barth. *The Global Positioning System & Inertial Navigation*. McGraw-Hill, New York, 1998.
28. Y. Yang, T. Sharpe, and R. Hatch. A Fast Ambiguity Resolution Technique for RTK Embedded Within a GPS Receiver. *Proceedings of ION GPS 2002*, September 2002.
29. J. Cheng, J. A. Farrell, L. Yu, and E. Thomas. Aided Integer Ambiguity Resolution Algorithm. *Proceedings of IEEE PLANS*, April 2004.
30. P. Hwang. Kinematic GPS: Resolving Integer Ambiguities On-the Fly. *Proceedings of IEEE PLANS*, March 1990.
31. R. R. Hatch. Instantaneous Ambiguity Resolution. *KIS-90 Symposium*, August 1990.
32. P. J. G. Teunissen. A New Method for Fast Carrier Phase Ambiguity Estimation. *Proceedings of IEEE PLANS*, April 1994.
33. R. Hatch. Synergism of GPS Code and Carrier Measurements. *Proceedings of the Third International Geodetic Symposium on Satellite Doppler Positioning*, February 1982.
34. W. Parkinson and J. Spilker Jr. *The Global Positioning System: Theory and Applications*, volume II. American Institute of Aeronautics and Astronautics, Inc., Washington, DC, 1996.
35. P. Misra and P. Enge. *Global Positioning System: Signals, Measurements, and Performance*. Ganga-Jumuna Press, Lincoln, MA, 2001.
36. J. A. Klobuchar. Ionospheric Time-Delay Algorithm for Single-Frequency GPS Users. *IEEE Transactions on Aerospace and Electronic Systems*, 23: 325–331, 1987.
37. Naser El-Sheimy and Klaus Peter Schwarz. Navigating Urban Areas by VISAT — A Mobile Mapping System Integrating GPS/INS/Digital Cameras for GIS Applications. *Journal of The Institute of Navigation*, 45: 275–285, 1998.
38. M. Rothblatt. Urban Area Performance of GPS Receiver with Simultrac Capability. *IEEE Aerospace Electronics Systems Magazine,* 161–174, 1992.

39. M. Tsakiri, A. Kealy, and M. Stewart. Urban Canyon Vehicle Navigation with Integrated GPS/GLONASS/DR Systems. *Journal of the Institute of Navigation*, 46, 1999.
40. M. H. Kao and D. H. Eller. Multiconfiguration Kalman Filter Design for High-Performance GPS Navigation. *IEEE Transactions on Automatic Control*, 28: 304–314, 1983.
41. D. H. Titterton and J. L. Weston. *Strapdown Inertial Navigation Technology*. Peter Peregrinus Ltd., London, 1997.
42. J. L. Farrell. *Integrated Aircraft Navigation*. Academic Press, New York, 1976.
43. K. R. Britting. *Inertial Navigation Systems Analysis*. Wiley-Interscience, New York, 1971.
44. Y. J. Cui, S. S. Ge, T. Goh, W. K. Tan, E. Sim, and K. Tan. Synchronization Solutions for a Loosely Coupled INS and GPS Navigation System. *Proceedings of the Asian Control Conference*, 2002.
45. N. A. Carlson. Federated Square Root Filter for Decentralized Parallel Processes. *IEEE Transactions on Aerospace and Electronic Systems*, 26: 517–525, 1990.
46. D. Gustafson and J. Dowdle. Deeply Integrated Code Tracking: Comparative Performance Analysis. *Proceedings of ION GPS 2003*, September 9–12, 2003.
47. B. Vik and T. I. Fossen. Nonlinear Analysis of GPS Aided by INS. *Proceedings of ION GPS 1999*, June 28–30, 1999.
48. S. Alban, D. M. Akos, and S. M. Rock. Performance Analysis and Architectures for INS-Aided GPS Tracking Loops. *Proceedings of ION GPS 2003*, January 22–24, 2003.

Biographies

Jingrong Cheng graduated from Harbin Engineering University with a Masters in electrical engineering in 2000. She is currently a Ph.D. candidate of University of California at Riverside. She specializes in the GPS/INS integration, integer ambiguity resolution, GPS signal processing and navigation.

Yu Lu received both his Bachelors and Masters degrees in electrical engineering from Peking University (P. R. China) in 1997 and 2000, respectively. Currently he is a Ph.D. candidate in electrical engineering at the University of California at Riverside. His research interests includes GPS signal processing, GPS software radio implementation, GPS/INS integration, and embedded system design.

Elmer R. Thomas graduated from the Bourns College of Engineering Computer Engineering Program at U.C. Riverside in December 2002. While an undergraduate he earned several fellowships, awards, and scholarships including: the Eugene Cota Robles fellowship, U.C. Leadership through Excellence and Advanced Degrees (UC LEADS) fellowship, and a certificate of recognition for excellence in leadership presented by Assemblyman John J. Benoit of the 64th Assembly District. His specialty is in Global Positioning Systems (GPS) and vehicle control. He has received an award for outstanding research in the area of GPS from the University of California president.

Jay A. Farrell received B.S. degrees (1986) in physics and electrical engineering from Iowa State University, and M.S. (1988) and Ph.D. (1989) degrees in electrical engineering from the University of Notre Dame. At Charles Stark Draper Lab (1989–1994), he was principal investigator on projects involving intelligent and learning control systems for autonomous vehicles. Dr. Farrell received the Engineering Vice President's Best Technical Publication Award in 1990, and Recognition Awards for Outstanding Performance and Achievement in 1991 and 1993. He is a professor and former chair of the Department of Electrical Engineering at the University of California, Riverside. His research interests include: identification and online control for nonlinear systems, integrated GPS/INS navigation, and artificial intelligence techniques for autonomous vehicles. He is author of the book *The Global Positioning System and Inertial Navigation* (McGraw-Hill, 1998) and over 120 additional technical publications.

4 Landmarks and Triangulation in Navigation

Huosheng Hu, Julian Ryde, and Jiali Shen

CONTENTS

4.1 INTRODUCTION

Landmarks are routinely used by biological systems as reference points during navigation. Their employment in robotic navigation requires the development of satisfactory sensor technologies for landmark selection and recognition, which poses a big challenge. During the last two decades, landmarks and triangulation techniques have been widely used in navigation of autonomous mobile robots in industry [1,2]. Such a navigation strategy relies on identification and subsequent recognition of distinctive environment features or objects that are either known a priori or extracted dynamically. This process has inherent difficulties in practice due to sensor noise and environment uncertainty [3]. This chapter outlines a number of landmark-based navigation algorithms that are able to locate the mobile robot and update landmarks autonomously.

Autonomous mobile robots need the capability to explore and navigate in dynamic or unknown environments in order to be useful in a wide range of real-world applications. Over the last few decades, many different types of sensing and navigation techniques have been developed in the field of mobile robots, some of which have achieved very promising results based on different sensors such as odometry, laser scanners, inertial sensors, gyro, sonar, and vision [4]. This trend has been mainly driven by the necessity of deployment of mobile robots in unstructured environments or coexisting with humans. However, since there is huge uncertainty in the real world and no sensor is perfect, it remains a great challenge today to build robust and intelligent navigation systems for mobile robots to operate safely in the real world.

In general, the methods for locating mobile robots in the real world are divided into two categories: relative positioning and absolute positioning. In relative positioning, odometry (or dead reckoning) [4] and inertial navigation (gyros and accelerometers) [5] are commonly used to calculate the robot positions from a starting reference point at a high updating rate. Odometry is one of the most popular internal sensor for position estimation because of its ease of use in real time. However the disadvantage of odometry and inertial navigation is that it has an unbounded accumulation of errors, and the mobile robot becomes lost easily. Therefore, frequent correction based on information obtained from other sensors becomes necessary.

In contrast, absolute positioning relies on detecting and recognizing different features in the robot environment in order for a mobile robot to reach a destination and implement specified tasks. These environment features are normally divided into four types [4] (i) active beacons that are fixed at known positions and actively transmit ultrasonic [6], IR or RF signals for the calculation of the absolute robot position from the direction of receiving incidence; (ii) artificial landmarks that are specially designed objects or markers placed at known locations in the environment; (iii) natural landmarks that are distinctive features in the environment and can be abstracted by robot sensors; and (iv) environment models that are built from prior knowledge about the environment and can be used for matching new sensor observations. Among these environment features, natural landmark-based navigation is flexible as no explicit artificial landmarks are needed, but may not function well when landmarks are sparse and often the environment must be known a priori. Although the artificial landmark and active beacon approaches are not flexible, the ability to find landmarks is enhanced and the process of map building is simplified. They have been widely adopted in many real-world applications, including Global Positioning Systems (GPSs) [7] and retro-reflective barcode targets [3]. This chapter only addresses the issues related to artificial landmarks and the associated navigation methods. More information on other landmarks can be found in Reference 8.

To make the use of mobile robots in daily deployment feasible, it is necessary to reach a trade-off between costs and benefits. Often, this calls for efficient landmark detection and triangulation algorithms that can guarantee real-time performance in the presence of insufficient or conflicting data from different types of sensors. Therefore, the use of multiple sensors (laser, sonar, and vision) and multiple landmarks (artificial and natural) for the position estimation of a mobile robot becomes absolutely necessary. Unlike odometry-based systems, landmark-based systems do not suffer from drift errors. However, how to select and recognize good landmarks in different circumstances is a nontrivial task since the different view angles of landmarks bring different errors into the measurements. Therefore, it is often the case that some landmarks are misidentified and this remains a challenging issue in many real-world applications. Moreover, the cooperative navigation of multiple mobile robots is a more flexible navigation method than navigation methods for a single robot.

The rest of the chapter is structured as follows. Section 4.2 presents an overview of our approach to landmark-based navigation, and proposes a multisensor system that can locate the robot and update different kinds of landmarks in the robot internal model concurrently. Section 4.3 describes a navigation system based on a rotating laser scanner and artificial landmarks, in which a triangulation method for calibrating the mobile robot position is also presented. Then the visual-based navigation is addressed in Section 4.4 for the mobile robot to recognize the digital and symbolic landmarks automatically. These landmarks are very common in office environments (name plates) and highway systems (road sign boards). Section 4.5 describes the localization system based on a SICK laser scanner and two cylinder landmarks, in which cylinder landmarks are fixed on two mobile robots and can change their relative position and distance for localization. Finally, a brief summary and potential future extension are given in Section 4.6.

4.2 Landmark-Based Navigation

In a landmark-based navigation system, the robot relies on its onboard sensors to detect and recognize landmarks in its environment to determine its position. This navigation system very much depends on the kind of sensors being used, the types of landmarks, and the number of landmarks available. For instance, Sugihara [9] used a single camera on a robot to detect the identical points in the environment and then adopted an $O(n^3 \lg n)$ algorithm to find the position and orientation of the robot such that each ray pierces at least one of the n points in the environment. An extended version was proposed in References 10 and 11, respectively. The localization based on distinguishable landmarks in the environment has been researched in Reference 12, in which the localization error varies depending on the configuration of landmarks. Apart from vision systems, other sensors have been widely used in position estimation, including laser [3], odometry [13], ultrasonic beacons [6], GPS [7], IR [12], and sonars [14]. Since no sensor is perfect and landmarks may change, none of these approaches is adequate for a mobile robot to operate autonomously in the real world. A landmark-based navigation system needs the integration of multiple sensors to achieve robustness and cope with uncertainties in both sensors and landmark positions. This motivates us to pursue a hybrid approach to the problem by integrating multiple sensors and different kinds of landmarks in a unified framework.

In general, the accuracy of the position estimation in a landmark-based navigation system is affected by two major problems. The first problem is that the navigation system cannot work well when landmarks accidentally change their positions. If natural landmarks are used in the navigation process, their positions must be prestored into the environment map so that it is possible for a mobile robot to localize itself during its operation. The second problem is that sensory measurements are noisy when the robot moves on an uneven floor surface or changes the speed frequently. The accuracy of robot positioning degrades gradually, and sometimes becomes unacceptable during a continuous operation. Therefore, re-calibration is needed from time to time and it becomes a burden for real-world applications.

To effectively solve these problems, we propose a novel landmark-based navigation system that is able to:

- Initialize its position through triangulation when necessary
- Update its internal landmark model when the position of landmarks is changed or new landmarks become available
- Localize the robot position by integrating data from odometry, laser scanner, sonar, and vision

Figure 4.1 shows the block diagram of our navigation system that is able to implement concurrent localization and map building automatically. It is

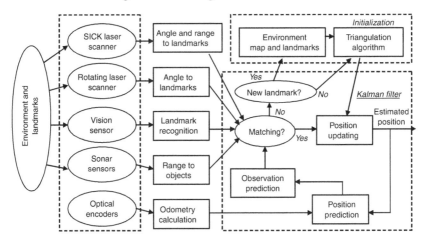

FIGURE 4.1   Landmark-based localization.

a closed-loop navigation process for position initialization, position updating, and map building. The system consists of three parts: an initialization part, a Kalman filter (KF) part, and a map-updating part:

- The initialization part includes a triangulation algorithm, which is based on angular measurements from the multiple sensors. Whenever the mobile robot is stationary, the triangulation algorithm is called to recalibrate the robot location so that the accumulative position errors can be corrected.
- The KF part aims to fuse measuring data from different sensors, and reduce individual sensor uncertainties. More details are presented in Section 4.3.3.
- The map building part is to update and maintain the internal world model of the mobile robot. A recursive least square algorithm is adopted to optimize the landmark position during operation. The key idea is to optimize the internal landmark model during the robot operation and add any new landmark that is consistently detected by the laser scanners and vision systems into the localization process. The choice of the least square criteria is of course based on the assumption that measurement errors have Gaussian distributions.

As can be seen in Figure 4.1, we have considered two types of laser scanners and one vision system for landmark detection and recognition. The proposed navigation system is especially aimed at service robots that operate in indoor environments such as offices and hospitals where the global map of the environment is two-dimensional. The position of the robot with respect to this map is unknown and needs to be determined.

There are three kinds of landmarks that are considered in our design (i) single strip retro-reflective landmarks, (ii) digital landmarks, and (iii) geometric landmarks. The positions of the first two kinds of landmarks are pre-input into a global map of the robot's environment. The positions of the geometric landmarks are abstracted and then registered into the global map dynamically. The next three sections describe their application in robot navigation respectively. Note that our navigation system is not restricted to these three kinds of landmarks, and can be easily extended into other kind of landmarks and their combination.

4.3 LASER SCANNER AND RETRO-REFLECTIVE LANDMARKS

4.3.1 Laser Scanner and Angle Observation

The localization system based on the laser scanner and retro-reflective landmarks is a promising absolute positioning technique in terms of performance and cost [8]. Using this technology, the coordinates of retro-reflective landmarks are prestored into an environment map. During its operation, the robot uses its onboard laser sensor to scan these landmarks in its surroundings and measure the bearing relative to each of them [1]. Then the position estimation of the mobile robot is normally calculated by using two distinctive methods: triangulation [12] and Kalman filtering [14].

Research here is based on a rotating laser scanner that is able to measure the angle between the robot base line and the beam line from either the leading or the falling edge of landmarks in the horizontal plane. As shown in Figure 4.2, the laser scanner is situated on top of the physical center of the robot, scanned 360° in azimuth up to 50 m range at a constant speed of 2 Hz. Note that an IR laser beam (870 nm) from a HeNe laser diode emits energy of 0.5 mW, which is eye-safe. As can be seen, there are six landmarks in this environment, namely $B_1, B_2, \ldots, B_6$. The landmarks are in the form of a single strip for easy detection from a large distance, instead of traditional bar-codes. All landmarks have an identical size of 50 cm in length and 10 cm in width. The positions of the landmarks are surveyed in advance and prestored into the robot memory as a look-up table, represented by the coordinates in the world frame:

$$\Re = [B_1, \ldots, B_i, \ldots, B_N] = [(b_{x1}, b_{y1}), \ldots, (b_{xi}, b_{yi}), \ldots] \qquad (4.1)$$

where (b_{xi}, b_{yi}) are the coordinates of the ith landmark and N is the total number of landmarks in use.

These landmarks can return strong reflective signals to the scanner, that is, the area inside dotted lines in Figure 4.2. The reflected light from these

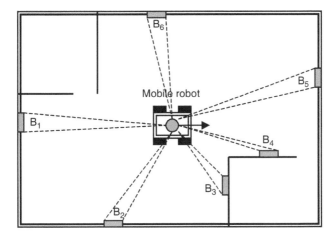

Figure 4.2 Landmarks and an onboard laser scanner.

landmarks is measured by a photo-detector inside the scanner. The scanner outputs the relative angles (with respect to the robot frame) measured by the scanner encoder at the falling edge of each landmark. The measurement variance would increase when the mobile robot moves around. This is because the vibration of the laser scanner would appear when the floor surface is not smooth.

4.3.2 Triangulation Algorithm

In the case of a stationary robot, the laser scanner senses all six landmarks, as shown in Figure 4.3, from a single location continuously. Then these data can be used to calculate the initial position and heading of the robot by the triangulation algorithms proposed in References 8 and 15. There are two ways to do triangulation. First, triangulation can be recursively implemented by choosing three landmarks in Figure 4.3 in turn when the mobile robot is stationary. It is actually identical to the "3-point problem" in land surveying. The laser scanner detects the falling edges of three landmarks and in turn provides three angle measurements, denoted by β_i ($i = 1, 2, 3$):

$$\beta_i = \tan^{-1} \frac{b_{yi} - y_l}{b_{xi} - x_l} - \theta \tag{4.2}$$

where θ is the robot orientation, (b_{xi}, b_{yi}) are the coordinates of the landmark β_i, and (x_l, y_l) are the coordinates of the laser scanner in a global frame.

Based on the trigonometric identity, the equations for calculating the robot position and orientation are easy to derive from Equation (4.2). More details can

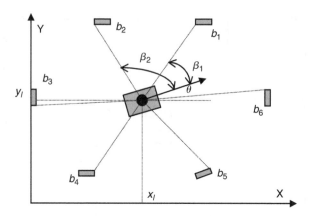

FIGURE 4.3   Triangulation example.

be seen from References 12 and 16. There are two problems in this triangulation process:

- First, the triangulation algorithm is normally sensitive to the positions of the three landmarks being used. When three targets are in an optimal position (about 120° apart), the results are very accurate. Otherwise, the robot position and orientation have big variances with respect to an optimal value.
- Second, it is very difficult to identify which landmark has been detected if all landmarks are identical. Mismatch is more likely to happen in practice when obstacles obscure one or more landmarks.

Alternatively, we can use all landmarks to make a least square solution with redundant observations so that the individual solutions do not depend on the specific choice of the landmarks. This solution is nonlinear, however, the equations can be readily linearized and used with the standard least square algorithm. The advantage of this approach is that the redundant observations can be used to check and, hopefully, eliminate blunders (misidentification of the targets, etc.) in the observation automatically. This approach can be readily automated and is, indeed, very popular in surveying. But it needs more computation time compared with the first approach.

Since the laser scanner can only measure the angles to the different landmarks, and cannot distinguish one landmark from another, a key problem is how to determine the correspondence between the measured angle and the landmark [1]. Therefore, the initialization of the robot position is normally done manually. Also, re-calibration is done manually when the mobile robot gets lost. This is inconvenient for real-world applications. It is necessary to find a feasible way to initialize the position of a mobile robot automatically, which can be found in Reference 3.

4.3.3 KF-Based Navigation Algorithm

The triangulation algorithm is difficult to implement when the robot moves around since it is necessary to compensate the time frame as each of three landmarks is detected at different robot positions. Skewis and Lumelsky [12] proposed a triangulation algorithm to attack this problem. However, there was no satisfactory result being obtained after the algorithm was tested. This is mainly due to the following reasons:

- Each of the landmarks is in a single strip and not encoded, that is, indistinguishable from one another.
- Noisy readings come from the laser scanner as some angle measurements are caused by random objects.
- In general, the robot environment is nonconvex. Therefore, not all landmarks can be seen by the laser scanner. Moreover some landmarks may be obscured by dynamic objects such as humans and other robots.

The KF algorithm is a natural choice for robot localization since it provides a convenient way to fuse the data from multiple sensors, for example, the laser scanner and odometry. However, it normally requires a linear dynamic model and a linear output model. However, in this research, both models are nonlinear as follows:

$$\mathbf{x}(k+1) = \mathbf{f}(\mathbf{x}(k), \mathbf{u}(k)) + \mathbf{w}(k) \tag{4.3}$$

$$\mathbf{z}(k+1) = \mathbf{h}(B_i, \mathbf{x}(k)) + \mathbf{v}(k) \tag{4.4}$$

where $\mathbf{f}(\mathbf{x}(k), \mathbf{u}(k))$ is the nonlinear state transition function of the robot. $\mathbf{w}(k) \sim N(\mathbf{0}, \mathbf{Q}(k))$ indicates a Gaussian noise with zero mean and covariance $\mathbf{Q}(k)$. $\mathbf{h}(B_i, \mathbf{x}(k))$ is the nonlinear observation model and $\mathbf{v}(k)$ is Gaussian noise with zero mean and covariance $\mathbf{R}(k)$.

The control vector is calculated by two optical encoders at each cycle time k:

$$\mathbf{u}(k) = [\Delta d, \Delta \theta]^{\mathrm{T}} \tag{4.5}$$

and the state transition function of the robot is

$$\mathbf{f}(\mathbf{x}(k), \mathbf{u}(k)) = \begin{bmatrix} x(k) + \Delta d \cos \theta \\ y(k) + \Delta d \sin \theta \\ \theta(k) + \Delta \theta \end{bmatrix} \tag{4.6}$$

For the laser scanner, the observation model is

$$\mathbf{h}(B_i, \mathbf{u}(k)) = \arctan \frac{b_{yi} - y(k)}{b_{xi} - x(k)} - \theta(k) \qquad (4.7)$$

Since the models (4.3) and (4.4) are nonlinear, the EKF [17] must be used here to integrate the laser measurements and encoder readings. Note that the EKF is recursively implemented as follows:

Step 1: Prediction — It predicts the next position of the robot using odometry.

$$\mathbf{x}(k + 1/k) = \mathbf{f}(\mathbf{x}(k), \mathbf{u}(k)) \qquad (4.8)$$

$$\mathbf{p}(k + 1/k) = \nabla \mathbf{f} \mathbf{P}(k/k) \nabla \mathbf{f}^{\mathrm{T}} + \mathbf{Q}(k) \qquad (4.9)$$

where $\nabla \mathbf{f}$ is the Jacobean matrix of the transition function, and is obtained by linearization

$$\nabla \mathbf{f} = \begin{bmatrix} 1 & 0 & -\Delta d(k) \sin \theta(k) \\ 0 & 1 & \Delta d(k) \cos \theta(k) \\ 0 & 0 & 1 \end{bmatrix} \qquad (4.10)$$

Step 2: Observation — It makes actual measurements.
The measurement of the laser scanner is

$$\mathbf{z}(k + 1) = \mathbf{h}(B_i, \mathbf{x}(k)) \qquad (4.11)$$

The predicted angular measurement is

$$\hat{\mathbf{z}}(k + 1) = \mathbf{h}(B_i, \hat{\mathbf{x}}(k + 1/k)) \qquad (4.12)$$

Step 3: Matching— It compares the real measurement with the predicted measurement.
To calculate the innovation, use

$$\mathbf{v}(k + 1) = \mathbf{z}(k + 1) - \hat{\mathbf{z}}(k + 1) \qquad (4.13)$$

The innovation covariance is:

$$\mathbf{S}(k + 1) = \nabla \mathbf{h} \mathbf{P}(k + 1) \nabla \mathbf{h}^{\mathrm{T}} + \mathbf{R}(k + 1) \qquad (4.14)$$

where $\nabla \mathbf{B}$ is the Jacobean matrix of the measurement function:

$$\nabla \mathbf{h} = \left[\frac{\partial \mathbf{h}}{\partial \mathbf{x}}, \; \frac{\partial \mathbf{h}}{\partial \mathbf{y}}, \; -1 \right] \qquad (4.15)$$

For each measurement, a validation gate is used to decide whether it is a match or not:

$$\mathbf{v}(k+1)\mathbf{S}(k+1)\mathbf{v}^\mathrm{T}(k+1) \leq \mathbf{G} \qquad (4.16)$$

If it is true, the current measurement is accepted. Otherwise, it is disregarded.

Step 4: Updating — It corrects the prediction error from odometry readings.

The filter gain is updated by:

$$\mathbf{W}(k+1) = \mathbf{P}(k+1/k)\nabla \mathbf{h}^\mathrm{T}\mathbf{S}^{-1}(k+1) \qquad (4.17)$$

The robot state is then calculated by:

$$\mathbf{x}(k+1/k+1) = \hat{\mathbf{x}}(k+1/k) + \mathbf{W}(k+1)\mathbf{v}(k+1) \qquad (4.18)$$

The covariance is updated by:

$$\mathbf{P}(k+1/k+1) = \mathbf{P}(k+1/k) - \mathbf{W}(k+1)\mathbf{S}(k+1)\mathbf{W}^\mathrm{T}(k+1) \qquad (4.19)$$

Step 5: Return to Step 1 to recursively implement the four steps earlier.

The algorithm is essentially very simple although it contains some very useful features. It produces the estimate of the current robot position at each cycle by integrating odometry data with only one angle measurement from the laser scanner. Recursively, it combines every new measurement with measurements made in the past to estimate the robot position, or "make a compromise." This can be seen as a pseudo "triangulation" technique in a dynamic sense.

4.3.4 Implementation and Results

The proposed navigation algorithm based on angle-only observations was implemented in our robotics research laboratory. The mobile robot equipped with a rotating laser scanner and single-stripe landmarks were fixed on the walls within the laboratory. The mobile robot was commanded to follow a close-loop route at a speed of 0.3 m/sec. The route is near circular with a diameter of 4 m.

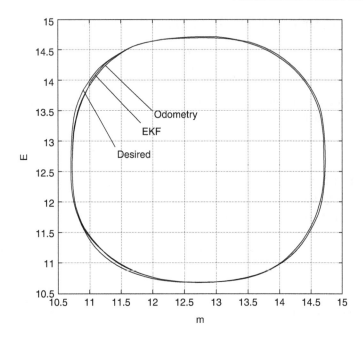

FIGURE 4.4 Navigating a close-loop route inside building.

Figure 4.4 presents the results gathered from the robot operation. There are three sets of data, namely a planned trajectory, a trajectory calculated from odometry, and a trajectory estimated by the EKF. As can be seen, the trajectory produced by odometry deviated further than the one generated by the EKF. Both odometry data and the EKF data look very close to the planned trajectory since the trajectory plotting is scaled down too much. However, the odometry data will deviate further away from the desired trajectory if the mobile robot travels continuously, which is due to the accumulative error of odometry.

4.4 VISION AND DIGITAL LANDMARKS

Visual robot navigation can be roughly classified into two major approaches: one is the iconic or appearance-based method that directly compares the raw data with the internal map and another is the feature-based method that focuses on the prominent features [18]. A feature-based navigation algorithm is often simpler and reliable, especially in dynamic environments. For instance, Atiya and Hager [19] used a stereo vision system to obtain vertical image edges in order to determine robot position. The observed landmark and stored map labeling problem is solved by a set-based method. Se et al. [20] proposed a random sample consensus (RANSAC) approach to determine the global position of the robot by matching the SIFT (scale invariant feature transform) features. Feature-based methods are often very efficient, and we have adopted it in our design.

However, the presence of nonunique feature landmarks causes the serious concern in feature-based visual navigation. Therefore, instead of undistinguishable landmarks addressed in the previous section, we propose a new type of artificial landmarks, which draws inspiration from wide applications of License Plate Recognition (LPR). These landmarks are embedded with characters and digit numbers that are similar to the name plates in offices and the license plates used in transport. A similar approach is presented in Reference 21, which proposed a visual landmark learning and recognition system for use in mobile robot navigation tasks that can read text inside well-defined landmarks such as nameplates, streets, and roads. However, there is no indication of its real-time performance.

Figure 4.5a presents the format of the proposed landmark, and Figure 4.5b shows a real landmark held by a person. Each landmark has the following features:

- Five characters, the letter L followed by four digits, are printed on the landmark.
- Each of the five characters has the same size, and the clearances between the characters are all the same (H, W, and D in Figure 4.1a). We currently select the parameters: L = 33, D = 200, H = 66, and W = 34 (mm), which may be changed in different application environments.
- The positions of the characters are also known (L in Figure 4.5a).

4.4.1 Landmark Recognition

The digits are the index of the landmark and the algorithm can identify the landmark with a digits recognition method. The standard size of the characters contains enough information for robot localization. Since the proposed landmark is similar to a license plate, many algorithms developed for license recognition can be used here directly, including the fuzzy-map method for locating the plate and the neural network for character recognition [22], and the fast plate location method based on vertical edges of the images [23]. Figure 4.6 shows a new landmark recognition algorithm that consists of three major modules: region finding, digits finding, and digits recognition.

4.4.1.1 Region finding module

This module is to find out all the probable regions that contain the landmark digits and exclude as much background as possible. Considering the features of the digits (sharply rising and falling edge in pairs in a horizontal scan line),

(a)

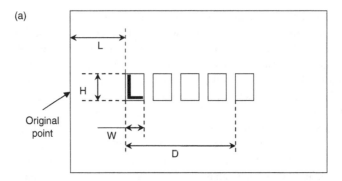

(b)

FIGURE 4.5 The proposed digital landmark. (a) The format of the landmark. (b) An example of the landmark.

we develop a simple region finding algorithm for extracting potential regions from images being captured.

While scanning the lines, the program will count the edge pairs (a pair is composed of a rising edge and a falling edge), and record the line sections that contain more than four edge pairs. In a scan line, the program may record more than one section if the pairs are further away from each other. The region extraction module analyzes the line sections recorded and finds out the probable regions based on the following assumptions:

- The line sections will gather closely in the digits region.
- The numbers of line sections in the digits region will not be less than 10.
- The clearance between line sections in a digit region will be limited.

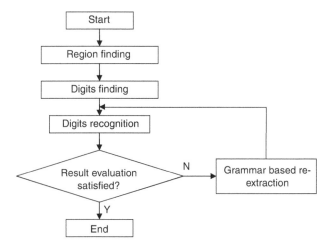

FIGURE 4.6   Landmark recognition algorithm.

4.4.1.2 Digits finding module

The digits finding module is mainly based on an edge following algorithm. The steps can be described as follows:

- Do top-down line scans of a potential region until an edge pixel of a digit is found.
- Follow the edge of the digit and record the parameters (position, width, height, etc.).
- Repeat steps above until all the digits in the region have been found.

4.4.1.3 Digits recognition module

The digits recognition module works in several steps as follows:

- Normalization — It divides the image areas and normalizes them to $64 * 64$ arrays regardless of the original size of the digits. If some noisy areas were found with the digits, this step will normalize them as well, in order to avoid losing information. Figure 4.7a is the result of normalization of the digits detected.
- Thinning — It is to extract out the skeleton (one-pixel-wide central line of a line). The skeleton is essential for texture analysis of a pattern. The end points, bifurcate points, etc. can be extracted from the skeleton. The program adopts an updated Hilditch algorithm to implement the thinning operation. Figure 4.7b is the result of thinning. There are often some noises in the thinning image, for

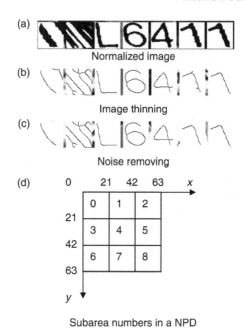

FIGURE 4.7   Normalization, thinning, and noise removing.

example, some odd pixels which will generate fake endpoints and bifurcate points.

- Noise removal — This step removes the noisy pixels according to the following rules (i) The isolated pixels are removed; (ii) short lines (the length is less then 60 pixels) are removed; and (iii) short odd lines are removed. An odd line is composed of the pixels from an endpoint to a bifurcate point. The bifurcate point pixel is preserved while processing. Figure 4.7c is an example of noise removed images.
- Feature extraction — It extracts a grid based 9-element feature vector $F = (f_1, f_2, \ldots, f_9)^{\mathrm{T}}$ for each of the normalized probable digits (NPD). The nine elements express the ratio of the number of black pixels in a subarea. The following figure gives the serial number of the subareas in a NPD. The borderlines of subareas are the four lines shown in Figure 4.7d, and the coordinate value of a NPD is from 0 to 63 in both x and y axes. The elements are defined by the following equation:

$$f_i = \frac{N_i}{\sum_{i=0,8} N_i} \qquad (4.20)$$

where N_i is the number of the black pixels in the ith subarea shown in Figure 4.7d. For instance the feature vector for the letter "L" in Figure 4.7c is (0.2343, 0, 0, 0.246, 0, 0, 0.246, 0.168, 0.144) T. We can find that in the NPD for "L," no black pixel is in subarea 1, 2, 4, and 5; and the black pixels in subareas 7 and 8 are relatively smaller than those of the subarea 0, 3, and 6. Because the features are relative values instead of absolute ones, the feature values are free from the different exposure level of the image, which causes the different width of the character strokes. The features are robust to the sloping digits, which may be due to a sloping camera. The image in Figure 4.5b is an example for sloping digits. We found in the NPD that the distribution of black pixels in each subarea of the NPD does not change due to the slope. The feature extracted from it also proves the same.

- Feature matching — It calculates the scalar products of the feature vector extracted in the earlier step and those from the features library; then it will give out the result according to the minimum scalar product. This step also contains a simple judgment of the results if more than one probable region is found. The following conditions are adopted to do this, if (i) five digits are found in a region; (ii) the first character of a region is recognized as "L"; (iii) the minimum scalar products are very small; and (iv) the region is more probable to be the right one. Another function of this step is to connect the characters recognized into a string according to the right region judgment and positions of digits and then output it.

4.4.2 Position Estimation

Assume that the robot position is expressed by the vector $p = (x, y, \theta)^T$, and three coordinate systems are adopted for our implementation:

- {W}: the global coordinates. The localization is to find out $^W P$, the position vector in {W}.
- {L}: the landmark coordinates. It is fixed on the current landmark which is being seen. The original point is fixed on the position shown in Figure 4.5a.
- {I}: the image coordinates. It is fixed on the image plane.

If a landmark is "seen" by the robot, it is able to identify the landmark and get the position information of the landmark in {W} from the database, and therefore the transformation matrix C_L^W is known. If the position in {L}, $^L P$, is deduced, the localization can be done by the transformation

$$^{W}P = C_{L}^{W} \cdot {}^{L}P \qquad (4.21)$$

The problem now is to calculate

$$^{L}P = (x_{c}^{L}, y_{c}^{L}, \theta_{c}^{L})^{T} \qquad (4.22)$$

In this section, two methods of localization, that is, triangulation, and least square estimation (LSE), are investigated in terms of two cases: single landmark and double landmarks.

4.4.2.1 Triangulation method

Figure 4.8 is the sketch map of landmark imaging, using a pinhole model. P1 and P2 are two of the vertical edges of the five characters (10 edges alltogether). The positions of P1 and P2 are known as $(p_{1}^{L}, 0)$ and $(p_{2}^{L}, 0)$. The parameters of the landmark are shown in Figure 4.5a.

According to the pinhole model, we get:

$$\frac{(y_{2}^{I} - y_{1}^{I})r}{f} = \frac{H}{l_{1}} \quad \text{and} \quad \frac{(y_{4}^{I} - y_{3}^{I})r}{f} = \frac{H}{l_{2}} \qquad (4.23)$$

where r is the resolution with the unit of MPD (millimeters per dot), and (x_{i}^{I}, y_{i}^{I}) is the coordinate value of the ith feature point, both endpoints of the selected vertical edges.

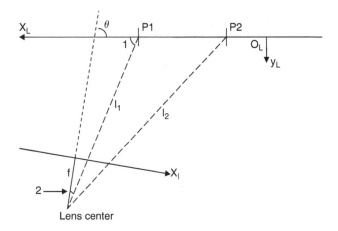

FIGURE 4.8 Landmark detection.

Assume that the position of the lens center is (x_c^L, y_c^L). According to Pythagoras theory, we have:

$$\begin{cases} (p_1^L - x_c^L)^2 + (y_c^L)^2 = l_1^2 \\ (p_2^L - x_c^L)^2 + (y_c^L)^2 = l_2^2 \end{cases} \qquad (4.24)$$

Considering that y_c^L is always a positive value, we have:

$$\begin{cases} x_c^L = \dfrac{l_1^2 - l_2^2 + p_2^{L^2} - p_1^{L^2}}{2(x_2^L - x_1^L)} \\ y_c^L = \sqrt{l_1^2 - (x_c^L - p_1^L)^2} \end{cases} \qquad (4.25)$$

In Figure 4.9, the direction may be easily obtained as:

$$\theta = \angle 1 + \angle 2 = \tan^{-1}\left(\frac{p_1^L - x_c^L}{y_c^L}\right) + \tan^{-1}\left(\frac{x_1^I r}{f}\right) \qquad (4.26)$$

By combining Equation (4.25) and Equation (4.26), we have

$$\begin{bmatrix} x_c^L \\ y_c^L \\ \theta_c^L \end{bmatrix} = \begin{bmatrix} \dfrac{l_1^2 - l_2^2 + p_2^{L^2} - p_1^{L^2}}{2(p_2^L - p_1^L)} \\ \sqrt{l_1^2 - (x_c^L - p_1^L)^2} \\ \tan^{-1}\left(\dfrac{p_1^L - x_c^L}{y_c^L}\right) + \tan^{-1}\left(\dfrac{x_1^I r}{f}\right) \end{bmatrix} \qquad (4.27)$$

We substitute $^L P$ into Equation (4.21), and then the localization is completed. Equation (4.27) is the result of localization. The coordinates given by the program are in {L}. The errors of this method are caused by the imprecise extraction of each character. Differentiating Equation (4.27), we have:

$$\begin{bmatrix} dx_c^L \\ dy_c^L \\ d\theta_c^L \end{bmatrix} = \begin{bmatrix} \dfrac{1}{(p_2^L - p_1^L)}(l_1 dl_1 - l_2 dl_2) \\ \dfrac{l_1}{\sqrt{l_1^2 - (x_c^L - p_1^L)^2}} dl_1 - \dfrac{x_c^L - p_1^L}{\sqrt{l_1^2 - (x_c^L - p_1^L)^2}} dx_c^L \\ \dfrac{dx_1^I}{1 + (x_1^I r/f)^2} - \dfrac{2y_c^L(p_1^L - x_c^L)}{y_c^{L^2} + (p_1^L - x_c^L)^2}\left(dx_c^L + \dfrac{dy_c^L}{y_c^{L^2}}\right) \end{bmatrix} \qquad (4.28)$$

(a)

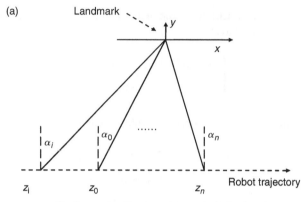

Position estimation based on a single landmark

(b)

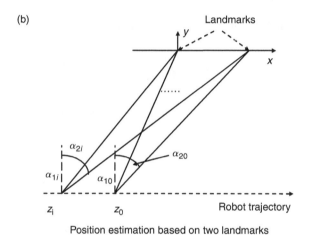

Position estimation based on two landmarks

FIGURE 4.9   Landmark-based localization of the robot.

Equation (4.28) shows the relationship between localization errors and character extraction errors. It can be seen that the localization error is relative to l_1 and l_2, as well as the distances between the robot and the features, which will be large when the observing angle is large. In the two-landmark case, the two features p_1 and p_2 can be selected from different landmarks, which can provide more accurate position results.

4.4.3   Least Square Estimator (LSE)

In a real application, the robot continuously samples data using its onboard camera. Errors may be reduced by fusing the data of individual samples. In this section, LSEs are used in terms of two different cases: single landmark case and dual landmark case. The method proposed here is the extension of Boley's LSE [24].

4.4.3.1 Single-landmark LSE (SLSE)

The single-landmark LSE algorithm is based on the coordinates shown in Figure 4.9a. The origin point is placed at the landmark, and the x-axis is set parallel to the robot moving trajectories. From each sampling point on a robot trajectory, the bearings to the landmark are measured as α_i ($i = 0, 1, \ldots, n$), by using the methods in Section 4.2. The position of each sample is noted in the vector $z_i = (x_i, y_i)^{\mathrm{T}}$, the distances between each position, and z_0, which can be obtained from the readings of odometry, are noted as $d_i = z_i - z_0 = (x_i - x_0, y_0)^{\mathrm{T}}$. It is easy to observe that:

$$\tan(\alpha_i) = \frac{x_0 + d_i}{y_0} \tag{4.29}$$

Rewriting Equation (4.29), we have:

$$x_0 \cos(\alpha_i) - y_0 \sin(\alpha_i) = -d_i \cos(\alpha_i) \tag{4.30}$$

Row-by-row collecting all the equations for $i = 1, \ldots, n$, we have over determined equations which can be expressed as:

$$A z_0 = b \tag{4.31}$$

where

$$A = \begin{pmatrix} \cos(\alpha_1) & -\sin(\alpha_1) \\ \cos(\alpha_2) & -\sin(\alpha_2) \\ \vdots & \vdots \\ \cos(\alpha_n) & -\sin(\alpha_n) \end{pmatrix}, \quad z_0 = \begin{pmatrix} x_0 \\ y_0 \end{pmatrix}, \quad b = \begin{pmatrix} -d_1 \cos(\alpha_1) \\ -d_2 \cos(\alpha_2) \\ \vdots \\ -d_n \cos(\alpha_n) \end{pmatrix}$$

Using the Least Square method, we can estimate z as follows:

$$z_0 = (A^{\mathrm{T}}A)^{-1}A^{\mathrm{T}}b \tag{4.32}$$

In this method, we adopt samples at the positions z_i ($i = 1, \ldots, n$) as the reference sample (RS) to estimate the robot position z_0.

4.4.3.2 Dual-landmark LSE (DLSE)

In this case, the coordinates are built on two landmarks. The original point is set to one landmark and the x-axis point to the other one. The coordinate values of two landmarks are known as $(0, 0)$ and $(D, 0)$, where D is the distance between two landmarks.

For each sampling point z_i, both landmarks are measured as $(\alpha_{1i}, \alpha_{2i})^T$, and two equations will be generated:

$$\begin{cases} \tan(\alpha_{1i}) = \dfrac{x_0 + d_{xi}}{y_0 + d_{yi}} \\[3mm] \tan(\alpha_{2i}) = \dfrac{x_0 + d_{xi} - D}{y_0 + d_{yi}} \end{cases} \tag{4.33}$$

where d_{xi} and d_{yi} are the displacement of the sampling points in two directions, which are obtained from the readings of the odometer.

Rewriting Equation (4.33), and collecting the equations for each RS, we have:

$$A z_0 = b \tag{4.34}$$

where

$$A = \begin{pmatrix} \cos(\alpha_{11}) & -\sin(\alpha_{11}) \\ \vdots & \vdots \\ \cos(\alpha_{1n}) & -\sin(\alpha_{1n}) \\ \cos(\alpha_{21}) & -\sin(\alpha_{21}) \\ \vdots & \vdots \\ \cos(\alpha_{2n}) & -\sin(\alpha_{2n}) \end{pmatrix}, \quad z_0 = \begin{pmatrix} x_0 \\ y_0 \end{pmatrix}$$

$$b = \begin{pmatrix} \sin\alpha_{11} d_{y1} - d_{x1}\cos\alpha_{11} \\ \vdots \\ \sin\alpha_{1n} d_{yn} - d_{xn}\cos\alpha_{1n} \\ \sin\alpha_{21} d_{y2} - (d_{x2} - D)\cos\alpha_{21} \\ \vdots \\ \sin\alpha_{2n} d2 - (d_{x2} - D)\cos\alpha_{2n} \end{pmatrix}$$

The position z_0 can also be deduced from Equation (4.32).

4.4.4 Implementation and Results

The experiments are carried out using a "Logitech QuickCam" web camera ($1/4''$ CCD sensor, 4.9 mm lens). In our implementation, the mobile robot is

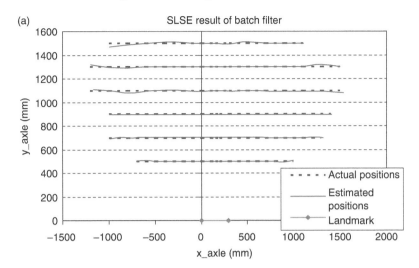

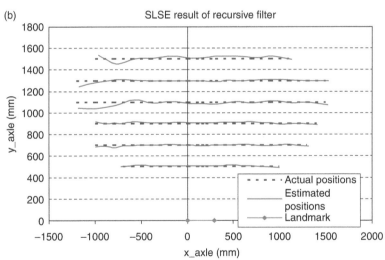

FIGURE 4.10   Single landmark–based LSE.

moving along a straight line at different distances to the landmark that was fixed on the wall. After landmark recognition, the robot's position is calculated through the triangulation method described in Section 4.2 [3,12,16].

Then LSE is implemented in two ways, namely batch processing and recursive processing. The experimental results for single landmark are presented in Figure 4.10. In contrast, the experimental results for dual landmarks are presented in Figure 4.11.

As we can see from Figure 4.10a and Figure 4.11a, the batch filtering algorithm gave better localization results than the recursive filtering algorithm, but it is not real time. Although the recursive filters have relatively large errors (the left side of each line) at the beginning, the estimated results converge rapidly when more data is available. The final result is therefore applicable in a real-time system.

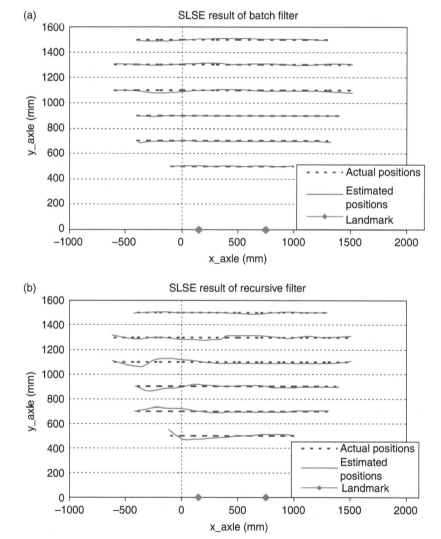

FIGURE 4.11   Dual landmark–based LSE.

4.5 SICK LASER SCANNER AND GEOMETRIC LANDMARKS

Geometric landmarks are widely used for robot navigation, which are normally static. Recently, Howard and his colleagues [25] proposed a new approach by equipping their robots with geometric landmarks that are easily found and movable within the environment. In their implementation, a large heterogeneous team of robots was adopted, each of which carried a SICK scanner and two geometric landmarks (cylinders). Motivated by their research, we have equipped each of our robots with a SICK scanner and a cylinder so that colocalization can be implemented.

Since indoor environments usually contain many straight lines, the detection process is greatly aided if the landmark always has identical range signatures regardless of relative position or orientation. This is the case for one shape only, the circle. This characteristic aids detection but is not helpful when determining relative positions between two or more robots because rotational changes cannot be perceived. Two distinguishable circles guarantee unique localization. If the circles are indistinguishable then localization is one of the two places. Figure 4.12 shows a typical mapping situation involving co-location. Two cylinders A and B are shown; these cylinders could be individual robots or one robot carrying two cylinders. The advantage of observing two robots is that large separations may be used, leading to more accurate localization, however, mounting both cylinders on one robot reduces the number of robots required, the observer and the mobile landmark robot. Figure 4.12 presents a cooperative localization and mapping scenario involving three robots R1, R2, and R3. R1 is equipped with a laser scanner and the remaining robots are mobile landmarks. The initial positions of R2 and R3 allow R1 to map the room on the left. Under the observation of R1, at position A, R2 and R3 move across the corridor to the

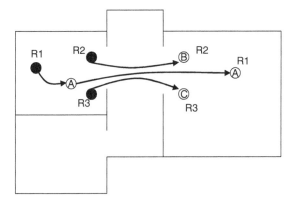

FIGURE 4.12   Cooperative localization scenario involving three robots.

second room where they adopt positions B and C. Now R1 can continue to D using R2 and R3 as artificial landmarks and map the second room.

Once the relative positions of the companion robots are known, map building is possible. The main difficulty is achieving fast and reliable detection of circles of known radius from noisy range data. The detection of shapes in images is a large area of research within the computer vision community and contains several relevant techniques such as the Hough Transform and least squares fitting approaches.

4.5.1   Circular Hough Transform

The Hough Transform [25] has been highly successful in the vision community, thanks to its tolerance of image noise and excellent straight-line detection. The Hough Transform may be generalized to any geometric primitive. However, the introduction of each new parameter adds another dimension to the Hough space. The geometric increase in storage and processing required for the accumulator grids have repercussions on performance. A typical high-resolution laser scan is given in Figure 4.13a which shows a relatively cluttered environment with two circular landmarks that are indicated with dashed lines. The standard Hough Transform is particularly ineffective for circle extraction from laser range data because of the uneven distribution of points in Cartesian space. The laser scanner samples at regular intervals of θ resulting in an increased density of readings from nearer objects. A nearby straight edge obstacle may be detected in preference to the circles.

This is rectified by a Range Weighted Hough Transform (RWHT) as discussed in Reference 26. The weight function applied is a simple linear increase from the origin of the scan. This linear increase negates the effect of the $1/r$ fall in point density. The improvement is immediately apparent in Figure 4.13b where the peaks of the circle centers can be distinguished from nearby walls. As can be seen, the two highest peaks correspond to the circular landmarks.

Only a 2D Hough parameter space is required for the circle search because the radii of the circles are known. The two parameters are the coordinates of the candidate circle center. The confusion of straight lines with circles is a serious problem that refuses to be resolved. A possible solution would be to first remove all points corresponding to straight lines and then perform the circular Hough Transform on the remaining points, however this is very time consuming. The process for Hough Transform circle detection is summarized in Figure 4.14.

There are a number of reasons why the circular Hough Transform is not particularly suited to this application. Range data are different from image data for which the Hough Transform was first devised. Another problem is that it always returns an answer even if the geometric primitive is not present in the data. The determination of peak significance by comparison with others and the kind of data expected requires a relatively complicated statistical analysis.

(a)

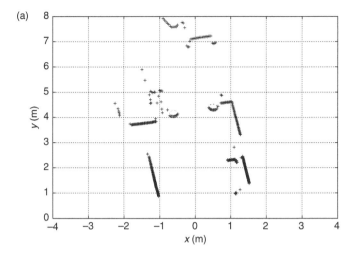

(b)

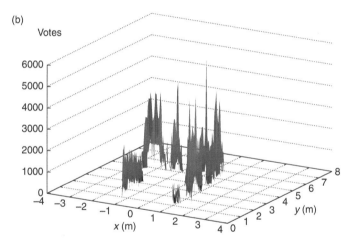

FIGURE 4.13 Circular Hough transform.

4.5.2 Least Squares Fitting of Circles

Poor performance of the Hough Transform approach prompted research into least squares curve fitting approaches. It is evident from Reference 27 that fitting circles to points is a nontrivial process, mainly because the resulting equations are highly nonlinear and circles cannot be elegantly expressed in Cartesian coordinate systems. No least squares algorithm suitable for range data could be found, therefore one was devised.

One of the problems with the circular Hough Transform is that there is much information specific to range scans that is not included in the search for

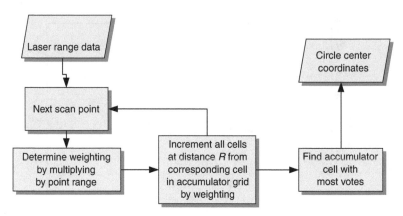

FIGURE 4.14   Flowchart of the circular Hough transform detection process.

circles. One important property of circles is that they are highly symmetric and so appear identical when viewed from any angle; this greatly eases the burden of detection. Also, the range data has an inherent sequence that is not obvious in Cartesian coordinates. Detection of a circle occurs when a sequence of adjacent points lie close to the circumference of that circle. Relaxing the requirement for the detection of occluded targets allows the following algorithm shown in Figure 4.15b.

The algorithm assumes the center of the circular target is at the scan angle of the current scan point being analyzed. The mean of the least squares differences is then calculated by Equation (4.38) and Equation (4.39). Scan angles with this quantity below a threshold (comparable to the accuracy of the laser scanner) are likely contenders for having the center of the target circle situated along them. Figure 4.15a illustrates the geometry involved with laser scan points depicted by crosses. Point A is the current scan point being evaluated and the circle represents the search target. The candidate circle for A is assumed to be positioned with center C, as shown on the line $\overline{OA}$ where O is the origin of the laser scan. Assuming the laser scan returns points evenly distributed over θ then the number of nearest neighbors to be incorporated is determined. Points that lie within an angle of $\hat{AOB}$ from A are candidate points where

$$A\hat{O}B = \arcsin \frac{R}{(R + \overline{OA})} \qquad (4.35)$$

and R is the radius of the circular landmark.

Care has to be taken regarding scan points lying near D and B, which are subject to glancing edge effects. The causes of these effects are specular reflection and pixel mixing which occurs when the laser spot spans an environmental

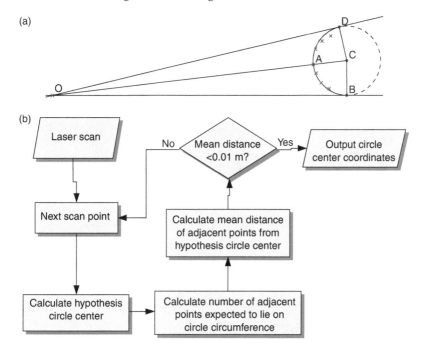

FIGURE 4.15 Least squares fitting of circles. (a) Geometric construction illustrating the least squares method for circle location. (b) Flowchart of the least-squares circle detection algorithm.

range discontinuity. The subset of laser range points processed is

$$S = \begin{pmatrix} r_1 & r_2 & \cdots & r_n \\ \theta_1 & \theta_2 & \cdots & \theta_n \end{pmatrix} \tag{4.36}$$

where r and θ are the polar coordinates of the scan points in the coordinate system of the robot. The position of the hypothesis circle in polar coordinates is

$$\begin{pmatrix} C_r \\ C_\theta \end{pmatrix} = \begin{pmatrix} r_{\frac{n+1}{2}} + R \\ \theta_{\frac{n+1}{2}} \end{pmatrix} \tag{4.37}$$

The distance of the ith point from circle circumference is

$$d_i = \sqrt{C_r^2 + r_i^2 - 2C_r r_i \cos(C_\theta - \theta_i)} - R \tag{4.38}$$

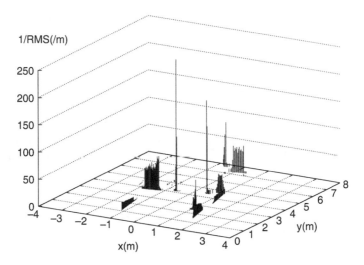

FIGURE 4.16 Reciprocal root mean least squares differences of the laser scan in Figure 4.13b.

Ultimately the mean least squares difference is calculated in the usual fashion as

$$\bar{d}^2 = \frac{1}{n} \sum_{i=1}^{n} d_i^2 \qquad (4.39)$$

This indicates how far, on average, the points are from the circumference of the hypothesis circle and the reciprocal is proportional to the likelihood of detection. This is repeated for each point in the scan. The points that exceed a threshold probability imply successful circle detection at that position. Figure 4.16 plots the reciprocal root mean least square differences for the example laser scan in Figure 4.13b. Note that the two prominent peaks correspond to the circular landmarks.

What is apparent from Figure 4.16 is the accurate detection and localization of the two circular targets with the smaller of the two circle peaks being nearly twice as big as the largest background peak. This ensures a superior performance of 98% reliability vs. 50% for a RWHT. A comparison of Figure 4.13b and Figure 4.16 emphasizes the effectiveness of the least squares algorithm over the RWHT for reliable circular target extraction from laser range data. The least squares algorithm takes advantage of range data specific characteristics like sequence and a single observation point. The more generic RWHT does not utilize this extra information and so the least squares method is not only 25 times more accurate but also faster and requires less memory.

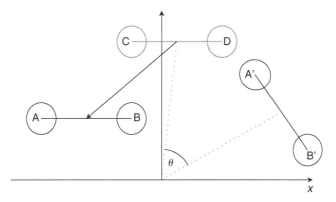

FIGURE 4.17   Pose change calculation from two observations.

4.5.3 Cooperative Position Estimation

The two cylindrical targets are observed from two different poses and the observations superimposed. This is shown in Figure 4.17 with the second observation cylinder positions indicated with an apostrophe, that is, A' and B'. The pose change consists of a rotation and translation. The rotation angle is the change in angle of the line joining the two circles. Once the rotation of the robot between the poses is known, the rotation effect can be reversed, that is, placing the cylinders at the positions C and D, as shown in Figure 4.17. The change in position or translation of the robot between observations is given by the difference in position of the midpoints of $\overline{CD}$ and $\overline{AB}$. Knowing the rotation θ and translation T of the robot between successive scans, enables the amalgamation of scan data to produce a global map. Scan data, L, is transformed point by point into the coordinate frame of the global map, L', by

$$ L'_i = \begin{pmatrix} T_x \\ T_y \end{pmatrix} + \begin{pmatrix} \cos\theta & -\sin\theta \\ \sin\theta & \cos\theta \end{pmatrix} L_i \qquad (4.40) $$

Given that a robot can observe other stationary robots, how may it determine changes in its pose? Changes in pose may be described as linear combinations of two geometric transforms, translation and rotation. An important consideration is if the observed robots are distinguishable; if they can be unambiguously identified then the determination of pose change between landmark observations is trivial. The rotation is calculated from the change in angle of the lines joining the landmarks, and the translation is the average displacement of each point to its image point. If the landmarks are indistinguishable then it is not so straightforward because each point cannot be associated with absolute certainty to the same point in the subsequent sensor update. Problems also arise with symmetric distributions of landmarks.

If the relative positional information of indistinguishable landmarks is available then three are sufficient to unambiguously determine pose. Initially two would appear sufficient, however the ambiguity of identity means that landmarks may be rotated 180°. Even though only three asymmetrically distributed indistinguishable landmarks are needed for unambiguous pose determination, the fewer landmarks required, the better. Is it possible to have reliable pose updates using only the relative positions of two landmarks? There are a number of ways that this may be achieved. The simplest is to use distinguishable landmarks, for instance circles of sufficiently different radii. If indistinguishable landmarks have to be used then they may be placed in such a configuration as localization is only required in one half plane. An example would be when they are against a wall then the robot cannot be localized in the half plane behind the wall and still be able to detect the landmarks. Use of odometry and fast updates means that the large pose changes that would cause ambiguity would never happen between updates or would be detected by the odometry sensors.

4.5.4 Implementation and Results

The experimental platform is a Magellan Pro-robot equipped with a SICK LMS 200 laser range finder. The range finder has a scanning angle width of 180° and a resolution of 0.5°. The laser range finder is almost an ideal sensor with unrivalled accuracy, acquisition time, and range. The main problems are cost, mass (4.5 kg), and power consumption (17.5 W). The characteristics of this LMS are detailed in References 28 and 29.

Experiments were performed to test the localization accuracy delivered and involved driving the robot along a straight line and in a square. The deviation of the colocation positions from this straight line give an indication of the localization error in the direction perpendicular to the line. This error depends approximately linearly on the angular resolution of the laser scanner, the range and separation of the geometric targets. The localization error was of the order of 0.02 m at ranges of 0 to 8 m with the laser scanner operating at a resolution of 0.25°.

Error in the range to the targets introduces error into the position estimation of the robot. Figure 4.18 illustrates the dependence of the pose uncertainty on the range error. The origin O is the true position of the robot and O' is its worst case perceived position if the range to the target A is over estimated and that to target B is underestimated. The error estimate is greatly simplified if a far field approximation is used which implies

$$\overline{AB} << \overline{OM} \qquad (4.41)$$

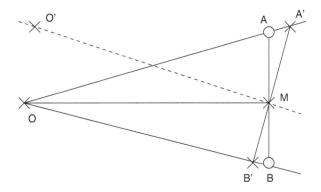

FIGURE 4.18 Geometric construction used to calculate the localization error.

in this approximation the following similarities prove useful

$$\sin\theta \approx \tan\theta \approx \theta \qquad (4.42)$$

for small angles of θ in radians. As $\overline{OM} = \overline{O'M}$ then for the displacement of O' the angle of rotation is

$$O\hat{M}O' = \arctan\left(\frac{2}{\sqrt{2}}\,\frac{\overline{AA'}}{\overline{AB}}\right) \approx \sqrt{2}\frac{\overline{AA'}}{\overline{AB}} \qquad (4.43)$$

Note the $\sqrt{2}$ factor is due to the addition of the errors in quadrature. Finally the position error can be expressed as

$$\overline{OO'} \approx \sqrt{2}\frac{\overline{AA'}}{\overline{AB}}\overline{OM} \qquad (4.44)$$

The far field approximation, expressed in Equation (4.41), falters if the targets are near and for large target separations, however in these situations the error is minimal. It should also be clear from Figure 4.19 that the dependence of position error σ_x on angular error σ_θ for the laser scanner is simply

$$\sigma_x \approx \overline{OM}\sigma_\theta \qquad (4.45)$$

The angular error for the SICK LMS 200 is ca. 0.5° so at a range of 4 m the position error due to angular error is around 0.03 m. Targets separated by 2 m with radii 0.1225 m at a range of 4 m observed with a range error of 0.01 m produced a position error of 0.03 m. This prediction is close enough to the error observed at this range in Figure 4.20.

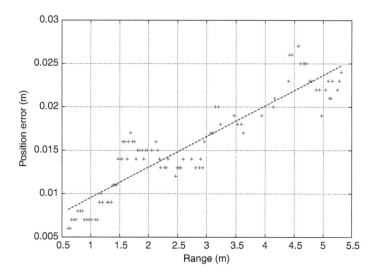

FIGURE 4.19   Plot showing the increase of position error with range to targets and line of best fit.

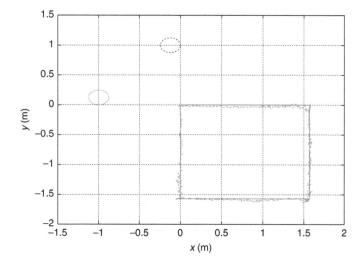

FIGURE 4.20   Localization along square path (solid line indicates true path taken and circles are the geometric targets used for localization).

A typical set of continuous localization results are displayed in Figure 4.20. The robot was moved one loop around a 1.57 m square at 0.2 m/sec. The laser scanner mounted on the robot has a maximum scan angle of 180° and so the robot had to reverse along some edges of the square in order to maintain tracking.

The target cylinders were located at $(0, 1)$ and $(-1, 0)$ because in these positions they can always be observed by the $180°$ scanner, allowing continuous position updates. The localization error can easily be extracted from Figure 4.20 and is of the order of 0.03 m. The position accuracy is better toward the origin of the graph because the robot is nearer to the target positions of $(0, 1)$ and $(-1, 0)$.

4.6 CONCLUSIONS

This chapter addresses the problem of landmarks and triangulation in navigation of mobile robots. A novel landmark-based navigation system is proposed, which consists of three types of landmarks (retro-reflective, digital, and geometric) and three types of sensors (laser, vision, and odometry), as well as sonar sensors. Some corresponding navigation and triangulation algorithms have been developed so that the robot is able to estimate its position and update its internal map continuously in a dynamic environment.

To improve the localization accuracy for mobile robots in continuous operation, the EKF algorithm has been adopted in the navigation process to integrate odometry data and angle observations from the laser scanner in order to provide a useful solution toward real-world applications. A triangulation module is embedded into the proposed architecture to re-calibrate the robot's position when the robot is stationary or gets lost. The experimental results are presented to show its applicability.

A digital landmark-based localization algorithm for mobile robots is demonstrated, which uses a fast digits recognition method. The algorithm provides an easy solution to landmark identification in complex environments, which is robust to slope images. Some advantages of the algorithm are the flexible extendibility of digital landmarks and the low computation cost of landmark recognition. We are currently investigating the following four issues (i) other information in the single landmark, for example, the edges of middle characters, may be used; (ii) other data fusion methods to use pre-known position information (from dead-reckoning or EKF); (iii) multiple landmarks may be seen in some conditions, and triangulation methods may be used; and (iv) localization algorithms based on texture landmarks.

The feasibility of cooperative localization based on one sensing robot and two landmark robots is also investigated. It is implemented by a least squares fitting approach optimized for the sequential natural of the range data and the highly symmetric aspect of the circular geometric targets. This colocation scheme allows fast position and orientation determination with bounded errors and reliability indicators in unknown indoor environments. The robust localization algorithm lays the foundation for mapping featureless and highly symmetric environments. Continuous localization was performed at 0.2 m/sec. Continuous localization can be provided, however these scans should not be incorporated into the global map, only the ones taken when stationary should be used to improve the quality of the global map. Improvements in colocation accuracy should be possible allowing either the extension of the range over which cooperative localization is possible or reducing the separation of the targets so that they may be mounted on one robot thus allowing cooperative mapping with only two robots. These improvements in colocation accuracy would primarily come from over-sampling the least squares fitting algorithm.

Although multiple sensors and multiple landmarks have been adopted in the proposed navigation system, they have been independently investigated and tested so far. A natural extension of future research is to investigate the integration of three landmark-based navigation algorithms. Moreover, the proposed navigation algorithms have potential applications for service robots in homes, offices, and hospitals. It can also be used for outdoor beacon-based navigation such as GPS navigation systems.

REFERENCES

1. Astrom K. (1992). A correspondence problem in laser guided navigation, in *Proceedings of Symposium on Image Analysis*, D. Eriksson and E. Benojsson (eds), Sweden, pp. 141–144.
2. Chee B.-Y. and Lang S. Y. T. (1996). A random sampling approach to landmark detection for mobile robot localization, *Proceedings of International Conference on Mechatronics*, 1, 41–44.
3. Hu H. and Gu D. (2000). Landmark-based navigation of industrial mobile robot, *International Journal of Industrial Robot*, 27, 458–467.
4. Everett H. R. (1995). *Sensors for Mobile Robots — Theory and Application*. A K Peters Ltd., Natick, MA.
5. Barshan B. and Durrant-Whyte H. F. (1995). Inertial sensing for mobile robotics, *IEEE Transactions on Robotics and Automation*, 11, 328–342.
6. Kleeman L. (1992). Optimal estimation of position and heading for mobile robots using ultrasonic beacons and dead reckoning, in *Proceedings of IEEE International Conference on Robotics and Automation*, Nice, France, pp. 2582–2587.
7. Cooper S. and Durrant-Whyte H. F. (1994). A Kalman filter model for GPS navigation of land vehicles, in *Proceedings of IEEE/RSJ/GI International Conference on Intelligent Robotic Systems*, Munich, Germany, pp. 157–163.
8. Borenstein J., Everett H. R., Feng L., and Wehe D. (1997). Mobile robot positioning — sensors and technologies, *Journal of Robotic Systems, Special Issue on Mobile Robots*, 14, 231–249.
9. Sugihara K. (1988). Some location problems for robot navigation using a single camera, *Journal of Computer Vision, Graphics and Image Processing*, 42, 112–129.
10. Avis D. and Imai H. (1990). Locating a robot with angle measurements, *Journal of Symbolic Computation*, 10, 311–326.
11. Krotov E. (1989). Mobile robot localisation using a single image, in *Proceedings of IEEE International Conference on Robotics and Automation*, Scottsdale, AZ, Vol. 2, pp. 978–983.

12. McGillem C. D. and Rappaport T. S. (1988). Infrared location system for navigation of autonomous vehicles, in *Proceedings of International Conference on Robotics and Automation*, pp. 1236–1238.

13. Borenstain J. (1994). The CLAPPER: a dual-drive mobile robot with internal correction of dead-reckoning errors, in *Proceedings of IEEE International Conference on Robotics and Automation*, San Diego, CA, pp. 3085–3090.

14. Durrant-Whyte H. F. (1996). An autonomous guided vehicle for cargo handling applications, *International Journal of Robotics Research*, 15, 407–440.

15. Madsen C. B. and Andersen C. S. (1998). Optimal landmark selection for triangulation of robot position, *International Journal of Robotics and Autonomous Systems*, 23, 277–292.

16. Skewis T. and Lumesly V. (1994). Experiments with a mobile robot operating in a cluttered unknown environment, *Journal of Robotic Systems*, 11, 281–300.

17. Leonard J. and Durrant-Whyte H. F. (1991). Mobile robot localization by tracking geometric beacons, *IEEE Transactions on Robotics and Automation*, 7, 376–382.

18. Yuen D. C. K. and MacDonald B. A. (2005). Vision-based localisation algorithm based on landmark matching, triangulation, reconstruction and comparison, *IEEE Transactions on Robotics*, 21, 217–226.

19. Atiya S. and Hager H. D. (1993). Real-time vision-based robot localisation, *IEEE Transactions on Robotics and Automation*, 9, 785–800.

20. Se S., Lowe D. G., and Little J. (2002). Global localisation using distinctive visual features, in *Proceedings of International Conference on Intelligent Robotic Systems*, Lausanne, Switzerland, pp. 226–231.

21. Matarić M. J. et al. (2002). Learning visual landmarks for mobile robot navigation, in *Proceedings of the 15th Triennial World Congress of Automatic Control (IFAC02)*, Barcelona, Spain, pp. 890–895.

22. Chage S., Chen L., Chung Y., and Chen S. (2004). Automatic license plate recognition, *IEEE Transactions on Intelligent Transportation Systems*, 5, 42–53.

23. Bai H., Zhu J., and Liu C. (2003). A fast license plate extraction method on complex background, intelligent transportation systems, *Proceedings of IEEE*, 2, 985–987.

24. Boley D. and Sutherland K. (1998). A rapidly converging recursive method for mobile robot localisation, *The International Journal of Robotics Research*, 17, 1027–1039.

25. Howard A., Mataric M. J., and Sukhatme G. S. (2003). Localization for mobile robot teams: a distributed MLE approach, *Experimental Robotics VIII*, Bruno Siciliano and Paulo Dario (eds), Springer-Verlag, Heidelberg, pp. 146–155.

26. Forsberg J., Larsson U., and Wernersson A. (1995). Mobile robot navigation using tange-weight Hough transforms, *IEEE Robotics and Automation Magazine*, 2, 18–26.

27. Chenov N. and Lesort C. (2003). Least squares fitting of circles and lines, http://arxiv.org/abs/cs/030100/, *Journal of Computer Vision and Pattern Recognition*, 1001–1027.

28. Aboshosha A. and Zell A. (2003). Robust mapping and path planning for indoor robots based on sensor integration of sonar and a 2D laser range finder, in

Proceedings of IEEE 7th International Conference on Intelligent Engineering Systems, Sssint-Luxor, Egypt.

29. Ye C. and Borenstein J. (2002). Characterisation of a 2D laser scanner for mobile robot obstacle negotiation, in *Proceedings of IEEE International Conference on Robotics and Automation*, Washington, DC, USA, pp. 2512–2518.
30. Kim J. J. and Cho H. S. (1992). Real-time determination of a mobile robot's position by linear scanning of a landmark, *International Journal of Robotics*, 10, 309–319.
31. Hough P. V. C. (1962). Methods and means for recognising complex patterns, U.S. Patent 3 069 654.

BIOGRAPHIES

Huosheng Hu earned the M.Sc. degree in industrial automation from the Central South University in China and the Ph.D. degree in robotics from the University of Oxford in the United Kingdom. Currently, he is a professor in Computer Science in the University of Essex, leading the Human Centred Robotics Group. His research interests include autonomous mobile robots, human-robot interaction, evolutionary robotics, multi-robot collaboration, embedded systems, pervasive computing, sensor integration, RoboCup, intelligent control and networked robotics. He has published over 200 papers in journals, books and conferences in these areas. He is one of the editors-in-chief for *International Journal of Automation and Computing*, and a member of program committees for many international conferences. He is a Chartered Engineer, a senior member of IEEE, and a member of IEE, AAAI, ACM, IASTED and IAS.

Julian Ryde obtained the M.Sc. and M.A. in Physics from the University of Cambridge in the United Kingdom. Currently he is studying for a Ph.D. degree in robotics at the University of Essex, United Kingdom. He has published research papers in a number of international conferences including ICRA, ICMA and IAS. He is a member of both the Institute of Physics and the IEEE. His research interests include autonomous mobile robots, multi-robot collaboration, 3D sensing and mapping.

Jiali Shen received the B.Sc. degree from Nankai University in China, in 1997, and the M.Eng. degree from Tsinghua University in China in 2000. Currently, he is a Ph.D. student in Department of Computer Science, University of Essex, United Kingdom. His research interests include autonomous mobile robots, human-robot interaction and machine learning. He has published a number of research papers in journals, book chapters and international conferences.

II

Modeling and Control

For robotic systems that are embodied, situated and mobile, intelligent inter-action with the environment and the successful operation in response to higher-level commands is crucial before such systems can qualify as autonom-ous and intelligent. This implies the ability of each robot to, at least, be capable of controlling the equipped hardware so as to take the action that is required of the robot, which ranges from moving between points, to changing the pose of equipment like robotic grippers and manipulators. Effective control of a robot's hardware faculties, and making use of sensor feedback, is therefore extremely important.

Due to Brockett's theorem, it is well known that nonholonomic systems with restricted mobility cannot be stabilized to a desired configuration (or posture) via differentiable, or even continuous, pure-state feedback. Therefore, different approaches have been proposed, which includes discontinuous, hybrid, and time varying control laws. Many elegant control strategies have been proposed for various nonholonomic systems. Among them, research results can generally be classified into two classes. The first class is kinematic control, which provides the solutions only on a pure kinematic level, where the systems are represented by their kinematic models and velocity acts as the control input. One commonly used approach for the controller design of nonholonomic systems is to convert, with appropriate state and input transformations, the original systems into some canonical forms for which the design can be carried out more easily. Chapter 5 explores the use of discontinuous control laws for the kinematic control of nonholonomic systems. The chapter also presents the design of a hybrid variable and a switching strategy to guarantee robust stability of the closed loop system in the presence of disturbances and measurement noise.

The second class of results on the control of nonholonomic systems is dynamic control, where the torque and force are taken as the control inputs. Both trajectory tracking and force control are manageable for a constrained robot if the exact robot dynamic model is available for controller design. In real applications, however, perfect cancellation of the robot dynamics is almost impossible. As such, adaptive control was proposed to deal with parameter uncertainties. Approximator-based adaptive control approaches have been extensively studied in the past decade using Lyapunov analysis for general nonlinear systems. Motivated by previous works on the control of nonholonomic constrained mechanical systems and the approximation-based adaptive control of nonlinear systems, the adaptive neuro-fuzzy (NF) control is developed in Chapter 6 for the control of nonholonomic constrained systems using the Lyapunov stability analysis in a unified procedure.

In addition, we should note that actuator dynamics constitute an important component of the complete robotic dynamics, especially in the case of high-velocity movement and highly varying loads. Many control methods have therefore been developed to take into account the effects of actuator dynamics. However, very few works in literature have considered the control of nonholonomic systems with actuator dynamics. To address this, Chapter 7 considers the stabilization problem for general nonholonomic mechanical systems at the actuator level, taking into account the uncertainties in dynamics and the actuators. The controller design consists of two stages. In the first stage, to facilitate control system design, the nonholonomic kinematic subsystem is transformed into a skew-symmetric form and the properties of the overall systems are discussed. Then, a virtual adaptive controller is presented to compensate for the parametric uncertainties of the kinematic and dynamic subsystems. In the second stage, an adaptive controller is designed at the actuator level and the controller guarantees that the configuration state of the system converges to the origin.

The last chapter of this part of the book considers the control of nonholonomic (specifically, car-like) robots for vehicle following. This is an important aspect of advanced autonomous mobile robot systems in which robots may very likely outnumber human operators. The nonholonomic nature of car-like mobile robot motion imposes intrinsic difficulties in control design. This chapter, hence, presents a unified control design for tracking maneuvers of two car-like mobile robots. The vehicle tracking maneuvers are formulated into an integrated framework, with forward tracking, backward tracking, driving, and steering, at the kinematics and dynamics levels. A nonlinear controller with a few design parameters is designed for maneuvers with simultaneous driving and steering for vehicle tracking — in both forward tracking and backward tracking maneuvers. Tracking stability is ensured by the proper design of a stable performance target dynamics with a set of sufficient conditions for selecting design parameters.

Together with the effective use of sensors, effective control of the configuration of the robots' hardware forms the basic and necessary capabilities that bring mobile robotic systems closer to autonomy. The possession of the sensing and control capabilities presented in the first two parts of the book is indispensable for autonomous mobile robots, and will fuse with higher level decision-making mechanisms, which focus on more abstract cognitive planning abilities, to bring forth truly autonomous and intelligent systems.

5 Stabilization of Nonholonomic Systems

Alessandro Astolfi

CONTENTS

5.1 INTRODUCTION

In this chapter we study the stabilization problem for nonholonomic systems.

Nonholonomic control systems are becoming increasingly important in research and industry as they present many interesting features and potentialities. From the researchers' point of view nonholonomic control systems are a prototype of strongly nonlinear systems, requiring a fully nonlinear analysis, since all first approximation methods are inadequate. Thus, the design of a *good* control law for a given nonholonomic system is a challenging task. On the other hand, from an industrial point of view, nonholonomic systems are extremely appealing for their efficiency and flexibility. They can be used as means of transport, inspection, and operation in free space and hostile environments.

Before moving to the technical discussion, it is worth pointing out why we deal with nonholonomic control systems and why we focus on noncontinuous, hybrid, or time-varying stabilizers. A possible answer to the first question can be found in the words of D. Edelen [1]: *"Real problems in the real world rarely exhibit themselves in those pleasant forms wherein one can model them in terms of systems with holonomic constraints. [...] The second law of thermodynamics tells us that such holonomic representations must ultimately degenerate from the domain of the real into ethereal flights of fancy."*

A more practical answer comes from everyday life. Consider the problem of parking a car, we can only drive forward or backward and steer to the left or to the right. Observe a falling cat, an astronaut, a gymnast, or a diver: they can change configuration requiring no contact with fixed objects. These examples seem to be *weakly* related but, from a mathematical point of view, they are all examples of nonholonomic control problems.

Consider now the second question. In the earlier examples an experienced operator is able to perform the proper succession of operations in order to *drive* a nonholonomic system from an initial configuration to a final one. However, when the ability of the operator is not enough or when we desire to automatically reconfigure a nonholonomic system, it is necessary to design a regulator. Hence the birth of the theory of *nonholonomic control or nonholonomic motion planning*. Unfortunately, one of the first results of such a theory was a *negative* one [2]: there exists no continuously differentiable, time invariant, control law able to asymptotically stabilize a controllable nonholonomic system. Therefore, many researchers have proposed and studied discontinuous, hybrid, or time varying control laws.

We now briefly review some of the existing results on the control of nonholonomic systems (see References 3 and 4 for further detail). The control strategies for nonholonomic systems can be divided into two main groups: open loop strategies and closed loop (feedback) strategies. In the latter group we can further distinguish between continuous and discontinuous control laws.[1]

---

[1] A third possible approach is the one based upon sampled-data control laws.

In open loop strategies (see e.g., [5–7]) the control signal is calculated off-line starting from the knowledge of the initial and the final configurations of the system. By their own nature, these strategies are not able to compensate for disturbances and model errors, therefore, in practise, the reached configuration may differ significantly from the desired final one. Nevertheless, it is possible to include open loop strategies in an iterative design method, which possesses some robustness properties. This approach is known as iterative state steering [8].

In closed loop strategies (see e.g., [9–16] for time-varying feedbacks, [17–21] for discontinuous ones, [12,22,23] for *middle* strategies [discontinuous and time varying], [24,25] for hybrid control laws, and [26–28] for multi-rate methods) the control signal is computed online, based on the knowledge of the actual configuration of the system and of the final one. They can potentially compensate for model errors and disturbances. However, the result of Reference 29 states that there does not exist a continuous homogeneous controller that robustly stabilizes nonholonomic systems against modeling uncertainties. This has motivated further research in this direction. Many researchers have been trying to solve this problem using discontinuous feedback (see [8,30–33]), or to find special instances in which a continuous feedback can yield robust stability (see e.g., [34]).

From the very brief discussion above it is apparent that several tools are available for the control of nonholonomic systems. However, to date, it is not possible to single-out a control strategy (or a set of tools) that *performs* better than the other ones. This is mainly due to the following facts. A good control law should have two basic features. First, it should drive the system from its initial state to the final one in a *simple* way, second it should be robust against model mismatches, noisy measurements, and the approximate knowledge of initial conditions. Open loop strategies are generally able to grant the first item, but nothing can be said on their robustness, although they can be exploited in robust iterative designs. On the other hand, closed loop strategies are potentially more robust, but the dynamics of the closed loop system may not be *natural*. In particular the closed loop system may show oscillatory response, which is not at all necessary or required to reach the desired final configuration. Note finally that closed loop strategies are potentially more robust than open loop ones. However, we will show that the robust stabilization problem for nonholonomic systems has very special properties, and it is intrinsically *hard*.

5.2 PRELIMINARIES AND DEFINITIONS

In this chapter, we discuss the problem of designing stabilizing control laws for nonholonomic systems described by equations of the form

$$\dot{x} = g(x)u \qquad (5.1)$$

with $x \in U \subset \mathbb{R}^n, u \in \mathbb{R}^m$, and $m < n$. Despite its simple formulation this problem does not possess a simple solution, as can be inferred from the Theorem of Brockett [2]. This theorem, yielding necessary conditions for smooth stabilizability for general nonlinear systems, provides necessary and sufficient conditions for feedback stabilizability of nonholonomic systems.

Theorem 5.1 *[2] Let $\dot{q} = g(q)u$ be given, with $q \in \mathbb{R}^n, u \in \mathbb{R}^m, g(q_0)u_0 = 0, g(\cdot)$ continuously differentiable in a neighborhood of q_0. Assume, moreover, that span$\{g(q)\}$ is a nonsingular distribution of dimension m in a neighborhood of q_0. Then:*

1. *There exists a continuously differentiable control law which makes (q_0, u_0) asymptotically stable iff $m \geq n$.*
2. *There exists a continuously differentiable and dynamic feedback law which makes[2] (q_0, ξ_0, u_0) asymptotically stable iff $m \geq n$.*

The first part of Theorem 5.1 is due to Brockett [2], while the second one to Pomet [10] (see also the work of Ryan [35] for a more general result in the framework of nonsmooth stabilizability). We will not present the proof of the above theorem, which can be found in the literature [2,10,36]; we simply mention that the provided obstruction to stabilizability has a topological nature. The essence of Theorem 5.1 is that the only *interesting* nonholonomic systems are those for which the distribution $g(q)$ drops dimension precisely at q_0, is not continuously differentiable at q_0, or is not defined at q_0. In such cases we cannot infer anything about the existence of C^1 (smooth), time invariant, static or dynamic, asymptotically stabilizing control laws. Motivated by the conclusions of Brockett's Theorem we focus on:

- State feedback control laws described by equations of the form

$$u = a(x) \qquad (5.2)$$

   where $\alpha : \mathbb{R}^n \to \mathbb{R}^m$ is a discontinuous function of its arguments.
- State feedback, hybrid, control laws described by equations of the form

$$u = k(x, s_d), \quad s_d = k_d(x, \bar{s}_d) \qquad (5.3)$$

---

[2] ξ denotes the state of the dynamic controller.

where s_d evolves in the finite set[3] $\{1, 2\}, k : \mathbb{R}^n \times \{1, 2\} \rightarrow \mathbb{R}^m$ is continuous in x for each fixed s_d; $k_d : \mathbb{R}^n \times \{1, 2\} \rightarrow \{1, 2\}$, and $\bar{s}_d$ is defined as $\bar{s}_d(t) = \lim_{s<t} s_d(s)$;

• Time varying, state feedback, sampled-data, control laws described by equations of the form

$$u = u_T(x(kT), kT) \qquad (5.4)$$

where $T > 0$ is the sampling time, and $u_T : \mathbb{R}^n \times \mathbb{R} \rightarrow \mathbb{R}^m$ is a continuous function of its arguments.

Remark 5.1 *Whenever we deal with discontinuous control laws, functions which are not defined at some points, for example, are unbounded at $x = 0$, are allowed. In particular the term discontinuous will be used throughout this chapter to denote functions which are unbounded, hence undefined, in a certain set; for example, the function $\frac{1}{x}$ is discontinuous at $x = 0$.*

The purpose of the control law is to guarantee that each initial state in a given set converges asymptotically to the origin. However, as we use different control laws, we will need different definitions of *stability*.

Definition 5.1 *[20] A control law described by equations of the form (5.2) almost stabilizes[4] the system (5.1) in the region Ω_0[5] if the following holds:*

(i) *For all initial states $x_0 \in \Omega_0$ the closed loop system admits a unique (forward) solution*

(ii) *For all initial states $x_0 \in \Omega_0$ one has, along the trajectories of the closed loop system, $\lim_{t \rightarrow \infty} \|x(t)\| = 0$*

Moreover, the control law almost exponentially stabilizes the system (5.1) in the region Ω_0 if in addition

(iii) *There exist positive constant c_0 and λ_0 such that for all initial states $x_0 \in \Omega_0$ and for all $t \geq 0$ one has, along the trajectories of the closed loop system, $\|x(t)\| \leq c_0 \exp^{-\lambda_0 t}$*

Hybrid and sampled-data control laws are discussed in relation with robust stabilization problems. To discuss the properties of hybrid control laws we need to introduce a notion of robustness with respect to small noise. To this end,

---

[3] For this controller to make sense we equip $\{1, 2\}$ with the discrete topology, that is, every set is an open set.

[4] This terminology differs from that introduced in Reference 37. Note also that stability has to be understood as Lagrange stability.

[5] The set Ω_0 does not need to be a neighborhood of the origin, but may be an open and dense set with the origin at its boundary.

consider two functions e and d satisfying the following *regularity assumptions*: e and d are in $\mathcal{L}_{\text{loc}}^{\infty}(\mathbb{R}^n \times [0, +\infty); R^n)$, and are continuous in x for each t. We introduce[6] these functions as a measurement noise e and an external noise d and define the perturbed system with u given by Equation (5.3), that is,

$$\dot{x} = g(x)k(x + e(x,t), s_d(t)) + d(x,t)$$
$$s_d = k_d(x + e(x,t), \bar{s}_d) \tag{5.5}$$

In this context the definition of global exponential stability is as follows.

Definition 5.2 *[32] Let e and d be two functions satisfying our standing regularity assumptions. The origin of the system (5.5) is said to be a globally exponentially stable equilibrium on $\mathbb{R}^n$ if the following two properties hold:*[7]

(i) *For every $(x_0, s_0) \in \mathbb{R}^n \times \{1, 2\}$, there exists a solution of (5.5) starting from (x_0, s_0). Moreover all maximal solutions of (5.5) are defined on $[0, +\infty)$.*

(ii) *There exists δ of class $\mathcal{K}_\infty$ and $C > 0$ such that, for all $r > 0$ and for all $(x_0, s_0) \in \mathbb{R}^n \times \{1, 2\}$ with $|x_0| \leq \delta(r)$ and for all maximal solutions (X, S_d) of (5.5) starting from (x_0, s_0), one has*

$$|X(t)| \leq re^{-Ct}, \quad \forall t \geq 0 \tag{5.6}$$

Finally, we characterize robustly stabilizing controllers.[8]

Definition 5.3 *[32] The controller (k, k_d) is a robustly globally exponentially stabilizing controller if there exists a continuous function $\rho : \mathbb{R}^n \to \mathbb{R}$ such that $\rho(x) > 0$, for all $x \neq 0$, and such that for any two functions e and d satisfying our standing regularity assumptions and*

$$\sup_{R_{\geq 0}} |e(x, \cdot)| \leq \rho(x), \quad \text{ess sup}_{R_{\geq 0}} |d(x, \cdot)| \leq \rho(x) \tag{5.7}$$

for all $x \in \mathbb{R}^n$, the origin of (5.5) is a globally exponentially stable equilibrium on $\mathbb{R}^n$.

---

[6] Using similar arguments we could also consider an actuator noise.

[7] A function $\gamma : R_{\geq 0} \to R_{\geq 0}$ is of class $\mathcal{K}$ if it is continuous, strictly increasing, and zero at zero. It is of class $\mathcal{K}_\infty$, if it is of class $\mathcal{K}$ and unbounded. A continuous function $\beta : R_{\geq 0} \times R_{\geq 0} \to R_{\geq 0}$ is of class $\mathcal{KL}$ if $\beta(\cdot, \tau)$ is of class $\mathcal{K}$ for each $\tau \geq 0$ and $\beta(s, \cdot)$ is decreasing to zero for each $s > 0$:

[8] Note that our notion of robust stability is closely related to the classical notion of Input-to-State stability [38].

To discuss generalized sampled-data control laws, consider the perturbed model

$$\dot{x} = g(x)u(x,t) + d(x,t) \tag{5.8}$$

where $d \in R^m$ is a disturbance. Assume the system is between a sampler and zero-order hold. Then it is possible to define a parameterized family[9] of discrete-time models of (5.8) described by

$$x(k+1) = F_T(k, x(k), u(k), d(k)) \tag{5.9}$$

where the free parameter $T > 0$ is the sampling period, and $x(k) = x(kT), u(k) = u(kT)$, and $d(k) = d(kT)$. If we use the approximate model (5.9) to design a discrete-time controller we obtain a discrete-time controller $u_T(x(k), k)$ that is also parameterized by T. Consider now the resulting closed loop system, namely

$$x(k+1) = F_T(k, x(k), u_T(x(k), k), d(k)) \tag{5.10}$$

Definition 5.4 *[39] The family of systems (5.10) is semiglobally practically input-to-state stable (SP-ISS) if there exist $\beta \in \mathcal{KL}$ and $\gamma \in \mathcal{K}$, such that for any strictly positive real numbers $\Delta_x, \Delta_d, \delta$ there exists $T^* > 0$ such that the solutions of the closed loop system satisfy*

$$|x(k, k_0, x_0, d)| \leq \beta(|x_0|, (k - k_0)) + \gamma(\|d\|_\infty) + \delta \tag{5.11}$$

for all $k \geq k_0, T \in (0, T^*), |x_0| \leq \Delta_x$, and $\|d\|_\infty \leq \Delta_d$. Moreover, if $d = 0$, and the above holds, the system is semiglobally practically asymptotically stable (SP-AS) and u_T is called a SP-AS controller.

We stress that, in practice, when designing a discrete-time controller for a continuous-time plant the final goal is to achieve stabilization for the sampled-data system. It is therefore important to note that, as discussed in References 40 and 42, SP-ISS (SP-AS) of the discrete time closed-loop systems implies, under the considered assumptions, SP-ISS (SP-AS) of the sampled-data controlled systems.

5.3 DISCONTINUOUS STABILIZATION

Discontinuous, time invariant, control laws have been dealt with in several research papers, see for example, References 17, 19, and 43; however, our

---

[9] The approximate model, to be useful for control design, has to satisfy the so-called one-step consistency property [40,41].

starting point is completely different. First of all we show that, under some technical assumptions, a nonholonomic system admits a smooth stabilizer only if a subset of the differential equations describing the system are not defined on a certain hyperplane passing through the origin of the coordinates system. Hence, we focus on such a class of systems and we give sufficient conditions for the existence of stabilizing control laws. Finally, we show that any smooth nonholonomic system can be always transformed into a system which is not defined on a certain hyperplane, say $\mathcal{P}$, passing through the origin of the coordinates system. Using the above results, we will propose in Section 5.3.4 a general procedure to design discontinuous almost asymptotically stabilizers for nonholonomic control systems. Such a procedure yields a control law which is not defined on $\mathcal{P}$; hence the closed loop system is not defined on $\mathcal{P}$. However, we will prove that every initial condition which lies outside $\mathcal{P}$ yields trajectories which converge asymptotically to the origin.

5.3.1 Stabilization of Discontinuous Nonholonomic Systems

In this section we discuss the issue of smooth asymptotic stabilizability for systems described by equations of the form (5.1) with $x \in \mathbb{R}^n$ and $u \in \mathbb{R}^{m+p}$. We exploit a few basic facts from geometric control theory, as presented in Reference 44. Note however that proper care has to be taken as we deal with discontinuous functions.

Lemma 5.1 *[20] Consider the system*

$$
\begin{aligned}
\dot{x}_1 &= g_{11}(x_1, x_2)u_1 \\
\dot{x}_2 &= g_{21}(x_1, x_2)u_1 + g_{22}(x_1, x_2)u_2
\end{aligned}
\tag{5.12}
$$

with $x_1 \in \mathbb{R}^p, x_2 \in \mathbb{R}^{n-p}, x = \mathrm{col}(x_1, x_2) \in \mathbb{R}^n, u_1 \in \mathbb{R}^p, u_2 \in \mathbb{R}^m, u = \mathrm{col}(u_1, u_2) \in \mathbb{R}^{m+p}$, and $m + p < n$ ($g_{ij}(x_1, x_2)$ are matrix functions of appropriate dimensions). Assume that the matrix function $g_{21}(x_1, x_2)$ is smooth in an open and dense set U, that the matrix functions $g_{11}(x_1, x_2)$ and $g_{22}(x_1, x_2)$ are smooth in $\bar{U}$,[10] and that the distribution

$$
G = \mathrm{span}\{g_1(x_1, x_2), \dots, g_{m+p}(x_1, x_2)\}
$$

---

[10] Let U be an open and dense set. We denote with $\bar{U}$ the smallest simply connected open set properly containing U.

where $g_i(x_1, x_2)$ denotes the ith column of the matrix

$$g(x_1, x_2) = \begin{bmatrix} g_{11}(x_1, x_2) & 0 \\ g_{21}(x_1, x_2) & g_{22}(x_1, x_2) \end{bmatrix}$$

is nonsingular in $\bar{U}$. Finally, assume, without lack of generality, that the set $\bar{U}$ contains the origin of $\mathbb{R}^n$.

   Then the following holds:

   1. *Set $u_1 = u_1(x_1, x_2)$ with*

$$u_1(0, x_2) = 0 \tag{5.13}$$

   for all x_2. Then, for every u_2, the $n - p$-dimensional manifold $\mathcal{M} = \{x \in \bar{U}: x_1 = 0\}$ is invariant for the system

$$\begin{aligned} \dot{x}_1 &= g_{11}(x_1, x_2)u_1(x_1, x_2) \\ \dot{x}_2 &= g_{21}(x_1, x_2)u_1(x_1, x_2) + g_{22}(x_1, x_2)u_2 \end{aligned} \tag{5.14}$$

   2. *If the matrix function $g_{11}(x_1, x_2)$ has constant rank equal to p in $\bar{U}$ and there exists a smooth scalar function $\phi(x_1)$ such that the matrix function $\phi(x_1)g_{21}(x_1, x_2)$ is smooth in $\bar{U}$, then the $n - p$-dimensional distribution*

$$\Delta = \text{span}\left\{ \begin{bmatrix} 0_{p \times (n-p)} \\ I_{n-p} \end{bmatrix} \right\}$$

   is controlled invariant.[11]

Remark 5.2 *As discussed in Reference 17, under mild hypotheses and with a proper choice of coordinates, it is always possible to write the kinematic equations of a nonholonomic system with equations having the form (5.12), with*

$$g_{11}(x_1, x_2) = I_p, \quad g_{21}(x_1, x_2) = \begin{bmatrix} 0 \\ *(x_1, x_2) \end{bmatrix}, \quad g_{22}(x_1, x_2) = \begin{bmatrix} I_m \\ *(x_1, x_2) \end{bmatrix}$$

This form is known as normal form [17].

   Lemma 5.1 is instrumental to yield a necessary condition and a certain number of sufficient conditions for asymptotic stabilizability of nonholonomic systems described by equations of the form (5.12).

---

[11] $0_{p \times (n-p)}$ denotes the zero matrix of dimensions $p \times (n - p)$ and I_s denotes the identity matrix of dimension s.

Theorem 5.2 *[20] Consider a system described by equations of the form (5.12). Let $\bar{U}$ be a neighborhood of the origin. Assume there exists a smooth control law*

$$u = u(x_1, x_2) = \begin{bmatrix} u_1(x_1, x_2) \\ u_2(x_1, x_2) \end{bmatrix}$$

defined on $\bar{U}$, which locally asymptotically stabilizes the resulting closed loop system. Moreover, assume that:

(i) *The control $u_1(x_1, x_2)$ satisfies the condition (5.13)*
(ii) *The vector field $g_{21}(x_1, x_2)u_1(x_1, x_2)$ is a smooth $n - p$-dimensional vector field defined in $\bar{U}$*
(iii) *The matrix functions $g_{11}(x_1, x_2)$ and $g_{22}(x_1, x_2)$ are smooth in $\bar{U}$.*

Then there exists a smooth matrix function $g^a(x_1, x_2)$, defined on $\bar{U}$, such that $g^a(0, x_2) \neq 0_{(n-p) \times p}$, and a smooth scalar function $g^b(x_1, x_2)$, defined on $\bar{U}$, such that $g^b(0, x_2) = 0$, having the property that

$$g_{21}(x_1, x_2) = \frac{g^a(x_1, x_2)}{g^b(x_1, x_2)} \tag{5.15}$$

that is, the matrix function $g_{21}(x_1, x_2)$ is not defined for $x_1 = 0$.

Remark 5.3 *Strictly speaking, it is not correct to discuss the asymptotic stability of the origin for a system described by equations of the form (5.12) with $g_{21}(x_1, x_2)$ fulfilling condition (5.15), as such a system is not defined at the origin. Hence, the origin is not an equilibrium. However, it is possible to overcome this problem using the following definition of asymptotic stability. We say that a smooth control law locally (globally) asymptotically stabilizes system (5.12) if the closed loop system is smooth in a neighborhood of the origin (in $\mathbb{R}^n$) and the origin is a locally (globally) asymptotically stable equilibrium of the closed loop system. For example, the system $\dot{x} = \frac{1}{x}u$ is globally asymptotically stabilized by the smooth control $u = -x^2$.*

We now discuss sufficient conditions for asymptotic stabilizability of systems described by equations of the form (5.12).

Theorem 5.3 *[20] Consider the system (5.12) defined in an open and dense set U, such that $\bar{U}$ contains the point $x = 0$. Consider the following hold.*

(i) *The matrix functions $g_{11}(x_1, x_2)$ and $g_{22}(x_1, x_2)$ are smooth in $\bar{U}$.*
(ii) *The matrix function $g_{21}(x_1, x_2)$ is smooth in U.*

(iii) *The matrix function $g_{22}(x_1, x_2)$ depends on x_2 only, that is,*
      $g_{22}(x_1, x_2) = \bar{g}_{22}(x_2)$ for some function $\bar{g}_{22}(\cdot)$.
(iv) *There exists a smooth vector function $u_1(x_1, x_2)$, zero for $x_1 = 0$*
     and for all x_2, that is, $u_1(0, x_2) = 0$, such that

$$-\infty < x_1' X g_{11}(x_1, x_2) u_1(x_1, x_2) < 0$$

   for some positive definite matrix X and for all nonzero x_1 in $\bar{U}$.
   Moreover $g_{21}(x_1, x_2) u_1(x_1, x_2)$ is smooth in $\bar{U}$ and it is a function
   of x_2 only, that is, $g_{21}(x_1, x_2) u_1(x_1, x_2) = \bar{f}_2(x_2)$, for some function
   $\bar{f}_2(\cdot)$ such that $\bar{f}_2(0) = 0$.
(v) *There exists a smooth function $u_2(x_2)$ that renders the equilibrium*
    $x_2 = 0$ of the system

$$\dot{x}_2 = \bar{f}_2(x_2) + \bar{g}_{22}(x_2) u_2(x_2)$$

   locally asymptotically stable.

Then, the smooth control law

$$u = u(x_1, x_2) = \begin{bmatrix} u_1(x_1, x_2) \\ u_2(x_2) \end{bmatrix}$$

locally asymptotically stabilizes the system (5.12).

As should be clear from Theorem 5.3, the possibility of rendering the manifold $x_1 = 0$ invariant for the closed loop system, allows the asymptotic stabilization problem to be solved in two successive steps. Hypothesis (iv) determines the component u_1 of the control law; whereas the component u_2 must be chosen to fulfill hypothesis (v). Observe that the choice of u_1 is crucial, as the existence of a smooth function $u_2(x_2)$ fulfilling hypothesis (v) depends on such a choice. The hypotheses of Theorem 5.3 may be easily strengthened to obtain a global result.

Theorem 5.4 *Consider the system (5.12) defined in an open and dense set U, such that $\bar{U} = \mathbb{R}^n$. Suppose (i), (ii), (iii), and (iv) of Theorem 5.3 hold. Moreover, suppose that the following holds:*

(v)′ *There exists a smooth function $u_2(x_2)$ which renders the equilibrium*
     $x_2 = 0$ of the system

$$\dot{x}_2 = \bar{f}_2(x_2) + \bar{g}_{22}(x_2) u_2(x_2)$$

   globally asymptotically stable.

Then the smooth control law

$$u = u(x_1, x_2) = \begin{bmatrix} u_1(x_1, x_2) \\ u_2(x_2) \end{bmatrix}$$

globally asymptotically stabilizes the system (5.12).

Example 5.1   The following simple example illustrates the obtained results. Consider the system

$$\dot{x}_1 = (x_1^2 + x_2^2)u_1$$
$$\dot{x}_2 = -\frac{x_2}{x_1}u_1 + u_2$$

defined on $U = \{x \in \mathbb{R}^2 | x_1 = 0\}$ and fulfilling hypotheses (i), (ii), and (iii) of Theorem 5.4. Setting $u_1 = -x_1$ we fulfill also (iv) and (v). Hence, simple calculations show that the smooth (linear) control law

$$u = \begin{bmatrix} -x_1 \\ -2x_2 \end{bmatrix}$$

yields a globally asymptotically stable closed loop system.

Before concluding this section we discuss another extension of Theorem 5.3.

Theorem 5.5   *[20] Consider the system (5.12) defined in an open and dense set U, such that $\bar{U}$ contains the point $x = 0$. Suppose (i), (ii), and (iii) of Theorem 5.3 hold. Suppose, moreover, that the following holds:*

   (iv)″   There exists a smooth vector function $u_1(x_1, x_2)$, zero for $x_1 = 0$, and for all x_2, that is, $u_1(0, x_2) = 0$, such that

$$-\infty < x_1' X g_{11}(x_1, x_2)u_1(x_1, x_2) < -x_1' Q x_1$$

   for some positive definite matrices X and Q and for all nonzero x_1 in $\bar{U}$. Moreover $g_{21}(x_1, x_2)u_1(x_1, x_2)$ is smooth in $\bar{U}$ and it is a function of x_2 only, that is, $g_{21}(x_1, x_2)u_1(x_1, x_2) = \bar{f}_2(x_2)$, for some function $\bar{f}_2(\cdot)$ such that $\bar{f}_2(0) = 0$.

   (v)″   There exists a smooth function $u_2(x_2)$ which renders the equilibrium $x_2 = 0$ of the system

$$\dot{x}_2 = \bar{f}_2(x_2) + \bar{g}_{22}(x_2)u_2(x_2)$$

   locally exponentially stable.

(v)″ Then the smooth control law

$$u = u(x_1, x_2) = \begin{bmatrix} u_1(x_1, x_2) \\ u_2(x_2) \end{bmatrix}$$

locally exponentially stabilizes the system (5.12).

The hypotheses of Theorems 5.3, 5.4, and 5.5 seem very restrictive. However, it is possible to transform several smooth nonholonomic systems in such a way that the aforementioned hypotheses are automatically fulfilled.

5.3.2 The σ Process

In this section we discuss the use of nonsmooth coordinates changes to transform continuous systems into discontinuous ones. We consider a choice of coordinates system in which, to a small displacement near a fixed point, there corresponds a great change in coordinates. The polar coordinates system possesses such a property; however the cartesian to polar transformation requires transcendental functions; therefore, when not needed, we avoid using the polar coordinates, using another procedure: the so-called σ process (see Reference 45, where the σ process is used to resolve singularities of vector fields).

Mainly, the σ process consists of a nonsmooth rational transformation, but, with abuse of notation, we denote with the term σ process every nonsmooth coordinates transformation possessing the property of *increasing the resolution* around a given point.

Example 5.2   Consider the two dimensional system with one control

$$\dot{x} = g_1(x, y)u, \quad \dot{y} = g_2(x, y)u$$

and perform the coordinates transformation

$$\begin{bmatrix} z \\ w \end{bmatrix} = \Phi(x, y) = \begin{bmatrix} x \\ y/x \end{bmatrix}$$

The resulting system is

$$\dot{z} = g_1(z, zw)u, \quad \dot{w} = \frac{g_2(z, zw) - wg_1(z, zw)}{z}u \qquad (5.16)$$

and it is discontinuous if one of the $g_i(z, zw)$ is such that $g_i(0, 0) \neq 0$. If the system (5.16) is not discontinuous we can further transform it with a second σ process.[12]

5.3.3 The Issue of Asymptotic Stability

Theorems 5.3, 5.4, and 5.5 yield sufficient conditions for stabilizability of discontinuous nonholonomic systems, while the σ process allows to *map a continuously differentiable* system into a *discontinuous* one. To have a practically useful result, we have to show that asymptotic stability of the transformed (discontinuous) system implies almost asymptotic stability of the original (continuously differentiable) system. Moreover, to implement a discontinuous control we must define it on the points of singularity.

Consider a continuously differentiable nonholonomic system described by equations of the form (5.1). Set $x = \mathrm{col}(x_1, x_2)$ with $x_1 \in \mathbb{R}$ and $x_2 = \mathrm{col}(x_{21}, \ldots, x_{2,n-1}) \in \mathbb{R}^{n-1}$ and define the σ process[13]

$$\xi = \begin{bmatrix} \xi_1 \\ \xi_2 \end{bmatrix} = \begin{bmatrix} x_1 \\ \sigma(x_1, x_2) \end{bmatrix} \qquad (5.17)$$

where $\xi_2 = \mathrm{col}(\xi_{21}, \ldots, \xi_{2,n-1})$, $\sigma(x_1, x_2) = \mathrm{col}(\sigma_1(x_1, x_2), \ldots, \sigma_{n-1}(x_1, x_2))$, and $\sigma_i(x_1, x_2) = x_{2i}^{\alpha_i} / x_1^{\beta_i}$ with $\alpha_i \geq 1$ and $\beta_i \geq 0$, for all $i = 1, \ldots, n-1$. The application of the σ process (5.17) to the system (5.1) yields a new system which is, in general, not defined for $\xi_1 = 0$. Suppose now that the transformed system, with state ξ, is *exponentially* stabilized by a control law $u = u(\xi)$, that is, $|\xi_1(t)| \leq c_1 \exp(-\lambda_1 t)$ and $|\xi_{2i}(t)| \leq c_{2i} \exp(-\lambda_{2i} t)$ for some positive $\lambda_1, \lambda_{2i}, c_1$, and c_{2i} and for all $i = 1, \ldots, n-1$. Then $|x_1(t)| \leq c_1 \exp^{-\lambda_1 t}$ and $|x_{2i}(t)| \leq (c_1 c_2)^{1/\alpha_i} \exp(\frac{-\lambda_1 \beta_i + \lambda_{2i}}{\alpha_i} t)$ for all $i = 1, \ldots, n-1$. We conclude that exponential convergence to zero of the state ξ of the transformed system implies exponential convergence to zero of the state x of the original system.

Remark 5.4 *The previous conclusions also remain valid if the stabilizer is dynamic. This fact is useful to design dynamic, output feedback, discontinuous stabilizers for nonholonomic systems [46].*

Remark 5.5 *Asymptotic stability of the system with state ξ does not imply asymptotic stability of the system with state x, as the inverse of the coordinates transformation (5.17) does not map neighborhood of $\xi = 0$ into neighborhood of $x = 0$, as illustrated in Figure 5.1. Therefore, exponential stability (in the sense of Lyapunov) of the closed loop system with state ξ implies only almost exponential stability of the closed loop system with state x.*

---

[12] Note that the composition of σ processes yields a σ process.
[13] The coordinates transformation (5.17) defines a σ process only if $\Sigma_i \beta_i \geq 1$.

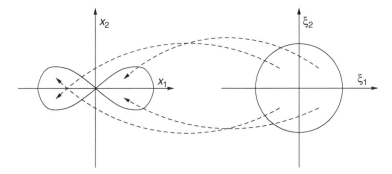

FIGURE 5.1   The anti-σ process $x_1 = \xi_1, x_2 = \xi_1\xi_2$ does not map the ball $\xi_1^2 + \xi_2^2 = R^2$ into a neighborhood of the origin in the $x_1 x_2$-plane.

The continuously differentiate control law which stabilizes a given discontinuous nonholonomic system needs to be transformed back to the original coordinates via inversion of the σ process (in Reference 45 such procedure is denoted anti-σ process). Note that the anti-σ process yields a discontinuous control law

$$u(\xi_1, \xi_{21}, \ldots, \xi_{2,n-1}) \xrightarrow{\text{anti-}\sigma} u\left(x_1, \frac{x_{21}^{\alpha_1}}{x_1^{\beta_1}}, \ldots, \frac{x_{2,n-1}^{\alpha_{n-1}}}{x_1^{\beta_{n-1}}}\right)$$

Such a control law cannot be directly implemented, because it is not defined at $x_1 = 0$. Nevertheless, it is *implementable* provided that some conditions are fulfilled.

Theorem 5.6   *[20] Consider a smooth nonholonomic system*

$$\dot{x} = g(x)u \tag{5.18}$$

with $x \in \mathbb{R}^n$, $u \in \mathbb{R}^m$, and $n > m$. Assume that $x_1(0) \neq 0$. Apply the σ process (5.17) and suppose there exists a continuously differentiable control law $u = u(\xi)$ globally exponentially stabilizing the transformed system, that is, $|\xi_1(t)| \leq c_1 \exp^{-\lambda_1 t}$ and $|\xi_{2i}(t)| \leq c_{2i} \exp^{-\lambda_{2i}t}$, for some positive $\lambda_1, \lambda_{2i}, c_1$, and c_{2i} and for all $i = 1, \ldots, n-1$. Assume moreover that there exist positive constants $c_0 \leq c_1$ and $\lambda_0 \geq \lambda_1$ such that[14] $c_0 \exp^{-\lambda_0 t} \leq |\xi_t(t)|$. Assume finally that

$$\beta_i \geq 0 \tag{5.19}$$

---

[14] This implies that the state ξ_1 does not converge to zero in finite time.

for all $i = 1, \ldots, n - 1$. Then for every $\epsilon > 0$ there exists a $\delta > 0$ (depending on ϵ) satisfying $\delta \ll c_0 \leq |x_1(0)| = |\xi_1(0)| \leq c_1$, such that the trajectories of the system in closed loop with the C^0 control law

$$
u =
\begin{cases}
u\left(x_1, \dfrac{x_{21}^{\alpha_1}}{x_1^{\beta_1}}, \ldots, \dfrac{x_{2,n-1}^{\alpha_{n-1}}}{x_1^{\beta_{n-1}}}\right) & \text{if } |x_1| > \delta \\
0 & \text{elsewhere}
\end{cases}
\tag{5.20}
$$

converge to the set $\Omega_\epsilon = \{x \in \mathbb{R}^n | \|x\| \leq \epsilon\}$ in some finite time T_* and remain therein for all $t \geq T_*$.

At this point the reader may argue whether it is possible or not to let δ go to zero, that is, what we can conclude about the (discontinuous) control law

$$
u =
\begin{cases}
u\left(x_1, \dfrac{x_{21}^{\alpha_1}}{x_1^{\beta_1}}, \ldots, \dfrac{x_{2,n-1}^{\alpha_{n-1}}}{x_1^{\beta_{n-1}}}\right) & \text{if } x_1 \neq 0 \\
0_{p \times 1} & \text{if } x_1 = 0
\end{cases}
\tag{5.21}
$$

Observe that the control law (5.21) is discontinuous at $x_1 = 0$ as a function of x, but it is continuous as a function of t, since $x_1(t) = 0$ only asymptotically (if $x_1(0) \neq 0$, which is without lack of generality). Moreover, by hypothesis, the variables $\xi_{2i} = x_{2i}^{\alpha_i} / x_1^{\beta_i}$ tend to zero when t goes to infinity. Thus

$$
\lim_{t \to \infty} u\left(x_1(t), \dfrac{x_{21}^{\alpha_1}(t)}{x_1^{\beta_1}(t)}, \ldots, \dfrac{x_{2,n-1}^{\alpha_{n-1}}(t)}{x_1^{\beta_{n-1}}(t)}\right) = u(0, 0, \ldots, 0) = 0
$$

As a consequence, the control law (5.21) is well defined and bounded, along the trajectories of the closed loop system, for all $t \geq 0$ and, viewed as a function of time, is even continuous (i.e., it is at least C^0) as t goes to infinity. Finally, using Theorem 5.6, with $\delta = 0$, and assuming that the conditions (5.19) hold, we conclude that the control law (5.21) almost exponentially stabilizes the system (5.18) on the open and dense set $\Omega = \{x \in \mathbb{R}^n | x_1 \neq 0\}$.

Remark 5.6 *The assumption $x_1(0) \neq 0$ is without lack of generality, as it is always possible to apply preventively an open loop control, for example, a constant control, driving the system away from the hyperplane $x_1 = 0$ [32,33,47].*

Remark 5.7 *By a general property of one dimensional dynamical systems, we conclude that the state variable $x_1 = \xi_1$ evolving from a nonzero initial condition approaches the equilibrium $x_1 = 0$ without ever crossing it, that is,* *there exists no finite time T such that $x_1(T) = 0$. Thus, the singular plane $x_1 = 0$ is never crossed, but is just approached asymptotically. Moreover, every trajectory starting in $\Omega^+ = \{x \in \mathbb{R}^n | x_1 > 0\}$ $(\Omega^- = \{x \in \mathbb{R}^n | x_1 < 0\})$ remains in $\Omega^+ (\Omega^-)$ for every finite t and approaches the border of $\Omega^+ (\Omega^-)$ as t goes to infinity.*

5.3.4 An Algorithm to Design Almost Stabilizers

In this section we propose a procedure to design discontinuous control laws for smooth nonholonomic systems described by equations of the form (5.12). The procedure is composed of the following steps.

(I) Transform a given smooth nonholonomic system, by means of a σ process, into a discontinuous system.

(II) Check if the discontinuous system admits a smooth control law yielding asymptotic stability. In case of positive answer proceed to step III, otherwise return to step I and use a different σ process.

(III) Build a smooth stabilizer for the transformed system.

(IV) Apply the anti-σ process to the obtained stabilizer to build a discontinuous control law for the original system.

The crucial points of the algorithm are the selection of the σ process (step I) and the design of the smooth asymptotically stabilizing control law for the transformed system (step III). In particular, step III can be easily solved for low dimensional systems; whereas there is no constructive or systematic way to perform step I successfully; that is, to select a σ process which allows to conclude positively the algorithm.

Finally, to obtain a discontinuous nonholonomic system described by equations of the form (5.12), with $g_{21}(x_1, x_2)$ fulfilling condition (5.15), the following simple result may be useful.

Proposition 5.1 *[20] Consider a nonholonomic system described by equations of the form (5.12). Assume that $g_{11}(x_1, x_2) = I_p$ and that the matrices $g_{21}(x_1, x_2)$ and $g_{22}(x_1, x_2)$ have smooth entries in $\mathbb{R}^n$. Consider a coordinates transformation (σ process) described by equations of the form*

$$\xi_1 = x_1, \quad \xi_2 = \frac{\Phi_2(x_1, x_2)}{\sigma(x_1)}$$

where $\Phi_2(x_1, x_2)$ is a smooth mapping such that $\Phi_2(0, x_2) \neq 0$ and $\sigma(x_1)$ is a smooth function which is zero at $x_1 = 0$. Then the transformed system is always described, in the new coordinates, by equations of the form (5.12) *but, in general, the matrix $g_{21}(\xi_1, \xi_2)$ is not defined at $\xi_1 = 0$, that is, fulfills condition (5.15).*

The presented discontinuous stabilization approach has been exploited in the control of underactuated spacecraft in Reference 36 and has been given an interesting geometric interpretation in Reference 48 and related references. Finally, in Reference 49 and related works, it has been shown that the proposed approach can be interpreted in terms of a state-dependent time-scaling.

5.4 CHAINED SYSTEMS AND POWER SYSTEMS

From this section onward, we focus on two special classes of nonholonomic systems: chained systems and power systems. They occupy a special place in the theory of nonholonomic control. Many nonholonomic mechanical systems can be represented by, or are *feedback equivalent* to, kinematic models in chained form or in power form. Chained systems have been introduced in Reference 7, where sufficient conditions for (local) feedback equivalence to chained forms have also been given. Power systems have been introduced in Reference 50. Therein, it has also been shown that chained systems and power systems are globally feedback equivalent. Chained systems[15] are described by equations of the form

$$\dot{x}_1 = u_1$$
$$\dot{x}_2 = u_2$$
$$\dot{x}_3 = x_2 u_1 \qquad\qquad (5.22)$$
$$\vdots$$
$$\dot{x}_n = x_{n-1} u_1.$$

Power systems are described by equations of the form

$$\dot{x}_1 = u_1$$
$$\dot{x}_2 = u_2$$
$$\dot{x}_3 = x_1 u_2 \qquad\qquad (5.23)$$
$$\vdots$$
$$\dot{x}_n = \frac{1}{(n-2)!} x_1^{n-2} u_2.$$

---

[15] In the terminology of Reference 7, Equations (5.22) describe a one-chain single generator system.

5.5   DISCONTINUOUS CONTROL OF CHAINED SYSTEMS

To begin with, we transform system (5.22) through the σ process

$$
\begin{aligned}
\xi_1 &= x_1 \\
\xi_2 &= x_2 \\
&\vdots \\
\xi_{n-1} &= \frac{x_{n-1}}{x_1^{(n-3)}} \\
\xi_n &= \frac{x_n}{x_1^{(n-2)}}
\end{aligned}
\tag{5.24}
$$

yielding a discontinuous system described by equations of the form

$$
\begin{aligned}
\dot{\xi}_1 &= u_1 \\
\dot{\xi}_2 &= u_2 \\
&\vdots \\
\dot{\xi}_{n-1} &= \frac{\xi_{n-2} - (n-3)\xi_{n-1}}{\xi_1} u_1 \\
\dot{\xi}_n &= \frac{\xi_{n-1} - (n-2)\xi_n}{\xi_1} u_1.
\end{aligned}
\tag{5.25}
$$

Remark 5.8   *The σ process (5.24) is a special case of (5.17) with $\alpha_i = 1$ for all $i = 1, \ldots, n-1$ and $\beta_i = i - 1$ for all $i = 1, \ldots, n-1$. Observe that such β_i fulfill the conditions (5.19).*

Consider now the system (5.25) and apply the control $u_1 = -k\xi_1$, with $k > 0$. A simple computation shows that the resulting system, described by equations of the form

$$
\dot{\xi} = A\xi + b_2 u_2
\tag{5.26}
$$

where $\xi = [\xi_1, \xi_2, \ldots, \xi_n]'$,

$$
A = \begin{bmatrix}
-k & 0 & 0 & 0 & \cdots & 0 \\
0 & 0 & 0 & 0 & \cdots & 0 \\
0 & -k & k & 0 & \cdots & 0 \\
0 & 0 & -k & 2k & \cdots & 0 \\
\vdots & \vdots & \vdots & \vdots & \ddots & \vdots \\
0 & 0 & 0 & 0 & \cdots & (n-2)k
\end{bmatrix}
\tag{5.27}
$$

and

$$b_2 = [0 \quad 1 \quad 0 \quad \cdots \quad 0]' \qquad (5.28)$$

is stabilizable with the second control input u_2. Therefore, recalling the results established in Section 5.3, we give the following statement.

Proposition 5.2 *[20] The discontinuous control law*

$$u = \begin{bmatrix} u_1 \\ u_2 \end{bmatrix} = \begin{bmatrix} -kx_1 \\ p_2x_2 + p_3\frac{x_3}{x_1} + \cdots + p_{n-1}\frac{x_{n-1}}{x_1^{n-3}} + p_n\frac{x_n}{x_1^{n-2}} \end{bmatrix} \qquad (5.29)$$

with $k > 0$ and $p = [0, p_2, p_3, \ldots, p_{n-1}, p_n]$ such that the eigenvalues of the matrix $A + b_2p$ have all negative real part, almost exponentially stabilizes the system (5.22) in the open and dense set $\Omega_1 = \{x \in \mathbb{R}^n | x_1 \neq 0\}$.

Remark 5.9 *If we rewrite the control law (5.29) as*

$$u = \begin{bmatrix} u_1 \\ u_2 \end{bmatrix} = \begin{bmatrix} -kx_1 \\ p_2x_2 + \frac{p_3}{x_1}x_3 + \cdots + \frac{p_{n-1}}{x_1^{n-3}}x_{n-1} + \frac{p_n}{x_1^{n-2}}x_n \end{bmatrix}$$

we can regard it as a linear control law with state dependent gains.

5.5.1 An Example: A Car-Like Vehicle

In this section we consider the problem of designing a discontinuous controller for a prototypical nonholonomic system: a car-like vehicle. For simplicity we consider an ideal system, that is, the wheels roll without slipping and all pairs of wheels are perfectly aligned and with the same radius. A thorough analysis of the phenomena caused by nonideal wheels can be found in Reference 51. The problem of stabilizing a car-like vehicle has been addressed with different techniques by several authors, see References 5 and 7 for open loop strategies and References 9, 12, and 52, for state feedback control laws. In what follows, exploiting the results in Section 5.3, we design a discontinuous state feedback controller. This control law, because of its singularity, is not directly implementable. However, as discussed in Section 5.3, and in Reference 33 and 53, and in Section 5.7, it is possible to build modifications yielding uniform ultimate boundedness or (robust) exponential stability.

The kinematic model of a car with rear tires aligned with the car and front tires allowed to spin about the vertical axis [7] is

$$\dot{x} = \cos\theta v_1$$

$$\dot{y} = \sin\theta v_1$$

$$\dot{\theta} = \frac{1}{l}\tan\theta v_1 \qquad\qquad (5.30)$$

$$\dot{\phi} = v_2$$

where (x, y) denotes the location of the center of the axle between the two rear wheels, θ the angle of the car body with respect to the x-axis, ϕ the steering angle with respect to the car body, and v_1 and v_2 the forward velocity of the rear wheels and the velocity of the steering wheels, respectively (see Figure 5.2). Applying the control transformation

$$\begin{bmatrix} v_1 \\ v_2 \end{bmatrix} \begin{bmatrix} \dfrac{u_1}{\cos\theta} \\ -\frac{3}{l}\sin^2\phi\tan\theta\sec\theta u_1 + l\cos^2\phi\cos^3\theta u_2 \end{bmatrix}$$

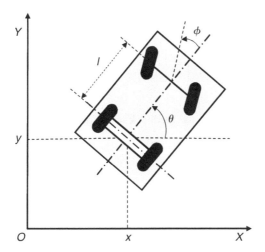

FIGURE 5.2 Model of an automobile.

and the σ process

$$\xi_1 = x$$

$$\xi_2 = \frac{1}{l}\sec^3\theta \tan\phi$$

$$\xi_3 = \frac{\tan\theta}{x}$$

$$\xi_4 = \frac{y}{x^2}$$

we obtain a system described by equations of the form

$$\dot{\xi}_1 = u_1$$

$$\dot{\xi}_2 = u_2$$

$$\dot{\xi}_3 = \frac{\xi_2 - \xi_3}{\xi_1}u_1$$

$$\dot{\xi}_4 = \frac{\xi_3 - 2\xi_4}{\xi_1}u_1$$

that is, by equations of the form (5.25) with $n = 4$. Thus, using Proposition (5.2), we design the state feedback control law

$$\begin{bmatrix} u_1 \\ u_2 \end{bmatrix} = \begin{bmatrix} -k\xi_1 \\ p_2\xi_2 + p_3\xi_3 + p_4\xi_4 \end{bmatrix}$$

In the original coordinates the feedback law is described by

$$v_1 = -k\frac{x}{\cos\theta}$$

$$v_2 = k\frac{3}{l}\sin^2\phi \tan\theta \sec\theta \frac{x}{\cos\theta}$$

$$+ l\cos^2\phi \cos^3\theta \left[p_2\left(\frac{1}{l}\sec^3\theta \tan\phi\right) + p_3\left(\frac{\tan\theta}{x}\right) + p_4\left(\frac{y}{x^2}\right) \right]$$

To have almost exponential stability of the closed loop system it is necessary to have $k > 0$ and to set p_2, p_3, and p_4 such that[16] $\sigma(A_4) \in \mathbb{C}^-$, where

$$A_4 = \begin{bmatrix} p_2 & p_3 & p_4 \\ -k & k & 0 \\ 0 & -k & 2k \end{bmatrix}$$

It must be noticed that the matrix A_4 is a submatrix of the matrix $A + b_2 p$ considered in Proposition 5.2. This is without lack of generality as only $n - 1$ eigenvalues of the matrix $A + b_2 p$ can be set with the vector p, whereas one eigenvalue is always equal to $-k$. Figure 5.3 and Figure 5.4 show the results of simulations carried out with the proposed controller.

5.5.2 Discussion

Discontinuous, state feedback, control laws to almost exponentially stabilize chained systems, have been presented. In contrast to other results, the given control laws are extremely simple and possess an intuitive interpretation in terms of linear feedback with state dependent gain scheduling. It is worth stressing that the design of the stabilizing control law involves mainly linear control tools, that is, stability of the closed loop system depends on the stability of some linear systems. A drawback of the proposed approach is the possibility for numerical problems to appear in real time implementations. In fact, most of the features of the closed loop system derive from the simplification in the product $\frac{1}{x_1} u_1$. If such a simplification takes place only approximately, for example, for the presence of measurement noise, the limit $\lim_{x_1 \to 0} \frac{1}{x_1} u_1(x_1^*)$, where x_1^* is the available measure on x_1, may be unbounded.

5.6 ROBUST STABILIZATION — PART I

The results in Section 5.4 can be interpreted as follows. For nominal and ideal conditions (e.g., exact integration, noise free measurements) and as long as $x_1(0) \neq 0$, the discontinuous controllers proposed therein are well defined and yield bounded control action, along the trajectories of the closed loop system. Moreover, as detailed in Reference 53, the analysis carried out in References 19–21, 43, and 54–56 is correct and yields an adequate picture of the ideal properties of this class of discontinuous controllers. However, a substantial difference is to be expected in a nonideal situation, as the control law blows up, that is, provides unbounded control action, whenever the discontinuity surface $x_1 = 0$ is intersected, for example, in the presence of external

---

[16] $\sigma(A)$ denotes the spectrum of the square matrix A and $\mathbb{C}^-$ denotes the open left-half of the complex plane.

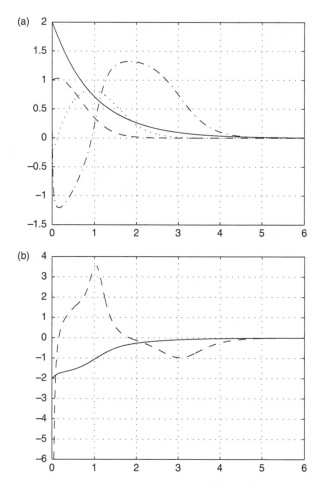

FIGURE 5.3   (a) Time histories of $x(t)$ (solid), $y(t)$ (dashed), $\phi(t)$ (dash-dotted), and $\theta(t)$ (dotted). (b) Translational (solid) and rotational (dashed) velocity controls.

disturbances. In what follows we perform a very simple robustness analysis, with reference to an interesting situation, namely in the presence of external disturbances and model errors, and for a prototype system.

Consider a three dimensional chained system perturbed by a constant nonzero disturbance entering the third equation,[17] that is,

$$\dot{x}_1 = u_1, \quad \dot{x}_2 = u_2, \quad \dot{x}_3 = x_2 u_1 + d \qquad (5.31)$$

---

[17] The disturbance models a violation of the nonholonomic constraint, that is, $x_2 \dot{x}_1 - \dot{x}_3 \neq 0$.

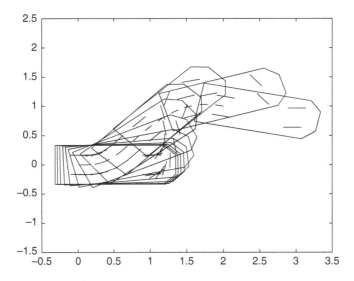

FIGURE 5.4   Parking maneuver. The dashed line describes the trajectory, in the xy-plane, of the center of the axle between the two rear wheels.

with $d \neq 0$. For such a system we point out some structural limitations, namely the nonexistence of sufficiently regular control laws yielding a closed loop system with *converging* solutions.

Proposition 5.3   *[53] Consider system (5.31) with $d \neq 0$ and a control law $u(x,t)$ such that, for any initial condition, $x_i(t)$ and $u_i(x(t),t)$ are absolutely continuous functions of time and $\lim_{t \to \infty} |x_1(t)| = x_{1,\infty}$. Then, $\lim_{t \to \infty} |x_3(t)| = \infty$.*

Proposition 5.3 points out a limitation of any *regular* control law applied to system (5.31) with $d \neq 0$. However, this limitation does not apply if we simply ask for boundedness (and not convergence) of the trajectories of the controlled system or if we use more general control signals.

Proposition 5.4   *[53] Consider the system (5.31) with $d \neq 0$ known. There exist absolutely continuous controls $u_i(t)$ such that $x(t)$ remains bounded for all $t \geq 0$. Moreover, if $x_2(0) = 0$ there exist impulsive controls $u_i(t)$ such that $x(t)$ remains bounded for all $t \geq 0$ and $x_3(t)$ converges to a constant value.*

Several points are left open by the above discussion. These will be partly addressed and solved in the next two sections, where we present robust hybrid and sampled-data stabilizers for chained and power systems.

5.7 ROBUST STABILIZATION — PART II

In this section we consider the robust stabilization problem for nonholonomic systems in the presence of measurement errors and exogenous disturbances. This problem has been only partly investigated, and several attempts have been made to study the robustness properties of existing control laws or to robustify given controllers [34,53,57]. Most of the robust stabilization results and investigations focus on the problems of parametric uncertainties or model errors, see for example, [58] where the problem of local robust stabilization by means of time-varying control laws have been studied; [57], where a similar problem has been addressed using the class of discontinuous control laws discussed in Section 5.4 and [8,24] where several types of hybrid control laws have been used to achieve local robustness against unknown parameters or unmodelled dynamics. On the other hand, the fundamental problems of robustness in the presence of sensor noise, external disturbances, and actuator disturbances have been only partially addressed, see for example, [33,53]. These problems are of special interest and relevance whenever discontinuous control laws are employed, as for such control laws classical robustness results and Lyapunov theory are not directly applicable, see however Reference 59, where a discontinuous control law, possessing a Lyapunov stability property, has been constructed. In what follows we make use of the class of discontinuous control laws presented in Section 5.4 and we show how, adding a proper modification together with a *hybrid variable,* it is possible to obtain a closed loop system with global stability properties and which is globally robust against measurement noises and exogenous disturbances. The proposed controller takes inspiration from the results in References 33, 60, and 61.

5.7.1 The Local Controller

Consider the system (5.22) and the control law $u_l : \mathbb{R}^n \to \mathbb{R}^2$ defined by

$$u_{1l}(x) = -x_1, \quad u_{2l}(x) = p_2 x_2 + p_3 \frac{x_3}{x_1} + \cdots + p_n \frac{x_n}{x_1^{n-2}} \qquad (5.32)$$

with the p_i such that the matrix

$$\bar{A} = \begin{bmatrix} p_2 + 1 & p_3 & \cdots & p_{n-1} & p_n \\ -1 & 2 & \cdots & 0 & 0 \\ 0 & -1 & \cdots & 0 & 0 \\ \vdots & \vdots & \vdots & \vdots & \vdots \\ 0 & 0 & \cdots & -1 & n-1 \end{bmatrix}$$

is Hurwitz.[18] Let $P = P' > 0$ be such that $\bar{A}'P + P\bar{A} < 0$, and let z be a variable in $\mathbb{R} \cup \{+\infty\}$ defined by

$$z = z(x) = \begin{cases} Y'PY & \text{if } x_1 \neq 0 \\ +\infty & \text{if } x_1 = 0 \end{cases} \tag{5.33}$$

with[19]

$$Y = Y(x) = \left[\frac{x_2}{x_1}, \frac{x_3}{x_1^2}, \ldots, \frac{x_n}{x_1^{n-1}} \right]', \quad \forall x \in \mathbb{R}^n, x_1 \neq 0$$

Consider now the perturbed closed loop system composed of the chained system (5.22) perturbed by an additive disturbance d and in closed loop with $u = u_l(x + e)$, where e represents a measurement noise. For such a perturbed system the following fact holds.

Lemma 5.2 *There exists a continuous function $\rho_l : \mathbb{R} \to \mathbb{R}$ satisfying $\rho_l(\xi) > 0, \forall \xi \neq 0$, such that, for all e and d satisfying the regularity assumptions in Section 5.2 and Equation (5.7) with $\rho = \rho_l(x_1)$, and for all x_0 satisfying $z(x_0) \leq M$, there exists a Carathéodory solution X starting from x_0 and all such Carathéodory solutions are maximally defined on $[0, +\infty)$. Moreover there exists a function δ_l of class $\mathcal{K}_\infty$ and $C > 0$ such that, for all r and M, and for all x_0 satisfying $|x_0| \leq \delta_l(r)$ and $z(x_0) \leq M$, we have $|X(t)| \leq r\sqrt{M}e^{-Ct}$ and $z(t) \leq Me^{-Ct}$, for all $t \geq 0$.*

Lemma 5.2 states that, for any $M > 0$, the region $z(x) \leq M$ is robustly forward invariant, that is, it is positively invariant in the presence of a class of measurement noise and external additive disturbances. Moreover, any trajectory in such a region converges exponentially to the origin.

5.7.2 The Global Controller

Let $\mu > 0$ and consider the control law u_g defined as $u_{1g} = 1$ and $u_{2g} = -\mu x_2$. Consider the perturbed closed loop system composed of the chained system (5.22) perturbed by an additive disturbance d and in closed loop with $u = u_g(x + e)$, where as before e represents a measurement noise. For such a perturbed system the following fact holds.

---

[18] The eigenvalues of the matrix $\bar{A}$ can be arbitrarily assigned by a proper selection of the coefficients p_i.

[19] The variable Y differs from the variable used in the σ-process in Section 5.4 and in References 20 and 57. It is not difficult to show that using the σ-process therein it is possible only to prove a weaker version of Theorem 5.7.

Lemma 5.3   *There exists a continuous function $\rho_g : \mathbb{R}^n \to \mathbb{R}$ satisfying $\rho_g(x) > 0, \forall x \neq 0$ such that, for any initial condition, the considered perturbed system where e and d satisfies the regularity assumptions in Section 5.2 and Equation (5.7) with $\rho = \rho_g(x)$, admit a unique Carathéodory solution, defined for all $t \geq 0$. Moreover there exists a function δ_g of class $\mathcal{K}_\infty$ such that, for any $r > 0$ and for any $M > 0$ there exists a time $T_g = T_g(M, \delta_g(r))$ such that, for all Carathéeodory solutions X with initial condition x_0 with $|x_0| \leq \delta_g(r)$, one has $z(X(t)) \leq M$ for all $t \geq T_g$, and $|X(t)| \leq r$ for all $t \leq T_g$.*

Lemma 5.3 states that, for any $M > 0$, the trajectories of the perturbed system enter the region $z(x) \leq M$ in finite time, while remaining bounded for all t.

5.7.3 Definition of the Hybrid Controller and Main Result

We are now ready to define the hybrid controller robustly stabilizing system (5.22). To this end, for any strictly positive number M, we define the subset Γ_M of $\mathbb{R}^n$ as $\Gamma_M = \{x, x_1 \neq 0, z < M\}$, where z is defined by (5.33). Let $M_2 > M_1 > 0$. The hybrid controller (k, k_d) is defined making a hysteresis between u_l and u_g on Γ_{M_2} and Γ_{M_1}, that is,

$$k(x, s_d) = \begin{cases} u_l(x) & \text{if } s_d = 1 \text{ and } x_1 \neq 0 \\ 0 & \text{if } s_d = 1 \text{ and } x_1 = 0 \\ u_g(x) & \text{if } s_d = 2 \end{cases} \tag{5.34}$$

$$k_d(x, s_d) = \begin{cases} 1 & \text{if } x \in \Gamma_{M_1} \cup \{0\} \\ s_d & \text{if } x \in \Gamma_{M_2} \backslash \Gamma_{M_1} \\ 2 & \text{if } x \notin \Gamma_{M_2} \cup \{0\} \end{cases} \tag{5.35}$$

Theorem 5.7   *[32] The hybrid controller (k, k_d), described in Section 5.7.1, Section 5.7.2, and Section 5.7.3 robustly globally exponentially stabilizes system (5.22).*

5.7.4 Discussion

A hybrid control law globally robustly exponentially stabilizing a chained system has been proposed. This controller retains the main features of the discontinuous controller proposed in Section 7.4, while allowing to counteract (small) exogenous disturbances and measurement noise. A similar, but local, result was developed in Proposition 3 of Reference 33. Note finally that the idea of *switching* between a *local* and a *global* controller to achieve stabilization in the large has been advocated in several papers, and typically in the context of stabilization of unstable equilibria of mechanical systems. However, what makes the present result interesting is that we aim at achieving robust asymptotic stability rather than asymptotic stability.

5.8 ROBUST STABILITAZION — PART III

In the aforementioned discussion, we have implicitly assumed that the control signals are continuous, that is, are generated by an analog device, and measurement signals are also continuous. In real applications, however, control signals are (in general) computed by a digital device, and measurements are obtained by sample and hold of physical signals. This implies that, from a realistic point of view, it is necessary to regard the system to be controlled as a sampled-data system. Control of nonlinear sampled-data systems has recently gained a lot of interest, see for example, References 40 and 62. The main issue in addressing and solving sampled-data control problems for nonlinear systems is the definition of an adequate discrete time model, which should describe (with a given accuracy) the behavior of the sampled-data system. This problem has been widely addressed in the numerical analysis literature, see References 40 and 41. In particular, it has been shown that approximate discrete time models obtained using standard Euler approximation are adequate for control, provided that one is ready to trade global properties with semi-global properties and asymptotic properties with practical properties.

5.8.1 Robust Sampled-Data Control of Power Systems

In this section we focus on systems in power form (see Equation (5.23)) and on their Euler approximate discrete time model given by

$$
\begin{aligned}
x_1(k+1) &= x_1(k) + Tu_1(k) + d_1(x(k), k)\\
x_2(k+1) &= x_2(k) + Tu_2(k) + d_2(x(k), k)\\
x_3(k+1) &= x_3(k) + Tx_1(k)u_2(k) + d_3(x(k), k)\\
&\vdots\\
x_n(k+1) &= x_n(k) + \frac{1}{(n-2)!}Tx_1^{(n-2)}u_2(k) + d_n(x(k), k)
\end{aligned}
\tag{5.36}
$$

where we have also included the additive disturbance $d(x(k), k) \in \mathbb{R}^n$.

Theorem 5.8 *[42] Consider the Euler approximate model in Equation (5.36) with $d(x(k), k) = 0$ for all k. Let $\rho(s) = g_0|s|^b$ with $b > 0$ and $g_0 > 0$ and*

$W(x) = \sum_{i=2}^{n} c_i |x_i|^{a_i}$, with $c_i > 0$ $a_i \in \{2, 3, \ldots\}$. Then there exists $T^* > 0$ such that for all $T \in (0, T^*)$ the controller $u_T := (u_{1T}, u_{2T})'$ where[20]

$$u_{1T} = - g_1 x_1 - \rho(W)(\cos((k+1)T) - \tfrac{\epsilon}{2} \sin((k+1)T))$$
$$\qquad + \tfrac{\epsilon}{2} \Delta_\rho \sin((k+1)T)$$
$$u_{2T} = - g_2 \mathrm{sign}(L_{f_2} W) |L_{f_2} W|^\alpha (2\rho(W) + 2(g_1 x_1 + \rho(W) \cos((k+1)T))$$
$$\qquad \times \cos((k+1)T) - \epsilon g_1 x_1 \sin((k+1)T)) \qquad (5.37)$$

with $g_1 > 0, g_2 > 0, a > 0$ and a sufficiently small $\epsilon > 0$, is a SP-AS controller for the system (5.36) and the function

$$V_T(k, x) = (g_1 x_1 + \rho(W) \cos(kT))^2 + \rho(W)^2 - \epsilon g_1 x_1 \rho(W) \sin(kT)$$
$$\qquad (5.38)$$

is a (strict) SP-AS Lyapunov function for the closed loop system (5.36), (5.37).

The control law is similar to the one proposed in Reference 50 and the Lyapunov function is a modification of the one proposed in Reference 10. The proposed result provides a discrete-time counterpart and to some extent a generalization of Theorem 2 of Reference 10. Theorem 5.8 states that V_T is a strict SP-AS Lyapunov function for the closed loop system. It is well known that the existence of a strict SP-AS negative Lyapunov function allows to address the stabilization problem in the presence of disturbances.

Proposition 5.5 *[42] There exist* $T^* > 0$ *such that for all* $T \in (0, T^*)$ *the controller (5.37) is a SP-ISS controller for system (5.36) and the function (5.38) is a SP-ISS Lyapunov function for the closed loop system (5.36), (5.37.)*

5.8.2 An Example: A Car-Like Vehicle Revisited

In this section we apply the proposed result to the model of a car-like vehicle introduced in Section 5.5.1. Consider the model (5.30) and the coordinate transformation [50,52]

$$x_1 = x$$
$$x_2 = \sec^3(\theta) \tan(\phi)$$
$$x_3 = x \sec^3(\theta) \tan(\phi) - l \tan(\theta) \qquad (5.39)$$
$$x_4 = ly + \tfrac{1}{2} x^2 \sec^3(\theta) \tan(\phi) - lx \tan(\theta)$$

---

[20] f_2 denotes the vector $[0, 1, x_1, \ldots, \frac{1}{(n-2)!} x_1^{n-2}]'$ and $L_{f_2} W = \frac{\partial W}{\partial x} f_2$.

yielding

$$\dot{x}_1 = u_1$$
$$\dot{x}_2 = u_2$$
$$\dot{x}_3 = x_1 u_2 \tag{5.40}$$
$$\dot{x}_4 = \tfrac{1}{2} x_1^2 u_2$$

where $u_1 = v_1 \cos\theta$ and $u_2 = \frac{d}{dt}(\sec^3(\theta)\tan(\phi))$. Applying Theorem 5.8 we construct the controller

$$u_{1T} = -3x_1 + \rho(W)(\cos((k+1)T) - \tfrac{\epsilon}{2}\sin((k+1)T))$$
$$u_{2T} = u_2(2\rho(W) - 3\epsilon x_1 \sin((k+1)T) + 2(3x_1 + \rho(W)) \tag{5.41}$$
$$\times \cos((k+1)T))\cos((k+1)T))$$

with $k = 1, \rho(W) = \frac{4}{10}\sqrt[6]{W(x)}$, and $u_2 = -\frac{3}{100}\text{sign}(L_{f_2}W(x))\sqrt[5]{|L_{f_2}W(x)|}$, which is a SP-AS controller for the Euler model (5.40). Figure 5.5 shows simulation results when the controller (5.41) is applied to control the plant (5.40). We have used $x_o = (0,0,0,1)', T = 0.2$, and $\epsilon = 0.35$.

5.8.3 Discussion

The problem of robust stabilization of nonholonomic systems in power form has been addressed and solved in the framework of nonlinear sampled-data control theory. It has been shown that, by modifying the periodic controller in Reference 10, SP-AS and SP-ISS can be achieved. The main drawback of the proposed controllers is the slow convergence rate, which is, however, intrinsic to smooth time-varying controllers [12].

5.9 CONCLUSIONS

The problem of (discontinuous) stabilization and robust stabilization for non-holonomic systems has been discussed from various perspectives. It has been shown that, in ideal situations, a class of discontinuous controllers allow to obtain fast convergence and *efficient* trajectories. This approach is, however, inadequate in the presence of disturbances and measurement noise, hence it is necessary to modify the proposed control by introducing a second controller, a hybrid variable, and a switching strategy, which together guarantee robust stability. Both these controllers have been designed in continuous time. It is therefore difficult to quantify the loss of performance arising from a sampled-data implementation. As a result, we have discussed the robust stabilization

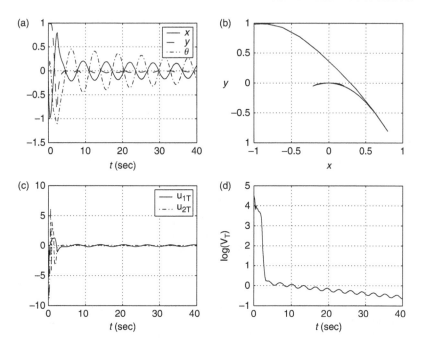

FIGURE 5.5 Response of the car model controlled using the controller (5.41): (a) variables x, y, and θ; (b) trajectory of the center of the axel between the two rear wheels; (c) control signals; (d) Lyapunov function.

problem in the framework of nonlinear sampled-data systems. The discussion in the chapter has highlighted main issues:

- For the class of nonholonomic system described by Equation (5.1) it is not possible to single out *the best* control strategy, that is, several control strategies with diverse and *conflicting* properties exist.
- It may be difficult to provide general stabilization results for nonholonomic systems described by Equation (5.1), hence it is convenient to consider special (canonical) forms, such as chained forms or power forms. The use of canonical forms allows the explicit construction of (robustly) stabilizing control laws, and the in-depth study of the asymptotic properties of closed loop systems.

Several issues have been left aside in this chapter. We mention the stabilization problem for systems with high-order nonholonomic constraints, the stabilization problems for systems which are not feedback equivalent to chained or power forms (e.g., the so-called ball and plate system, and all systems arising in dextrous manipulation), the stabilization of dynamic models of nonholonomic systems, and the adaptive stabilization of nonholonomic systems with unknown parameters. Finally, the important problem of trajectory tracking for nonholonomic systems has not been discussed at all. We believe that the list of reference (although by no means complete) provides adequate pointers to investigate and study the above issues.

ACKNOWLEDGMENTS

The present chapter is the result of collaborative works with D.S. Laila, M.C. Laiou, F. Mazenc, C. Prieur, E. Valtolina, and W. Schaufelberger.

REFERENCES

1. Edelen, D. G. B., *Lagrangian Mechanics of Nonconservative Nonholonomic Systems*. Noordhoff International Publishing, Leyden, 1977.
2. Brockett, R. W., Asymptotic stability and feedback stabilization, in *Differential Geometry Control Theory*, Birkhauser, Boston, p. 181, 1983.
3. Kolmanovsky, I. and McClamroch, N. H., Developments in nonholonomic control problems, *IEEE Control Systems Magazine*, 15, 20, 1995.
4. Morin, P., Pomet, J.-B., and Samson, C., Developments in time-varying feedback stabilization of nonlinear systems, in *Symposium on Nonlinear Control System Design*, Twente, NL, 1998.
5. Lafferriere, G. and Sussmann, H. J., *A Differential Geometric Approach to Motion Planning*, Nonholonomic Motion Planning, Kluwer Academic, Dordrecht, 1992.
6. Brockett, R. W. and Dai, L., *Non-Holonomic Kinematics and the Role of Elliptic Functions in Constructive Controllability*, Nonholonomic Motion Planning. Kluwer Academic, Dordrecht, 1992.
7. Murray, R. M. and Sastry, S. S., Nonholonomic motion planning: steering using sinusoids, *IEEE Transactions on Automatic Control*, 38, 700, 1993.
8. Lucibello, P. and Oriolo, G., Robust stabilization via iterative state steering with an application to chained form systems, *Automatica*, 37, 71, 2001.
9. Samson, C., Velocity and torque feedback control of a nonholonomic cart, in *International Workshop on Adaptive and Nonlinear Control*. Grenoble, FR, p. 125, 1990.
10. Pomet, J.-B., Explicit design of time-varying stabilizing control laws for a class of controllable systems without drift, *Systems and Control Letters*, 18, 147, 1992.
11. Coron, J. M., Global asymptotic stabilization for controllable systems without drift, *Mathematics of Control Signals and Systems*, 295, 1992.
12. M'Closkey, R. T. and Murray, R. M., Exponential stabilization of driftless nonlinear control systems using homogeneous feedback. *IEEE Transactions on Automatic Control*, 42, 614, 1997.

13. Samson, C., Control of chained systems. Application to path following and time-varying point-stabilization of mobile robots, *IEEE Transactions on Automatic Control*, 40, 64, 1995.

14. Coron, J. M., Stabilizing time-varying feedback, in *Symposium on Non-linear Control System Design*, Lake Tahoe, CA, p. 176, 1995.

15. Kolmanovsky, I. and McClamroch, N. H., Application of integrator backstepping to nonholonomic control problems, in *Symposium on Nonlinear Control System Design*, Lake Tahoe, CA, p. 753, 1995.

16. Morin, P. and Samson, C., Practical stabilization of driftless systems on Lie groups: the transverse function approach, *IEEE Transactions on Automatic Control*, 48, 1496, 2003.

17. Bloch, A. M., Reyhanoglu, M., and McClamroch, N. H., Control and stabilization of nonholonomic dynamic systems, *IEEE Transactions on Automatic Control*, 37, 1746, 1992.

18. Canudas de Wit, C. and Sørdalen, O. J., Example of piecewise smooth stabilization of driftless NL systems with less inputs than states, in *Symposium on Nonlinear Control System Design*, Bordeaux, FR, p. 57, 1992.

19. Khennouf, H. and Canudas de Wit, C., On the construction of stabilizing discontinuous controllers for nonholonomic systems, in *Symposium on Nonlinear Control System Design*, Lake Tahoe, CA, p. 747, 1995.

20. Astolfi, A., Discontinuous control of nonholonomic systems, *Systems and Control Letters*, 27, 37, 1996.

21. Laiou, M. C. and Astolfi, A., Discontinuous control of high-order generalized chained systems, *Systems and Control Letters*, 37, 309, 1999.

22. Sørdalen, O. J. and Egeland, O., Exponential stabilization of nonholonomic chained systems, *IEEE Transactions on Automatic Control*, 40, 35, 1995.

23. Godhavn, J. M. and Egeland, O., A Lyapunov approach to exponential stabilization of nonholonomic systems in power form, *IEEE Transactions on Automatic Control*, 42, 1028, 1997.

24. Hespanha, J. P. and Morse, A. S., Stabilization of nonholonomic integrators via logic-based switching, *Automatica*, 35, 385, 1998.

25. Casagrande, D., Astolfi, A., and Parisini, T., Control of nonholonomic systems: a simple stabilizing time-switching strategy, in *16th IFAC World Congress*, Praha, CR, 2005.

26. Nam, T. K. et al., Control of high order chained form systems, in *Proceedings of 41st SICE Annual Conference*, Osaka, TP, p. 2196, 2002.

27. Monaco, S. and Normand-Cyrot, D., An introduction to motion planning under multirate digital control, in *Proceedings of 31st IEEE Conference on Decision Control*, Tucson, AZ, p. 1780, 1992.

28. Tilbury, D. M. and Chelouah, A., Steering a three-input nonholonomic system using multi-rate controls, in *2nd European Control Conference*, p. 1428, 1993.

29. Lizarraga, D. A., Morin, P., and Samson, C., Non-robustness of continuous homogeneous stabilizers for affine control systems, in *Proceedings of 38th IEEE Conference on Decision Control*, Phoenix, AZ, p. 855, 1999.

30. Lizarraga, D. A., Anneke, N., and Nijmeijer, H., Robust exponential stabilization for the extended chained form via hybrid control, in *Proceedings of 41st IEEE Conference on Decision Control*, Las Vegas, NV, p. 2798, 2002.

31. Morin, P. and Samson, C., Exponential stabilization of nonlinear driftless systems with robustness to unmodelled dynamics, *Control, Optimization and Calculus Variations*, 4, 1, 1999.

32. Prieur, C. and Astolfi, A., Robust stabilization of chained systems via hybrid control, *IEEE Transactions on Automatic Control*, 48, 1768, 2003.

33. Valtolina, E. and Astolfi, A., Local robust regulation of chained systems, *Systems and Control Letters*, 49, 231, 2003.

34. Maini, M. et al., On the robust stabilization of chained systems by continuous feedback, in *Proceedings of Conference on Decision Control IEEE*, Phoenix, AZ, p. 3472, 1999.

35. Ryan, E. P., On Brockett's condition for smooth stabilizability and its necessity in a context of nonsmooth feedback, *SIAM Journal of Control and Optimization*, 32, 1597, 1994.

36. Bacciotti, A., *Local Stabilizability of Nonlinear Control Systems*, Series on Advances in Mathematics for Applied Sciences, World Scientific, Singapore, 1991.

37. Sontag, E. D., Feedback stabilization of nonlinear systems, in *Robust Control of Linear Systems and Nonlinear Control*, M.A. Kaashoek, J.H. van Schuppen, and A.C.M. Ran (eds). Birkhäuser, Basel, Boston, p. 61, 1990.

38. Sontag, E. D., On characterizations of the input-to-state stability property, *Systems and Control Letters*, 24, 351, 1995.

39. Jiang, Z. P. and Wang, Y., Input-to-state stability for discrete-time nonlinear systems, *Automatica*, 37, 857, 2001.

40. Nešić, D. and Laila, D. S., A note on input-to-state stabilization for nonlinear sampled-data systems, *IEEE Transactions on Automatic Control*, 47, 1153, 2002.

41. Stuart, A. M. and Humphries, A. R., *Dynamical Systems and Numerical Analysis*, Cambridge University Press, New York, 1996.

42. Laila, D. S. and Astolfi, A., Input-to-state stability for discrete-time time-varying systems with applications to robust stabilization of systems in power form, *Automatica*, 41, 1891–1903, 2005.

43. Tsiotras, P., Corless, M. J., and Longuski, J. M., A novel approach to the attitude control of axisymmetric spacecraft, *Automatica*, 31, 1099, 1995.

44. Isidori, A., *Nonlinear Control Systems*, 2nd ed., Springer-Verlag, Berlin, 1989.

45. Arnold, V. I., *Geometrical Methods in the Theory of Ordinary Differential Equations*, 2nd ed., Springer-Verlag, New York, 1987.

46. Astolfi, A. and Schaufelberger, W., State and output feedback stabilization of multiple chained systems with discontinuous control, *Systems and Control Letters*, 32, 49, 1997.

47. Luo, J. and Tsiotras, P., Control design for chained-form systems with bounded inputs, *Systems and Control Letters*, 29, 123, 2000.

48. Sun, Z. and Ge, S. S., Nonregular feedback linearization: a nonsmooth approach, *IEEE Transactions on Automatic Control*, 48, 1772, 1998

49. Date, H. et al., Simultaneous control of position and orientation for ball-plate manipulation problem based on time-state control form, *IEEE Transactions on Robotics and Automatics*, 20, 465, 2004.

50. Pomet, J.-B. and Samson, C., Time-varying exponential stabilization of nonholonomic systems in power form, in *Proceedings of IFAC Symposium of Robust Control Design*, p. 447, 1994.

51. Alexander, J. C. and Maddocks, J. H., On the maneuvering of vehicles, *SIAM Journal of Applied Mathematics*, 48, 38, 1993.

52. Teel, A. R., Murray, R. M., and Walsh, G., Nonholonomic control systems: from steering to stabilization with sinusoids, in *31st Conference on Decision and Control*, Tucson, AZ, p. 1603, 1992.

53. Astolfi, A., Laiou, M. C., and Mazenc, F., New results and examples on a class of discontinuous controllers, in *European Control Conference*, 1999.

54. Bloch, A. M. and Drakunov, S. V., Stabilization and tracking in the non-holonomic integrator via sliding modes, *Systems and Control Letters*, 29, 91, 1996.

55. Reyhanoglu, M., Exponential stabilization of an underactuated autonomous surface vessel, *Automatica*, 33, 2249, 1997.

56. Escobar, G., Ortega, R., and Reyhanoglu, M., Regulation and tracking of the nonholonomic double integrator: a field-oriented control approach, *Automatica*, 34, 125 ,1998.

57. Jiang, Z. P., Robust exponential regulation of nonholonomic systems with uncertainties, *Automatica*, 36, 189, 2000.

58. Morin, P. and Samson, C., Robust stabilization of driftless systems with hybrid open-loop/feedback control, in *American Control Conference*, Chicago, IL, 2000.

59. Marchand, N. and Alamir, M., Discontinuous exponential stabilization of chained form systems, *Automatica*, 39, 343, 2003.

60. Prieur, C., A robust globally asymptotically stabilizing feedback: the example of the Artstein's circles, In *Nonlinear Control in the Year 2002, LNCIS 258*, A. Isidori et al. (eds), p. 279, Springer-Verlag, London, 2000.

61. Prieur, C., Uniting local and global controllers with robustness to vanishing noise, *Mathematics Control Signals and Systems*, 14, 143, 2001.

62. Nešić, D. and Teel, A. R., A framework for stabilization of nonlinear sampled-data systems based on their approximate discrete-time models, *IEEE Transactions on Automatic Control*, 49, 1103, 2004.

BIOGRAPHY

Alessandro Astolfi was born in Rome, Italy, in 1967. He graduated in electrical engineering from the University of Rome in 1991. In 1992, he joined ETH-Zurich where he obtained an M.Sc. in information theory in 1995 and the Ph.D. degree with Medal of Honour in 1995 with a thesis on discontinuous stabilization of nonholonomic systems. In 1996 he was awarded a PhD from the University of Rome for his work on nonlinear robust control.

Since 1996, he has been with the Electrical and Electronic Engineering Department of Imperial College, London (UK), where he is currently professor in nonlinear control theory. In 1998, he was appointed associate professor at the Department of Electronics and Information of the Politecnico of Milano (Italy).

He has been a visiting lecturer in "Nonlinear Control" in several universities, including ETH-Zurich (1995–1996); Terza University of Rome (1996); Rice University, Houston (1999); Kepler University, Linz (2000); SUPELEC, Paris (2001).

His research interests are focused on mathematical control theory and control applications, with special emphasis for the problems of discontinuous stabilization, robust control, and adaptive control.

He is the author of more than 40 journal papers, 15 book chapters, and over 100 papers in refereed conference proceedings. He is co-editor of a book *Modeling and Control of Mechanical Systems.*

He is associate editor of *Systems and Control Letters, Automatica,* the *International Journal of Control,* the *European Journal of Control, and the IEEE Transactions Automatic Control.* He has also served in the IPC of various international conferences.

6 Adaptive Neural-Fuzzy Control of Nonholonomic Mobile Robots

Fan Hong, Shuzhi Sam Ge, Frank L. Lewis, and Tong Heng Lee

CONTENTS

6.1 INTRODUCTION

In recent years, control and stabilization of mechanical systems with nonholonomic constraints has been an area of active research. Due to Brockett's theorem [1], it is well known that nonholonomic systems with restricted mobility cannot be stabilized to a desired configuration (or posture) via differentiable, or even continuous, pure-state feedback, although it is controllable. A number of approaches including (i) discontinuous time-invariant stabilization [2,3], (ii) time-varying stabilization [4], and (iii) hybrid stabilization [5] have been proposed for the problem (see the Survey Paper 6 and the references therein for more details).

For the controller design of nonholonomic systems, there are efforts focused on the kinematic control problem, where the systems are represented by their kinematic models and the velocity acts as the control input. One commonly

229

used approach for the controller design of nonholonomic systems is to convert, with appropriate state and input transformations, the original systems into some canonical forms for which the design can be carried out more easily [7,8]. Using the special algebraic structures of the canonical forms, various feedback strategies have been proposed to stabilize nonholonomic systems in the literature [9–12]. The majority of these constructive methods have been developed based on exact system models. However, it is more practical to design the controller against possible existence of modeling errors and external disturbances. A hybrid feedback algorithm based on supervisory adaptive control was presented to globally asymptotically stabilize a wheeled mobile robot [13]. Output feedback tracking and regulation controllers were presented in Reference 14 for practical wheeled mobile robots. Robustness issues with regard to disturbances in the kinematic model have also been investigated.

In practice, however, it is more realistic to formulate the nonholonomic system control problem at the dynamic level, where the torque and force are taken as the control inputs. In actual applications, however, exact knowledge of the robot dynamics is almost impossible. Adaptive control strategies were proposed to stabilize dynamic nonholonomic systems [15]. Sliding mode control was applied to guarantee the uniform ultimate boundedness of tracking error in Reference 16. In Reference 17, stable adaptive control was investigated for dynamic nonholonomic chained systems with uncertain constant parameters. Using geometric phase as a basis, control of Caplygin dynamical systems was studied in Reference 18, and the closed-loop system was proved to achieve the desired local asymptotic stabilization of a single equilibrium solution. Thanks to the research in References 19 and 20, the motion control part of the problem can be reduced to a problem similar to the free-motion control of a robot with less degrees of freedom. Robust adaptive motion controllers were proposed in References 21 and 22 using the linear-in-the-parameter property of the system dynamics and the bound of the robot parameters.

The difficulty in precise dynamic modeling has invoked the development of approximator-based control approaches, using Lyapunov synthesis for the general nonlinear system [23–28]. Neural networks (NNs) are well known for its ability to extend adaptive control techniques to systems in nonlinear-in-the-parameters. The universal approximation properties of NNs in the feedback control systems successfully avoid the use of regression matrices, and assumptions such as certainty equivalence. It requires no persistence of excitation conditions by using the robustifying terms. For a comprehensive study of the subject, readers are referred to Reference 29 and the references therein. For fuzzy logic systems, it provides natural and linguistic representation of human's (or expert's) knowledge, reasoning about vague rules that describe the imprecise and qualitative relationship between the system's input and output. The combination of NNs and fuzzy logic systems can overcome some of the individual weaknesses and offer some appealing features. It offers an architecture that uses fuzzy algorithms to represent the knowledge in a natural and interpretable manner, while preserving the learning ability of NNs as well as the associated convergence and stability. The neuro-fuzzy (NF) system is a NN-based fuzzy logic control and decision system, and is suitable for online systems identification and control.

For adaptive NF control system design, the parameterized NF approximators are generally expressed as a series of the commonly used radial basis function (RBF) because of its nice approximation properties, that is, $y = \sum_j w_j \phi(\sigma_j, \|x - c_j\|)$, where w_j is the connection weight, and c_j and σ_j are the center and width respectively that decide the shape of the function ϕ. The major challenge in the RBF approximation problem lies in the selection of the receptive center and width, that is, c_j and σ_j as they both appear nonlinearly. In general, there are three kinds of methods to determine c_j and σ_j. The first is the grid-type partition method, which uses a grid partitioning of the multidimensional space and defines a number of fuzzy sets or nodes for each variable. This is the most intuitive approach but the problem is the exponential growth of fuzzy rules or nodes in relation to the dimension of the input space. The second kind is the clustering algorithm, such as fuzzy C-means (FCM) [30] and the nearest-neighborhood cluster algorithm [31]. These methods are found to be useful in choosing parameters, but require off-line learning. In addition, the gradient descent method is usually employed for fine tuning the parameters c_j and σ_j by clustering algorithm so that the approximation accuracy is improved. The last type consists of optimization approaches such as genetic algorithms (GA). However, the problem with either the gradient descent method or GA is that the learning and the adaptation speeds are slow. On the other hand, most of the adaptive control schemes using RBF as an approximator only consider the updating law of weights w_j to simplify the design [32]. However, it is obvious that the parameters, c_j and σ_j are important in capturing the fast-changing system dynamics, reducing the approximation error, and improving the control performance [33]. An adaptive scheme of tuning both the weights w_j and the center and width, c_j and σ_j, was presented in Reference 34.

Motivated by previous works on the control of nonholonomic constrained mechanical systems and the approximation-based adaptive control of nonlinear systems, adaptive NF control is developed in this chapter for nonholonomic constrained mobile robotic systems using Lyapunov stability analysis in a unified procedure. Despite the differences between the NNs and fuzzy logic systems, they actually can be unified at the level of the universal function approximator, termed as the NF networks which are multilayer feedforward networks that integrate the TSK-type fuzzy system and RBF NN into a connectionist structure. Indeed, for simple systems, the rules are fairly easy to derive with physical insight, however, they become unreasonably difficult for systems with strong nonlinear couplings yet without a good physical understanding. Because of the difficulty in deriving the rules in fuzzy systems for systems with little physical insights, we present the adaptive laws to design the outputs of the "rules" numerically using adaptive (NN) control techniques. It is shown that the motion tracking error converges to zero, the force tracking error is uniformly bounded, and the closed-loop stability is guaranteed without the requirement of the PE condition.

The rest of the chapter is organized as follows. The dynamics of mobile robot systems subject to nonholonomic constraints are briefly described in Section 6.2. Multilayer NF systems as the key design tool are introduced in Section 6.3. The main results of the adaptive NF control design are presented in Section 6.4, and a simulation example is provided in Section 6.5. Concluding remarks are given in Section 6.6.

6.2 DYNAMICS OF NONHOLONOMIC MOBILE ROBOTS

In general, a nonholonomic mobile robot system having an n-dimensional configuration space with generalized coordinates $q = [q_1, \ldots, q_n]^T$ and subject to $(n - m)$ constraints can be described by [35]

$$M(q)\ddot{q} + C(q, \dot{q})\dot{q} + G(q) = B(q)\tau + f + \tau_d \tag{6.1}$$

where $M(q) \in \mathbb{R}^{n \times n}$ is the inertia matrix and $M(q)^T = M(q) > 0$, $C(q, \dot{q}) \in \mathbb{R}^{n \times n}$ is the centripetal and coriolis matrix, $G(q) \in \mathbb{R}^n$ is the gravitation force vector, $B(q) \in \mathbb{R}^{n \times r}$ is the full-rank input transformation matrix and is assumed to be known, as it is a function of fixed geometry of the system, $\tau \in \mathbb{R}^r$ is the input vector of forces and torques, $f \in \mathbb{R}^n$ is the constrained force vector, and $\tau_d \in \mathbb{R}^n$ denotes bounded unknown disturbances including unstructured unmodeled dynamics. The dynamic system (6.1) has the following properties [32,36]:

Property 6.1 *Matrices $M(q)$, $G(q)$ are uniformly bounded and uniformly continuous if q is uniformly bounded and continuous, respectively. Matrix $C(q, \dot{q})$ is uniformly bounded and uniformly continuous if $\dot{q}$ is uniformly bounded and continuous.*

Property 6.2 *Matrix $\dot{M} - 2C$ is skew-symmetric, that is, $x^T(\dot{M} - 2C)x = 0$, $\forall x \neq 0$.*

When the system is subjected to nonholonomic constraints, the $(n - m)$ nonintegrable and independent velocity constraints can be expressed as

$$J(q)\dot{q} = 0 \tag{6.2}$$

where $J(q) \in \mathbb{R}^{(n-m) \times n}$ is the matrix associated with the constraint.

The constraint (6.2) is referred to as the classical nonholonomic constraint when it is not integrable. In the chapter, constraint (6.2) is assumed to be completely nonholonomic and exactly known. The effect of the constraints can be viewed as restricting the dynamics on the manifold Ω_{nh} as

$$\Omega_{nh} = \{(q, \dot{q})|J(q)\dot{q} = 0\}$$

It is noted that since the nonholonomic constraint (6.2) is nonintegrable, there is no explicit restriction on the values of the configuration variables.

Based on the nonholonomic constraint (6.2), the generalized constraint forces in the mechanical system (6.1) can be given by

$$f = J^{T}(q)\lambda \tag{6.3}$$

where $\lambda \in \mathbb{R}^{n-m}$ is known as friction force on the contact point between the rigid body and environmental surfaces.

Since $J(q) \in \mathbb{R}^{(n-m)\times n}$, it is always possible to find an m rank matrix $R(q) \in \mathbb{R}^{n\times m}$ formed by a set of smooth and linearly independent vector fields spanning the null space of $J(q)$, that is,

$$R^{T}(q)J^{T}(q) = 0 \tag{6.4}$$

Denote $R(q) = [r_1(q), \ldots, r_m(q)]$ and define an auxiliary time function $\dot{z}(t) = [\dot{z}_1(t), \ldots, \dot{z}_m(t)]^{T} \in \mathbb{R}^m$ such that

$$\dot{q} = R(q)\dot{z}(t) = r_1(q)\dot{z}_1(t) + \cdots + r_m(q)\dot{z}_m(t) \tag{6.5}$$

Equation (6.5) is the so-called kinematic model of nonholonomic systems in the literature. Usually, $\dot{z}(t)$ has physical meaning, consisting of the linear velocity v and the angular velocity ω, that is, $\dot{z}(t) = [v \ \omega]^{T}$. Equation (6.5) describes the kinematic relationship between the motion vector $q(t)$ and the velocity vector $\dot{z}(t)$.

Differentiating (6.5) yields

$$\ddot{q} = \dot{R}(q)\dot{z} + R(q)\ddot{z} \tag{6.6}$$

From (6.5), $\dot{z}$ can be obtained from q and $\dot{q}$ as

$$\dot{z} = [R^{T}(q)R(q)]^{-1}R^{T}(q)\dot{q} \tag{6.7}$$

The dynamic equation (6.1), which satisfies the nonholonomic constraint (6.2), can be rewritten in terms of the internal state variable $\dot{z}$ as

$$M(q)R(q)\ddot{z} + [M(q)\dot{R}(q) + C(q,\dot{q})R(q)]\dot{z} + G(q) = B(q)\tau + J^{T}(q)\lambda + \tau_d \tag{6.8}$$

Substituting (6.5) and (6.6) into (6.1), and then premultiplying (6.1) by $R^{T}(q)$, the constraint matrix $J^{T}(q)\lambda$ can be eliminated by virtue of (6.4). As a consequence, we have the transformed nonholonomic system

$$\dot{q} = R(q)\dot{z} = r_1(q)\dot{z}_1 + \cdots + r_m(q)\dot{z}_m \tag{6.9}$$

$$M_1(q)\ddot{z} + C_1(q,\dot{q})\dot{z} + G_1(q) = B_1(q)\tau + \tau_{d1} \tag{6.10}$$

where

$$M_1(q) = R^{T}M(q)R$$

$$C_1(q,\dot{q}) = R^{T}[M(q)\dot{R} + C(q,\dot{q})R]$$

$$G_1(q) = R^{T}G(q)$$

$$B_1(q) = R^{T}B(q)$$

$$\tau_{d1} = R^{T}\tau_d$$

which is more appropriate for the controller design as the constraint λ has been eliminated from the dynamic equation.

Exploiting the structure of the dynamic equation (6.10), some properties are listed as follows.

Property 6.3 *Matrix $D_1(q)$ is symmetric and positive-definite.*

Property 6.4 *Matrix $\dot{D}_1(q) - 2C_1(q,\dot{q})$ is skew-symmetric.*

Property 6.5 *$D(q)$, $G(q)$, $J(q)$, and $R(q)$ are bounded and continuous if z is bounded and uniformly continuous. $C(q,\dot{q})$ and $\dot{R}(q)$ are bounded if $\dot{z}$ is bounded. $C(q,\dot{q})$ and $\dot{R}(q)$ are uniformly continuous if $\dot{z}$ is uniformly continuous [37].*

In the following, the kinematic nonholonomic subsystem (6.5) is converted into the chained canonical form. The nonholonomic chained system considered in this chapter is the m-input, $(m-1)$-chain, single generator chained form given by Walsh and Bushnell [38]

$$\dot{x}_1 = u_1$$

$$\dot{x}_{j,i} = u_1 x_{j,i+1} \quad (2 \leq i \leq n_j - 1)(1 \leq j \leq m - 1) \qquad (6.11)$$

$$\dot{x}_{j,n_j} = u_{j+1}$$

Note that in Equation (6.11), $X = [x_1, X_2, \ldots, X_m]^T \in \mathbb{R}^n$ with $X_j = [x_{j-1,2}, \ldots, x_{j-1,n_{j-1}}]$ $(2 \leq j \leq m)$ are the states and $u = [u_1, u_2, \ldots, u_m]^T$ are the inputs of the kinematic subsystem.

The class of nonholonomic systems in chained form was first introduced in Reference 7 and has been studied as a benchmark example in the literature. It is the most important canonical form that is commonly used in the study of nonholonomic control systems. The necessary and sufficient conditions for transforming system (6.5) into the chained form are given in Reference 39. Theoretical challenges and practical interests have provided substantial motivation for the extensive study of nonholonomic systems in chained form. The following assumption is made.

Assumption 6.1 *The kinematic model of the nonholonomic system given by (6.5) can be converted into the chained form (6.11) by some diffeomorphic coordinate transformation $X = T_1(q)$ and state feedback $v = T_2(q)u$ where u is a new control input.*

The existence and construction of these systems have been established in References 38 and 40. For the notations on the differential geometry used below, readers are referred to Reference 41.

Proposition 6.1 *Consider the drift-free nonholonomic system*

$$\dot{q} = r_1(q)\dot{z}_1 + \cdots + r_m(q)\dot{z}_m$$

where $r_i(q)$ are smooth, linearly independent input vector fields. There exist state transformation $X = T_1(q)$ and feedback $\dot{z} = T_2(q)u$ on some open set $\mathbb{U} \subset \mathbb{R}^n$ to transform the system into an $(m-1)$-chain, single-generator chained form, if and only if there exists a basis $f_1, \ldots, f_m$ for $\Delta_0 := \text{span}\{r_1, \ldots, r_m\}$

which has the form

$$f_1 = (\partial/\partial q_1) + \sum_{i=2}^{n} f_1^i(q)\partial/\partial q_i$$

$$f_j = \sum_{i=1}^{n} f_j^i(q)\partial/\partial q_i, \quad 2 \le j \le m$$

such that the distributions

$$G_j = \text{span}\{\text{ad}_{f_1}^i f_2, \ldots, \text{ad}_{f_1}^i f_m : 0 \le i \le j\}, \quad 0 \le j \le n-1$$

have constant dimension on $\mathbb{U}$ *and are all involutive, and* G_{n-1} *has dimension* $n-1$ *on* $\mathbb{U}$ *[38,40].*

For a two-input controllable system, a constructive method was reproduced in Reference 10 and it is given here for completeness. Consider

$$\dot{q} = r_1(q)\dot{z}_1 + r_2(q)\dot{z}_2 \tag{6.12}$$

where $r_1(q)$, $r_2(q)$ are linearly independent and smooth, $q \in \mathbb{R}^n$, and $\dot{z} = [\dot{z}_1, \dot{z}_2]^{\mathsf{T}}$.

Define

$$\Delta_0 := \text{span}\{r_1, r_2, \text{ad}_{r_1} r_2, \ldots, \text{ad}_{r_1}^{n-2} r_2\}$$

$$\Delta_1 := \text{span}\{r_2, \text{ad}_{r_1} r_2, \ldots, \text{ad}_{r_1}^{n-2} r_2\}$$

$$\Delta_2 := \text{span}\{r_2, \text{ad}_{r_1} r_2, \ldots, \text{ad}_{r_1}^{n-3} r_2\}$$

If $\Delta_0(q) = \mathbb{R}^n$, $\forall q \in \mathbb{U}$ (where $\mathbb{U}$ is some open set of $\mathbb{R}^n$), Δ_1 and Δ_2 are involutive on $\mathbb{U}$, and $r_1(q)$ satisfies $[r_1, \Delta_1] \subset \Delta_1$, then there exist two independent functions $h_1 : \mathbb{U} \to \mathbb{R}$ and $h_2 : \mathbb{U} \to \mathbb{R}$ which satisfy the following relationships:

$$dh_1 \cdot \Delta_1 = 0, \quad dh_1 \cdot r_1 = 1$$

$$dh_2 \cdot \Delta_2 = 0, \quad dh_2 \cdot \text{ad}_{r_1}^{n-2} r_2 \ne 0$$

Let $T_1(q) : q \to X$ as

$$x_1 = h_1$$

$$x_2 = L_{r_1}^{n-2} h_2$$

$$\vdots$$

$$x_{n-1} = L_{r_1} h_2$$

$$x_n = h_2$$

It may be verified that $T_1(q)$ is a valid change of coordinates by evaluating the Jacobian of $T_1(q)$ at the origin.

Since $L_{r_2} L_{r_1}^{n-2} h_2 \neq 0$, let $T_2(q) : \dot{z} \to u$ as

$$\dot{z}_1 := u_1$$

$$\dot{z}_2 := \frac{1}{L_{r_2} L_{r_1}^{n-2} h_2} [u_2 - (L_{r_1}^{n-1} h_2) u_1]$$

Then, the local coordinate transformation $X = T_1(q)$ and state feedback $\dot{z} = T_2(q)u$ render system (6.12) into the chained form

$$\dot{x}_1 = u_1$$

$$\dot{x}_2 = u_2$$

$$\dot{x}_3 = x_2 u_1$$

$$\vdots$$

$$\dot{x}_n = x_{n-1} u_1$$

Remark 6.1 *Under certain conditions which has been stated in Proposition 6.1, the kinematic model (6.5) can be converted into a chained form driven by integrators.*

6.3 MULTI-LAYER NF SYSTEMS

Despite the differences between the NNs and fuzzy logic systems, they actually can be unified at the level of the universal function approximator, which are multilayer feedforward networks that integrate the TSK-type fuzzy system and RBF NN into a connectionist structure.

Typically, fuzzy logic systems are rule-based systems, which consists of the fuzzifier, the fuzzy rule base, the fuzzy inference engine, and the defuzzifier.

The purpose of the fuzzifier is to provide scale mapping of the crisp input to corresponding linguistic forms noted as labels of fuzzy set. The fuzzy rule base stores knowledge base for linguistic data and is expressed as a collection of fuzzy IF–THEN rules. The typical fuzzy rule used in the Takagi–Sugeno–Kang (TSK) model [42] is in the following form:

$$R^l: \text{IF } z_1 \text{ is } F_1^l \text{ AND } z_2 \text{ is } F_2^l \cdots \text{ AND } z_n \text{ is } F_n^l$$

$$\text{THEN } y^l = k_0^l + k_1^l z_1 + \cdots + k_n^l z_n$$

where F_i^l ($i = 1, 2, \ldots, n$) are fuzzy sets, k_j^l ($j = 0, 1, \ldots, n$) are real-valued parameters, $z = [z_1, z_2, \ldots, z_n]^T$ is the system input, y^l is the system output due to rule R^l, and $l = 1, 2, \ldots, N$. For the zero-order TSK-fuzzy system, we have $y^l = k_0^l$. The fuzzy inference engine is the kernel of the fuzzy system and uses the fuzzy IF–THEN rules to determine a mapping from the input universe to the output universe based on fuzzy logic policies. The role of the defuzzifier is the scale mapping of the linguistic value to a corresponding crisp output value. For simple systems, the rules are fairly easy to derive with physical insight. However, they become unreasonably difficult for systems with strong nonlinear couplings yet without a good physical understanding.

On the other hand, the NNs can build up a very nice mapping between system's inputs and outputs. Due to its great learning capability, it can be used to approximate any continuous function to any desired accuracy. Despite the differences between the NNs and fuzzy logic systems, they can, in fact, be unified at the level of the universal function approximator which integrates the TSK-type fuzzy system and RBF NN into a connectionist structure. Nodes in the first layer are called input linguistic nodes and corresponds to input variables. These nodes only transmit input values to the next layer directly. Nodes in the second layer play the role of membership functions specifying the degree to which an input value belongs to a fuzzy set. The nodes in the third layer are called rule nodes which represent fuzzy rules. The fourth layer is the output layer. The links in the third layer act as the precondition of fuzzy rules and the links in the fourth layer act as the consequence of fuzzy rules.

The output of the whole NF system is then given by

$$y(x) = \sum_{l=1}^{n_r} w_l \left[\frac{\prod_{i=1}^{n_i} \mu_{A_i^l}(x_i)}{\sum_{k=1}^{n_r} \prod_{i=1}^{n_i} \mu_{A_i^k}(x_i)} \right] \tag{6.13}$$

where $x = [x_1, x_2, \ldots, x_{n_i}]^T$, $\mu_{A_i^k}(x_i)$ is the membership function of linguistic variable x_i with

$$\mu_{A_i^k}(x_i) = \exp\left[-\frac{(x_i - c_{ik})^2}{\sigma_{ik}^2} \right] \tag{6.14}$$

For clarity, let us define the weight vector and fuzzy basis function vector respectively as

$$W = [w_1, w_2, \ldots, w_{n_r}]^{\mathrm{T}}$$

$$S(x, c, \sigma) = [s_1, s_2, \ldots, s_{n_r}]^{\mathrm{T}}$$

where $s_l = \prod_{i=1}^{n_i} \mu_{A_i^l}(x_i) / [\sum_{k=1}^{n_r} \prod_{i=1}^{n_i} \mu_{A_i^k}(x_i)]$, $c = [c_1^{\mathrm{T}}, c_2^{\mathrm{T}}, \ldots, c_{n_r}^{\mathrm{T}}]^{\mathrm{T}}$, and $\sigma = [\sigma_1^{\mathrm{T}}, \sigma_2^{\mathrm{T}}, \ldots, \sigma_{n_r}^{\mathrm{T}}]^{\mathrm{T}}$. Then, Equation (6.13) can be represented as

$$y = W^{\mathrm{T}} S(x, c, \sigma) \qquad (6.15)$$

Remark 6.2   *For Equation (6.15), W and $S(x, c, \sigma)$ are the weights and the (normalized) basis functions in NN terminology, while they are the outputs of the rules and the weighted firing strength in fuzzy logic terminology. Because of the difficulty in deriving the rules in fuzzy systems for systems with little physical insights, we would hereby like to present the adaptive laws to design the outputs of the "rules" numerically using adaptive (NN) control techniques.*

It has been proven that, if the number of the fuzzy rules n_r is sufficiently large, a fuzzy logic system (6.15) is capable of uniformly approximating any given real continuous function, $h(x)$, over a compact set $\Omega_x \subset \mathbb{R}^{n_i}$ to any arbitrary degree of accuracy in the form

$$h(x) = W^{*\mathrm{T}} S(x, c^*, \sigma^*) + \epsilon(x), \quad \forall x \in \Omega_x \subset \mathbb{R}^{n_i} \qquad (6.16)$$

where W^*, c^*, and σ^* are the ideal constant vectors, and $\epsilon(x)$ is the approximation error. The following assumption is made for W^*, c^*, σ^*, and $\epsilon(x)$.

Assumption 6.2   *The ideal NF vectors W^*, c^*, σ^*, and the NF approximation error are bounded over the compact set, that is,*

$$\|W^*\| \leq w_m, \quad \|c^*\| \leq c_m, \quad \|\sigma^*\| \leq \sigma_m, \quad |\epsilon(x)| \leq \epsilon^*$$

$\forall x \in \Omega_x$ with w_m, c_m, σ_m, and ϵ^* being unknown positive constants.

Remark 6.3   *The optimal weight vector W^*, c^*, and σ^* in (6.16) is an "artificial" quantity required only for analytical purposes. Typically, W^*, c^*, and σ^* are chosen as the value of W that minimizes $\epsilon(x)$ for all $x \in \Omega_x \subset \mathbb{R}^{n_i}$, that is,*

$$(W^*, c^*, \sigma^*) := \arg \min_{W, c, \sigma} \left\{ \sup_{x \in \Omega_x} |h(x) - W^{\mathrm{T}} S(x, c, \sigma)| \right\}$$

Remark 6.4   *The approximation error $\epsilon(x)$, is a critical quantity and can be reduced by increasing the number of the fuzzy rules n_r. According to the universal approximation theorem, it can be made as small as possible if the number of fuzzy rules n_r is sufficiently large.*

From the analysis given above, we see that the system uncertainties are converted to the estimation of unknown parameters W^*, c^*, σ^*, and unknown bounds ϵ^*.

As the ideal vectors/constants W^*, c^*, σ^*, and ϵ^* are usually unknown, we use their estimates $\hat{W}$, $\hat{c}$, $\hat{\sigma}$, and $\hat{\epsilon}$ instead. The following lemma gives the properties of the approximation errors $\hat{W}_0^\mathrm{T} S(x,\hat{c},\hat{\sigma}) - W^{*\mathrm{T}} S(x,c^*,\sigma^*)$. The definition of induced norm of matrices is given here first.

Definition 6.1   *For an $m \times n$ matrix $A = \{a_{ij}\}$, the induced p-norm, $p = 1, 2$ of A is defined as*

$$\|A\|_1 = \max_j \left\{ \sum_{i=1}^{m} |a_{ij}| \right\} \quad \text{column sum}$$

$$\|A\|_2 = \max_i \left\{ \sqrt{\lambda_i(A^\mathrm{T} A)} \right\}$$

Usually, $\|A\|_2$ is abbreviated to $\|A\|$.

The Frobenius norm is defined as the root of the sum of the squares of all elements

$$\|A\|_\mathrm{F}^2 = \sum a_{ij}^2 = \mathrm{tr}(A^\mathrm{T} A)$$

with $\mathrm{tr}(\cdot)$ the matrix trace, that is, sum of diagonal elements.

Lemma 6.1   *[34, 43] The approximation error can be expressed as*

$$\hat{W}^\mathrm{T} S(x,\hat{c},\hat{\sigma}) - W^{*\mathrm{T}} S(x,c^*,\sigma^*)$$
$$= \tilde{W}^\mathrm{T}(\hat{S} - \hat{S}_c' \hat{c} - \hat{S}_\sigma' \hat{\sigma}) + \hat{W}^\mathrm{T}(\hat{S}_c' \tilde{c} + \hat{S}_\sigma' \tilde{\sigma}) + d_u \qquad (6.17)$$

where $\hat{S} = S(x,\hat{c},\hat{\sigma})$, $\tilde{W} = \hat{W} - W^*$, $\tilde{c} = \hat{c} - c^*$, and $\tilde{\sigma} = \hat{\sigma} - \sigma^*$ are defined as approximation error, and $\hat{S}_c' = [\hat{s}_{1c}', \hat{s}_{2c}', \ldots, \hat{s}_{n_r c}']^\mathrm{T} \in \mathbb{R}^{n_r \times (n_i \times n_r)}$ with

$$\hat{s}_{ic}' = \left. \frac{\partial s_i}{\partial c} \right|_{c=\hat{c},\sigma=\hat{\sigma}} \in \mathbb{R}^{(n_i \times n_r) \times 1}, \quad i = 1, \ldots, n_r$$

and $\hat{S}'_\sigma = [\hat{s}'_{1\sigma}, \hat{s}'_{2\sigma}, \ldots, \hat{s}'_{n_r\sigma}]^T \in \mathbb{R}^{n_r \times (n_i \times n_r)}$ *with*

$$\hat{s}'_{i\sigma} = \frac{\partial s_i}{\partial \sigma}\bigg|_{c=\hat{c},\sigma=\hat{\sigma}} \in \mathbb{R}^{(n_i \times n_r) \times 1}, \quad i = 1, \ldots, n_r$$

and the residual term d_u is bounded by

$$|d_u| \leq \|c^*\| \cdot \|\hat{S}'^T_c \hat{W}\| + \|\sigma^*\| \cdot \|\hat{S}'^T_\sigma \hat{W}\| + \|W^*\| \cdot \|\hat{S}'_c \hat{c}\|$$
$$+ \|W^*\| \cdot \|\hat{S}'_\sigma \hat{\sigma}\| + \|W^*\|_1 \tag{6.18}$$

Proof   The Taylor series expansion of $S(x, c^*, \sigma^*)$ with respect to $(x, \hat{c}, \hat{\sigma})$ can be expressed as

$$S(x, c^*, \sigma^*) = S(x, \hat{c}, \hat{\sigma}) - \hat{S}'_c \tilde{c} - \hat{S}'_\sigma \tilde{\sigma} + O(x, \tilde{c}, \tilde{\sigma}) \tag{6.19}$$

where $O(x, \tilde{c}, \tilde{\sigma})$ denotes the sum of the high order terms in the Taylor series expansion.

Using (6.19), we obtain

$$\hat{W}^T S(x, \hat{c}, \hat{\sigma}) - W^{*T} S(x, c^*, \sigma^*)$$
$$= (\tilde{W} + W^*)^T S(x, \hat{c}, \hat{\sigma}) - W^{*T}[S(x, \hat{c}, \hat{\sigma}) - \hat{S}'_c \tilde{c} - \hat{S}'_\sigma \tilde{\sigma} + O(x, \tilde{c}, \tilde{\sigma})]$$
$$= \tilde{W}^T \hat{S} + (\hat{W} - \tilde{W})^T \hat{S}'_c \tilde{c} + (\hat{W} - \tilde{W})^T \hat{S}'_\sigma \tilde{\sigma} - W^{*T} O(x, \tilde{c}, \tilde{\sigma})$$
$$= \tilde{W}^T \hat{S} + \hat{W}^T \hat{S}'_c \tilde{c} - \tilde{W}^T \hat{S}'_c (\hat{c} - c^*) + \hat{W}^T \hat{S}'_\sigma \tilde{\sigma} - \tilde{W}^T \hat{S}'_\sigma (\hat{\sigma} - \sigma^*)$$
$$\quad - W^{*T} O(x, \tilde{c}, \tilde{\sigma})$$
$$= \tilde{W}^T (\hat{S} - \hat{S}'_c \hat{c} - \hat{S}'_\sigma \hat{\sigma}) + \hat{W}^T (\hat{S}'_c \tilde{c} + \hat{S}'_\sigma \tilde{\sigma}) + d_u \tag{6.20}$$

where the residual term d_u is given by

$$d_u = \tilde{W}^T (\hat{S}'_c c^* + \hat{S}'_\sigma \sigma^*) - W^{*T} O(x, \tilde{c}, \tilde{\sigma})$$

Noting that $\tilde{W} = \hat{W} - W^*$, $\tilde{c} = \hat{c} - c^*$, and $\tilde{\sigma} = \hat{\sigma} - \sigma^*$, Equation (6.20) implies that

$$d_u = \hat{W}^T \hat{S} - W^{*T} S^* - (\hat{W} - W^*)^T (\hat{S} - \hat{S}'_c \hat{c} - \hat{S}'_\sigma \hat{\sigma})$$
$$\quad - \hat{W}^T [\hat{S}'_c (\hat{c} - c^*) + \hat{S}'_\sigma (\hat{\sigma} - \sigma^*)]$$
$$= \hat{W}^T \hat{S}'_c c^* + \hat{W}^T \hat{S}'_\sigma \sigma^* - W^{*T} \hat{S}'_c \hat{c} - W^{*T} \hat{S}'_\sigma \hat{\sigma} + W^{*T} (\hat{S} - S^*)$$

with $S^* \triangleq S(x, c^*, \sigma^*)$.

Since every element of the vector $(\hat{S} - S^*)$ is bounded in $[-1, +1]$, we have

$$W^{*^{\mathrm{T}}}(\hat{S} - S^*) \leq \sum_{i=1}^{n_r} |w_i^*| \overset{\triangle}{=} \|W^*\|_1$$

Considering $\hat{W}^{\mathrm{T}} \hat{S}_c' c^* = \mathrm{tr}\{\hat{W}^{\mathrm{T}} \hat{S}_c' c^*\} \leq \|\hat{W}^{\mathrm{T}} \hat{S}_c'\|_{\mathrm{F}} \cdot \|c^*\|_{\mathrm{F}} = \|\hat{S}_c'^{\mathrm{T}} \hat{W}\| \cdot \|c^*\|$, we have

$$|d_u| \leq \|c^*\| \cdot \|\hat{S}_c'^{\mathrm{T}} \hat{W}\| + \|\sigma^*\| \cdot \|\hat{S}_\sigma'^{\mathrm{T}} \hat{W}\| + \|W^*\| \cdot \|\hat{S}_c' \hat{c}\|$$
$$+ \|W^*\| \cdot \|\hat{S}_\sigma' \hat{\sigma}\| + \|W^*\|_1$$

Thus, we have shown that (6.18) holds.

6.4 ADAPTIVE NF CONTROL DESIGN

In this section, the adaptive NF control is presented for nonholonomic mobile robots with uncertainties and external disturbances.

The following lemmas are useful in the controller design.

Lemma 6.2 *Let $e = H(s)r$ with $H(s)$ representing an $(n \times m)$-dimensional strictly proper exponentially stable transfer function, r and e denoting its input and output, respectively. Then $r \in L_2^m \cap L_\infty^m$ implies that $e, \dot{e} \in L_2^n \cap L_\infty^n$, e is continuous, and $e \to 0$ as $t \to \infty$. If, in addition, $r \to 0$ as $t \to \infty$, then $\dot{e} \to 0$ [32].*

Lemma 6.3 *Given a differentiable function $\phi(t) \colon \mathbb{R}^+ \to \mathbb{R}$, if $\phi(t) \in L_2$ and $\dot{\phi}(t) \in L_\infty$, then $\phi(t) \to 0$ as $t \to \infty$, where L_∞ and L_2 denote bounded and square integrable function sets, respectively.*

Consider the constrained dynamic equation (6.1) together with $(n-m)$ independent nonholonomic constraints (6.2). For simplicity of design, the following assumptions are made throughout this section.

Assumption 6.3 *Matrix $R^{\mathrm{T}}(q)B(q)$ is of full rank, which guarantees all m degrees of freedom can be (independently) actuated.*

It has been proven that the nonholonomic system (6.1) and (6.2) cannot be stabilized to a single point using smooth state feedback [18]. It can only be stabilized to a manifold of dimension $(n - m)$ due to the existence of $(n - m)$ nonholonomic constraints. Though the nonsmooth feedback laws [44] or time-varying feedback laws [4] can be used to stabilize these systems to a point, it is worth mentioning that different control objectives may also be pursued, such as stabilization to manifolds of equilibrium points (as opposed to a single equilibrium position) or to trajectories.

By appropriate selection, a set of vector $\dot{z}(t) \in \mathbb{R}^m$, the control objective can be specified as: given a desired $z_d(t)$, $\dot{z}_d(t)$, and desired constraint λ_d, determine a control law such that for any $(q(0), \dot{q}(0)) \in \Omega$, $z(t)$ and $\dot{q}$ asymptotically converge to a manifold Ω_{nhd} specified as

$$\Omega_{nhd} = \{(q, \dot{q}) | z(t) = z_d, \ \dot{q} = R(q)\dot{z}_d(t)\} \tag{6.21}$$

while the constraint force error $(\lambda - \lambda_d)$ is bounded in a certain region. The variable $z(t)$ can be thought as m "output equations" of the nonholonomic system.

Assumption 6.4 *The desired reference trajectory $z_d(t)$ is assumed to be bounded and uniformly continuous, and has bounded and uniformly continuous derivatives up to the second order. The desired $\lambda_d(t)$ is bounded and uniformly continuous.*

Let us define the following notations as

$$e_z = z - z_d \tag{6.22}$$

$$e_\lambda = \lambda - \lambda_d \tag{6.23}$$

$$\dot{z}_r = \dot{z}_d - \rho_1 e_z \tag{6.24}$$

$$s = \dot{e}_z + \rho_1 e_z \tag{6.25}$$

where $\dot{z}_r$ is the reference trajectory described in internal state space.

Apparently, we have

$$\dot{z} = \dot{z}_r + s \tag{6.26}$$

For force control, define μ as

$$\dot{\mu} = -\rho_2 \mu - \rho_3^{-1} J^T \lambda \tag{6.27}$$

where $\mu \in \mathbb{R}^n$. For the convenience of controller design, combining s and μ to form the following new hybrid variables

$$\sigma = Rs + \mu \tag{6.28}$$

$$v = R\dot{z}_r - \mu \tag{6.29}$$

From (6.26), (6.28), and (6.29), we have

$$\sigma + v = R\dot{z} \tag{6.30}$$

The time derivatives of v and σ are given by

$$\dot{v} = \dot{R}\dot{z}_r + R\ddot{z}_r - \dot{\mu} \tag{6.31}$$

$$\dot{\sigma} = \dot{R}\dot{z} + R\ddot{z} - \dot{v} \tag{6.32}$$

From the dynamic equation (6.8) together with (6.30) and (6.32), we have

$$M(q)\dot{\sigma} + C(q,\dot{q})\sigma + M(q)\dot{v} + C(q,\dot{q})v + G(q) = B(q)\tau + J^{\mathrm{T}}(q)\lambda + \tau_d \tag{6.33}$$

Consider the control law as

$$B\tau = \hat{M}(q)\dot{v} + \hat{C}(q,\dot{q})v + \hat{G}(q) - K_\sigma \sigma - J^{\mathrm{T}}\lambda_d + k_\lambda J^{\mathrm{T}}e_\lambda - K_s \mathrm{sgn}(\sigma)$$

$$- \hat{b}_m \sum_{i=1}^{n}\sum_{j=1}^{n}\bar{\phi}_{m_{ij}}|\sigma_i \dot{v}_j| - \hat{b}_c \sum_{i=1}^{n}\sum_{j=1}^{n}\bar{\phi}_{c_{ij}}|\sigma_i v_j| - \hat{b}_g \sum_{i=1}^{n}\bar{\phi}_{g_i}|\sigma_i| \tag{6.34}$$

where matrix $K_\sigma > 0$, constant $k_\lambda > 0$, matrix $K_s = \mathrm{diag}\{k_{sii}\}$ with $k_{sii} \geq |E_i|$ and E_i is the element of vector E (defined later), $\hat{M}(q)$, $\hat{C}(q,\dot{q})$, and $\hat{G}(q)$ are the estimates of $M(q)$, $C(q,\dot{q})$, and $G(q)$, respectively, the elements of which, that is, $m_{ij}(q)$, $c_{ij}(q,\dot{q})$, and $g_i(q)$ can be expressed by NF networks as

$$m_{ij}(q) = W_{m_{ij}}^{*\mathrm{T}}S(q,c_{m_{ij}}^*,\sigma_{m_{ij}}^*) + \epsilon_{m_{ij}}(q) \tag{6.35}$$

$$c_{ij}(q,\dot{q}) = W_{c_{ij}}^{*\mathrm{T}}S(q,\dot{q},c_{c_{ij}}^*,\sigma_{c_{ij}}^*) + \epsilon_{c_{ij}}(q,\dot{q}) \tag{6.36}$$

$$g_i(q) = W_{g_i}^{*\mathrm{T}}S(q,c_{g_i}^*,\sigma_{g_i}^*) + \epsilon_{g_i}(q) \tag{6.37}$$

where $W_{m_{ij}}^*$, $W_{c_{ij}}^*$, $W_{g_i}^*$ are ideal constant weight vectors, $c_{m_{ij}}^*$, $c_{c_{ij}}^*$, $c_{g_i}^*$ are the ideal constant center vectors, $\sigma_{m_{ij}}^*$, $\sigma_{c_{ij}}^*$, $\sigma_{g_i}^*$ are the ideal constant width vectors, and $\epsilon_{m_{ij}}(q)$, $\epsilon_{c_{ij}}(q,\dot{q})$, $\epsilon_{g_i}(q)$ are the approximation errors.

In addition, $\hat{b}_m$, $\hat{b}_c$, and $\hat{b}_g$ are the estimates of constants b_m^*, b_c^*, and b_g^*, respectively, which are defined by

$$b_m^* \stackrel{\triangle}{=} \max_{i,j}\{b_{m_{ij}}^*\} > 0, \quad b_{m_{ij}}^* \stackrel{\triangle}{=} \max\{w_{m_{ij}}, c_{m_{ij}}, \sigma_{m_{ij}}\} \tag{6.38}$$

$$b_c^* \stackrel{\triangle}{=} \max_{i,j}\{b_{c_{ij}}^*\} > 0, \quad b_{c_{ij}}^* \stackrel{\triangle}{=} \max\{w_{c_{ij}}, c_{c_{ij}}, \sigma_{c_{ij}}\} \tag{6.39}$$

$$b_g^* \stackrel{\triangle}{=} \max_{i}\{b_{g_i}^*\} > 0, \quad b_{g_i}^* \stackrel{\triangle}{=} \max\{w_{g_i}, c_{g_i}, \sigma_{g_i}\} \tag{6.40}$$

and $\bar{\phi}_{m_{ij}}$, $\bar{\phi}_{c_{ij}}$, and $\bar{\phi}_{g_i}$ are known positive functions defined by

$$\bar{\phi}_{m_{ij}} = \|\hat{S}'^{\mathrm{T}}_{c_{m_{ij}}} \hat{W}_{m_{ij}}\| + \|\hat{S}'^{\mathrm{T}}_{\sigma_{m_{ij}}} \hat{W}_{m_{ij}}\| + \|\hat{S}'_{c_{m_{ij}}} \hat{c}_{m_{ij}}\| + \|\hat{S}'_{\sigma_{m_{ij}}} \hat{\sigma}_{m_{ij}}\| + n_{rm_{ij}} \tag{6.41}$$

$$\bar{\phi}_{c_{ij}} = \|\hat{S}'^{\mathrm{T}}_{c_{c_{ij}}} \hat{W}_{c_{ij}}\| + \|\hat{S}'^{\mathrm{T}}_{\sigma_{c_{ij}}} \hat{W}_{c_{ij}}\| + \|\hat{S}'_{c_{c_{ij}}} \hat{c}_{c_{ij}}\| + \|\hat{S}'_{\sigma_{c_{ij}}} \hat{\sigma}_{c_{ij}}\| + n_{rc_{ij}} \tag{6.42}$$

$$\bar{\phi}_{g_i} = \|\hat{S}'^{\mathrm{T}}_{c_{g_i}} \hat{W}_{g_i}\| + \|\hat{S}'^{\mathrm{T}}_{\sigma_{g_i}} \hat{W}_{g_i}\| + \|\hat{S}'_{c_{g_i}} \hat{c}_{g_i}\| + \|\hat{S}'_{\sigma_{g_i}} \hat{\sigma}_{g_i}\| + n_{rg_i} \tag{6.43}$$

Using the "GL" matrix (denoted by upright and bold symbol with curly bracket) and operator (denoted by "•") introduced in Reference 32, the function emulators (6.35)–(6.37) can be collectively expressed as

$$M(q) = [\{\mathbf{W}_M^*\}^{\mathrm{T}} \bullet \{\mathbf{S}_M\}] + E_M \tag{6.44}$$

$$C(q,\dot{q}) = [\{\mathbf{W}_C^*\}^{\mathrm{T}} \bullet \{\mathbf{S}_C\}] + E_C \tag{6.45}$$

$$G(q) = [\{\mathbf{W}_G^*\}^{\mathrm{T}} \bullet \{\mathbf{S}_G\}] + E_G \tag{6.46}$$

where $[\{\mathbf{W}_M^*\}, \{\mathbf{S}_M\}]$, $[\{\mathbf{W}_C^*\}, \{\mathbf{S}_C\}]$, and $[\{\mathbf{W}_G^*\}, \{\mathbf{S}_G\}]$ are the desired weights and basis function GL matrices pairs of the NF emulation of $M(q)$, $C(q,\dot{q})$, and $G(q)$, respectively; and E_M, E_C, E_G are the collective NF reconstruction errors, respectively.

The estimates $\hat{M}(q)$, $\hat{C}(q,\dot{q})$, $\hat{G}(q)$, can, accordingly, be expressed as

$$\hat{M}(q) = [\{\hat{\mathbf{W}}_M\}^{\mathrm{T}} \bullet \{\hat{\mathbf{S}}_M\}] \tag{6.47}$$

$$\hat{C}(q,\dot{q}) = [\{\hat{\mathbf{W}}_C\}^{\mathrm{T}} \bullet \{\hat{\mathbf{S}}_C\}] \tag{6.48}$$

$$\hat{G}(q) = [\{\hat{\mathbf{W}}_G\}^{\mathrm{T}} \bullet \{\hat{\mathbf{S}}_G\}] \tag{6.49}$$

Note that in real implementation, the actual control torque τ must be provided rather than $B\tau$ given in (6.34). There are various approaches available in the literature to solve τ from (6.34), either analytically or numerically. In this chapter, the following scheme is applied to compute the control torque τ with rigor and rationality.

Define

$$u = B\tau \qquad (6.50)$$

Premultiplying both sides of (6.50) by R^T, we obtain

$$R^T u = R^T B \tau$$

From Assumption 6.3, it is known that $R^T B$ is nonsingular. Thus, τ is obtained as

$$\tau = (R^T B)^{-1} R^T u \qquad (6.51)$$

Substituting (6.51) and (6.47)–(6.49) into the dynamic equation (6.33) yields the closed-loop system error equation as

$$
\begin{aligned}
M\dot{\sigma} + C\sigma = {}& ([\{\hat{\mathbf{W}}_M\}^T \bullet \{\hat{\mathbf{S}}_M\}] - [\{\mathbf{W}_M^*\}^T \bullet \{\mathbf{S}_M\}])\dot{v} \\
& + ([\{\hat{\mathbf{W}}_C\}^T \bullet \{\hat{\mathbf{S}}_C\}] - [\{\mathbf{W}_C^*\}^T \bullet \{\mathbf{S}_C\}])v \\
& + ([\{\hat{\mathbf{W}}_G\}^T \bullet \{\hat{\mathbf{S}}_G\}] - [\{\mathbf{W}_G^*\}^T \bullet \{\mathbf{S}_G\}]) \\
& - K_\sigma \sigma + J^T \lambda - E - K_s \mathrm{sgn}(\sigma) \\
& - \hat{b}_m \sum_{i=1}^{n}\sum_{j=1}^{n} \bar{\phi}_{mij}|\sigma_i \dot{v}_j| - \hat{b}_c \sum_{i=1}^{n}\sum_{j=1}^{n} \bar{\phi}_{cij}|\sigma_i v_j| - \hat{b}_g \sum_{i=1}^{n} \bar{\phi}_{gi}|\sigma_i|
\end{aligned}
$$
$$(6.52)$$

where $E = E_M \dot{v} + E_C v + E_G - \tau_d$.

The stability of the closed-loop system will be illustrated in the following theorem.

Theorem 6.1 *Consider the nonholonomic mobile robot system described by dynamic equation (6.1) and the $(n-m)$ independent nonholonomic constraints (6.2). If the control law is chosen by (6.34), and the parameter adaptation laws* *are chosen by*

$$\dot{\hat{\mathbf{W}}}_{Mi} = -\Gamma_{Mi} \bullet (\{\hat{\mathbf{S}}_{Mi}\} - \{\hat{\mathbf{S}}_{Mci}\} - \{\hat{\mathbf{S}}_{M\sigma i}\})\dot{v}\sigma_i \qquad (6.53)$$

$$\dot{\hat{\mathbf{W}}}_{Ci} = -\Gamma_{Ci} \bullet (\{\hat{\mathbf{S}}_{Ci}\} - \{\hat{\mathbf{S}}_{Cci}\} - \{\hat{\mathbf{S}}_{C\sigma i}\})v\sigma_i \qquad (6.54)$$

$$\dot{\hat{\mathbf{W}}}_{Gi} = -\Gamma_{Gi}(\hat{\mathbf{S}}_{Gi} - \hat{\mathbf{S}}_{Gci} - \hat{\mathbf{S}}_{G\sigma i})\sigma_i \qquad (6.55)$$

$$\dot{\hat{\mathbf{C}}}_{Mi} = -\mathbf{\Theta}_{Mi} \bullet \{\widehat{\mathbf{SW}}_{Mci}\}\dot{v}\sigma_i \qquad (6.56)$$

$$\dot{\hat{\mathbf{C}}}_{Ci} = -\mathbf{\Theta}_{Ci} \bullet \{\widehat{\mathbf{SW}}_{Cci}\}v\sigma_i \qquad (6.57)$$

$$\dot{\hat{\mathbf{C}}}_{Gi} = -\mathbf{\Theta}_{Gi}\widehat{\mathbf{SW}}_{Gci}\sigma_i \qquad (6.58)$$

$$\dot{\hat{\mathbf{\Sigma}}}_{Mi} = -\mathbf{\Xi}_{Mi} \bullet \{\widehat{\mathbf{SW}}_{M\sigma i}\}\dot{v}\sigma_i \qquad (6.59)$$

$$\dot{\hat{\mathbf{\Sigma}}}_{Ci} = -\mathbf{\Xi}_{Ci} \bullet \{\widehat{\mathbf{SW}}_{C\sigma i}\}v\sigma_i \qquad (6.60)$$

$$\dot{\hat{\mathbf{\Sigma}}}_{Gi} = -\mathbf{\Xi}_{Gi}\widehat{\mathbf{SW}}_{G\sigma i}\sigma_i \qquad (6.61)$$

$$\dot{\hat{b}}_m = \gamma_{bm} \sum_{i=1}^{n} \sum_{j=1}^{n} \bar{\phi}_{mij}|\sigma_i \dot{v}_j| \qquad (6.62)$$

$$\dot{\hat{b}}_c = \gamma_{bc} \sum_{i=1}^{n} \sum_{j=1}^{n} \bar{\phi}_{cij}|\sigma_i v_j| \qquad (6.63)$$

$$\dot{\hat{b}}_g = \gamma_{bg} \sum_{i=1}^{n} \bar{\phi}_{gi}|\sigma_i| \qquad (6.64)$$

where matrices $\Gamma_{Mi}, \Gamma_{Ci}, \Gamma_{Gi}, \mathbf{\Theta}_{Mi}, \mathbf{\Theta}_{Ci}, \mathbf{\Theta}_{Gi}, \mathbf{\Xi}_{Mi}, \mathbf{\Xi}_{Ci}, \mathbf{\Xi}_{Gi}$ *are symmetric positive definite, and constants* $\gamma_{bm}, \gamma_{bc}, \gamma_{bg} > 0,$ *the signals* e_z *and* $\dot{e}_z$ *asymptotically converge to zero, and all the other closed loop signals are semiglobally uniformly ultimately bounded.*

Proof   The time derivative of $\frac{1}{2}\sigma^{\mathrm{T}}M\sigma$ along (6.52) is

$$
\begin{aligned}
\sigma^{\mathrm{T}}M^{\mathrm{T}}\dot{\sigma} ={}& -\sigma^{\mathrm{T}}K_{\sigma}\sigma - \sigma^{\mathrm{T}}E - \sigma^{\mathrm{T}}K_{s}\mathrm{sgn}(\sigma) + \sigma^{\mathrm{T}}J^{\mathrm{T}}\lambda - \sigma^{\mathrm{T}}C\sigma \\
&+ \sigma^{\mathrm{T}}([\{\hat{\mathbf{W}}_{M}\}^{\mathrm{T}} \bullet \{\hat{\mathbf{S}}_{M}\}] - [\{\mathbf{W}_{M}^{*}\}^{\mathrm{T}} \bullet \{\mathbf{S}_{M}\}])\dot{v} \\
&+ \sigma^{\mathrm{T}}([\{\hat{\mathbf{W}}_{C}\}^{\mathrm{T}} \bullet \{\hat{\mathbf{S}}_{C}\}] - [\{\mathbf{W}_{C}^{*}\}^{\mathrm{T}} \bullet \{\mathbf{S}_{C}\}])v \\
&+ \sigma^{\mathrm{T}}([\{\hat{\mathbf{W}}_{G}\}^{\mathrm{T}} \bullet \{\hat{\mathbf{S}}_{G}\}] - [\{\mathbf{W}_{G}^{*}\}^{\mathrm{T}} \bullet \{\mathbf{S}_{G}\}]) \\
&- \hat{b}_{m}\sum_{i=1}^{n}\sum_{j=1}^{n}\bar{\phi}_{m_{ij}}|\sigma_{i}\dot{v}_{j}| - \hat{b}_{c}\sum_{i=1}^{n}\sum_{j=1}^{n}\bar{\phi}_{c_{ij}}|\sigma_{i}v_{j}| - \hat{b}_{g}\sum_{i=1}^{n}\bar{\phi}_{g_{i}}|\sigma_{i}|
\end{aligned}
$$

$$(6.65)$$

Using the properties (6.17) and (6.18) given in Lemma 6.1, we have the following property for the NF approximation error:

$$
\begin{aligned}
\sigma^{\mathrm{T}}([\{\hat{\mathbf{W}}_{M}\}^{\mathrm{T}} \bullet \{\hat{\mathbf{S}}_{M}\}] &- [\{\mathbf{W}_{M}^{*}\}^{\mathrm{T}} \bullet \{\mathbf{S}_{M}\}])\dot{v} \\
={}& \sigma^{\mathrm{T}}([\{\tilde{\mathbf{W}}_{M}\}^{\mathrm{T}} \bullet (\{\hat{\mathbf{S}}_{M}\} - \{\hat{\mathbf{S}}_{Mc}\} - \{\hat{\mathbf{S}}_{M\sigma}\}))] \\
&+ [\{\tilde{\mathbf{C}}_{M}\}^{\mathrm{T}} \bullet \{\widehat{\mathbf{SW}}_{Mc}\}] + [\{\tilde{\mathbf{\Sigma}}_{M}\}^{\mathrm{T}} \bullet \{\widehat{\mathbf{SW}}_{M\sigma}\}] + D_{Mu})\dot{v}
\end{aligned}
$$

$$(6.66)$$

where GL matrices $\{\hat{\mathbf{S}}_{Mc}\}$, $\{\hat{\mathbf{S}}_{M\sigma}\}$, $\{\widehat{\mathbf{SW}}_{Mc}\}$, $\{\widehat{\mathbf{SW}}_{M\sigma}\}$, and matrix D_{Mu} are defined respectively as:

$$
\{\hat{\mathbf{S}}_{Mc}\} = \begin{Bmatrix} \{\hat{\mathbf{S}}_{Mc1}\} \\ \vdots \\ \{\hat{\mathbf{S}}_{Mcn}\} \end{Bmatrix}, \quad
\{\hat{\mathbf{S}}_{M\sigma}\} = \begin{Bmatrix} \{\hat{\mathbf{S}}_{M\sigma 1}\} \\ \vdots \\ \{\hat{\mathbf{S}}_{M\sigma n}\} \end{Bmatrix}
$$

$$
\{\widehat{\mathbf{SW}}_{Mc}\} = \begin{Bmatrix} \{\widehat{\mathbf{SW}}_{Mc1}\} \\ \vdots \\ \{\widehat{\mathbf{SW}}_{Mcn}\} \end{Bmatrix}, \quad
\{\widehat{\mathbf{SW}}_{M\sigma}\} = \begin{Bmatrix} \{\widehat{\mathbf{SW}}_{M\sigma 1}\} \\ \vdots \\ \{\widehat{\mathbf{SW}}_{M\sigma n}\} \end{Bmatrix}
$$

with

$$
\{\hat{\mathbf{S}}_{Mci}\} = \{\hat{\mathbf{S}}_{Mci1} \quad \cdots \quad \hat{\mathbf{S}}_{Mcin}\}, \quad \hat{\mathbf{S}}_{Mcij} = \hat{S}'_{c_{m_{ij}}}\hat{c}_{m_{ij}}
$$

$$
\{\hat{\mathbf{S}}_{M\sigma i}\} = \{\hat{\mathbf{S}}_{M\sigma i1} \quad \cdots \quad \hat{\mathbf{S}}_{M\sigma in}\}, \quad \hat{\mathbf{S}}_{M\sigma ij} = \hat{S}'_{\sigma_{m_{ij}}}\hat{\sigma}_{m_{ij}}
$$

$$
\{\widehat{\mathbf{SW}}_{Mci}\} = \{\widehat{\mathbf{SW}}_{Mci1} \quad \cdots \quad \widehat{\mathbf{SW}}_{Mcin}\}, \quad \widehat{\mathbf{SW}}_{Mcij} = \hat{S}'^{\mathrm{T}}_{c_{m_{ij}}}\hat{W}_{m_{ij}}
$$

$$
\{\widehat{\mathbf{SW}}_{M\sigma i}\} = \{\widehat{\mathbf{SW}}_{M\sigma i1} \quad \cdots \quad \widehat{\mathbf{SW}}_{M\sigma in}\}, \quad \widehat{\mathbf{SW}}_{M\sigma ij} = \hat{S}'^{\mathrm{T}}_{\sigma_{m_{ij}}}\hat{W}_{m_{ij}}
$$

and $D_{Mu} = [d_{muij}]$ with

$$|d_{muij}| \leq \|c^*_{m_{ij}}\| \cdot \|\hat{S}'^{\mathrm{T}}_{c_{m_{ij}}} \hat{W}_{m_{ij}}\| + \|\sigma^*_{m_{ij}}\| \cdot \|\hat{S}'^{\mathrm{T}}_{\sigma_{m_{ij}}} \hat{W}_{m_{ij}}\| + \|W^*_{m_{ij}}\| \cdot \|\hat{S}'_{c_{m_{ij}}} \hat{c}_{m_{ij}}\|$$

$$+ \|W^*_{m_{ij}}\| \cdot \|\hat{S}'_{\sigma_{m_{ij}}} \hat{\sigma}_{m_{ij}}\| + \|W^*_{m_{ij}}\|_1 \tag{6.67}$$

Noting Assumption 6.2, the following can be obtained:

$$\|c^*_{m_{ij}}\| \cdot \|\hat{S}'^{\mathrm{T}}_{c_{m_{ij}}} \hat{W}_{m_{ij}}\| + \|\sigma^*_{m_{ij}}\| \cdot \|\hat{S}'^{\mathrm{T}}_{\sigma_{m_{ij}}} \hat{W}_{m_{ij}}\| + \|W^*_{m_{ij}}\| \cdot \|\hat{S}'_{c_{m_{ij}}} \hat{c}_{m_{ij}}\|$$

$$+ \|W^*_{m_{ij}}\| \cdot \|\hat{S}'_{\sigma_{m_{ij}}} \hat{\sigma}_{m_{ij}}\| + \|W^*_{m_{ij}}\|_1$$

$$\leq c_{m_{ij}} \|\hat{S}'^{\mathrm{T}}_{c_{m_{ij}}} \hat{W}_{m_{ij}}\| + \sigma_{m_{ij}} \|\hat{S}'^{\mathrm{T}}_{\sigma_{m_{ij}}} \hat{W}_{m_{ij}}\| + w_{m_{ij}}$$

$$\times (\|\hat{S}'_{c_{m_{ij}}} \hat{c}_{m_{ij}}\| + \|\hat{S}'_{\sigma_{m_{ij}}} \hat{\sigma}_{m_{ij}}\| + n_{rm_{ij}})$$

$$\leq b^*_{m_{ij}} (\|\hat{S}'^{\mathrm{T}}_{c_{m_{ij}}} \hat{W}_{m_{ij}}\| + \|\hat{S}'^{\mathrm{T}}_{\sigma_{m_{ij}}} \hat{W}_{m_{ij}}\| + \|\hat{S}'_{c_{m_{ij}}} \hat{c}_{m_{ij}}\| + \|\hat{S}'_{\sigma_{m_{ij}}} \hat{\sigma}_{m_{ij}}\| + n_{rm_{ij}})$$

$$= b^*_{m_{ij}} \bar{\phi}_{m_{ij}} \tag{6.68}$$

Thus, (6.66) becomes

$$\sigma^{\mathrm{T}}([\{\hat{\mathbf{W}}_M\}^{\mathrm{T}} \bullet \{\hat{\mathbf{S}}_M\}] - [\{\mathbf{W}_M^*\}^{\mathrm{T}} \bullet \{\mathbf{S}_M\}])\dot{v}$$

$$= \sigma^{\mathrm{T}}([\{\tilde{\mathbf{W}}_M\}^{\mathrm{T}} \bullet (\{\hat{\mathbf{S}}_M\} - \{\hat{\mathbf{S}}_{Mc}\} - \{\hat{\mathbf{S}}_{M\sigma}\})] + [\{\tilde{\mathbf{C}}_M\}^{\mathrm{T}} \bullet \{\widehat{\mathbf{SW}}_{Mc}\}]$$

$$+ [\{\tilde{\mathbf{\Sigma}}_M\}^{\mathrm{T}} \bullet \{\widehat{\mathbf{SW}}_{M\sigma}\}])\dot{v} + \sum_{i=1}^{n}\sum_{j=1}^{n} \sigma_i d_{uij}\dot{v}_j$$

$$\leq \sigma^{\mathrm{T}}([\{\tilde{\mathbf{W}}_M\}^{\mathrm{T}} \bullet (\{\hat{\mathbf{S}}_M\} - \{\hat{\mathbf{S}}_{Mc}\} - \{\hat{\mathbf{S}}_{M\sigma}\})] + [\{\tilde{\mathbf{C}}_M\}^{\mathrm{T}} \bullet \{\widehat{\mathbf{SW}}_{Mc}\}]$$

$$+ [\{\tilde{\mathbf{\Sigma}}_M\}^{\mathrm{T}} \bullet \{\widehat{\mathbf{SW}}_{M\sigma}\}])\dot{v} + \sum_{i=1}^{n}\sum_{j=1}^{n} b^*_{m_{ij}} \bar{\phi}_{m_{ij}} |\sigma_i \dot{v}_j|$$

$$\leq \sigma^{\mathrm{T}}([\{\tilde{\mathbf{W}}_M\}^{\mathrm{T}} \bullet (\{\hat{\mathbf{S}}_M\} - \{\hat{\mathbf{S}}_{Mc}\} - \{\hat{\mathbf{S}}_{M\sigma}\})] + [\{\tilde{\mathbf{C}}_M\}^{\mathrm{T}} \bullet \{\widehat{\mathbf{SW}}_{Mc}\}]$$

$$+ [\{\tilde{\mathbf{\Sigma}}_M\}^{\mathrm{T}} \bullet \{\widehat{\mathbf{SW}}_{M\sigma}\}])\dot{v} + b^*_m \sum_{i=1}^{n}\sum_{j=1}^{n} \bar{\phi}_{m_{ij}} |\sigma_i \dot{v}_j| \tag{6.69}$$

Similarly, we have the following inequalities for other approximation errors as

$$\sigma^{\mathrm{T}}([\{\hat{\mathbf{W}}_C\}^{\mathrm{T}} \bullet \{\hat{\mathbf{S}}_C\}] - [\{\mathbf{W}_C^*\}^{\mathrm{T}} \bullet \{\mathbf{S}_C\}])v$$

$$\leq \sigma^{\mathrm{T}}([\{\tilde{\mathbf{W}}_C\}^{\mathrm{T}} \bullet (\{\hat{\mathbf{S}}_C\} - \{\hat{\mathbf{S}}_{Cc}\} - \{\hat{\mathbf{S}}_{C\sigma}\})] + [\{\tilde{\mathbf{C}}_C\}^{\mathrm{T}} \bullet \{\widehat{\mathbf{SW}}_{Cc}\}]$$

$$+ [\{\tilde{\mathbf{\Sigma}}_C\}^{\mathrm{T}} \bullet \{\widehat{\mathbf{SW}}_{C\sigma}\}])v + b_c^* \sum_{i=1}^{n}\sum_{j=1}^{n} \bar{\phi}_{cij}|\sigma_i v_j| \tag{6.70}$$

$$\sigma^{\mathrm{T}}([\{\hat{\mathbf{W}}_G\}^{\mathrm{T}} \bullet \{\hat{\mathbf{S}}_G\}] - [\{\mathbf{W}_G^*\}^{\mathrm{T}} \bullet \{\mathbf{S}_G\}])$$

$$\leq \sigma^{\mathrm{T}}([\{\tilde{\mathbf{W}}_G\}^{\mathrm{T}} \bullet (\{\hat{\mathbf{S}}_G\} - \{\hat{\mathbf{S}}_{Gc}\} - \{\hat{\mathbf{S}}_{G\sigma}\})] + [\{\tilde{\mathbf{C}}_G\}^{\mathrm{T}} \bullet \{\widehat{\mathbf{SW}}_{Gc}\}]$$

$$+ [\{\tilde{\mathbf{\Sigma}}_G\}^{\mathrm{T}} \bullet \{\widehat{\mathbf{SW}}_{G\sigma}\}]) + b_g^* \sum_{i=1}^{n} \bar{\phi}_{gi}|\sigma_i| \tag{6.71}$$

where definition for GL matrices $\{\hat{\mathbf{S}}_{Cc}\}$, $\{\hat{\mathbf{S}}_{C\sigma}\}$, $\{\widehat{\mathbf{SW}}_{Cc}\}$, $\{\widehat{\mathbf{SW}}_{C\sigma}\}$, $\{\hat{\mathbf{S}}_{Gc}\}$, $\{\hat{\mathbf{S}}_{G\sigma}\}$, $\{\widehat{\mathbf{SW}}_{Gc}\}$, and $\{\widehat{\mathbf{SW}}_{G\sigma}\}$, which is omitted here for conciseness, can be similarly made.

Consider the Lyapunov function candidate

$$V = \frac{1}{2}\sigma^{\mathrm{T}}M\sigma + \frac{1}{2}\sum_{i=1}^{n}\tilde{\mathbf{W}}_{Mi}^{\mathrm{T}}\mathbf{\Gamma}_{Mi}^{-1}\tilde{\mathbf{W}}_{Mi} + \frac{1}{2}\sum_{i=1}^{n}\tilde{\mathbf{W}}_{Ci}^{\mathrm{T}}\mathbf{\Gamma}_{Ci}^{-1}\tilde{\mathbf{W}}_{Ci} + \frac{1}{2}\sum_{i=1}^{n}\tilde{\mathbf{W}}_{Gi}^{\mathrm{T}}$$

$$\times \mathbf{\Gamma}_{Gi}^{-1}\tilde{\mathbf{W}}_{Gi} + \frac{1}{2}\sum_{i=1}^{n}\tilde{\mathbf{C}}_{Mi}^{\mathrm{T}}\mathbf{\Theta}_{Mi}^{-1}\tilde{\mathbf{C}}_{Mi} + \frac{1}{2}\sum_{i=1}^{n}\tilde{\mathbf{C}}_{Ci}^{\mathrm{T}}\mathbf{\Theta}_{Ci}^{-1}\tilde{\mathbf{C}}_{Ci}$$

$$+ \frac{1}{2}\sum_{i=1}^{n}\tilde{\mathbf{C}}_{Gi}^{\mathrm{T}}\mathbf{\Theta}_{Gi}^{-1}\tilde{\mathbf{C}}_{Gi} + \frac{1}{2}\sum_{i=1}^{n}\tilde{\mathbf{\Sigma}}_{Mi}^{\mathrm{T}}\mathbf{\Xi}_{Mi}^{-1}\tilde{\mathbf{\Sigma}}_{Mi} + \frac{1}{2}\sum_{i=1}^{n}\tilde{\mathbf{\Sigma}}_{Ci}^{\mathrm{T}}\mathbf{\Xi}_{Ci}^{-1}\tilde{\mathbf{\Sigma}}_{Ci}$$

$$+ \frac{1}{2}\sum_{i=1}^{n}\tilde{\mathbf{\Sigma}}_{Gi}^{\mathrm{T}}\mathbf{\Xi}_{Gi}^{-1}\tilde{\mathbf{\Sigma}}_{Gi} + \frac{1}{2}\gamma_{bm}^{-1}\tilde{b}_m^2 + \frac{1}{2}\gamma_{bc}^{-1}\tilde{b}_c^2 + \frac{1}{2}\gamma_{bg}^{-1}\tilde{b}_g^2 + \frac{1}{2}\rho_3\mu^{\mathrm{T}}\mu$$

$$\tag{6.72}$$

with $\tilde{(\cdot)} = \hat{(\cdot)} - (\cdot)^*$.

By virtue of (6.52), (6.69) to (6.71), the time derivative of V is given by

$$
\dot{V} = \sigma^{\mathrm{T}} M \dot{\sigma} + \frac{1}{2} \sigma^{\mathrm{T}} \dot{M} \sigma + \sum_{i=1}^{n} \tilde{\mathbf{W}}_{Mi}^{\mathrm{T}} \mathbf{\Gamma}_{Mi}^{-1} \dot{\hat{\mathbf{W}}}_{Mi} + \sum_{i=1}^{n} \tilde{\mathbf{W}}_{Ci}^{\mathrm{T}} \mathbf{\Gamma}_{Ci}^{-1} \dot{\hat{\mathbf{W}}}_{Ci}
$$

$$
+ \sum_{i=1}^{n} \tilde{\mathbf{W}}_{Gi}^{\mathrm{T}} \mathbf{\Gamma}_{Gi}^{-1} \dot{\hat{\mathbf{W}}}_{Gi} + \sum_{i=1}^{n} \tilde{\mathbf{C}}_{Mi}^{\mathrm{T}} \mathbf{\Theta}_{Mi}^{-1} \dot{\hat{\mathbf{C}}}_{Mi} + \sum_{i=1}^{n} \tilde{\mathbf{C}}_{Ci}^{\mathrm{T}} \mathbf{\Theta}_{Ci}^{-1} \dot{\hat{\mathbf{C}}}_{Ci}
$$

$$
+ \sum_{i=1}^{n} \tilde{\mathbf{C}}_{Gi}^{\mathrm{T}} \mathbf{\Theta}_{Gi}^{-1} \dot{\hat{\mathbf{C}}}_{Gi} + \sum_{i=1}^{n} \tilde{\boldsymbol{\Sigma}}_{Mi}^{\mathrm{T}} \boldsymbol{\Xi}_{Mi}^{-1} \dot{\hat{\boldsymbol{\Sigma}}}_{Mi} + \sum_{i=1}^{n} \tilde{\boldsymbol{\Sigma}}_{Ci}^{\mathrm{T}} \boldsymbol{\Xi}_{Ci}^{-1} \dot{\hat{\boldsymbol{\Sigma}}}_{Ci}
$$

$$
+ \sum_{i=1}^{n} \tilde{\boldsymbol{\Sigma}}_{Gi}^{\mathrm{T}} \boldsymbol{\Xi}_{Gi}^{-1} \dot{\hat{\boldsymbol{\Sigma}}}_{Gi} + \gamma_{bm}^{-1} \tilde{b}_m \dot{\hat{b}}_m + \gamma_{bc}^{-1} \tilde{b}_c \dot{\hat{b}}_c + \gamma_{bg}^{-1} \tilde{b}_g \dot{\hat{b}}_g + \rho_3 \mu^{\mathrm{T}} \dot{\mu}
$$

$$
\leq \frac{1}{2} \sigma^{\mathrm{T}} \dot{M} \sigma - \sigma^{\mathrm{T}} C \sigma - \sigma^{\mathrm{T}} K_\sigma \sigma - \sigma^{\mathrm{T}} E - \sigma^{\mathrm{T}} K_s \mathrm{sgn}(\sigma) + \sigma^{\mathrm{T}} J^{\mathrm{T}} \lambda
$$

$$
+ \sigma^{\mathrm{T}} ([\{\tilde{\mathbf{W}}_M\}^{\mathrm{T}} \bullet (\{\hat{\mathbf{S}}_M\} - \{\hat{\mathbf{S}}_{Mc}\} - \{\hat{\mathbf{S}}_{M\sigma}\})]
$$

$$
+ [\{\tilde{\mathbf{C}}_M\}^{\mathrm{T}} \bullet \{\widehat{\mathbf{SW}}_{Mc}\}] + [\{\tilde{\boldsymbol{\Sigma}}_M\}^{\mathrm{T}} \bullet \{\widehat{\mathbf{SW}}_{M\sigma}\}]) \dot{\nu}
$$

$$
+ \sigma^{\mathrm{T}} ([\{\tilde{\mathbf{W}}_C\}^{\mathrm{T}} \bullet (\{\hat{\mathbf{S}}_C\} - \{\hat{\mathbf{S}}_{Cc}\} - \{\hat{\mathbf{S}}_{C\sigma}\})]
$$

$$
+ [\{\tilde{\mathbf{C}}_C\}^{\mathrm{T}} \bullet \{\widehat{\mathbf{SW}}_{Cc}\}] + [\{\tilde{\boldsymbol{\Sigma}}_C\}^{\mathrm{T}} \bullet \{\widehat{\mathbf{SW}}_{C\sigma}\}]) \nu
$$

$$
+ \sigma^{\mathrm{T}} ([\{\tilde{\mathbf{W}}_G\}^{\mathrm{T}} \bullet (\{\hat{\mathbf{S}}_G\} - \{\hat{\mathbf{S}}_{Gc}\} - \{\hat{\mathbf{S}}_{G\sigma}\})]
$$

$$
+ [\{\tilde{\mathbf{C}}_G\}^{\mathrm{T}} \bullet \{\widehat{\mathbf{SW}}_{Gc}\}] + [\{\tilde{\boldsymbol{\Sigma}}_G\}^{\mathrm{T}} \bullet \{\widehat{\mathbf{SW}}_{G\sigma}\}])
$$

$$
+ \sum_{i=1}^{n} \tilde{\mathbf{W}}_{Mi}^{\mathrm{T}} \mathbf{\Gamma}_{Mi}^{-1} \dot{\hat{\mathbf{W}}}_{Mi} + \sum_{i=1}^{n} \tilde{\mathbf{W}}_{Ci}^{\mathrm{T}} \mathbf{\Gamma}_{Ci}^{-1} \dot{\hat{\mathbf{W}}}_{Ci} + \sum_{i=1}^{n} \tilde{\mathbf{W}}_{Gi}^{\mathrm{T}} \mathbf{\Gamma}_{Gi}^{-1} \dot{\hat{\mathbf{W}}}_{Gi}
$$

$$
+ \sum_{i=1}^{n} \tilde{\mathbf{C}}_{Mi}^{\mathrm{T}} \mathbf{\Theta}_{Mi}^{-1} \dot{\hat{\mathbf{C}}}_{Mi} + \sum_{i=1}^{n} \tilde{\mathbf{C}}_{Ci}^{\mathrm{T}} \mathbf{\Theta}_{Ci}^{-1} \dot{\hat{\mathbf{C}}}_{Ci} + \sum_{i=1}^{n} \tilde{\mathbf{C}}_{Gi}^{\mathrm{T}} \mathbf{\Theta}_{Gi}^{-1} \dot{\hat{\mathbf{C}}}_{Gi}
$$

$$
+ \sum_{i=1}^{n} \tilde{\boldsymbol{\Sigma}}_{Mi}^{\mathrm{T}} \boldsymbol{\Xi}_{Mi}^{-1} \dot{\hat{\boldsymbol{\Sigma}}}_{Mi} + \sum_{i=1}^{n} \tilde{\boldsymbol{\Sigma}}_{Ci}^{\mathrm{T}} \boldsymbol{\Xi}_{Ci}^{-1} \dot{\hat{\boldsymbol{\Sigma}}}_{Ci} + \sum_{i=1}^{n} \tilde{\boldsymbol{\Sigma}}_{Gi}^{\mathrm{T}} \boldsymbol{\Xi}_{Gi}^{-1} \dot{\hat{\boldsymbol{\Sigma}}}_{Gi}
$$

$$
- \bar{b}_m \sum_{i=1}^{n} \sum_{j=1}^{n} \bar{\phi}_{mij} |\sigma_i \dot{\nu}_j| - \bar{b}_c \sum_{i=1}^{n} \sum_{j=1}^{n} \bar{\phi}_{cij} |\sigma_i \nu_j| - \bar{b}_g \sum_{i=1}^{n} \bar{\phi}_{gi} |\sigma_i|
$$

$$
+ \gamma_{bm}^{-1} \tilde{b}_m \dot{\hat{b}}_m + \gamma_{bc}^{-1} \tilde{b}_c \dot{\hat{b}}_c + \gamma_{bg}^{-1} \tilde{b}_g \dot{\hat{b}}_g + \rho_3 \mu^{\mathrm{T}} \dot{\mu} \tag{6.73}
$$

As matrix $\dot{M} - 2C$ is skew-symmetric, $\sigma^{\mathrm{T}}(\dot{M} - 2C)\sigma = 0, \forall x \neq 0$.

Noting that

$$\sigma^{\mathrm{T}}[\{\tilde{\mathbf{W}}_M\}^{\mathrm{T}} \bullet \{\hat{\mathbf{S}}_M\}]\dot{v} = \begin{bmatrix} \sigma_1 & \sigma_2 & \cdots & \sigma_n \end{bmatrix} \begin{bmatrix} \{\tilde{\mathbf{W}}_{M1}\}^{\mathrm{T}} \bullet \{\hat{\mathbf{S}}_{M1}\}\dot{v} \\ \vdots \\ \{\tilde{\mathbf{W}}_{Mn}\}^{\mathrm{T}} \bullet \{\hat{\mathbf{S}}_{Mn}\}\dot{v} \end{bmatrix}$$

$$= \sum_{i=1}^{n}\{\tilde{\mathbf{W}}_{Mi}\}^{\mathrm{T}} \bullet \{\hat{\mathbf{S}}_{Mi}\}\dot{v}\sigma_i \qquad (6.74)$$

and similarly

$$\sigma^{\mathrm{T}}[\{\tilde{\mathbf{W}}_M\}^{\mathrm{T}} \bullet (\{\hat{\mathbf{S}}_M\} - \{\hat{\mathbf{S}}_{Mc}\} - \{\hat{\mathbf{S}}_{M\sigma}\})]\dot{v}$$

$$= \sum_{i=1}^{n}\{\tilde{\mathbf{W}}_{Mi}\}^{\mathrm{T}} \bullet (\{\hat{\mathbf{S}}_{Mi}\} - \{\hat{\mathbf{S}}_{Mci}\} - \{\hat{\mathbf{S}}_{M\sigma i}\})\dot{v}\sigma_i$$

$$\sigma^{\mathrm{T}}[\{\tilde{\mathbf{C}}_M\}^{\mathrm{T}} \bullet \{\widehat{\mathbf{SW}}_{Mc}\}]\dot{v} = \sum_{i=1}^{n}\{\tilde{\mathbf{C}}_{Mi}\}^{\mathrm{T}} \bullet \{\widehat{\mathbf{SW}}_{Mci}\}\dot{v}\sigma_i$$

$$\sigma^{\mathrm{T}}[\{\tilde{\boldsymbol{\Sigma}}_M\}^{\mathrm{T}} \bullet \{\widehat{\mathbf{SW}}_{M\sigma}\}]\dot{v} = \sum_{i=1}^{n}\{\tilde{\boldsymbol{\Sigma}}_{Mi}\}^{\mathrm{T}} \bullet \{\widehat{\mathbf{SW}}_{M\sigma i}\}\dot{v}\sigma_i$$

$$\sigma^{\mathrm{T}}[\{\tilde{\mathbf{W}}_C\}^{\mathrm{T}} \bullet (\{\hat{\mathbf{S}}_C\} - \{\hat{\mathbf{S}}_{Cc}\} - \{\hat{\mathbf{S}}_{C\sigma}\})]v$$

$$= \sum_{i=1}^{n}\{\tilde{\mathbf{W}}_{Ci}\}^{\mathrm{T}} \bullet (\{\hat{\mathbf{S}}_{Ci}\} - \{\hat{\mathbf{S}}_{Cci}\} - \{\hat{\mathbf{S}}_{C\sigma i}\})v\sigma_i$$

$$\sigma^{\mathrm{T}}[\{\tilde{\mathbf{C}}_C\}^{\mathrm{T}} \bullet \{\widehat{\mathbf{SW}}_{Cc}\}]v = \sum_{i=1}^{n}\{\tilde{\mathbf{C}}_{Ci}\}^{\mathrm{T}} \bullet \{\widehat{\mathbf{SW}}_{Cci}\}v\sigma_i$$

$$\sigma^{\mathrm{T}}[\{\tilde{\boldsymbol{\Sigma}}_C\}^{\mathrm{T}} \bullet \{\widehat{\mathbf{SW}}_{C\sigma}\}]v = \sum_{i=1}^{n}\{\tilde{\boldsymbol{\Sigma}}_{Ci}\}^{\mathrm{T}} \bullet \{\widehat{\mathbf{SW}}_{C\sigma i}\}v\sigma_i$$

$$\sigma^{\mathrm{T}}[\{\tilde{\mathbf{W}}_G\}^{\mathrm{T}} \bullet (\{\hat{\mathbf{S}}_G\} - \{\hat{\mathbf{S}}_{Gc}\} - \{\hat{\mathbf{S}}_{G\sigma}\})] = \sum_{i=1}^{n}\tilde{\mathbf{W}}_{Gi}^{\mathrm{T}}(\hat{\mathbf{S}}_{Gi} - \hat{\mathbf{S}}_{Gci} - \hat{\mathbf{S}}_{G\sigma i})\sigma_i$$

$$\sigma^{\mathrm{T}}[\{\tilde{\mathbf{C}}_G\}^{\mathrm{T}} \bullet \{\widehat{\mathbf{SW}}_{Gc}\}] = \sum_{i=1}^{n}\tilde{\mathbf{C}}_{Gi}^{\mathrm{T}}\widehat{\mathbf{SW}}_{Gci}\sigma_i$$

$$\sigma^{\mathrm{T}}[\{\tilde{\boldsymbol{\Sigma}}_G\}^{\mathrm{T}} \bullet \{\widehat{\mathbf{SW}}_{G\sigma}\}] = \sum_{i=1}^{n}\tilde{\boldsymbol{\Sigma}}_{Gi}^{\mathrm{T}}\widehat{\mathbf{SW}}_{G\sigma i}\sigma_i$$

Equation (6.73) becomes

$$\dot{V} \leq \frac{1}{2}\sigma^T\dot{M}\sigma - \sigma^T C\sigma - \sigma^T K_\sigma\sigma - \sigma^T E - \sigma^T K_s\mathrm{sgn}(\sigma) + \sigma^T J^T\lambda$$

$$+ \sum_{i=1}^{n}\{\tilde{\mathbf{W}}_{Mi}\}^T \bullet (\{\hat{\mathbf{S}}_{Mi}\} - \{\hat{\mathbf{S}}_{Mci}\} - \{\hat{\mathbf{S}}_{M\sigma i}\})\dot{v}\sigma_i$$

$$+ \sum_{i=1}^{n}\{\tilde{\mathbf{C}}_{Mi}\}^T \bullet \{\widehat{\mathbf{SW}}_{Mci}\}\dot{v}\sigma_i + \sum_{i=1}^{n}\{\tilde{\boldsymbol{\Sigma}}_{Mi}\}^T \bullet \{\widehat{\mathbf{SW}}_{M\sigma i}\}\dot{v}\sigma_i$$

$$+ \sum_{i=1}^{n}\{\tilde{\mathbf{W}}_{Ci}\}^T \bullet (\{\hat{\mathbf{S}}_{Ci}\} - \{\hat{\mathbf{S}}_{Cci}\} - \{\hat{\mathbf{S}}_{C\sigma i}\})v\sigma_i$$

$$+ \sum_{i=1}^{n}\{\tilde{\mathbf{C}}_{Ci}\}^T \bullet \{\widehat{\mathbf{SW}}_{Cci}\}v\sigma_i + \sum_{i=1}^{n}\{\tilde{\boldsymbol{\Sigma}}_{Ci}\}^T \bullet \{\widehat{\mathbf{SW}}_{C\sigma i}\}v\sigma_i$$

$$+ \sum_{i=1}^{n}\tilde{\mathbf{W}}_{Gi}^T(\hat{\mathbf{S}}_{Gi} - \hat{\mathbf{S}}_{Gci} - \hat{\mathbf{S}}_{G\sigma i})\sigma_i + \sum_{i=1}^{n}\tilde{\mathbf{C}}_{Gi}^T\widehat{\mathbf{SW}}_{Gci}\sigma_i + \sum_{i=1}^{n}\tilde{\boldsymbol{\Sigma}}_{Gi}^T\widehat{\mathbf{SW}}_{G\sigma i}\sigma_i$$

$$+ \sum_{i=1}^{n}\tilde{\mathbf{W}}_{Mi}^T\boldsymbol{\Gamma}_{Mi}^{-1}\dot{\hat{\mathbf{W}}}_{Mi} + \sum_{i=1}^{n}\tilde{\mathbf{W}}_{Ci}^T\boldsymbol{\Gamma}_{Ci}^{-1}\dot{\hat{\mathbf{W}}}_{Ci} + \sum_{i=1}^{n}\tilde{\mathbf{W}}_{Gi}^T\boldsymbol{\Gamma}_{Gi}^{-1}\dot{\hat{\mathbf{W}}}_{Gi}$$

$$+ \sum_{i=1}^{n}\tilde{\mathbf{C}}_{Mi}^T\boldsymbol{\Theta}_{Mi}^{-1}\dot{\hat{\mathbf{C}}}_{Mi} + \sum_{i=1}^{n}\tilde{\mathbf{C}}_{Ci}^T\boldsymbol{\Theta}_{Ci}^{-1}\dot{\hat{\mathbf{C}}}_{Ci} + \sum_{i=1}^{n}\tilde{\mathbf{C}}_{Gi}^T\boldsymbol{\Theta}_{Gi}^{-1}\dot{\hat{\mathbf{C}}}_{Gi}$$

$$+ \sum_{i=1}^{n}\tilde{\boldsymbol{\Sigma}}_{Mi}^T\boldsymbol{\Xi}_{Mi}^{-1}\dot{\hat{\boldsymbol{\Sigma}}}_{Mi} + \sum_{i=1}^{n}\tilde{\boldsymbol{\Sigma}}_{Ci}^T\boldsymbol{\Xi}_{Ci}^{-1}\dot{\hat{\boldsymbol{\Sigma}}}_{Ci} + \sum_{i=1}^{n}\tilde{\boldsymbol{\Sigma}}_{Gi}^T\boldsymbol{\Xi}_{Gi}^{-1}\dot{\hat{\boldsymbol{\Sigma}}}_{Gi}$$

$$- \tilde{b}_m\sum_{i=1}^{n}\sum_{j=1}^{n}\bar{\phi}_{m_{ij}}|\sigma_i\dot{v}_j| - \tilde{b}_c\sum_{i=1}^{n}\sum_{j=1}^{n}\bar{\phi}_{c_{ij}}|\sigma_i v_j| - \tilde{b}_g\sum_{i=1}^{n}\bar{\phi}_{g_i}|\sigma_i|$$

$$+ \gamma_{bm}^{-1}\tilde{b}_m\dot{\hat{b}}_m + \gamma_{bc}^{-1}\tilde{b}_c\dot{\hat{b}}_c + \gamma_{bg}^{-1}\tilde{b}_g\dot{\hat{b}}_g + \rho_3\mu^T\dot{\mu} \tag{6.75}$$

Substituting the weight vectors updating laws (6.53)–(6.55), the center vectors updating laws (6.56)–(6.58), the width vectors updating laws (6.59)–(6.61), and the constant parameters updating laws (6.62)–(6.64) into (6.75) yields

$$\dot{V} \leq -\sigma^T K_\sigma\sigma - \sigma^T E - \sigma^T K_s\mathrm{sgn}(\sigma) + \sigma^T J^T\lambda + \rho_3\mu^T\dot{\mu} \tag{6.76}$$

Noting that $k_{sii} \geq |E_i| > 0$, it is obvious that $[-\sigma^T E - \sigma^T K_s\mathrm{sgn}(\sigma)] \leq 0$. In addition, from (6.27), we know that $\dot{\mu} = -\rho_2\mu - \rho_3^{-1}J^T\lambda$ and from (6.28),

$\sigma^T = s^T R^T + \mu^T$. Thus, we have

$$\sigma^T J^T \lambda + \rho_3 \mu^T (1 + k_\lambda) \dot{\mu} = -\rho_2 \rho_3 \mu^T \mu + s^T R^T J^T \lambda \qquad (6.77)$$

Noting $R^T J^T = 0$ from (6.4), we then have

$$\dot{V} \leq -\sigma^T K_\sigma \sigma - \rho_2 \rho_3 \mu^T \mu \leq 0 \qquad (6.78)$$

As $V \geq 0$ and $\dot{V} \leq 0$, $V \in L_\infty$. From the definition of V, it follows that σ, $\mu \in L_\infty^n$, $\hat{\mathbf{W}}_{Mi}, \hat{\mathbf{W}}_{Ci}, \hat{\mathbf{W}}_{Gi}, \hat{\mathbf{C}}_{Mi}, \hat{\mathbf{C}}_{Ci}, \hat{\mathbf{C}}_{Gi}, \hat{\boldsymbol{\Sigma}}_{Mi}, \hat{\boldsymbol{\Sigma}}_{Ci}, \hat{\boldsymbol{\Sigma}}_{Gi} \in L_\infty^{n_i}$, $i = 1, \ldots, n$ with n_i denoting the compatible size of the vectors, and $\hat{b}_m, \hat{b}_c, \hat{b}_g \in L_\infty$.

Integrating both sides of (6.78), we have

$$\int_0^t \sigma^T K_\sigma \sigma \leq V(0) - V(t) \leq V(0) \qquad (6.79)$$

Hence $\sigma \in L_2^n$.

From (6.28), we have $s = (R^T R)^{-1} R^T (\sigma - \mu)$, hence $s \in L_\infty^m$ since R is bounded. From Lemma 6.2, it can be concluded that $e_z, \dot{e}_z \in L_\infty^m$.

From (6.27), (6.29), and (6.31), we have

$$\hat{M}\dot{v} + \hat{C}v + \hat{G} = \hat{M}(\dot{R}\dot{z}_r + R\ddot{z}_r - \dot{\mu}) + \hat{C}(R\dot{z}_r - \mu) + \hat{G}$$
$$= \hat{M}(\dot{R}\dot{z}_r + R\ddot{z}_r) + \hat{C}R\dot{z}_r + \hat{G} - \hat{C}\mu + \rho_3^{-1}\hat{M}J^T\lambda \qquad (6.80)$$

From (6.26), it is known that

$$\dot{q} = R\dot{z}_r + Rs \qquad (6.81)$$
$$\ddot{q} = \dot{R}\dot{z}_r + R\ddot{z}_r + \dot{R}s + R\dot{s} \qquad (6.82)$$

Replacing τ by (6.51) in dynamic equation (6.1) by noting $f = J^T(q)\lambda$, Equations (6.80)–(6.82), the closed-loop system becomes

$$M\dot{R}s + MR\dot{s} + CRs - (\hat{M} - M)(\dot{R}\dot{z}_r + R\ddot{z}_r) - (\hat{C} - C)R\dot{z}_r - (\hat{G} - G)$$

$$+ \hat{C}\mu + K_\sigma \sigma + K_s \mathrm{sgn}(\sigma) + \hat{b}_m \sum_{i=1}^n \sum_{j=1}^n \bar{\phi}_{m_{ij}} |\sigma_i \dot{v}_j|$$

$$+ \hat{b}_c \sum_{i=1}^n \sum_{j=1}^n \bar{\phi}_{c_{ij}} |\sigma_i v_j| + \hat{b}_g \sum_{i=1}^n \bar{\phi}_{g_i} |\sigma_i| - \tau_d$$

$$= (\rho_3^{-1}\hat{M} + I_n)J^T\lambda \qquad (6.83)$$

Invoking (6.44)–(6.46) and (6.47)–(6.49), Equation (6.83) then becomes

$$M\dot{R}s + MR\dot{s} + CRs - ([\{\hat{\mathbf{W}}_M\}^{\mathrm{T}} \bullet \{\hat{\mathbf{S}}_M\}] - [\{\mathbf{W}_M^*\}^{\mathrm{T}} \bullet \{\mathbf{S}_M\}])(R\dot{z}_r + R\ddot{z}_r)$$

$$- ([\{\hat{\mathbf{W}}_C\}^{\mathrm{T}} \bullet \{\hat{\mathbf{S}}_C\}] - [\{\mathbf{W}_C^*\}^{\mathrm{T}} \bullet \{\mathbf{S}_C\}])R\dot{z}_r$$

$$- ([\{\hat{\mathbf{W}}_G\}^{\mathrm{T}} \bullet \{\hat{\mathbf{S}}_G\}] - [\{\mathbf{W}_G^*\}^{\mathrm{T}} \bullet \{\mathbf{S}_G\}]) + \hat{C}\mu$$

$$+ K_\sigma \sigma + K_s \mathrm{sgn}(\sigma) + \hat{b}_m \sum_{i=1}^{n}\sum_{j=1}^{n} \bar{\phi}_{m_{ij}}|\sigma_i \dot{v}_j| + \hat{b}_c \sum_{i=1}^{n}\sum_{j=1}^{n} \bar{\phi}_{c_{ij}}|\sigma_i v_j|$$

$$+ \hat{b}_g \sum_{i=1}^{n} \bar{\phi}_{g_i}|\sigma_i| + E_M(R\dot{z}_r + R\ddot{z}_r) + E_C R\dot{z}_r + E_G - \tau_d$$

$$= (\rho_3^{-1}\hat{M} + I_n)J^{\mathrm{T}}\lambda \tag{6.84}$$

Since $M(q)$ is nonsingular, multiplying $J(q)M^{-1}(q)$ on both sides of (6.84) yields

$$J\dot{R}s + JM^{-1}\Bigg[CRs - ([\{\hat{\mathbf{W}}_M\}^{\mathrm{T}} \bullet \{\hat{\mathbf{S}}_M\}] - [\{\mathbf{W}_M^*\}^{\mathrm{T}} \bullet \{\mathbf{S}_M\}])(R\dot{z}_r + R\ddot{z}_r)$$

$$- ([\{\hat{\mathbf{W}}_C\}^{\mathrm{T}} \bullet \{\hat{\mathbf{S}}_C\}] - [\{\mathbf{W}_C^*\}^{\mathrm{T}} \bullet \{\mathbf{S}_C\}])R\dot{z}_r$$

$$- ([\{\hat{\mathbf{W}}_G\}^{\mathrm{T}} \bullet \{\hat{\mathbf{S}}_G\}] - [\{\mathbf{W}_G^*\}^{\mathrm{T}} \bullet \{\mathbf{S}_G\}]) + \hat{C}\mu$$

$$+ K_\sigma \sigma + K_s \mathrm{sgn}(\sigma) + \hat{b}_m \sum_{i=1}^{n}\sum_{j=1}^{n} \bar{\phi}_{m_{ij}}|\sigma_i \dot{v}_j| + \hat{b}_c \sum_{i=1}^{n}\sum_{j=1}^{n} \bar{\phi}_{c_{ij}}|\sigma_i v_j|$$

$$+ \hat{b}_g \sum_{i=1}^{n} \bar{\phi}_{g_i}|\sigma_i| + E_M(R\dot{z}_r + R\ddot{z}_r) + E_C R\dot{z}_r + E_G - \tau_d \Bigg]$$

$$= JM^{-1}(\rho_3^{-1}\hat{M} + I_n)J^{\mathrm{T}}\lambda \tag{6.85}$$

Since we have established that $e_z, \dot{e}_z \in L_\infty^m$, from Assumption 6.4 and (6.24), it can be concluded that $\dot{z}_r(t), \ddot{z}_r(t) \in L_\infty^m$. As r is shown to be bounded, so is $\dot{z}$ from (6.26). Hence, $\dot{q}(t) = R\dot{z}(t) \in L_\infty^n$. It follows that $M(q), \hat{M}(q), C(q, \dot{q}), \hat{C}(q, \dot{q}) \in L_\infty^{n \times n}$, and $G(q), \hat{G}(q) \in L_\infty^n$. Thus, the left hand side of (6.85) is bounded. In fact, ρ_3 can be properly chosen to keep $(\rho_3^{-1}\hat{M} + I_n)$ on the right hand side of (6.85) from being singular. Hence, we have $\lambda \in L_\infty^{n-m}$. As λ_d is bounded, so are e_λ and $B\tau$.

From (6.1), we can conclude that $\ddot{q} \in L_\infty^n$.

As $\lambda \in L_\infty^{n-m}$ and $\mu \in L_\infty^n$, from Equation (6.27), it is obvious that $\dot{\mu} \in L_\infty^n$. Thus, from (6.31), we have $\dot{\nu} \in L_\infty^n$. Since $\dot{z}, \ddot{z} \in L_\infty^m$ have been established before, we can conclude from (6.32) that $\dot{\sigma} \in L_\infty^n$. Now, with $\sigma, \mu \in L_2^n, \dot{\sigma}, \dot{\mu} \in L_\infty^n$, according to Lemma 6.3, we can conclude that σ and μ asymptotically converge to zero. Hence, from (6.28), it can be concluded that $s \to 0$ as $t \to \infty$. According to Lemma 6.3, we can also obtain $e_z, \dot{e}_z \to 0$ as $t \to \infty$.

Since $\dot{q}, \ddot{q} \in L_\infty^n$, q and $\dot{q}$ are uniformly continuous. Therefore, from Property 6.1, we can conclude that matrices $M(q)$, $C(q, \dot{q})$, $G(q)$, $S(q)$, $J(q)$, $\hat{D}(q)$, $\hat{C}(q, \dot{q})$, and $\hat{G}(q)$ are uniformly continuous.

Remark 6.5 *If $B\tau$ is directly replaced by (6.34) in the dynamic equation (6.1) without considering the real implementation issue, a wrong conclusion may be drawn.*

Substituting (6.34) and (6.47) to (6.49) into the dynamic equation (6.33) yields the closed-loop system error equation as

$$M\dot{\sigma} + C\sigma = ([\{\hat{\mathbf{W}}_M\}^{\mathrm{T}} \bullet \{\hat{\mathbf{S}}_M\}] - [\{\mathbf{W}_M^*\}^{\mathrm{T}} \bullet \{\mathbf{S}_M\}])\dot{\nu}$$

$$+ ([\{\hat{\mathbf{W}}_C\}^{\mathrm{T}} \bullet \{\hat{\mathbf{S}}_C\}] - [\{\mathbf{W}_C^*\}^{\mathrm{T}} \bullet \{\mathbf{S}_C\}])\nu$$

$$+ ([\{\hat{\mathbf{W}}_G\}^{\mathrm{T}} \bullet \{\hat{\mathbf{S}}_G\}] - [\{\mathbf{W}_G^*\}^{\mathrm{T}} \bullet \{\mathbf{S}_G\}])$$

$$- K_\sigma\sigma + (1 + k_\lambda)J^{\mathrm{T}}e_\lambda - E - K_s\mathrm{sgn}(\sigma)$$

$$- \hat{b}_m \sum_{i=1}^n \sum_{j=1}^n \bar{\phi}_{m_{ij}}|\sigma_i\dot{\nu}_j| - \hat{b}_c \sum_{i=1}^n \sum_{j=1}^n \bar{\phi}_{c_{ij}}|\sigma_i\nu_j| - \hat{b}_g \sum_{i=1}^n \bar{\phi}_{g_i}|\sigma_i|$$

$$(6.86)$$

which is misleading as it seems there is control effort applied to force error e_λ and the wrong conclusion of asymptotic convergence of e_λ may be drawn. This is due to the ignorance of the inherent property $R^{\mathrm{T}}J^{\mathrm{T}} = 0$. Thus, for the proposed scheme in this chapter, one can only guarantee the boundedness of e_λ, which will be confirmed in the simulation study.

6.5 SIMULATION STUDIES

Consider a mobile robot moving on a horizontal plane, driven by two rear wheels mounted on the same axis, and having one front passive wheel. The dynamic model can be expressed in the matrix form (6.1) with

$$
M(q) = \begin{bmatrix} m & 0 & mL\sin\theta \\ 0 & m & -mL\cos\theta \\ mL\sin\theta & -mL\cos\theta & I \end{bmatrix}
$$

$$
C(q,\dot{q}) = \begin{bmatrix} 0 & 0 & mL\dot{\theta}\cos\theta \\ 0 & 0 & mL\dot{\theta}\sin\theta \\ 0 & 0 & 0 \end{bmatrix}, \quad G(q) = 0, \quad B(q) = \frac{1}{R_1}\begin{bmatrix} \cos\theta & \cos\theta \\ \sin\theta & \sin\theta \\ R_2 & -R_2 \end{bmatrix}
$$

$$(6.87)$$

where $q = [x_c \; y_c \; \theta]^T \in R^3$ is the generalized coordinate with (x_c, y_c) being the coordinates of the center of mass of the vehicle, and θ being the orientation angle of the vehicle with respect to the X-axis, $\tau = [\tau_r \; \tau_l]^T \in R^2$ is the input vector with τ_r and τ_l being the torques provided by the motors mounted on the right and left respectively, m is the mass of the vehicle, I is its inertial moment around the vertical axis at the center of mass, L denotes the distance between the mid-distance of the rear wheels to the center of mass, $2R_1$ denotes the radius of the rear wheels, and $2R_2$ is the distance between the two rear wheels. The constraint forces are $f = J^T(q)\lambda$.

The nonholonomic constraints confine the vehicle to move only in the direction normal to the axis of the driving wheels, that is, the mobile bases satisfying the conditions of pure rolling and nonslipping

$$
\dot{x}_c \sin\theta - \dot{y}_c \cos\theta + L\dot{\theta} = 0 \tag{6.88}
$$

From (6.88), it is known that $J(q)$ and $R(q)$ are in the form

$$
J^T(q) = \begin{bmatrix} \sin\theta \\ -\cos\theta \\ L \end{bmatrix}, \quad R(q) = \begin{bmatrix} \cos\theta & -L\sin\theta \\ \sin\theta & L\cos\theta \\ 0 & 1 \end{bmatrix} \tag{6.89}
$$

Thus, the constraint forces can be written as $f = J^T(q)\lambda$ with

$$
\lambda = m\ddot{x}_c \sin\theta - m\ddot{y}_c \cos\theta + mL\ddot{\theta} \tag{6.90}
$$

In addition, the kinematic model (6.5) of the nonholonomic systems in terms of linear velocity v and angular velocity ω can be written as

$$\dot{z} = \begin{bmatrix} v \\ \omega \end{bmatrix} = \begin{bmatrix} \dot{z}_1 \\ \dot{z}_2 \end{bmatrix}, \quad \begin{bmatrix} \dot{x}_c \\ \dot{y}_c \\ \dot{\theta} \end{bmatrix} = \begin{bmatrix} \cos\theta & -L\sin\theta \\ \sin\theta & L\cos\theta \\ 0 & 1 \end{bmatrix} \begin{bmatrix} v \\ \omega \end{bmatrix} \tag{6.91}$$

The desired manifold Ω_{nhd} is chosen as

$$\Omega_{\text{nhd}} = \{(q, \dot{q}, \lambda) | z(t) = z_d(t), \ \dot{q} = S(q)\dot{z}_d(t), \ \lambda = \lambda_d\}$$

with $z_d = \dot{z}_d = 0$, $\lambda_d = 10$.

The existence of sgn-function in the controller (6.34) may inevitably lead to chattering in control torques. To avoid such a phenomenon, a sat-function is used to replace the sgn-function. The sat-function is given by

$$\text{sat}(\sigma) = \begin{cases} 1 & \text{if } \sigma > \epsilon \\ -1 & \text{if } \sigma < -\epsilon \\ \dfrac{1}{\epsilon}\sigma & \text{otherwise} \end{cases}$$

where $\epsilon = 0.01$ and $K_s = 5$ are chosen in the simulation.

The simulation is carried out using NF networks which are essentially the TSK-type fuzzy system with its membership function being chosen as the Gaussian function. Each element of the unknown system matrices $M(q)$ and $C(q, \dot{q})$ is modeled by the NF networks, which makes it different from conventional adaptive control design, where a relatively large amount of a prior knowledge about the system dynamics and the linear parametrization condition are required. The proposed adaptive NF controller, on the other hand, can be treated as an indirect adaptive scheme or partitioned NF systems [29,45], and does not require any precise knowledge on the system dynamics. The parameters in each NF subsystem can be separately tuned, which yield a faster updating speed, as can be seen from the simulation results.

In the simulation, the parameters of the system are taken as: $m = 10$ kg, $I = 5$ kgm^2, $R_1 = 0.05$ m, $R_2 = 0.5$ m, $L = 0.4$ m, $\tau_d(t) = [0.5\sin t, \ 0.1\sin t, \ 0.2\cos t]^{\text{T}}$, $q(0) = [2.0, \ 0.5, \ 0.785]^{\text{T}}$, $\dot{q}(0) = [0.2, \ 0.2, \ 0]^{\text{T}}$, and $\rho_1 = \text{diag}(5, 5)$, $\rho_2 = 1$, $\rho_3 = 10$. The control gain K_σ and force control gain K_λ are selected as $K_\sigma = \text{diag}(1, 1)$, $K_\lambda = 1$. The neural weights adaptation gains are chosen as $\Gamma_M = 0.1 I_{N_1}$, $\Gamma_C = 0.1 I_{N_2}$, with $N_1 = 100$ and $N_2 = 200$ being the number of rules of the NF system to estimate matrices M and C, respectively.

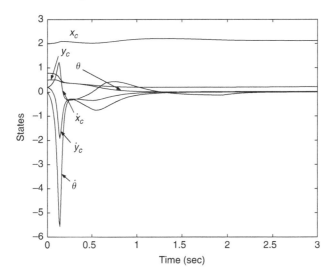

FIGURE 6.1   Responses of the states of the system.

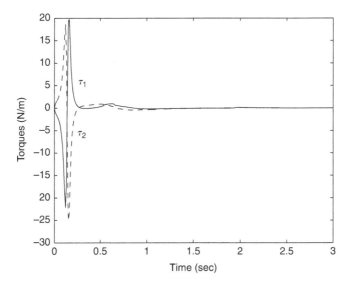

FIGURE 6.2   Control torques of the mobile robot.

The simulation results are shown in Figure 6.1 to Figure 6.5, among which, Figure 6.1 shows that the system's states response, including x_c, y_c, θ, $\dot{x}_c$, $\dot{y}_c$, and $\dot{\theta}$, are all bounded, and the control torques are bounded as can be seen in Figure 6.2. The estimates of the NN weights are shown to be bounded

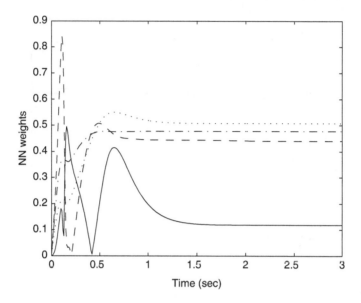

FIGURE 6.3   Responses of the norm of the NN weights.

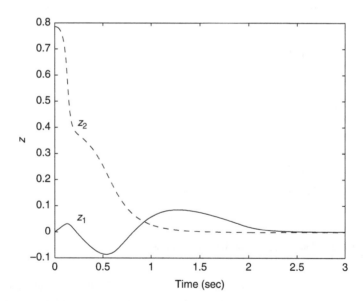

FIGURE 6.4   Responses of the internal states z.

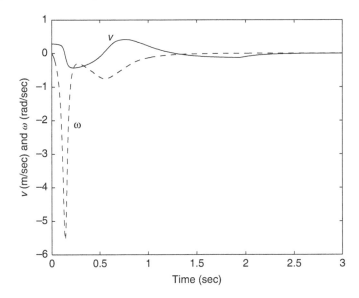

FIGURE 6.5   Responses of the linear velocity v and angular velocity ω.

in Figure 6.3 using some norms of the estimates for illustration. Figure 6.4 confirms that the stabilization of internal state z is achieved, while the linear velocity v and angular velocity ω are shown to converge asymptotically to zero in Figure 6.5.

In the simulations, the parameters have been selected at will to demonstrate the effectiveness of the proposed method. Different control performance can be achieved by adjusting parameter adaptation gains and other factors, such as the size of the networks, and the exploration of the knowledge of the systems. In fact, the control method has been developed as a turn-key solution without the need for much detailed analysis of the physical systems. For the best perform-ance, the physical properties should be explored and implemented in control system design. By examining the exact expressions for $D(q)$ and $C(q, \dot{q})$, we know that many of their elements are constants, such as m, I, and 0. In actual control system design, there is no need to estimate the 0s, while adaptive laws can be used to update the unknown m and I more elegantly.

6.6 CONCLUSION

In this chapter, adaptive NF control has been investigated for uncertain nonholonomic mobile robots in the presence of unknown disturbances. Despite the differences between the NNs and fuzzy logic systems, a unified adaptive NF control has been presented for function approximation. Because of the difficulty in deriving the rules in fuzzy systems for systems with little physical insights, the outputs of the "rules" are updated numerically using adaptive control techniques. It is shown that the controller can drive the system motion to converge to the desired manifold and at the same time guarantee the asymptotic convergence of the force tracking error without the requirement of the PE condition. By using NF approximation, the proposed controller is indeed a turned key solution for control system design as it requires little information on the system dynamics. Numerical simulation has been carried out to show the effectiveness of the proposed method for uncertain mobile robots.

REFERENCES

1. R. Brockett, "Asymptotic stability and feedback stabilization," *Diff. Geom. Control Theory* (Basel, Birkhauser), 181–208, 1983.
2. J. Guldner and V. I. Utkin, "Stabilization of nonholonomic mobile robots using Lyapunov function for navigation and sliding mode control," in *Proceedings of the 33rd IEEE Conference on Decision and Control* (Lake Buena, FL), pp. 2967–2972, 1994.
3. A. Astolfi, "Discontinuous control of nonholonomic systems," *Syst. Control Lett.*, 27, 37–45, 1996.
4. C. Samson, "Time-varying feedback stabilization of a nonholonomic wheeled mobile robot," *Int. J. Robotics Res.*, 12, 55–66, 1993.
5. O. J. Sødalen and O. Egeland, "Exponential stabilization of nonholonomic chained systems," *IEEE Trans. Automat. Contr.*, 40, 35–49, 1995.
6. I. Kolmanovsky and N. McClamroch, "Development in nonholonomic control problems," *IEEE Control Syst. Mag.*, 15, 20–36, 1995.
7. R. Murray and S. Sastry, "Nonholonomic motion planning: steering using sinusoids," *IEEE Trans. Automat. Contr.*, 38, 700–716, 1993.
8. W. Huo and S. S. Ge, "Exponential stabilization of nonholonomic systems: an ENI approach," *Int. J. Control*, 74, 1492–1500, 2001.
9. S. S. Ge, Z. Sun, T. H. Lee, and M. W. Spong, "Feedback linearization and stabilization of second-order nonholonomic chained systems," *Int. J. Control*, 74, 219–245, 2001.
10. R. Murray, "Control of nonholonomic systems using chained form," *Fields Inst. Commun.*, 1, 219–245, 1993.
11. Z. Sun, S. S. Ge, W. Huo, and T. H. Lee, "Stabilization of nonholonomic chained systems via nonregular feedback linearization," *Systems Control Lett.*, 44, 279–289, 2001.
12. Z. Jiang and H. Nijmeijer, "A recursive technique for tracking control of nonholonomic systems in chained form," *IEEE Trans. Automat. Contr.*, 44, 265–279, 1999.
13. J. P. Hespanha, S. Liberzon, and A. S. Morse, "Towards the supervisory control of uncertain nonholonomic systems," in *Proceedings of the American Control Conference* (San Diego, CA), pp. 3520–3524, 1999.

14. W. E. Dixon, D. M. Dawson, E. Zergeroglu, and A. Behal, *Nonlinear Control of Wheeled Mobile Robots*, London: Springer-Verlag, 2001.

15. S. S. Ge, J. Wang, T. H. Lee, and G. Y. Zhou, "Adaptive robust stabilization of dynamic nonholonomic chained systems," *J. Robot. Syst.*, 18, 119–133, 2001.

16. C. Y. Su and Y. Stepanenko, "Robust motion/force control of mechanical systems with classical nonholonomic constraints," *IEEE Trans. Automat. Contr.*, 39, 609–614, 1994.

17. W. Dong and W. Huo, "Adaptive stabilization of uncertain dynamic nonholonomic systems," *Int. J. Control*, 72, 1689–1700, 1999.

18. A. Bloch, M. Reyhanoglu, and N. H. McClamroch, "Control and stabilization of non-holonomic dynamic systems," *IEEE Trans. Automat. Contr.*, 37, 1746–1757, 1992.

19. G. Campion, B. d'Andrea Nobel, and G. Bastin, "Controllability and state feedback stability of nonholonomic mechanical systems," in *Advanced Robot Control* (C. C. de Wit, ed.), pp. 106–124, New York: Springer-Verlag, 1991.

20. C. Y. Su, T. P. Leung, and Q. J. Zhou, "Force/motion control of constrained robots using sliding mode," *IEEE Trans. Automat. Contr.*, 37, 668–672, 1992.

21. Z. P. Wang, S. S. Ge, and T. H. Lee, "Robust motion/force control of uncertain holonomic/nonholonomic mechanical systems," *IEEE/ASME Trans. Mechatronics*, 9, 118–123, 2004.

22. F. Hong, S. S. Ge, and T. H. Lee, "Robust adaptive fuzzy control of uncertain nonholonomic systems," in *Proceedings of IEEE International Symposium on Intelligent Control* (Taipei, Taiwan), pp. 192–197, 2004.

23. S. S. Ge, C. C. Hang, T. H. Lee, and T. Zhang, *Stable Adaptive Neural Network Control*, Boston, MA: Kluwer Academic Publisher, 2002.

24. A. Yesildirek and F. L. Lewis, "Feedback linearization using neural networks," *Automatica*, 31, 1659–1664, 1995.

25. E. B. Kosmatopoulos, "Universal stabilization using control Lyapunov functions, adaptive derivative feedback, and neural network approximator," *IEEE Trans. Syst., Man, Cybern. B*, 28, 472–477, 1998.

26. G. A. Rovithakis, "Stable adaptive neuro-control design via Lyapunov function derivative estimation," *Automatica*, 37, 1213–1221, 2001.

27. S. S. Ge, C. C. Hang, and T. Zhang, "Adaptive neural network control of nonlinear systems by state and output feedback," *IEEE Trans. Syst., Man, Cybern. B*, 29, 818–828, 1999.

28. N. Hovakimyan, F. Nardi, A. Calise, and K. Nakwan, "Adaptive output feedback control of uncertain nonlinear systems using single-hidden-layer neural networks," *IEEE Trans. Neural Networks*, 13, 1420–1431, 2002.

29. F. L. Lewis and S. S. Ge, "Neural networks in feedback control systems," in *Mechanical Engineer's Handbook*, New York: John Wiley, 2005.

30. J. Jang, C. Sun, and E. Mizutani, *Neuro-Fuzzy and Soft Computing*, Upper Saddle River, NJ: Prentice Hall, 1997.

31. L. X. Wang, *Adaptive Fuzzy Systems and Control: Design and Stability Analysis*, Englewood Cliffs, NJ: Prentice Hall, 1994.

32. S. S. Ge, T. H. Lee, and C. J. Harris, *Adaptive Neural Network Control of Robotic Manipulators*, London: World Scientific, 1998.

33. H. Han, C. Su, and Y. Stepanenko, "Adaptive control of a class of nonlinear systems with nonlinearly parameterized fuzzy approximators," *IEEE Trans. Fuzzy Syst.*, 9, 315–323, 2001.

34. L. Jia, S. S. Ge, and M.-S. Chiu, "Adaptive neuro-fuzzy control of nonaffine nonlinear systems," in *Proceedings of IEEE International Symposium on Intelligent Control* (Limassol, Cyprus), pp. 286–291, 2005.

35. A. Bloch and S. Drakunov, "Stabilization and tracking in the nonholonomic integrator via sliding modes," *Syst. Control Lett.*, 29, 91–99, 1996.

36. F. L. Lewis, C. T. Abdallah, and D. M. Dawson, *Control of Robot Manipulators*, New York: Macmillan, 1993.

37. Y. C. Chang and B. S. Chen, "Robust tracking designs for both holonomic and nonholonomic constrained mechanical systems: adaptive fuzzy approach," *IEEE Trans. Fuzzy Syst.*, 8, 46–66, 2000.

38. G. C. Walsh and L. G. Bushnell, "Stabilization of multiple input chained form control systems," *Syst. Control Lett.*, 25, 227–234, 1995.

39. R. M. Murray, "Nilpotent bases for a class of nonintegrable distributions with applications to trajectory generation for nonholonomic ssytems." *Mathematics of Control, Signals, and systems*, 7, 58–75, 1994.

40. L. Bushnell, D. Tilbury, and S. S. Sastry, "Steering three-input chained form non-holonomic systems using sinusoids: the firetruck example," in *Proceedings of the the European Control Conference* (Groningen, The Netherlands), pp. 1432–1437, 1993.

41. A. Isidori, *Nonlinear Control Systems*, Berlin; New York: Springer, 3rd ed., 1995.

42. T. Takagi and M. Sugeno, "Fuzzy identification of systems and its application to modeling and control," *IEEE Trans. Syst., Man, Cybern.*, 15, 116–132, 1985.

43. S. S. Ge, C. C. Hang, and T. Zhang, "Design and performance analysis of a direct adaptive controller for nonlinear systems," *Automatica*, 35, 1809–1817, 1999.

44. A. Astolfi, "On the stabilization of nonholonomic systems," in *Proceedings of the 33rd IEEE Conference on Decision and Control* (Lake Buena, FL), pp. 3481–3486, 1994.

45. F. L. Lewis, K. Liu, and A. Yesildirek, "Neural net robot controller with guaranteed tracking performance," *IEEE Trans. Neural Networks*, 6, 703–715, 1995.

BIOGRAPHIES

Fan Hong earned the Bachelor's degree from Xiamen University, China, in 1996, and the Ph.D. degree from the National University of Singapore in 2004. After her postdoctoral research in the National University of Singapore, she joined the A*STAR Data Storage Institute, Singapore, as a senior research fellow. Her research interests include adaptive control, neural networks, and nonlinear time-delay systems.

Shuzhi Sam Ge, IEEE Fellow, is a full professor with the Electrical and Computer Engineering Department at the National University of Singapore. He earned the B.Sc. degree from the Beijing University of Aeronautics and Astronautics (BUAA) in 1986, and the Ph.D. degree and the Diploma of Imperial College (DIC) from the Imperial College of Science, Technology and Medicine in 1993. His current research interests are in the control of nonlinear systems, hybrid systems, neural/fuzzy systems, robotics, sensor fusion, and real-time implementation. He has authored and co-authored over 200 international journal and conference papers, three monographs and co-invented three patents. He was the recipient of a number of prestigious research awards, and has been serving as the editor and associate editor of a number of flagship international journals. He is also serving as a technical consultant for the local industry.

Frank L. Lewis, IEEE Fellow, PE Texas, is a distinguished scholar professor and Moncrief-O'Donnell chair at the University of Texas at Arlington. He earned the B.Sc. degree in Physics and Electrical Engineering and the M.S.E.E. at Rice University, the MS in aeronautical engineering from the University of West Florida, and the Ph.D. at Georgia Institute of Technology. He works in feedback control and intelligent systems. He is the author of 4 U.S. patents, 160 journal papers, 240 conference papers, and 9 books. He received the Fulbright Research Award, the NSF Research Initiation Grant, and the ASEE Terman Award. He was selected as Engineer of the Year in 1994 by the Fort Worth IEEE Section and is listed in the Fort Worth Business Press *Top 200 Leaders in Manufacturing*. He was appointed to the NAE Committee on Space Station in 1995. He is an elected guest consulting professor at both Shanghai Jiao Tong University and South China University of Technology.

Tong Heng Lee earned the B.A. degree with First Class Honours in the engineering tripos from Cambridge University, England, in 1980, and the Ph.D. degree from Yale University in 1987. He is a professor in the Department of Electrical and Computer Engineering at the National University of Singapore. He is also currently Head of the Drives, Power, and Control Systems Group in this department. Professor Lee's research interests are in the areas of adaptive systems, knowledge-based control, intelligent mechatronics, and computational intelligence. He currently holds associate editor appointments in *Automatica; the IEEE Transactions in Systems, Man and Cybernetics; Control Engineering Practice* (an IFAC journal); the *International Journal of Systems Science* (Taylor & Francis, London); and *Mechatronics* journal (Oxford, Pergamon Press). Professor Lee was a recipient of the Cambridge University Charles Baker Prize in engineering. He has also co-authored three research monographs, and holds four patents (two of which are in the technology area of adaptive systems, and the other two are in the area of intelligent mechatronics).

7 Adaptive Control of Mobile Robots Including Actuator Dynamics

Zhuping Wang, Chun-Yi Su, and Shuzhi Sam Ge

CONTENTS

7.1 INTRODUCTION

A number of typical mobile robots can be described by the chained form or more general nonholonomic systems. Due to Brockett's theorem [1], it is well known that nonholonomic systems with restricted mobility cannot be stabilized to a desired configuration (or posture) via differentiable, or even continuous, pure-state feedback [2]. The design of stabilizing control laws for these systems is a challenging problem which has attracted much attention in the control community. A number of approaches have been proposed for the problem, which can be classified as (i) discontinuous time-invariant stabilization [3], (ii) time-varying stabilization [4], and (iii) hybrid stabilization [5, 6]. In References 7 and 8, an elegant approach to constructing piecewise continuous controllers has been developed. A nonsmooth state transformation is used to overcome the obstruction to stabilizability due to Brockett's theorem, and a smooth time-invariant feedback is used to stabilize the transformed system. In the original coordinates, the resulting feedback control is discontinuous. Various time varying controllers have been proposed in the literature [4,9]. The kinematic nonholonomic control systems can be asymptotically stabilized to an equilibrium point by smooth time-periodic static state feedback. However, the convergence rate for this method is comparatively slow. Hybrid controllers combine continuous time features with either discrete event features or discrete time features [6,10].

Among the many control strategies that have been proposed for various nonholonomic systems, research results can generally be classified into two classes. The first class is kinematic control, which provides the solutions only at the pure kinematic level, where the systems are represented by their kinematic models and velocity acts as the control input. Based on exact system kinematics, different control strategies have been proposed [4,5,8]. Recently, a few research works have been carried out to design controllers against possible existence of modeling uncertainties and external disturbances [11–13]. Robust exponential regulation is proposed in Reference 11 by assuming known bounds of the nonlinear drifts. It is also required that the x_0-subsystem is Lipschitz. To relax this condition, adaptive state feedback control is proposed in Reference 12 for systems with strong nonlinear drifts.

It is noted that one commonly used approach for control system design of nonholonomic systems is to convert, with appropriate state and input transformations, the original systems into some canonical forms for which controller design can be carried out more easily [14–17]. The chained form [14] and the power form [15] are two of the most important canonical forms of nonholonomic control systems. The class of nonholonomic systems in chained form was first introduced by Murray and Sastry [14] and has been studied as a benchmark example in the literatures. It is well known that many mechanical systems with nonholonomic constraints can be locally, or globally, converted to the chained form under coordinate change and state feedback [5,14]. The typical examples include tricycle-type mobile robots and cars towing several trailers. A new canonical form, called extended nonholonomic integrators (ENI) was presented in Reference 17, and it was shown that nonholonomic systems in ENI form, chained and power forms are equivalent, and can thus be dealt with in a unified framework. Using the special algebraic structures of the canonical forms, various feedback strategies have been proposed to stabilize nonholonomic systems in the literature [16–21].

The second class is dynamic control, taking inertia and forces into account, where the torque and force are taken as the control inputs. Different researchers have investigated this problem. Sliding mode control is applied to guarantee the uniform ultimate boundedness of tracking error in Reference 24. In Reference 23, stable adaptive control is investigated for dynamic nonholonomic chained systems with uncertain constant parameters. In Reference 24, adaptive robust stabilization is considered for dynamic nonholonomic chained systems with external disturbances. Using geometric phase as a basis, control of Caplygin dynamical systems was studied in Reference 2, and the closed-loop system was proved to achieve the desired local asymptotic stabilization of a single equilibrium solution. The principal limitation associated with these schemes is that controllers are designed at the velocity input level or torque input level and the actuator dynamics are excluded.

As demonstrated in Reference 25, actuator dynamics constitute an important component of the complete robot dynamics, especially in the case of high-velocity movement and highly varying loads. Many control methods have therefore been developed to take into account the effects of actuator dynamics (see, for instance, References 26–29). However, the literature is sparse on the control of the nonholonomic systems including the actuator dynamics.

In this chapter, the stabilization problem is considered for general nonholonomic mobile robots at the actuator level, taking into account the uncertainties in dynamics and the actuators. The controller design consists of two stages. In the first stage, to facilitate control system design, the nonholonomic kinematic subsystem is transformed into a skew-symmetric form and the properties of the overall systems are discussed. Then, a virtual adaptive controller is presented to compensate for the parametric uncertainties of the kinematic and dynamic subsystems. In the second stage, an adaptive controller is designed at the actuator level and the controller guarantees that the configuration state of the system converges to the origin.

This chapter is organized as follows: the model and model transformation of the system including actuator dynamics are presented in Section 7.2. The adaptive control law and stability analysis are presented in Section 7.3. Simulation studies are presented in Section 7.4 to show that the proposed method is effective. The conclusions are given in Section 7.5.

7.2 DYNAMIC MODELING AND PROPERTIES

In general, a nonholonomic system including actuator dynamics, having an n-dimensional configuration space with generalized coordinates $q = [q_1, \ldots, q_n]^\mathrm{T}$ and subject to $n - m$ constraints can be described by [30]

$$J(q)\dot{q} = 0 \tag{7.1}$$

$$M(q)\ddot{q} + C(q, \dot{q})\dot{q} + G(q) = B(q)K_N I + J^\mathrm{T}(q)\lambda \tag{7.2}$$

$$L\frac{\mathrm{d}I}{\mathrm{d}t} + RI + K_a \omega = v \tag{7.3}$$

where $M(q) \in R^{n \times n}$ is the inertia matrix which is symmetric positive definite, $C(q, \dot{q}) \in R^{n \times n}$ is the centripetal and coriolis matrix, $G(q) \in R^n$ is the gravitation force vector, $B(q) \in R^{n \times r}$ is the input transformation

matrix, $K_N \in R^{r \times r}$ is a positive definite diagonal matrix which characterizes the electromechanical conversion between current and torque, I denotes
an r-element vector of armature current, $J(q) \in R^{(n-m) \times n}$ is the matrix
associated with the constraint, and $\lambda \in R^{n-m}$ is the vector of constraint
forces. The terms $L = \text{diag}[L_1, L_2, L_3, \ldots, L_r]$, $R = \text{diag}[R_1, R_2, R_3, \ldots, R_r]$,
$K_a = \text{diag}[K_{a1}, K_{a2}, K_{a3}, \ldots, K_{ar}]$, $\omega = [\omega_1, \omega_2, \ldots, \omega_r]^T$, and $v \in R^r$ represent the equivalent armature inductances, resistances, back emf constants,
angular velocities of the driving motors, and the control input voltage vector,
respectively. Constraint 7.1 is assumed to be completely nonholonomic for all
$q \in \Re^n$ and $t \in \Re$. To completely actuate the nonholonomic system, $B(q)$ is
assumed to be a full-rank matrix and $r \geq m$.

Dynamic subsystem (7.2) has the following properties [31,32]:

Property 7.1   *There exists a so-called inertial parameter p and vector θ with
components depending on the mechanical parameters (mass, moment of inertia,
etc.,) such that*

$$M(q)\dot{v} + C(q, \dot{q})v + G(q) = \Phi(q, \dot{q}, v, \dot{v})\theta \tag{7.4}$$

*where Φ is a matrix of known functions of $q, \dot{q}, v,$ and $\dot{v}$; and θ is a vector of
inertia parameters and assumed completely unknown in this chapter.*

Property 7.2 $\dot{M} - 2C$ *is skew-symmetric.*

If matrix $N \in R^{n \times n}$ is skew-symmetric, then $N = -N^T$ and $Y^T N Y = 0$ for
all $Y \in R^n$.

Since $J(q) \in R^{(n-m) \times n}$, it is always realizable to find an m rank matrix
$S(q) \in R^{n \times m}$ formed by a set of smooth and linearly independent vector fields
spanning the null space of $J(q)$, that is,

$$S^T(q)J^T(q) = 0 \tag{7.5}$$

Since $S(q) = [s_1(q), \ldots, s_m(q)]$ is formed by a set of smooth and linearly
independent vector fields spanning the null space of $J(q)$, define an auxiliary
time function $v = [v_1, \ldots, v_m]^T \in R^m$ such that

$$\dot{q} = S(q)v(t) = s_1(q)v_1 + \cdots + s_m(q)v_m \tag{7.6}$$

Equation (7.6) is the so-called kinematic model of nonholonomic systems in
the literature.

Differentiating Equation (7.6) yields

$$\ddot{q} = \dot{S}(q)v + S(q)\dot{v} \tag{7.7}$$

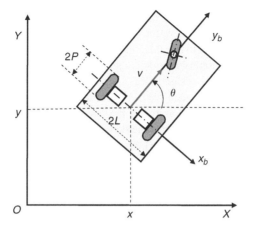

FIGURE 7.1   Differential drive wheeled mobile robot.

Substituting (7.6) and (7.7) into Equation (7.2), we have the transformed kinematic and dynamic subsystems of the whole nonholonomic system

$$\dot{q} = S(q)v = s_1(q)v_1 + \cdots + s_m(q)v_m \qquad (7.8)$$

$$M(q)S(q)\dot{v} + C_1(q,\dot{q})v + G(q) = B(q)K_N I + J^{\mathrm{T}}\lambda \qquad (7.9)$$

where

$$C_1(q,\dot{q}) = M(q)\dot{S} + C(q,\dot{q})S$$

In the actuator dynamics (7.3), the relationship between ω and v is dependent on the type of mechanical system and can be generally expressed as

$$\omega = \mu v \qquad (7.10)$$

The structure of μ depends on the mechanical systems to be controlled. For instance, in the simulation example, a type (2,0) differential drive mobile robot is used to illustrate the controller design, where μ can be derived as

$$\mu = \frac{1}{P}\begin{bmatrix} 1 & L \\ 1 & -L \end{bmatrix} \qquad (7.11)$$

where P and L are shown in Figure 7.1.

Eliminating ω from the actuator dynamics (7.3) by substituting (7.10), one obtains

$$L\frac{\mathrm{d}I}{\mathrm{d}t} + RI + K_a\mu v = v \qquad (7.12)$$

Until now we have brought the kinematics (7.1), dynamics (7.2), and actuator dynamics (7.3) of the considered nonholonomic system from the generalized coordinate system $q \in \mathfrak{R}^n$ to feasible independent generalized velocities $v \in \mathfrak{R}^m$ without violating the nonholonomic constraint (7.1).

For ease of controller design in this chapter, the existing results for the control of nonholonomic canonical forms in the literature are exploited. In the following, the kinematic nonholonomic subsystem (7.8) is first converted into the chained canonical form, and then to the skew-symmetric chained form for which a very nice controller structure [18] exists in the literature and can be utilized. This will be detailed later. The nonholonomic chained subsystem considered in this chapter is m-input, $(m - 1)$-chain, single-generator chained form given by [9,24]

$$\dot{x}_1 = u_1$$
$$\dot{x}_{j,i} = u_1 x_{j,i+1} \quad (2 \le i \le n_j - 1) \; (1 \le j \le m - 1) \qquad (7.13)$$
$$\dot{x}_{j,n_j} = u_{j+1}$$

Note that, in Equation (7.13), $X = [x_1, X_2, \ldots, X_m]^T \in R^n$ with $X_j = [x_{j-1,2}, \ldots, x_{j-1,n_{j-1}}]$ $(2 \le j \le m)$ are the states and $u = [u_1, u_2, \ldots, u_m]^T$ are the inputs of the kinematic subsystem.

The chained form is one of the most important canonical forms of nonholonomic control systems. It has been shown in References 5 and 14 and references therein that many nonlinear mechanical systems with nonholonomic constraints on velocities can be transformed, either locally or globally, to the chained form system via coordinates and state feedback transformation. The necessary and sufficient conditions for transforming system (7.8) into the chained form are given in Reference 33. The following assumption is made in this chapter.

Assumption 7.1 *The kinematic model of a nonholonomic system given by Equation (7.8) can be converted into chained form (7.13) by some diffeomorphic coordinate transformation $X = T_1(q)$ and state feedback $v = T_2(q)u$ where u is a new control input.*

The existence and construction of the transformation for these systems have been established in the literature [9,34]. It is given here for completeness of the presentation. For detailed explanations of the notations on differential geometry used below, readers are referred to Reference 35.

Proposition 7.1 *Consider the drift-free nonholonomic system*

$$\dot{q} = s_1(q)v_1 + \cdots + s_m(q)v_m$$

where $s_i(q)$ are smooth, linearly independent input vector fields. There exist state transformation $X = T_1(q)$ and feedback $v = T_2(q)u$ on some open set $U \subset R^n$ to transform the system into an $(m-1)$-chain, single-generator chained form, if and only if there exists a basis $f_1, \ldots, f_m$ for $\Delta_0 := \mathrm{span}\{s_1, \ldots, s_m\}$ which has the form

$$f_1 = (\partial/\partial q_1) + \sum_{i=2}^{n} f_1^i(q)\partial/\partial q_i$$

$$f_j = \sum_{i=2}^{n} f_j^i(q)\partial/\partial q_i, \quad 2 \le j \le m$$

such that the distributions

$$G_j = \mathrm{span}\{\mathrm{ad}_{f_1}^i f_2, \ldots, \mathrm{ad}_{f_1}^i f_m : 0 \le i \le j\},$$

$$0 \le j \le n-1$$

have constant dimension on U, are all involutive, and G_{n-1} has dimension $n-1$ on U [9, 34].

Using the constructive method given in Reference 14, a two input controllable system, that is,

$$\dot{q} = s_1(q)v_1 + s_2(q)v_2 \tag{7.14}$$

where $s_1(q), s_2(q)$ are linearly independent and smooth, $q \in R^n$, $v = [v_1, v_2]^T$, can be transformed into chained form (7.13) as

$$\dot{x}_1 = u_1$$
$$\dot{x}_2 = u_2$$
$$\dot{x}_3 = x_2 u_1 \tag{7.15}$$
$$\vdots$$
$$\dot{x}_n = x_{n-1} u_1$$

Under Assumption 7.1, that is, the existence of transformations $X = T_1(q)$, $v = T_2(q)u$, dynamic subsystem (7.9) is correspondingly converted into

$$M_2(X)S_2(X)\dot{u} + C_2(X, \dot{X})u + G_2(X) = B_2(X)K_N I + J_2^T(X)\lambda \tag{7.16}$$

where

$$M_2(X) = M(q)|_{q=T_1^{-1}(X)}$$
$$S_2(X) = S(q)T_2(q)|_{q=T_1^{-1}(X)}$$
$$C_2(X, \dot{X}) = C_1(q, \dot{q})T_2 + M(q)S(q)\dot{T}_2(q)|_{q=T_1^{-1}(X)}$$
$$G_2(X) = G(q)|_{q=T_1^{-1}(X)}$$
$$B_2(X) = B(q)|_{q=T_1^{-1}(X)}$$
$$J_2(X) = J(q)|_{q=T_1^{-1}(X)}$$

The actuator dynamics is transformed to

$$L\frac{dI}{dt} \mid RI + K_a Q(u, \mu, X) = v \qquad (7.17)$$

where

$$Q = \mu T_2(q)u|_{q=T_1^{-1}(X)}$$

Next, let us further transform the chained form into skew-symmetric chained form for the convenience of controller design. This transformation is the simple extension of the transformation of the one-generation, two-inputs, single-chained system given by Samson [18]. As shown in References 18, 23, and 24 by introducing the skew-symmetric chained form, via Lyapunov-like analysis, it is easier to design $U_2 = [u_2, \ldots, u_m]^T$ and a time-varying control u_1 to globally stabilize $[x_1, X_2, \ldots, X_m]^T$ of the kinematic subsystem, as will be detailed later.

The kinematic model of chained form (7.13) can be equivalently written as

$$\dot{X} = h_1(X)u_1 + \sum_{j=2}^{m} h_{2,j}u_j = h_1(X)u_1 + h_2 U_2 \qquad (7.18)$$

where

$$h_1(X) = [1, x_{1,3}, \ldots, x_{1,n1}, 0, \ldots, x_{m-1,3}, \ldots, x_{m-1,n_{m-1}}, 0]^T$$

$$h_2 = [h_{2,2}, \ldots, h_{2,m}]^T$$

and $h_{2,j}, j = 2, \ldots, m$ is an n-dimensional vector with the $1 + \sum_{i=1}^{j}(n_i - 1)$th element being 1 and other elements being zero.

Consider the following coordinates transformation

$$z_1 = x_1$$
$$z_{j,2} = x_{j,2}$$
$$z_{j,3} = x_{j,3}$$
$$z_{j,i+3} = \rho_{j,i} z_{j,i+1} + L_{h_1} z_{j,i+2} \quad (1 \leq i \leq n_j - 3) \, (1 \leq j \leq m - 1)$$
(7.19)

where $\rho_{j,i}$ are real positive numbers, and $L_{h1} z_{j,i} = (\partial z_{j,i}/\partial X) h_1(X)$ are the Lie derivatives of $z_{j,i}$ along $h_1(X)$. This transformation can convert the original chained system into the skew-symmetric chained form.

Define $Z = [z_1, z_{1,2}, \ldots, z_{1,n_1}, \ldots, z_{m-1,2}, \ldots, z_{m-1,n_{m-1}}]^T \in R^n$. Coordinate transformation (7.19) can also be written in a matrix form as below:

$$Z = \Psi X$$

where $\Psi = \text{diag}[1, \Psi_1, \ldots, \Psi_{m-1}]^T$ with $\Psi_k = [\psi_{j,i}] \in R^{n_k - 1 \times n_k - 1}$ being

$$\psi_{j,j} = 1 \qquad\qquad (j = 1, 2, \ldots, n_k - 1)$$
$$\psi_{j,i} = 0 \qquad\qquad (j < i; i, j = 1, 2, \ldots, n_k - 1)$$
$$\psi_{j,i} = 0 \qquad\qquad ((i + j) \bmod 2 \neq 0)$$
$$\psi_{j,i} = \rho_{j,i-3}\psi_{j,i-2} + \psi_{j-1,i-1} \quad (j = 3, 4, \ldots, n_k - 1; i = 1, 2, \ldots, n_k - 1)$$
(7.20)

It is explicit that matrix Ψ is of full rank. Moreover, $L_{h_2} z_{j,i} U_2 = 0$ $(1 \leq i \leq n_j - 1)$, and $L_{h_2} z_{j,n_j} U_2 = u_{j+1}$. Taking the time derivative of $z_{j,i+3}$ and using (7.18), we have

$$\dot{z}_{j,i+3} = \frac{\partial z_{j,i+3}}{\partial X} \dot{X} = (L_{h_1} z_{j,i+3}) u_1 + (L_{h_2} z_{j,i+3}) U_2$$
(7.21)

From (7.19), we know that for $0 \leq i \leq n_j - 4$, there is

$$L_{h_1} z_{j,i+3} = -\rho_{j,i+1} z_{j,i+2} + z_{j,i+4}$$
(7.22)

Hence, for $0 \leq i \leq n_j - 4$, Equation (7.21) becomes

$$\dot{z}_{j,i+3} = -\rho_{j,i+1} u_1 z_{j,i+2} + u_1 z_{j,i+4} \quad (1 \leq j \leq m - 1)$$
(7.23)

while for $i = n_j - 3$

$$\dot{z}_{j,i+3} = L_{h_1} z_{j,n_j} u_1 + u_{j+1} \quad (1 \leq j \leq m - 1)$$
(7.24)

Thus the original system has been converted into the following skew-symmetric chained form with actuator dynamics:

$$\dot{z}_1 = u_1$$

$$\dot{z}_{j,2} = u_1 z_{j,3}$$

$$\dot{z}_{j,i+3} = -\rho_{j,i+1} u_1 z_{j,i+2} + u_1 z_{j,i+4} \quad (1 \le j \le m - 1)(0 \le i \le n_j - 4)$$

$$\dot{z}_{j,n_j} = L_{h1} z_{j,n_j} u_1 + u_{j+1} \tag{7.25}$$

$$M_3(Z)S_3(Z)\dot{u} + C_3(Z,\dot{Z})u + G_3(Z) = B_3(Z)K_N I + J_3^T(Z)\lambda \tag{7.26}$$

$$L\frac{dI}{dt} + RI + K_a Q_3(u, \mu, Z) = v \tag{7.27}$$

where

$$M_3(Z) = M_2(X)|_{X=\Psi^{-1}(Z)}$$

$$C_3(Z,\dot{Z}) = C_2(X,\dot{X})|_{X=\Psi^{-1}(Z)}$$

$$G_3(Z) = G_2(X)|_{X=\Psi^{-1}(Z)}$$

$$B_3(Z) = B_2(X)|_{X=\Psi^{-1}(Z)}$$

$$J_3(Z) = J_2(X)|_{X=\Psi^{-1}(Z)}$$

$$Q_3(u, \mu, Z) = Q_3(u, \mu, X)|_{X=\Psi^{-1}(Z)}$$

Multiplying S_3^T to both sides of (7.26), we have

$$S_3^T M_3 S_3 \dot{u} + S_3^T C_3 u + S_3^T G_3 = S_3^T B_3 K_N I \tag{7.28}$$

which is more appropriate for controller design as the constraint λ has been eliminated from the dynamic equation.

To facilitate controller design, the properties of dynamic model (7.26) are listed below.

Property 7.3 $M_4 = S_3^T M_3 S_3$ *is symmetric positive definite and bounded.*

Property 7.4 $\dot{M}_4 - 2S_3^T C_3$ *is a skew-symmetric matrix. This property will be fully exploited for control system design.*

Property 7.5 *The dynamics can be expressed in the linear-in-parameters form*

$$M_3(Z)S_3(Z)\dot{\xi} + C_3(Z,\dot{Z})\xi + G_3(Z) = \Phi_1(Z,\dot{Z},\xi,\dot{\xi})\theta \tag{7.29}$$

where $\Phi_1(Z, \dot{z}, \xi, \dot{\xi}) \in R^{n \times p}$ is the known regressor matrix and $\theta \in R^p$ is the unknown parameters vector of system. For any physical system, we know that $\|\theta\|$ is always bounded.

Assumption 7.2   *$B_3(Z)$ is assumed known because it is a function of fixed geometry of the system. Accordingly, $B_3(Z)$ is assumed to be known exactly for subsequent discussion.*

7.3   CONTROL SYSTEM DESIGN

Consider the nonholonomic systems described by Equations (7.25–7.27). An adaptive controller is designed to stabilize the system states Z to the origin. Since $Z = \Psi X$ is of global diffeomorphism, the stabilization problem of X is the same as the stabilization problem of Z.

The controller design will consist of two stages (i) a virtual adaptive control input I_d is designed so that the subsystems (7.25) and (7.26) converge to the origin, and (ii) the actual control input v is designed in such a way that $I \rightarrow I_d$. In turn, this allows Z to be stabilized to the origin.

The following theorems are useful for the controller design. They are given here for completeness.

Corollary 7.1   *(Corollary of Barbalat's theory [31]): If $f(t)$, $\dot{f}(t) \in L_\infty$, and $f(t) \in L_p$, for some $p \in [1, \infty)$, then $f(t) \rightarrow 0$ as $t \rightarrow \infty$.*

Theorem 7.1   *(The extended version of Barbalat's theorem [18]): If a differentiable function $f(t) : R^+ \rightarrow R$ converges to a limit value as t tends to infinity, and if its derivative $\mathrm{d}/\mathrm{d}t(f(t))$ is the sum of two terms, one being uniformly continuous and another tending to zero as t tends to infinity, then $\mathrm{d}/\mathrm{d}t(f(t))$ tends to zero when t tends to infinity.*

7.3.1   Kinematic and Dynamic Subsystems

Define an auxiliary vector $u_d \in R^m$ as

$$u_d = \begin{bmatrix} -k_{u1}z_1 + h(Z_2, t) \\ -(\rho_{1,n1-2}z_{1,n1-1} + L_{h1}z_{1,n1})u_{d1} - k_{u2}z_{1,n1} \\ \vdots \\ -(\rho_{m-1,n_{m-1}-2}z_{m-1,n_{m-1}-1} + L_{h1}z_{m-1,n_{m-1}})u_{d1} - k_{um}z_{m-1,n_{m-1}} \end{bmatrix}$$

$$(7.30)$$

where $Z = [z_1, Z_2^T]^T$, $Z_2 = [z_{1,2}, \ldots, z_{1,n_1}, z_{2,2}, \ldots, z_{2,n_2}, \ldots, z_{m-1,2}, \ldots,$ $z_{m-1,n_{m-1}}]^T$, k_{uj} $(1 \leq j \leq m)$, and ρ_{j,n_j-2} $(1 \leq j \leq m)$ are positive constants, and $h(Z_2, t)$ satisfies the following assumption as given in References 18 and 23.

Assumption 7.3 $h(Z_2, t)$ *is a function of class* $C^{p+1}(p \geq 1)$*, uniformly bounded with respect to t, with all successive partial derivatives also uniformly bounded with respect to t, and such that:*

(i) $h(Z_2, t)$ *is a function of class* $C^{j+1}, j \geq 1$*, and all its successive partial derivatives are uniformly bounded with respect to t, and* $h(0, t) = 0, \forall t$

(ii) $z_{j,i}$ *being bounded and* $\dot{z}_{j,i}, z_{j,i}\dot{h}(Z_2, t), 2 \leq j \leq m - 1, 1 \leq i \leq n_j$ *tending to zero imply that* $z_{j,i}, 2 \leq j \leq m - 1, 1 \leq i \leq n_j$ *tending to zero*

Remark 7.1   *Note that it is not difficult to find* $h(Z_2, t)$ *satisfying the required conditions just as shown in the simulation. In fact, function* $h(Z_2, t)$ *is referred to as the heat function, and its primary role is to force the system in motion as long as the system has not reached the desired equilibrium point, thus preventing the system's state from converging to other equilibrium points. The conditions imposed upon the heat function in Assumption 7.3 are not severe and can easily be met. For example, the following three functions all satisfy the conditions [18, 24]:*

$$h(Z_2, t) = \|Z_2\|^2 \sin(t)$$

$$h(Z_2, t) = \sum_{j=0}^{n-2} a_j \sin(\beta_j t) z_{2+j}$$

$$h(Z_2, t) = \sum_{j=0}^{n-2} a_j \frac{\exp(b_j z_{2+j}) - 1}{\exp(b_j z_{2+j}) + 1} \sin(\beta_j t)$$

with $a_j \neq 0, b_j \neq 0, \beta_j \neq 0,$ *and* $\beta_i \neq \beta_j$ *when* $i \neq j$.

Considering the parameter vector θ to be uncertain, a virtual control input I_d has to be designed in such a way that the outputs of the dynamic subsystem (the inputs of the kinematic system) u tend to the auxiliary signals u_d, and the controller design at the dynamic level is achieved. As has been shown in Reference 18, when u_1 tends to u_{d1}, $u_1 Z_2$ and $\dot{Z}_2$ converge to zero, the definition of $h(Z_2, t)$ will guarantee Z goes to zero as well.

Defining $\tilde{u} = u - u_d$ the control objective at the dynamic level is to synthesis I_d to make $\tilde{u} \to 0$ so that $u \to u_d$.

From Property 7.5, we have

$$M_3(Z)S_3(Z)\dot{u}_d + C_3(Z, \dot{Z})u_d + G_3(Z) = \Phi_1(Z, \dot{Z}, u_d, \dot{u}_d)\theta \qquad (7.31)$$

where θ is the unknown parametric vector and Φ_1 is the regressor matrix of known kinematic functions.

Define

$$
\Lambda = \begin{bmatrix}
\sum\limits_{i=1}^{m-1} \dfrac{z_{i,n_i-1}z_{i,n_i}}{\rho_{i,1}\cdots\rho_{i,n_i-3}} + \sum\limits_{j=1}^{m-1} \dfrac{z_{j,n_j}L_{h1}z_{j,n_j}}{\rho_{j,1}\cdots\rho_{j,n_j-2}} \\[3mm]
\dfrac{z_{1,n_1}}{\rho_{1,1}\cdots\rho_{1,n_1-2}} \\
\vdots \\
\dfrac{z_{m-1,n_{m-1}}}{\rho_{m-1,1}\cdots\rho_{m-1,n_{m-1}-2}}
\end{bmatrix}
\qquad (7.32)
$$

The virtual control input I_d is designed as

$$
I_d = \hat{K}_{N\text{Inv}}\,\tau_{md} \qquad (7.33)
$$

where

$$
\tau_{md} = [S_3^T B_3]^{-1} S_3^T [\Phi_1(Z,\dot{Z},u_d,\dot{u}_d)\hat{\theta} - K_e S_3(u-u_d) - S_3(S_3^T S_3)^{-1}\Lambda] \quad (7.34)
$$

where $\hat{\theta}$ is the estimate of the unknown inertia parameter θ, and the adaptive law for $\hat{\theta}$ is given by

$$
\dot{\hat{\theta}} = -\Gamma\Phi_1^T S_3\tilde{u} \qquad (7.35)
$$

where Γ is a symmetric positive definite constant matrix.

In the design of the control law, K_N is considered as an unknown parameter and $\hat{K}_{N\text{Inv}} \in R^{r\times r}$ is an estimation of the parameter K_N^{-1} and is given as below:

$$
\dot{\hat{K}}_{N\text{Inv}} = \text{diag}[\dot{\hat{K}}_{N\text{Inv}i}] = \text{diag}[-f_i\tau_{mdi}] \qquad (7.36)
$$

where $F^T = [f_1, f_2, \ldots, f_r] = \tilde{u}^T S_3^T B_3$.

7.3.2 Control Design at the Actuator Level

Until now, we have designed a virtual controller I_d and embedded controller u_d for kinematic and dynamic subsystems, respectively. u tends to u_d can be guaranteed, if the actual input control signal of the dynamic system I be of the form I_d which can be realized from the actuator dynamics by the design of the actual control input v. On the basis of the above statements, we can conclude that if v is designed in such a way that I tends to I_d then $Z \to 0$ and $\tilde{u} \to 0$.

Define $I = e_I + I_d$ and substituting in (7.27) one gets

$$L\dot{e}_I + RI + K_aQ = -L\dot{I}_d + v$$

Adding K_Ie_I on both sides in the equation given above, one gets

$$L\dot{e}_I + RI + K_aQ + K_Ie_I = K_Ie_I - L\dot{I}_d + v \qquad (7.37)$$

When the actuator parameters L, R, and K_a are available for controller design, the control input v can be easily given as

$$v = L\dot{I}_d + RI - K_Ie_I + K_aQ + H \qquad (7.38)$$

where

$$H = -\hat{K}_NF \qquad (7.39)$$

with $\hat{K}_N \in R^{r \times r}$ is an estimate of the unknown parameter K_N, and the adaptation law for $\hat{K}_N$ is given by

$$\dot{\hat{K}}_N = \text{diag}[\dot{\hat{K}}_{Ni}] = \text{diag}[\gamma_N e_{Ii} f_i] \qquad (7.40)$$

where $\gamma_N > 0$ is a design constant, and determines the rate of adaptation.

When the actuator parameters L, R, and K_a are considered unknown for controller design, they are estimated online adaptively. Consider the adaptive control

$$v = \hat{L}\dot{I}_d + \hat{R}I - K_Ie_I + \hat{K}_aQ + H \qquad (7.41)$$

where $\hat{L} \in R^{r \times r}$, $\hat{R} \in R^{r \times r}$, and $\hat{K}_a \in R^{r \times r}$ are the estimates of the unknown parameters L, R, and K_a respectively, which are adaptively tuned as follows:

$$\dot{\hat{L}} = \text{diag}[\dot{\hat{L}}_i] = \text{diag}[-\gamma_L \dot{I}_{di} e_{Ii}] \qquad (7.42)$$

$$\dot{\hat{R}} = \text{diag}[\dot{\hat{R}}_i] = \text{diag}[-\gamma_R I_i e_{Ii}] \qquad (7.43)$$

$$\dot{\hat{K}}_a = \text{diag}[\dot{\hat{K}}_{ai}] = \text{diag}[-\gamma_a Q_i e_{Ii}] \qquad (7.44)$$

where γ_L, γ_R, $\gamma_a > 0$ are design constants, which determine rates of the adaptation for $\hat{L}_i$, $\hat{R}_i$, and $\hat{K}_{ai}$, respectively, and H is given by Equation (7.39).

Substituting (7.33) into (7.26) and using $I = e_I + I_d$, the error dynamics for the dynamic subsystem can be obtained

$$S_3^T M_3 S_3 \dot{\tilde{u}} = S_3^T \Phi_1 \tilde{\theta} - S_3^T C_3 \tilde{u} - S_3^T K_e S_3 \tilde{u} - \Lambda$$
$$+ S_3^T B_3 K_N \tilde{K}_{N\text{Inv}} \tau_{md} + S_3^T B_3 K_N e_I \qquad (7.45)$$

where $\tilde{K}_{N\text{Inv}} = \hat{K}_{N\text{Inv}} - K_{N\text{Inv}}$.

Denote $\tilde{L} = L - \hat{L}$, $\tilde{R} = R - \hat{R}$, $\tilde{K}_a = K_a - \hat{K}_a$, and $\tilde{K}_N = K_N - \hat{K}_N$.

Substituting (7.38) into (7.27), we have the error dynamics for the actuator dynamic subsystem

$$L\dot{e}_I = -K_I e_I + \tilde{L}\dot{I}_d + \tilde{R}I + \tilde{K}_a Q + H \tag{7.46}$$

The closed-loop stability is summarized in Theorem 7.2.

Theorem 7.2   *For a nonholonomic system described by (7.25)–(7.27), using the control law (7.41) with the virtual control (7.33) and the parameter adaptation laws (7.35), (7.40), (7.42)–(7.44), Z is globally asymptotically stabilizable at the origin $Z = 0$.*

Proof   For the convenience of proof, define the following three functions:

$$V_1 = \frac{1}{2}\tilde{u}^{\mathrm{T}}M_4(Z)\tilde{u} + \frac{1}{2}\tilde{\theta}^{\mathrm{T}}\Gamma^{-1}\tilde{\theta} + \frac{1}{2}\sum_{i=1}^{r}\gamma_N^{-1}K_{Ni}\tilde{K}_{N\mathrm{Inv}i}^2 \tag{7.47}$$

$$V_2 = \sum_{j=1}^{m-1}\frac{1}{2}\left[z_{j,2}^2 + \frac{1}{\rho_{j,1}}z_{j,3}^2 + \cdots + \frac{1}{\prod_{i=1}^{n_j-2}\rho_{j,i}}z_{j,n_j}^2\right] \tag{7.48}$$

$$V_3 = \frac{1}{2}e_I^{\mathrm{T}}Le_I + \frac{1}{2}\sum_{i=1}^{r}\gamma_L^{-1}\tilde{L}_i^2 + \frac{1}{2}\sum_{i=1}^{r}\gamma_R^{-1}\tilde{R}_i^2$$

$$+ \frac{1}{2}\sum_{i=1}^{r}\gamma_a^{-1}\tilde{K}_{ai}^2 + \frac{1}{2}\sum_{i=1}^{r}\gamma_N^{-1}\tilde{K}_{Ni}^2 \tag{7.49}$$

where $\tilde{\theta} = \hat{\theta} - \theta$. Since θ is a constant vector, we have $\dot{\tilde{\theta}} = \dot{\hat{\theta}}$

The derivative of V_1 along Equation (7.45) is given as

$$\dot{V}_1 = \tilde{u}^{\mathrm{T}}M_4(Z)\dot{\tilde{u}} + \frac{1}{2}\tilde{u}^{\mathrm{T}}\dot{M}_4(Z)\tilde{u} + \tilde{\theta}^{\mathrm{T}}\Gamma^{-1}\dot{\tilde{\theta}}$$

$$= \tilde{u}^{\mathrm{T}}(S_3^{\mathrm{T}}\Phi_1\tilde{\theta} - S_3^{\mathrm{T}}K_e S_3\tilde{u} - \Lambda + S_3^{\mathrm{T}}B_3 K_N \tilde{K}_{N\mathrm{Inv}}\tau_{md}) + \tilde{\theta}^{\mathrm{T}}\Gamma^{-1}\dot{\tilde{\theta}}$$

$$+ \sum_{i=1}^{r}\gamma_N^{-1}K_{Ni}\tilde{K}_{N\mathrm{Inv}i}\dot{\tilde{K}}_{N\mathrm{Inv}i} + \tilde{u}^{\mathrm{T}}S_3^{\mathrm{T}}B_3 K_N e_I \tag{7.50}$$

where the property that $\dot{M}_4 - 2S_3^{\mathrm{T}}C_3$ is skew-symmetric has been used.

The time derivative of V_2 is given by

$$\dot{V}_2(Z_2) = \sum_{j=1}^{m-1} \left[z_{j,2}\dot{z}_{j,2} + \frac{1}{\rho_{j,1}}z_{j,3}\dot{z}_{j,3} + \cdots + \frac{1}{\prod_{i=1}^{n_j-2}\rho_{j,i}}z_{j,n_j}\dot{z}_{j,n_j} \right] \tag{7.51}$$

Substituting (7.25) into (7.51), we have

$$\dot{V}_2 = \sum_{j=1}^{m-1} \left[z_{j,2}u_1 z_{j,3} - \frac{1}{\rho_{j,1}}z_{j,3}\rho_{j,1}u_1 z_{j,2} + \frac{1}{\rho_{j,1}}z_{j,3}u_1 z_{j,4} + \cdots \right.$$

$$- \frac{1}{\prod_{i=1}^{n_j-3}\rho_{j,i}}z_{j,n_j-1}\rho_{j,n_j-3}u_1 z_{j,n_j-2} + \frac{1}{\prod_{i=1}^{n_j-3}\rho_{j,i}}z_{j,n_j-1}u_1 z_{j,n_j}$$

$$\left. + \frac{1}{\prod_{i=1}^{n_j-2}\rho_{j,i}}z_{j,n_j}(L_{h1}z_{j,n_j}u_1 + u_{j+1}) \right]$$

$$= \sum_{j=1}^{m-1} \left[\frac{1}{\prod_{i=1}^{n_j-2}\rho_{j,i}}z_{j,n_j}((\rho_{j,n_j}-2z_{j,n_j-1}+L_{h1}z_{j,n_j})u_1 + u_{j+1}) \right] \tag{7.52}$$

The time derivative of V_3 is given by

$$\dot{V}_3 = -e_I^\mathsf{T}K_I e_I + e_I^\mathsf{T}\tilde{L}\dot{I}_d + e_I^\mathsf{T}\tilde{R}I + e_I^\mathsf{T}\tilde{K}_a Q$$

$$+ e_I^\mathsf{T}H + \sum_{i=1}^{r}\gamma_L^{-1}\tilde{L}_i\dot{\tilde{L}}_i + \sum_{i=1}^{r}\gamma_R^{-1}\tilde{R}_i\dot{\tilde{R}}_i$$

$$+ \sum_{i=1}^{r}\gamma_a^{-1}\tilde{K}_{ai}\dot{\tilde{K}}_{ai} + \sum_{i=1}^{r}\gamma_N^{-1}\tilde{K}_{Ni}\dot{\tilde{K}}_{Ni} \tag{7.53}$$

For stability analysis, let us consider the following Lyapunov function candidate:

$$V = V_1 + V_2 + V_3 \tag{7.54}$$

Combining Equation (7.50), Equation (7.52), and Equation (7.53), and using the adaptation laws (7.35), (7.40), (7.42) to (7.44), the derivative of V can be obtained as

$$\dot{V} = -\sum_{j=1}^{m-1} \frac{1}{\prod_{i=1}^{n_j-2}\rho_{j,i}}k_{u_{j+1}}z_{j,n_j}^2 - \tilde{u}^\mathsf{T}S_3^\mathsf{T}K_e S_3\tilde{u} - e_I^\mathsf{T}K_I e_I \tag{7.55}$$

We have $\dot{V} \leq 0$. Accordingly, $\tilde{u}$, $\tilde{\theta}$, Z_2, e_I, $\tilde{L}$, $\tilde{R}$, $\tilde{K}_a$, $\tilde{K}_N$ are all bounded in the sense of Lyapunov.

From Equation (7.55), using the Corollary of Baralart's theory [31], $\tilde{u} \to 0$ as $t \to \infty$, $z_{j,n_j} \to 0$ $(1 \leq j \leq m-1)$ as $t \to \infty$, and $e_I \to 0$ as $t \to \infty$.

Next, let us prove the asymptotic stability of Z.

The first equation of the controlled system is

$$\dot{z}_1 = -k_{u1}z_1 + h(Z_2, t) + \tilde{u}_1 \tag{7.56}$$

From Assumption 7.3, we know that $h(Z_2, t)$ is uniformly bounded. In addition, with $\tilde{u}_1$ converging to zero, (7.56) is a stable linear system subjected to the bounded additive perturbation $h(Z_2, t) + e_{u1}$. Therefore, $z_1(t)$ is also bounded uniformly.

Because z_1 and $h(Z_2, t)$ are bounded, it is clear that u_{d1} is bounded from (7.30). Together with $\tilde{u}$ converging to zero, u_1 is bounded. Since u_1 and Z_2 are bounded, $\tilde{u}$ goes to zero, u_{dj} and u_j $(2 \leq j \leq m)$ are bounded. Under the condition that Z_2, u_1, and u_j $(2 \leq j \leq m)$ are bounded, $\dot{z}_{j,n_j}$ and $\dot{z}_{j,i}$ $(1 \leq j \leq m-1, 2 \leq i \leq n_j - 1)$, from (7.25), are bounded.

In the following, let us show that $u_{d1}Z_2$ tends to zero. For $1 \leq j \leq m-1$, since u_{d1} is bounded and z_{j,n_j} tends to zero, $u_{d1}^2 z_{j,n_j}$ tends to zero. Taking the time derivative of $u_{d1}^2 z_{j,n_j}$, we have

$$\frac{\mathrm{d}}{\mathrm{d}t}(u_{d1}^2 z_{j,n_j}) = u_{d1}^2(-k_{u_j+1}z_{j,n_j} - \rho_{j,n_j-2}u_{d1}z_{j,n_j-1}$$

$$+ \tilde{u}_1 L_{h1}z_{j,n_j} + \tilde{u}_2) + z_{j,n_j}\frac{\mathrm{d}}{\mathrm{d}t}(u_{d1})^2$$

$$= -\rho_{j,n_j-2}u_{d1}^3 z_{j,n_j-1} + (2\dot{u}_{d1}u_{d1}z_{j,n_j} - k_{u_j+1}u_{d1}^2 z_{j,n_j}$$

$$+ u_{d1}^2 \tilde{u}_1 L_{h1}z_{j,n_j} + u_{d1}^2 \tilde{u}_2) \tag{7.57}$$

Since

$$\frac{\mathrm{d}}{\mathrm{d}t}u_{d1}^3 z_{j,n_j-1} = u_{d1}^3 \dot{z}_{j,n_j-1} + 3u_{d1}^2 \dot{u}_{d1}z_{j,n_j-1}$$

is bounded, the first term in (7.57) is uniformly continuous. Together with the fact that all other terms in (7.57) tend to zero (since $u_{d1}z_{j,n_j}$ and $\tilde{u}$ tend to zero), from the extended version of Barbalat's Lemma, $(\mathrm{d}/\mathrm{d}t)(u_{d1}^2 z_{j,n_j})$ tends to zero. Therefore, $u_{d1}^3 z_{j,n_j-1}$ also tends to zero. So $u_{d1}z_{j,n_j-1}$ also tends to zero.

Differentiating $u_{d1}^2 z_{j,n_j-1}$ yields

$$\frac{d}{dt}(u_{d1}^2 z_{j,n_j-1}) = 2u_{d1}\dot{u}_{d1}z_{j,n_j-1} + u_{d1}^2(-\rho_{j,n_j-3}u_{d1}z_{j,n_j-2} + u_{d1}z_{j,n_j}$$

$$- \rho_{j,n_j-3}z_{j,n_j-2}\tilde{u}_1 + \tilde{u}_1 z_{j,n_j})$$

$$= -\rho_{j,n_j-3}u_{d1}^3 z_{j,n_j-2} + (2\dot{u}_{d1}u_{d1}z_{j,n_j-1} + u_{d1}^2(-\rho_{j,n_j-3}z_{j,n_j-2}\tilde{u}_1$$

$$+ \tilde{u}_1 z_{j,n_j}) + u_{d1}^3 z_{j,n_j}) \tag{7.58}$$

where the first term is uniformly continuous since its time derivative is bounded, the other terms tend to zero. From the extended version of Barbalat's Lemma, $(d/dt)(u_{d1}^2 z_{j,n_j-1})$ tends to zero. Therefore, $u_{d1}^3 z_{j,n_j-2}$ and $u_{d1}z_{j,n_j-2}$ tend to zero.

Taking the time derivative of $u_{d1}^2 z_{j,i}, 2 \leq i \leq n_j - 2$ and repeating the above procedure iteratively, one obtains that $u_{d1}z_{j,i}, 2 \leq i \leq n_j$ tends to zero. From (7.25) and considering $L_{h1}z_{j,n_j}$ being a linear combination of $z_{j,i}, 2 \leq i \leq n_j$, we know $u_{d2}, \dot{Z}_2$ tends to zero.

Differentiating $u_{d1}z_{j,i}, 2 \leq i \leq n_j - 1$ yields

$$\frac{d}{dt}(u_{d1}z_{j,i}) = \dot{u}_{d1}z_{j,i} + u_{d1}\dot{z}_{j,i}$$

$$= z_{j,i}\dot{h} + (-k_{u_1}u_{d1}z_{j,i} - k_{u_1}\tilde{u}_1 z_{j,i} + u_{d1}\dot{z}_{j,i})$$

where the first term is uniformly continuous, the other terms tend to zero. From the extended version of Barbalat's Lemma, $z_{j,i}\dot{h}$ tends to zero. By condition (ii) in Assumption 7.3 on h, it can be concluded that $z_{j,i}$ tends to zero, which leads to h tending to zero. By examining (7.30), noting $\tilde{u}_1$ tending to zero and condition (i) in Assumption 7.3, z_1 tends to zero. From (7.25) and condition (i), u_{d1} tends to zero, therefore $u = u_d + \tilde{u}$ tends to zero. The theorem is proved.

Remark 7.2 *In Theorem 7.2, it has been proven that Z is globally asymptotically stabilizable, and all the signals in the closed-loop are bounded. Accordingly, we can only claim the boundedness of the estimated parameters and no conclusion can be made on its convergence. In general, to guarantee the convergence of the parameter estimation errors, persistently exciting trajectories are needed [31, 36], which is hard to meet in practice. Therefore, for the globally asymptotical stability of Z, it is an advantage to remove the stringent requirement of persistent excitation conditions for parameter convergence in actual implementation.*

7.4 SIMULATION

Consider a wheeled mobile robot moving on a horizontal plane, as shown in Figure 7.1, which has three wheels (two are differential drive wheels, one is a caster wheel), and is characterized by the configuration $q = [x, y, \theta]^T$. We assume that the robot does not contain flexible parts, all steering axes are perpendicular to the ground, the contact between wheels and the ground satisfies the condition of pure rolling and nonslipping.

The complete nonholonomic dynamic model of the wheeled mobile robot is given by

$$J(q)\dot{q} = 0 \tag{7.59}$$

$$M(q)\ddot{q} + C(q, \dot{q})\dot{q} + G(q) = B(q)K_N I + J^T(q)\lambda \tag{7.60}$$

$$L\dot{I} + RI + K_a \omega = v \tag{7.61}$$

The constraint of the nonslipping condition can be written as

$$\dot{x}\cos\theta + \dot{y}\sin\theta = 0$$

From the constraint, we have

$$J(q) = [\cos\theta, \sin\theta, 0]$$

which leads to

$$S(q) = \begin{bmatrix} -\sin\theta & 0 \\ \cos\theta & 0 \\ 0 & 1 \end{bmatrix}$$

Lagrange formulation can be used to derive the dynamic equations of the wheeled mobile robot. Because the mobile base is constrained to the horizontal plane, its potential energy remain constant, and accordingly $G(q) = 0$. The kinematic energy K is given by [37]

$$K = \tfrac{1}{2}\dot{q}^T M(q)\dot{q}$$

where

$$M(q) = \begin{bmatrix} m_0 & 0 & 0 \\ 0 & m_0 & 0 \\ 0 & 0 & I_0 \end{bmatrix}$$

with m_0 being the mass of the wheeled mobile robot, and I_0 being its inertia moment around the vertical axis at point Q. As a consequence, we obtain

$$C(q, \dot{q})\dot{q} = \dot{M}(q)\dot{q} - \frac{\partial K}{\partial q} = 0$$

From Figure 7.1, we have

$$B(q) = 1/P \begin{bmatrix} -\sin\theta & -\sin\theta \\ \cos\theta & \cos\theta \\ L & -L \end{bmatrix}$$

where P is the radius of the wheels and $2L$ is the length of the axis of the two fixed differential drive wheels as shown in Figure 7.1. The matrices $K_N = \mathrm{diag}[K_{N1}, K_{N2}]$, $L = \mathrm{diag}[L_1, L_2]$, $R = \mathrm{diag}[R_1, R_2]$, $K_a = \mathrm{diag}[K_{a1}, K_{a2}]$, and ω is given by (7.10) and (7.11).

Following the description in Section 7.2, the dynamics of the wheeled mobile robot can be written as

$$\dot{x} = v_1 \cos\theta$$

$$\dot{y} = v_1 \sin\theta$$

$$\dot{\theta} = v_2$$

$$M(q)S(q)\dot{v} + C_1(q)v + G = BK_N I + J^T \lambda \tag{7.62}$$

$$L\frac{dI}{dt} + RI + K_a \mu v = v \tag{7.63}$$

where

$$C_1 = \begin{bmatrix} -m_0 \cos\theta\,\dot{\theta} & 0 \\ -m_0 \sin\theta\,\dot{\theta} & 0 \\ 0 & 0 \end{bmatrix}, \quad v = [v_1, v_2]^T$$

with v_1, v_2 the linear and angular velocities of the robot.

Considering the coordinates transformation $X = T_1(q)$ and state feedback $u = T_2^{-1}(q)v$ given by [38]

$$\begin{bmatrix} x_1 \\ x_2 \\ x_3 \end{bmatrix} = \begin{bmatrix} 0 & 0 & 1 \\ \cos\theta & \sin\theta & 0 \\ -\sin\theta & \cos\theta & 0 \end{bmatrix} \begin{bmatrix} x \\ y \\ \theta \end{bmatrix}$$

$$u_1 = v_2$$

$$u_2 = v_1 - v_2 x_2$$

together with the transform matrix $\Psi = I$ in this special case, system (7.63) is
converted to

$$\dot{z}_1 = u_1 \tag{7.64}$$

$$\dot{z}_2 = z_3 u_1 \tag{7.65}$$

$$\dot{z}_3 = u_2 \tag{7.66}$$

$$M_3(Z)S_3(Z)\dot{u} + C_3(Z,\dot{Z})u + G_3(Z) = B_3(Z)K_N I + J_3^{\mathrm{T}}(Z)\lambda \tag{7.67}$$

$$L\frac{dI}{dt} + RI + K_a Q_3(u,\mu,Z) = v \tag{7.68}$$

where

$$M_3(Z) = \begin{bmatrix} m_0 & 0 & 0 \\ 0 & m_0 & 0 \\ 0 & 0 & I_0 \end{bmatrix}$$

$$S_3(Z) = \begin{bmatrix} -z_2 \sin z_1 & -\sin z_1 \\ z_2 \cos z_1 & \cos z_1 \\ 1 & 0 \end{bmatrix}$$

$$C_3(Z,\dot{Z}) = \begin{bmatrix} -m_0\dot{z}_2 \sin z_1 - m_0 z_2 \cos z_1 \dot{z}_1 & -m_0 \cos z_1 \dot{z}_1 \\ -m_0\dot{z}_2 \cos z_1 - m_0 z_2 \sin z_1 \dot{z}_1 & -m_0 \sin z_1 \dot{z}_1 \\ 0 & 0 \end{bmatrix}$$

$$B_3(Z) = 1/P \begin{bmatrix} -\sin z_1 & -\sin z_1 \\ \cos z_1 & \cos z_1 \\ L & -L \end{bmatrix}$$

$$G_3 = 0$$

$$J_3(Z) = [\cos z_1 \ \sin z_1 \ 0]$$

$$Q_3(u,\mu,Z) = \mu T_2 u|_{q=T_1^{-1}(Z)}$$

and we have the following property for the system dynamics:

$$S_3^{\mathrm{T}} M_3 S_3 \dot{u}_d + S_3^{\mathrm{T}} C_3 u_d + S_3^{\mathrm{T}} G_3 = \Phi(Z,\dot{Z},u_d,\dot{u}_d)\theta$$

with the inertia parameters vector $\theta = [m_0, I_0]^{\mathrm{T}}$ and

$$\Phi(Z,\dot{Z},u_d,\dot{u}_d) = \begin{bmatrix} z_2^2\dot{u}_{d1} + z_2\dot{u}_{d2} + z_2\dot{z}_2 u_{d1} & \dot{u}_{d1} \\ z_2\dot{u}_{d1} + \dot{u}_{d2} + \dot{z}_2 u_{d1} & 0 \end{bmatrix} \tag{7.69}$$

The auxiliary signal u_d is chosen as

$$u_d = \begin{bmatrix} u_{d1} \\ u_{d2} \end{bmatrix} = \begin{bmatrix} -k_{u1}z_1 + h(Z_2, t) \\ -\rho_1 z_2 u_{d1} - k_{u2}z_3 \end{bmatrix}$$

where $h(Z_2, t)$ is chosen as

$$h(Z_2, t) = (z_2^2 + z_3^2)\sin t$$

It is easy to see that the selected $h(Z_2, t)$ satisfies Assumption 7.3.

In the simulation, the parameters of the system are assumed to be $m_0 = I_0 = 1.0$, $P = 0.1$, $L = 1.0$, and $L_1 = L_2 = 1, R_1 = R_2 = 1, K_{N1} = K_{N2} = 1, K_{a1} = K_{a2} = 1$. The initial estimate $\hat{\theta}(0) = [0.5, 0.5]^T$ which is different from the true value. The design parameters are chosen as $k_{u1} = 0.2, k_{u2} = 1.0$, $\rho_1 = 1.0, \Gamma = \mathrm{diag}[10, 10]$, $\gamma_R = \gamma_L = \gamma_a = \gamma_N = 1$, and $K_e = \mathrm{diag}[5, 5]$.

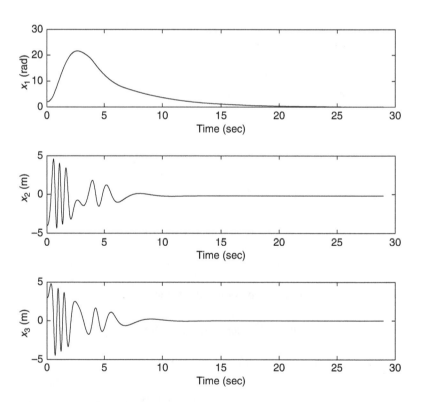

FIGURE 7.2   Responses of states x_1, x_2, and x_3.

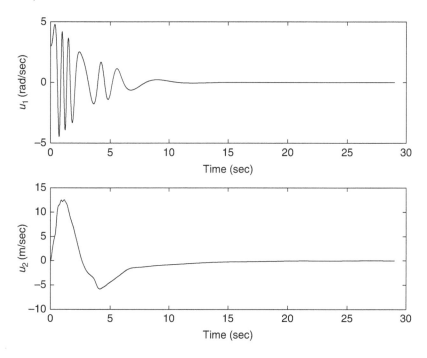

FIGURE 7.3   Responses of u_1 and u_2.

Simulation results are shown in Figure 7.2 to Figure 7.4. From Figure 7.2 and Figure 7.3, we can see that the responses of states x_1, x_2, x_3, u_1, and u_2 of the considered system asymptotically tend to zero. From Figure 7.4, the control sequence v_1 and v_2 remain bounded and tend to zero as well. The results of the simulation verify the validity of proposed algorithm.

7.5   CONCLUSION

In this chapter, stabilization of uncertain nonholonomic mobile robotic systems has been investigated with unknown constant inertia parameters and actuator dynamics. The controller design consists of two stages. In the first stage, for the convenience of controller design, the nonholonomic chained subsystems were first converted to the skew-symmetric chained form. Then a virtual adaptive controller was proposed where parametric uncertainties were compensated for by adaptive control techniques. In the second stage, an adaptive controller was designed at the actuator level to incorporate the actuator dynamics. The controller guarantees that the configuration state of the system converges to the origin. Throughout this chapter, feedback control design and stability analysis are performed via explicit Lyapunov techniques. Simulation studies on the

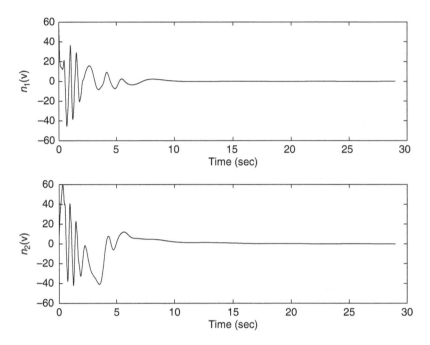

FIGURE 7.4   Control signals v_1 and v_2.

stabilization of unicycle wheeled mobile robot have been used to show the effectiveness of the proposed scheme.

REFERENCES

1. R. Brockett, "Asymptotic stability and feedback stabilization," in *Differential Geometry Control Theory*. Basel: Birkhauser, pp. 181–208, 1983.
2. A. Bloch, M. Reyhanoglu, and N. McClamroch, "Control and stabilization of nonholonomic dynamic systems," *IEEE Transactions on Automatic Control*, 37, 1746–1757, 1992.
3. J. Guldner and V. I. Utkin, "Stabilization of nonholonomic mobile robots using Lyapunov function for navigation and sliding mode control," in *Proceedings of the 33rd IEEE Conference on Decision & Control* (Lake Buena, FL), pp. 2967–2972, 1994.
4. C. Samson, "Time-varying feedback stabilization of a nonholonomic wheeled mobile robot," *International Journal of Robotics Research*, 12, 55–66, 1993.
5. I. Kolmanovsky and N. McClamroch, "Development in nonholonomic control problems," *IEEE Control System Magazine*, 15, 20–36, 1995.
6. O. J. Sordalen and O. Egeland, "Exponential stabilization of nonholonomic chained systems," *IEEE Transactions on Automatic Control*, 40, 35–49, 1995.

7. A. Astolfi, "On the stabilization of nonholonomic systems," in *Proceedings of the 33rd IEEE Conference on Decision & Control* (Lake Buena Vista, FL, USA), pp. 3481–3486, December 1994.

8. A. Astolfi, "Discontinuous control of nonholonomic systems," *Systems and Control Letters*, 27, 37–45, 1996.

9. G. C. Walsh and L. G. Bushnell, "Stabilization of multiple input chained form control systems," *System and Control Letters*, 25, 227–234, 1995.

10. C. Canudas de Wit, H. Berghuis, and H. Nijmeijer, "Practical stabilization of nonlinear systems in chained form," in *Proceedings of the 33rd IEEE Conference on Decision & Control* (Lake Buena Vista, FL, USA), pp. 3475–3480, 1994.

11. Z. P. Jiang, "Robust exponential regulation of nonholonomic systems with uncertainties," *Automatica*, 36, 189–209, 2000.

12. S. S. Ge, Z. P. Wang, and T. H. Lee, "Adaptive stabilization of uncertain nonholonomic systems by state and output feedback," *Automatica*, 39, 1451–1460, 2003.

13. W. E. Dixon, D. M. Dawson, E. Zergeroglu, and A. Behal, *Nonlinear Control of Wheeled Mobile Robots*. LNCIS 262, London: Springer-Verlag, 2001.

14. R. Murray and S. Sastry, "Nonholonomic motion planning: steering using sinusoids," *IEEE Transactions on Automatic Control*, 38, 700–716, 1993.

15. R. M'Closkey and R. Murray, "Convergence rate for nonholonomic systems in power form," in *Proceedings of the American Control Conference* (Chicago, USA), pp. 2489–2493, June 1992.

16. Z. Sun, S. S. Ge, W. Huo, and T. H. Lee, "Stabilization of nonholonomic chained systems via nonregular feedback linearization," *System and Control Letters*, 44, 279–289, 2001.

17. W. Huo and S. S. Ge, "Exponential stabilization of nonholonomic systems: an ENI approach," *International Journal of Control*, 74, 1492–1500, 2001.

18. C. Samson, "Control of chained systems: application to path following and time-varying point-stabilization of mobile robots," *IEEE Transactions on Automatic Control*, 40, 64–77, 1995.

19. S. S. Ge, Z. Sun, T. H. Lee, and M. W. Spong, "Feedback linearization and stabilization of second-order nonholonomic chained systems," *International Journal of Control*, 74, 1383–1392, 2001.

20. R. Murray, "Control of nonholonomic systems using chained form," *Fields Institute of Communication*, 1, 219–245, 1993.

21. Z. P. Wang, S. S. Ge, and T. H. Lee, "Robust adaptive neural network control of uncertain nonholonomic systems with strong nonlinear drifts," *IEEE Transactions on Systems, Man, and Cybernetics, Part B: Cybernetics*, 34, 2048–2059, 2004.

22. C. Y. Su and Y. Stepanenko, "Robust motion/force control of mechanical systems with classical nonholonomic constraints," *IEEE Transactions on Automatic Control*, 39, 609–614, 1994.

23. W. Dong and W. Huo, "Adaptive stabilization of uncertain dynamic nonholonomic systems," *International Journal of Control*, 72, 1689–1700, 1999.

24. S. S. Ge, J. Wang, T. H. Lee, and G. Y. Zhou, "Adaptive robust stabilization of dynamic nonholonomic chained systems," *Journal of Robotic Systems*, 18, 119–133, 2001.

25. M. C. Good, L. M. Sweet, and K. L. Strobel, "Dynamic models for control system design of integrated robot and drive systems," *Journal of Dynamic Systems, Measurement, and Control*, 107, 53–59, 1985.

26. J. H. Yang, "Adaptive robust tracking control for compliant-joint mechanical arms with motor dynamics," in *Proceedings of IEEE Conference on Decision & Control* (Phoenix, AZ), pp. 3394–3399, December 1999.

27. D. M. Dawson, Z. Qu, and J. Carroll, "Tracking control of rigid link electrically-driven robot manipulators," *International Journal of Control*, 56, 991–1006, 1992.

28. R. Colbaugh and K. Glass, "Adaptive regulation of rigid-link electrically-driven manipulators," in *Proceedings of IEEE International Conference on Robotics & Automation* (Nagoya, Japan), pp. 293–299, 1995.

29. C. Y. Su and Y. Stepanenko, "Hybrid adaptive/robust motion control of rigid-link electrically-driven robot manipulators," *IEEE Transactions on Robotics & Automation*, 11, 426–432, 1995.

30. A. Bloch and S. Drakunov, "Stabilization and tracking in the nonholonomic integrator via sliding mode," *Systems and Control Letters*, 29, 91–99, 1996.

31. S. S. Ge, T. H. Lee, and C. J. Harris, *Adaptive Neural Network Control of Robot Manipulators*. River Edge, NJ: World Scientific, 1998.

32. F. L. Lewis, C. T. Abdallah, and D. M. Dawson, *Control of Robots Manipulators*. New York: Macmillan, 1993.

33. R. Murray, "Nilpotent bases for a class of nonintergrable distributions with applications to trajectory generation for nonholonomic systems," *Mathematics of Control, Signals, and Systems*, 7, 58–75, 1994.

34. L. Bushnell, D. Tilbury, and S. S. Sastry, "Steering three-input chained form nonholonomic systems using sinusoids: the firetruck example," in *European Controls Conference* (Groningen, The Netherlands), pp. 1432–1437, 1993.

35. A. Isidori, *Nonlinear Control Systems*, 3rd edn. Berlin: Springer-Verlag, 1995.

36. J. J. E. Slotine and W. Li, *Applied Nonlinear Control*. Englewood Cliffs, NJ: Prentice Hall, 1991.

37. F. L. Lewis, S. Jagannathan, and A. Yesildirek, *Neural Network Control of Robots Manipulators and Nonlinear Systems*. 1 Gunpowder Square, London: Taylor & Francis, 1999.

38. C. C. de Wit, B. Siciliano, and G. Bastin, *Theory of Robot Control*. New York: Springer-Verlag, 1996.

BIOGRAPHIES

Zhuping Wang earned the B.E. and the M.E. degrees in automatic control from Northwestern Polytechnic University, China; and the Ph.D. degree from the National University of Singapore in 1994, 1997, and 2003, respectively. She is currently a professional officer in the Department of Electrical and Computer Engineering at the National University of Singapore, Singapore.

Dr. Wang's research interests include control of flexible link robots and smart materials robot control, control of nonholonomic systems, adaptive nonlinear control, neural network control, and robust control.

Chun-Yi Su earned his B.E. degree in control engineering from Xian University of Technology in 1982, his M.S. and Ph.D. degrees in control engineering from South China University of Technology, China, in 1987 and 1990, respectively. After a long stint at the University of Victoria, he joined Concordia University in 1998, where he is currently an associate professor and holds the Concordia Research Chair (Tier II) in intelligent control of nonsmooth dynamic systems.

Dr. Su's main research interests are in nonlinear control theory and robotics. He is the author and co-author of over 100 publications, which have appeared in journals, as book chapters, and in conference proceedings. He was the general co-chair of the Fourth International Conference on Control and Automation (ICCA'03).

Shuzhi Sam Ge IEEE Fellow, is a full professor with the Electrical and Computer Engineering Department at the National University of Singapore. He earned the B.Sc. degree from the Beijing University of Aeronautics and Astronautics (BUAA) in 1986, and the Ph.D. degree and the Diploma of Imperial College (DIC) from the Imperial College of Science, Technology and Medicine in 1993. His current research interests are in the control of nonlinear systems, hybrid systems, neural/fuzzy systems, robotics, sensor fusion, and real-time implementation. He has authored and co-authored over 200 international journal and conference papers, three monographs and co-invented three patents. He was the recipient of a number of prestigious research awards, and has been serving as the editor and associate editor of a number of flagship international journals. He is also a technical consultant for local industry.

8   Unified Control Design for Autonomous Car-Like Vehicle Tracking Maneuvers

Danwei Wang and Minhtuan Pham

CONTENTS

8.1   INTRODUCTION

Vehicle tracking has been one keen research topic on autonomous vehicles and mobile robotics in recent years. A vehicle tracking system consists of at least two vehicles. One vehicle leads a platoon while others autonomously track and follow the leader vehicle. An autonomous tracking controller is required for each of the follower vehicles. Based on the relative distance, orientation, velocity, and even acceleration of the leader vehicle, the controller generates corresponding control input for the follower vehicle. Many controllers for vehicle tracking have been proposed. In a planar configuration of vehicle tracking, the relative position between two vehicles is basically composed of two parts: longitudinal relative distance and lateral deviation.

Longitudinal control systems [1–5] concentrate on the longitudinal relative distance, also called intervehicular spacing, with the assumption that the vehicle following runs on a practically straight path or a fixed path without concerns with steering. Thus, the tracking error is the difference between the relative position and a predetermined spacing l. To further improve the tracking performance and stability, the relative velocity of the two vehicles is also taken into account. Using this additional tracking error, different control laws have been proposed, for example, a simple proportional integral differential (PID) controller [5] or with an additional quadratic term (PIQ controller) as in Reference 3, and an acceleration controller with a variable feedback gain as in Reference 2; or using adaptive control as in References 3 and 4. Lateral control, on the other hand, is used in two applications. The first one is lane following where all vehicles follow the center of the road or a sequence of landmarks [6–8]. The second application, which is of our interest, concerns the path traveled by the preceding vehicle or the leader vehicle. The only information that can be directly measured is the relative position and orientation between two consecutive vehicles. PID controllers are used for lateral control in References 9 and 10. A steering controller with nonlinear feedback is presented in Reference 11. This controller is based on a sliding mode observer and a linearized model of the truck to issue the steering command. To achieve better performance in the lane following method, part of the recently traveled path of the preceding vehicle is estimated and steering control can be obtained based on linearized [12] or nonlinear [13] dynamic/kinematic vehicle models.

Most of the above-mentioned controllers guarantee good tracking performances only when the leader vehicle moves forward in front of the follower vehicle. Backward tracking is still a challenge due to difficulties in backward driving as pointed out in Reference 14. Reference 14 presents a controller that imitates the human driving of a boat with the rudder. Some preliminary results on backward tracking for trailer systems have also been presented in References 15–17.

Lately, a tracking control method based on output feedback theory has been introduced in Reference 18, referred as the full-state tracking control for wheeled mobile robots. This nonlinear tracking method ensures exponential stability and convergence, and integrates both longitudinal control and lateral control into one controller. In this chapter, we present a unified control design for tracking maneuvers of two car-like mobile robots. The vehicle tracking maneuvers are formulated into an integrated framework with forward tracking, backward tracking, driving, and steering at kinematics and dynamics levels, respectively. A nonlinear controller with a few design parameters is designed for maneuvers with simultaneous driving as well as steering for vehicle tracking: in both forward tracking and backward tracking maneuvers. Tracking stability is ensured by the proper design of a stable performance target dynamics with a set of sufficient conditions for selecting design parameters. Simulation results show the effectiveness of the control scheme in both tracking cases. Tracking performance is evaluated with respect to the selection of parameters: the desired intervehicular spacing l and the desired steering angle multiplier p. The effects of the parameter value selections on the tracking performance are also examined via extensive simulations.

8.2 DYNAMICS OF TRACKING MANEUVERS

8.2.1 Vehicle Kinematics and Dynamics

Consider a car-like mobile robot with front wheels for steering and rear wheels for driving. Its kinematic model can be described by the following equation [18, 19]:

$$\dot{q} = G(\theta, \gamma)\mu \tag{8.1}$$

where $q = [x \quad y \quad \theta \quad \gamma]^{\mathrm{T}}$ is the state configuration of the vehicle with (x, y) being the generalized coordinates of the reference point located at the center of the rear axle, θ the heading angle of the vehicle with respect to the x-axis, and γ the steering angle of the front wheels; $\mu = [v \quad \omega]^{\mathrm{T}}$ contains the velocity v and the steering rate ω; and

$$G(\theta, \gamma) = \begin{bmatrix} \cos\theta & 0 \\ \sin\theta & 0 \\ \frac{1}{a}\tan\gamma & 0 \\ 0 & 1 \end{bmatrix} \tag{8.2}$$

with a being the length of the vehicle.

A dynamic model of the vehicle is as in (8.3)

$$\begin{cases} \dot{q} = G(\theta, \gamma)\mu \\ \dot{\mu} = u \end{cases} \tag{8.3}$$

where $u = [u_m \quad u_s]^{\mathrm{T}}$ consists of the driving acceleration u_m and the steering acceleration u_s homogenous to the driving and steering torques.

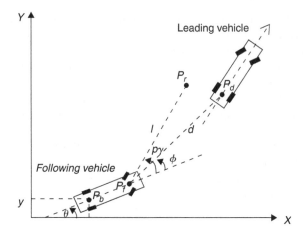

FIGURE 8.1   Forward tracking configuration.

8.2.2 Dynamics of Tracking Maneuvers

In the vehicle tracking system considered in this chapter, two car-like vehicles are moving in a horizontal plane. This vehicle tracking system can be executed in one of the two modes: forward tracking and backward tracking.

Forward Tracking. The leader vehicle moves forward in front of the follower vehicle as in Figure 8.1 and both vehicles move with positive velocities. In this case, the tracked point is the center point P_d at the rear axle of the leader vehicle. The relative distance between two vehicles is measured by the length $d > 0$ of $P_f P_d$ and the relative orientation angle ϕ is formed by the longitudinal axis $P_b P_f$ and $P_f P_d$ $(-\pi/2 \leq \phi \leq \pi/2)$.

Backward Tracking. The leader vehicle moves backward behind the follower vehicle as in Figure 8.2 and both vehicles move with negative velocities. In this case, the tracked point is the center point P_d at the front axle of the leader vehicle. The relative distance between two vehicles is measured by the length $d > 0$ of $P_b P_d$ and the relative orientation angle ϕ formed by the longitudinal axis $P_f P_b$ and $P_b P_d$ $(-\pi/2 \leq \phi \leq \pi/2)$.

For both cases, the point P_d of the leader vehicle is related to the point P_b of the follower vehicle by the unified function, referred as the virtual intervehicular connection as follows

$$P_d = z_d = \begin{bmatrix} x + \frac{1+f}{2}a\cos\theta + fd\cos(\theta + \phi) \\ y + \frac{1+f}{2}a\sin\theta + fd\sin(\theta + \phi) \end{bmatrix} \tag{8.4}$$

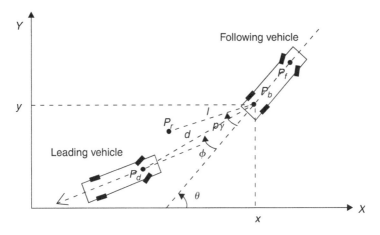

FIGURE 8.2   Backward tracking configuration.

where

$$f = \begin{cases} 1 & \text{for forward tracking} \\ -1 & \text{for backward tracking} \end{cases}$$

Vehicle tracking motion is defined as the collective motions of both vehicles as well as the relative distance and orientation angle between two vehicles. A good performance of tracking maneuvers is ensured only if the follower vehicle can follow the leader vehicle at a specified spacing and with a tracking error bounded or going to zero.

8.3   A Unified Tracking Controller

The objective of tracking control is to drive the follower vehicle automatically to follow the leader vehicle and maintain a predetermined intervehicular spacing. In this section, a nonlinear output feedback controller is developed. The idea is motivated by what a human driver does in car following maneuvers. The driver keeps his eye focused on the leader vehicle at a comfortable distance. He drives the vehicle so that his eye focus point is able to follow the leader vehicle with the same distance. With this motivation, we develop a look-ahead controller for forward tracking maneuver and a look-behind controller for backward tracking. In the following, these two controllers are developed as a unified nonlinear controller.

The focus point P_r is defined l meters away from the vehicle with l being the desired spacing between two vehicles ($P_f P_r = l$ in forward tracking and

$P_b P_r = l$ in backward tracking). The focus point P_r has a directional angle defined by the longitudinal axis of the vehicle ($P_b P_f$) and the focus line $P_f P_r$ (in the forward tracking as shown in Figure 8.1) or $P_b P_d$ (in the backward tracking as shown in Figure 8.2). It is p times as much as the steering angle γ. This focus point P_r, in both cases, can be expressed with respect to P_b as follows

$$
P_r = z = \begin{bmatrix} x + \frac{1+f}{2} a \cos\theta + l \cos(\theta + p\gamma) \\ y + \frac{1+f}{2} a \sin\theta + l \sin(\theta + p\gamma) \end{bmatrix} \tag{8.5}
$$

where l and p are two system parameters that will affect the performance of the vehicle tracking system. The focus point as defined in (8.5) can be viewed as a nonlinear output function of the posture of the follower vehicle.

An output tracking error can be defined as the difference between the output of the follower vehicle (8.5) and the virtual intervehicular connection (8.4) as follows

$$
\tilde{z} = z - z_d = \begin{bmatrix} l \cos(\theta + p\gamma) - fd \cos(\theta + \phi) \\ l \sin(\theta + p\gamma) - fd \sin(\theta + \phi) \end{bmatrix}
$$

$$
= R^{\mathrm{T}}(\theta) \begin{bmatrix} l \cos p\gamma - fd \cos\phi \\ l \sin p\gamma - fd \sin\phi \end{bmatrix} \tag{8.6}
$$

where $R(\theta)$ is a standard rotation matrix of θ as follows

$$
R(\theta) = \begin{bmatrix} \cos\theta & \sin\theta \\ -\sin\theta & \cos\theta \end{bmatrix} \tag{8.7}
$$

Lemma 8.1   *Consider a car-like vehicle with restricted steering angle, $|\gamma| \le \gamma_{max} < \pi/2$, and a vehicle tracking problem formulated as forward tracking or backward tracking. If the parameter p is chosen such that $|p| < \pi/(2\gamma_{max})$ and l is chosen as a finite constant, then the following two statements are equivalent:*

1. *The vehicle tracking error converges to zero, that is,*

$$
\lim_{t \to \infty} \|\tilde{z}(t)\| = 0
$$

2. *The relative orientation angle ϕ converges to $p\gamma$, that is,*

$$
\lim_{t \to \infty} (\phi - p\gamma) = 0
$$

and the intervehicular spacing d converges to fl, that is,

$$\lim_{t \to \infty} (d - fl) = 0$$

Proof   From the definition of the tracking error (8.6), we have

$$\|\tilde{z}\|^2 = l^2 + d^2 - 2lfd \cos(p\gamma - \phi)$$
$$= [d - fl \cos(p\gamma - \phi)]^2 + [fl \sin(p\gamma - \phi)]^2 \qquad (8.8)$$

Note that l is a finite constant, Statement 1 has an equivalent statement as follows

$$\lim_{t \to \infty} \|\tilde{z}(t)\| = 0 \Leftrightarrow \begin{cases} \lim_{t \to \infty} \sin(\phi - p\gamma) = 0 & \text{(i)} \\ \lim_{t \to \infty} [d - fl \cos(\phi - p\gamma)] = 0 & \text{(ii)} \end{cases} \qquad (8.9)$$

(a) If Statement 2 is true, it is easy to check that Statement 2 ensures both (8.9-i) and (8.9-ii) satisfied. Hence, $\lim_{t \to \infty} \|\tilde{z}(t)\| = 0$, that is, Statement 1 is true.

(b) If Statement 1 is true, we now prove that Statement 2 is true. Since $|\gamma| \le \gamma_{max}$ and $|p| < \pi/(2\gamma_{max})$, we have $|p\gamma| < \pi/2$. Furthermore, the relative orientation angle ϕ is also bounded, $|\phi| \le \pi/2$. Thus, we have

$$|p\gamma - \phi| \le |p\gamma| + |\phi| < \pi$$

As a result, (8.9-i) leads to

$$\lim_{t \to \infty} (\phi - p\gamma) = 0 \qquad (8.10)$$

Combining (8.10) with (8.9-ii) produces

$$\lim_{t \to \infty} (d - fl) = 0 \qquad (8.11)$$

Lemma 8.1 implies that a control law that ensures the convergence of the tracking error $\tilde{z}(t)$ can guarantee that the intervehicular spacing ultimately converges to the desired distance $|l|$. In practice, sensing range is limited, $0 < d < d_{max}$ and parameter l must be chosen such that fl lies in the valid range of the sensor

$$0 < fl < d_{max} \qquad (8.12)$$

Condition (8.12) shows that l must be positive in the look-ahead tracking, that is, $f = 1$, and negative in the look-behind tracking, that is, $f = -1$. To ensure robust and reliable performance, fl should be chosen well away from the boundaries, $0 \ll |l| \ll d_{max}$, so that d can be effectively kept within the valid range of the sensor. Equation (8.10) gives an interpretation of the parameter p. At steady state, $\phi = p\gamma$, and p is a multiplier relating the steering angle, γ, of the following vehicle and the relative orientation angle, ϕ.

To obtain the dynamic relationship between the output function $z(t)$ and the control input μ, take time derivative of (8.5)

$$\dot{z} = \frac{\partial z}{\partial q}\dot{q} = \frac{\partial z}{\partial q}G\mu = E(\theta, \gamma)\mu \tag{8.13}$$

where

$$E(\theta, \gamma) = R^{\mathrm{T}}(\theta)\bar{E}(\gamma) \tag{8.14}$$

with

$$\bar{E}(\gamma) = \begin{bmatrix} 1 - \dfrac{l}{a}\tan\gamma\sin p\gamma & -lp\sin p\gamma \\[2mm] \left(\dfrac{1+f}{2} + \dfrac{l}{a}\cos p\gamma\right)\tan\gamma & lp\cos p\gamma \end{bmatrix} \tag{8.15}$$

To ensure the existence of a feedback control, the matrix $E(\theta, \gamma)$ has to be nonsingular and the following lemma presents such a set of sufficient conditions.

Lemma 8.2   *Consider a car-like vehicle with restricted steering angle, $|\gamma| \leq \gamma_{max} < \pi/2$, and a vehicle tracking problem formulated as the forward tracking or the backward tracking. A control input μ exists for (8.13) if the design parameters l and p are chosen so that the following two conditions are satisfied:*

1. $lp \neq 0$

2. $\left| p - \dfrac{1+f}{2} \right| < \dfrac{\pi}{2\gamma_{max}}$

Proof   The existence of the input μ is guaranteed iff the matrix $E(\theta, \gamma)$ or, equivalently, the matrix $\bar{E}(\gamma)$ is nonsingular. This is equivalent to the determinant of matrix $\bar{E}(\gamma)$ being nonzero, that is,

$$\det(\bar{E}) = lp \cos p\gamma + \frac{1+f}{2} lp \tan \gamma \sin p\gamma \neq 0 \tag{8.16}$$

Since f only takes two values -1 or 1, we have the following equality:

$$\frac{1+f}{2} \tan \gamma = \tan\left(\frac{1+f}{2}\gamma\right) \tag{8.17}$$

and (8.16) becomes

$$\det(\bar{E}) = lp \cos p\gamma + lp \tan\left(\frac{1+f}{2}\gamma\right) \sin p\gamma$$

$$= lp \frac{\cos[(p - ((1+f)/2))\gamma]}{\cos((1+f/2)\gamma)} \neq 0 \tag{8.18}$$

Condition (8.18) is satisfied if the following two conditions are satisfied:

$$\begin{cases} 1. & lp \neq 0 \\ 2. & \left|p - \frac{1+f}{2}\right||\gamma| \leq \left|p - \frac{1+f}{2}\right|\gamma_{max} < \frac{\pi}{2} \\ & \Rightarrow \left|p - \frac{1+f}{2}\right| < \frac{\pi}{2\gamma_{max}} \end{cases} \tag{8.19}$$

For practical car-like wheeled mobile robots, the steering γ is restricted by $|\gamma| \leq \gamma_{max} < \pi/2$. Condition 1 in Lemma 8.2 requires (1) $(l \neq 0)$, that is, the focus point P_r cannot be fixed at the front center point P_f of the follower vehicle in forward tracking or at the back point P_b in backward tracking; and (2) $(p \neq 0)$, that is, P_r cannot be fixed on the longitudinal center axis. Condition 2 in Lemma 8.2 indicates that the selectable range of parameter p is bounded.

Lemmas 8.1 and 8.2 provide some sufficient conditions in choosing the design parameters l and p. It can be expected that vehicle tracking stability require more conditions on l and p. By examining the basic maneuvers, we can gain some insights and necessary conditions on l and p for tracking stability. Vehicle tracking along a straight path is a basic maneuver and its requirement on stability will offer some insight and a set of necessary conditions. In the following Lemma 8.3, a set of such conditions are derived for this purpose.

Lemma 8.3 *Consider a basic maneuver of vehicle following along a straight path $(\gamma_d = 0)$ at a speed $(v_d \neq 0)$. Suppose there exists a feedback vehicle-following controller that guarantees the convergence of the tracking error $\tilde{z}(t)$*

and its derivative $\ddot{\bar{z}}(t)$ as well as the steering rate ω to zero, that is,

$$\lim_{t\to\infty} \|\bar{z}(t)\| = \lim_{t\to\infty} \|\dot{\bar{z}}(t)\| = \lim_{t\to\infty} \omega = 0$$

In addition to the conditions in Lemmas 8.1 and 8.2, the parameters l, p, and f are necessary to satisfy $lp > 0$ and $fv_d > 0$.

Proof When the leader vehicle moves on a straight path ($\gamma_d = 0$) at a speed $v_d \neq 0$, its heading angle θ_d will stay as constant ($\dot{\theta}_d = 0$), we have

$$\dot{z}_d = \begin{bmatrix} v_d \cos\theta_d \\ v_d \sin\theta_d \end{bmatrix} = R^{\mathrm{T}}(\theta_d) \begin{bmatrix} v_d \\ 0 \end{bmatrix} \tag{8.20}$$

We also define the following tracking errors:

$$\tilde{\eta} = \eta - \eta_d = \begin{bmatrix} \tilde{\theta} \\ \tilde{\gamma} \end{bmatrix} = \begin{bmatrix} \theta - \theta_d \\ \gamma - \gamma_d \end{bmatrix} \tag{8.21}$$

We now prove Lemma 8.3 in two steps. First, we prove that the convergence of the tracking error $\bar{z}(t)$ to zero implies that $\tilde{\eta}$ converges to zero, that is,

$$\lim_{t\to\infty} \|\tilde{\eta}\| = \lim_{t\to\infty} \tilde{\theta} = \lim_{t\to\infty} \tilde{\gamma} = 0$$

Second, we derive the necessary conditions of parameter l, p, and f for the tracking stability of $\tilde{\eta}$ defined in (8.21).

1. *The convergence of $\bar{z}$ implies the convergence of $\tilde{\eta}$.*

Suppose $\gamma = \tilde{\gamma}$ converges to $\gamma_d = 0$, that is, the follower vehicle eventually moves on a straight path. This implies θ converges to a constant and

$$\lim_{t\to\infty} \dot{z} = \lim_{t\to\infty} R^{\mathrm{T}}(\theta)\bar{E}(\gamma)\mu = \lim_{t\to\infty} R^{\mathrm{T}}(\theta) \begin{bmatrix} 1 & 0 \\ 0 & lp \end{bmatrix} \begin{bmatrix} v \\ \omega \end{bmatrix} = R^{\mathrm{T}}(\theta) \begin{bmatrix} v \\ 0 \end{bmatrix}$$

where we have used the assumption $\lim_{t\to\infty} \omega = 0$. Furthermore

$$\lim_{t\to\infty} \dot{\bar{z}} = \lim_{t\to\infty} (\dot{z} - \dot{z}_d) = R^{\mathrm{T}}(\theta) \begin{bmatrix} v \\ 0 \end{bmatrix} - R^{\mathrm{T}}(\theta_d) \begin{bmatrix} v_d \\ 0 \end{bmatrix}$$

$$= R^{\mathrm{T}}(\theta) \begin{bmatrix} v - v_d \cos\tilde{\theta} \\ v_d \sin\tilde{\theta} \end{bmatrix} = 0$$

implies $\lim_{t\to\infty}\tilde{\theta} = \lim_{t\to\infty}(\theta - \theta_d) = 0$, given the constraint $|\tilde{\theta}| < \pi/2$, and $\lim_{t\to\infty}(v - v_d) = 0$. This shows $\lim_{t\to\infty}\|\tilde{\eta}\| = 0$.

On the other hand, suppose that γ diverges from $\gamma_d = 0$. The assumption $\lim_{t\to\infty}\omega = 0$ implies $\lim_{t\to\infty}\gamma = c$, with c being a nonzero angle. The instantaneous turning radius of the follower vehicle

$$r = \frac{a}{\tan\gamma}$$

converges to a finite constant. The follower vehicle eventually moves on a circular path while the leader vehicle moves on a straight path. This contradicts the assumption of $\lim_{t\to\infty}\|\tilde{z}\| = 0$, or, equivalently, $\lim_{t\to\infty}d = fl$, by Lemma 8.1.

These show that $\lim_{t\to\infty}\|\tilde{z}\| = 0$ implies $\lim_{t\to\infty}\|\tilde{\eta}\| = 0$. The reverse may not be true because two vehicles may be moving on two separate and parallel straight paths ($\tilde{\eta} = 0$), whereas the tracking error $\tilde{z}$ is not zero.

2. *Necessary conditions on l, p, and f for $\tilde{\eta}$ to converge to zero.*

Since $\dot{\theta}_d = 0$ and $\gamma = \gamma_d + \tilde{\gamma} = \tilde{\gamma}$, the error $\tilde{\eta}$ is computed as follows:

$$\dot{\tilde{\eta}} = \dot{\eta} - \dot{\eta}_d = \dot{\eta} = \begin{bmatrix} \dfrac{v}{a}\tan\gamma \\ \omega \end{bmatrix} = \begin{bmatrix} \dfrac{1}{a}\tan\gamma & 0 \\ 0 & 1 \end{bmatrix}\mu$$

$$= \begin{bmatrix} \dfrac{1}{a}\tan\tilde{\gamma} & 0 \\ 0 & 1 \end{bmatrix}\mu = Q(\tilde{\gamma})\mu \tag{8.22}$$

When the convergence of $\tilde{z}(t)$ and $\dot{\tilde{z}}(t)$ to zero are achieved, we have

$$0 \equiv \dot{\tilde{z}} = \dot{z} - \dot{z}_d = E(\theta, \gamma)\mu - \dot{z}_d$$

Since the matrix E is nonsingular (Lemma 8.2), we obtain

$$\mu = E^{-1}(\theta, \gamma)\dot{z}_d = \bar{E}^{-1}(\gamma)R(\theta)\dot{z}_d \tag{8.23}$$

Noting $\dot{z}_d$ in (8.20) and $\gamma = \tilde{\gamma}$, (8.23) becomes

$$\mu = \bar{E}^{-1}(\tilde{\gamma})R(\theta)R^{\mathsf{T}}(\theta_d)\begin{bmatrix} v_d \\ 0 \end{bmatrix} = \bar{E}^{-1}(\tilde{\gamma})R(\tilde{\theta})\begin{bmatrix} v_d \\ 0 \end{bmatrix} \tag{8.24}$$

Then the system (8.22) becomes

$$\dot{\eta} = Q(\tilde{\gamma})\mu = Q(\tilde{\gamma})\bar{E}^{-1}(\tilde{\gamma})R(\tilde{\theta})\begin{bmatrix} v_d \\ 0 \end{bmatrix} = \Gamma(\tilde{\eta}, v_d) \qquad (8.25)$$

It is easy to check that the system $\dot{\eta} = \Gamma(\tilde{\eta}, v_d)$ in (8.25) has an equilibrium at $\tilde{\eta} = 0$ and a linear approximation as follows [18]

$$\dot{\tilde{\eta}} = \left[\frac{\partial \Gamma(\tilde{\eta}, v_d)}{\partial \tilde{\eta}} \right]\bigg|_{\tilde{\eta}=0} \tilde{\eta} = A(v_d)\tilde{\eta} \qquad (8.26)$$

If (8.26) is exponentially stable in the neighborhood of $\tilde{\eta} = 0$, then (8.25) is uniformly asymptotically stable. Furthermore, for a linear system like that in (8.26), the condition to be exponentially stable is that all eigenvalues have negative real part. With some calculations, it is straightforward to yield

$$A(v_d) = \begin{bmatrix} 0 & \dfrac{1}{a} \\ -\dfrac{1}{lp} & -\dfrac{1}{p}\left(\dfrac{1+f}{2l} + \dfrac{1}{a} \right) \end{bmatrix} v_d \qquad (8.27)$$

with eigenvalues

$$\lambda_{1,2} = \frac{v_d}{2alp}\left\{ -\left(\frac{1+f}{2}a + l \right) \pm \sqrt{\Delta} \right\} \qquad (8.28)$$

$$\Delta = \left(\frac{1+f}{2}a + l \right)^2 - 4alp$$

With these eigenvalues, we have

$$\lambda_1 \lambda_2 = \frac{v_d^2}{alp}$$

$$\frac{\lambda_1 + \lambda_2}{2} = -\frac{v_d}{2alp}\left(\frac{1+f}{2}a + l \right) = -\frac{fv_d}{2alp}\left(\frac{1+f}{2}a + fl \right)$$

If both eigenvalues λ_1 and λ_2 are real numbers, the conditions for them to be negative real numbers are

$$\Delta \geq 0 \quad \text{and} \quad \lambda_1\lambda_2 > 0 \quad \text{and} \quad \frac{\lambda_1 + \lambda_2}{2} < 0$$

Since Lemma 8.1 implies $fl > 0$, which lead to $(((1+f)/2)a + fl) > 0$, the above conditions are equivalent to

$$0 < lp \leq \frac{(((1+f)/2)a + l)^2}{4a} \quad \text{and} \quad fv_d > 0$$

Likewise, if both eigenvalues are a pair of complex conjugates, then the conditions are

$$\Delta < 0 \quad \text{and} \quad \frac{\lambda_1 + \lambda_2}{2} < 0$$

which lead to

$$lp > \frac{(((1+f)/2)a + l)^2}{4a} > 0 \quad \text{and} \quad fv_d > 0$$

In summary, matrix $A(v_d)$ has both eigenvalues with negative real part iif $lp > 0$ and $fv_d > 0$. Under these conditions, the system (8.22) is uniformly asymptotically stable. In other words, the tracking error $\tilde{\eta}$ converges to zero.

The condition $fv_d > 0$ in Lemma 8.3 implies that vehicle-following maneuver is feasible and successful only if the leader vehicle moves forward ($v_d > 0$) in the look-ahead tracking mode ($f = 1$) and moves backward ($v_d < 0$) in the look-behind tracking mode ($f = -1$). This condition is satisfied automatically based on the formulations of the forward tracking and backward tracking defined earlier in Section 8.2.2. Condition $lp > 0$ implies that the vehicle tracking must be in the formations defined in Figure 8.1 and Figure 8.2 for forward tracking and backward tracking, respectively. In other words, look-ahead control can only be used for forward tracking formation and look-behind control can only be used for backward tracking control.

8.3.1 Kinematics-Based Tracking Controller

The target performance of the vehicle tracking maneuvers can be specified by a first-order system for the closed-loop output tracking error

$$\dot{\tilde{z}} + \lambda \tilde{z} = 0 \tag{8.29}$$

where the convergence rate $\lambda > 0$ can be specified for a desired target performance.

Equation (8.29) can then be rewritten equivalently as

$$\dot{z} = \dot{z}_d - \lambda \tilde{z} \tag{8.30}$$

Time differentiation of (8.4) leads to

$$\dot{z}_d = R^{\mathrm{T}}(\theta) \begin{bmatrix} v + f\dot{d}\cos\phi - fd(\dot{\theta} + \dot{\phi})\sin\phi \\ \dfrac{1+f}{2}v\tan\gamma + f\dot{d}\sin\phi + fd(\dot{\theta} + \dot{\phi})\cos\phi \end{bmatrix} \tag{8.31}$$

By substituting (8.6), (8.13), and (8.31) into (8.30), we have

$$E(\theta, \gamma)\mu = F_{\mathrm{kin}}(\theta, v, \gamma, d, \dot{d}, \phi, \dot{\phi}) \tag{8.32}$$

where

$$F_{\mathrm{kin}} = \dot{z}_d - \lambda \tilde{z} = R^{\mathrm{T}}(\theta)\bar{F}_{\mathrm{kin}}(v, \gamma, d, \dot{d}, \phi, \dot{\phi}) \tag{8.33}$$

with

$$\bar{F}_{\mathrm{kin}} = \begin{bmatrix} v - \lambda l \cos p\gamma \\ \dfrac{1+f}{2}v\tan\gamma - \lambda l \sin p\gamma \end{bmatrix} + fR^{\mathrm{T}}(\phi)\begin{bmatrix} \dot{d} + \lambda d \\ d(\dot{\theta} + \dot{\phi}) \end{bmatrix} \tag{8.34}$$

Multiplying the orthogonal matrix $R(\theta)$ to both sides of (8.32) produces

$$\bar{E}(\gamma)\mu = \bar{F}_{\mathrm{kin}}(v, \gamma, d, \dot{d}, \phi, \dot{\phi}) \tag{8.35}$$

With the parameters l and p chosen satisfying Lemma 8.2, the resultant nonlinear kinematics-based controller can be obtained from (8.35) by using (8.15) and (8.34)

$$\mu_{\mathrm{input}} = \bar{E}^{-1}(\gamma)\bar{F}_{\mathrm{kin}}\left(v, \gamma, d, \dot{d}, \phi, \dot{\phi}\right) \tag{8.36}$$

where $\mu_{input} = [v_{input} \quad \omega_{input}]^T$,
with

$$
\begin{aligned}
v_{input} = v &+ \{\cos(((1+f)/2)\gamma)\{-\lambda l + f(\dot{d} + \lambda d) \\
&\times \cos(p\gamma - \phi) + fd(\dot{\theta} + \dot{\phi})\sin(p\gamma - \phi)\}\} \\
&\times \{\cos[(p-(1+f)/2)\gamma]\}^{-1}
\end{aligned} \tag{8.37}
$$

ω_{input}

$$
\begin{aligned}
= &-\frac{\dot{\theta}}{p} - \frac{\lambda}{p}\tan\left[\left(p - \frac{1+f}{2}\right)\gamma\right] - \frac{\tan\gamma\cos(((1+f)/2)\gamma)}{ap\cos[(p-((1+f)/2))\gamma]} \\
&\times \{-\lambda l + f(\dot{d}+\lambda d)\cos(p\gamma - \phi) + fd(\dot{\theta}+\dot{\phi})\sin(p\gamma - \phi)\} \\
&+ \frac{f(\dot{d}+\lambda d)\sin(\phi - ((1+f)/2)\gamma) + fd(\dot{\theta}+\dot{\phi})\cos(\phi - ((1+f)/2)\gamma)}{lp\cos[(p-((1+f)/2))\gamma]}
\end{aligned} \tag{8.38}
$$

The above development of the kinematics-based vehicle-following controller is summarized in the following theorem.

Theorem 8.1 *Consider the car-like vehicle tracking maneuvers of forward tracking, shown in Figure 8.1, and backward tracking, shown in Figure 8.2. The kinematic motion of these tracking maneuvers is defined collectively as the kinematics (8.1) of both vehicles and the virtual intervehicular connection (8.4). Define the tracking error $\tilde{z}$ in (8.5) as the difference between the output of the follower vehicle (8.5) and the virtual intervehicular connection (8.4). The tracking target performance in $\tilde{z}$ is defined by the stable first order system (8.29) and can be ensured if the nonlinear control laws (8.37) for driving and (8.38) for steering are applied, and the following conditions are satisfied:*

- *Forward tracking: $f = 1$*

$$
\begin{cases}
v_d > 0 \\
\lambda > 0 \\
0 < l < d_{max} \\
0 < p < \frac{\pi}{2\gamma_{max}}
\end{cases} \tag{8.39}
$$

- *Backward tracking:* $f = -1$

$$
\begin{cases}
v_d < 0 \\
\lambda > 0 \\
-d_{\max} < l < 0 \\
-\dfrac{\pi}{2\gamma_{\max}} < p < 0
\end{cases}
\tag{8.40}
$$

Proof   The conditions in Lemmas 8.1 and 8.2 guarantee the existence of the control laws (8.37) and (8.38).

Combining the target performance specification, the conditions for tracking convergence equivalence in Lemma 8.1, the conditions for the existence of control input in Lemma 8.2, and the necessary conditions for the tracking stability in Lemma 8.3, we obtain

$$
\begin{cases}
\text{Target dynamics} & \Rightarrow & \{\lambda > 0 \\[4pt]
\text{Lemma 8.1} & \Rightarrow & \begin{cases} |p| < \dfrac{\pi}{2\gamma_{\max}} \\ 0 < fl < d_{\max} \end{cases} \\[12pt]
\text{Lemma 8.2} & \Rightarrow & \begin{cases} lp \neq 0 \\ \left| p - \dfrac{1+f}{2} \right| < \dfrac{\pi}{2\gamma_{\max}} \end{cases} \\[12pt]
\text{Lemma 8.3} & \Rightarrow & \begin{cases} lp > 0 \\ fv_d > 0 \end{cases}
\end{cases}
\Rightarrow
\begin{cases}
\lambda > 0 \\
|p| < \dfrac{\pi}{2\gamma_{\max}} \\
0 < fl < d_{\max} \\
\left| p - \dfrac{1+f}{2} \right| < \dfrac{\pi}{2\gamma_{\max}} \\
lp > 0 \\
fv_d > 0
\end{cases}
\tag{8.41}
$$

For $f = 1$, conditions (8.41) will lead to (8.39)
For $f = -1$, conditions (8.41) will lead to (8.40)

8.3.2 Dynamics-Based Tracking Controller

If the access to the torques/forces, or their corresponding convertible accelerations, of the vehicle control is available, the controller can be developed based on the dynamic model (8.3).

In this case, the target performance of the vehicle tracking maneuvers can be specified by a second-order system for the closed-loop output tracking error

$$
\ddot{\tilde{z}} + 2\xi\lambda\dot{\tilde{z}} + \lambda^2 \tilde{z} = 0
\tag{8.42}
$$

where the natural frequency $\lambda > 0$ and the damping ratio $\xi > 0$.

Equation (8.42) can then be rewritten equivalently as

$$\ddot{z} = \ddot{z}_d - 2\xi\lambda\dot{\tilde{z}} - \lambda^2\tilde{z} \tag{8.43}$$

Taking the differentiation of (8.13) yields

$$\ddot{z} = \frac{\partial(E\mu)}{\partial q}G\mu + E\dot{\mu} = H(\theta,\gamma)\mu + E(\theta,\gamma)u \tag{8.44}$$

where

$$H(\theta,\gamma) = R^{\mathrm{T}}(\theta)\bar{H}(\gamma) \tag{8.45}$$

and

$$\bar{H}(\gamma) = \begin{bmatrix} -\tan\gamma\left\{\dfrac{1+f}{2}\dot{\theta} \right. & -\dfrac{l}{a}\dfrac{v}{\cos^2\gamma}\sin p\gamma \\ \left. +\dfrac{l}{a}(\dot{\theta}+p\omega)\cos p\gamma\right\} & -lp(\dot{\theta}+p\omega)\cos p\gamma \\ \dot{\theta} - \dfrac{l}{a}(\dot{\theta}+p\omega)\tan\gamma\sin p\gamma & \left(\dfrac{1+f}{2}+\dfrac{l}{a}\cos p\gamma\right)\dfrac{v}{\cos^2\gamma} \\ & -lp(\dot{\theta}+p\omega)\sin p\gamma \end{bmatrix} \tag{8.46}$$

Subsequently, differentiation of (8.31) leads to

$$\ddot{z}_d = R^{\mathrm{T}}(\theta)\left\{\begin{bmatrix} \dot{v} - \dfrac{1+f}{2}a\dot{\theta}^2 \\ v\dot{\theta} + \dfrac{1+f}{2}a\ddot{\theta} \end{bmatrix} + fR^{\mathrm{T}}(\phi)\begin{bmatrix} \{\ddot{d} - d(\dot{\theta}+\dot{\phi})^2\} \\ \{2\dot{d}(\dot{\theta}+\dot{\phi}) + d(\ddot{\theta}+\ddot{\phi})\} \end{bmatrix}\right\} \tag{8.47}$$

where

$$\dot{\theta} = \frac{v\tan\gamma}{a} \tag{8.48}$$

$$\ddot{\theta} = \frac{\dot{v}\tan\gamma}{a} + \frac{v\omega}{a\cos^2\gamma} \tag{8.49}$$

Likewise, taking differentiation of $\tilde{z}$ in (8.6) yields

$$\dot{\tilde{z}} = R^{\mathrm{T}}(\theta)\begin{bmatrix} -l(\dot{\theta}+p\omega)\sin p\gamma - \dot{d}\cos\phi + d(\dot{\theta}+\dot{\phi})\sin\phi \\ l(\dot{\theta}+p\omega)\cos p\gamma - \dot{d}\sin\phi - d(\dot{\theta}+\dot{\phi})\cos\phi \end{bmatrix} \tag{8.50}$$

Having substituted $\tilde{z}$ in (8.6), $\dot{\tilde{z}}$ in (8.44), $\ddot{z}_d$ in (8.47), and $\dot{\ddot{z}}$ in (8.50) into (8.43), we obtain

$$E(\theta, \gamma)u = F_{\text{dyn}}(\theta, v, \dot{v}, \gamma, \omega, d, \dot{d}, \ddot{d}, \phi, \dot{\phi}, \ddot{\phi}) \tag{8.51}$$

where

$$F_{\text{dyn}} = \ddot{z}_d - 2\xi\lambda\dot{\tilde{z}} - \lambda^2\tilde{z} - H(\theta, \gamma)\mu$$
$$= R^{\text{T}}(\theta)\bar{F}_{\text{dyn}}(v, \dot{v}, \gamma, \omega, d, \dot{d}, \ddot{d}, \phi, \dot{\phi}, \ddot{\phi}) \tag{8.52}$$

with

$$\bar{F}_{\text{dyn}} = \begin{bmatrix} \dot{v} \\ \dfrac{1+f}{2}\dot{v}\tan\gamma \end{bmatrix} + lR^{\text{T}}(p\gamma)\begin{bmatrix} (\dot{\theta}+p\omega)^2 - \lambda^2 \\ -\dfrac{v\omega}{a\cos^2\gamma} - 2\xi\lambda(\dot{\theta}+p\omega) \end{bmatrix}$$
$$+ fR^{\text{T}}(\phi)\begin{bmatrix} \ddot{d} + 2\xi\lambda\dot{d} + \lambda^2 d - d(\dot{\theta}+\dot{\phi})^2 \\ d(\ddot{\theta}+\ddot{\phi}) + 2(\dot{d}+\xi\lambda d)(\dot{\theta}+\dot{\phi}) \end{bmatrix} \tag{8.53}$$

Multiplying the orthogonal matrix $R(\theta)$ to both sides of (8.51) produces

$$\bar{E}(\gamma)u = \bar{F}_{\text{dyn}}(v, \dot{v}, \gamma, \omega, d, \dot{d}, \ddot{d}, \phi, \dot{\phi}, \ddot{\phi}) \tag{8.54}$$

Conditions that satisfy (8.39) for the look-ahead tracking mode or (8.40) for the look-behind tracking mode guarantee that the decoupling matrices $\bar{E}(\gamma)$ and $E(\theta, \gamma)$ are invertible. Under those conditions, the dynamics-based vehicle-following controller can be achieved

$$u_{\text{input}} - \bar{E}^{-1}(\gamma)\bar{F}_{\text{dyn}}(v, \dot{v}, \gamma, \omega, d, \dot{d}, \ddot{d}, \phi, \dot{\phi}, \ddot{\phi}) \tag{8.55}$$

where $u_{\text{input}} = [u_m \quad u_s]^{\text{T}}$, with

$$u_m = \dot{v} + \frac{\cos(((1+f)/2)\gamma)}{\cos[(p-((1+f)/2))\gamma]}\{l[(\dot{\theta}+p\omega)^2 - \lambda^2]$$
$$+ f[\ddot{d} + 2\xi\lambda\dot{d} + \lambda^2 d - d(\dot{\theta}+\dot{\phi})^2]\cos(p\gamma-\phi)$$
$$+ f[d(\ddot{\theta}+\ddot{\phi}) + 2(\dot{d}+\xi\lambda d)(\dot{\theta}+\dot{\phi})]\sin(p\gamma-\phi)\} \tag{8.56}$$

$$u_s = -\frac{1}{p}(\ddot{\theta} + 2\xi\lambda(\dot{\theta} + p\omega)) + \frac{1}{p}((\dot{\theta} + p\omega)^2 - \lambda^2)\tan\left[\left(p - \frac{1+f}{2}\right)\gamma\right]$$

$$-\frac{\cos(((1+f)/2)\gamma)}{p\cos[(p - ((1+f)/2))\gamma]}\left\{\frac{l}{a}((\dot{\theta} + p\omega)^2 + \lambda^2)\tan\gamma\right.$$

$$+f\left[\frac{\sin(\phi - ((1+f)/2)\gamma)}{l\cos(((1+f)/2)\gamma)} - \frac{\tan\gamma}{a}\cos(p\gamma - \phi)\right]$$

$$\times[\ddot{d} + 2\xi\lambda\dot{d} + \lambda^2 d - d(\dot{\theta} + \dot{\phi})^2]$$

$$+f\left[\frac{\cos(\phi - ((1+f)/2)\gamma)}{l\cos(((1+f)/2)\gamma)} - \frac{\tan\gamma}{a}\sin(p\gamma - \phi)\right]$$

$$\left.\times[d(\ddot{\theta} + \ddot{\phi}) + 2(\dot{d} + \xi\lambda d)(\dot{\theta} + \dot{\phi})]\right\} \tag{8.57}$$

The above development can be summarized as follows.

Theorem 8.2 *Consider the car-like mobile robot performing forward tracking, shown in Figure 8.1, and backward tracking, shown in Figure 8.2. The dynamic motion of these tracking maneuvers is defined collectively as the dynamics (8.3) of both vehicles and the virtual intervehicular connection (8.4).*

Define the tracking error $\tilde{z}$ in (8.5) as the difference between the output of the follower vehicle (8.5) and the virtual intervehicular connection (8.4). The tracking target performance in $\tilde{z}$ is defined by the stable second-order system (8.42) and can be ensured if the nonlinear controls (8.56) for driving and (8.57) for steering are applied and the following necessary conditions are satisfied.

- *Forward tracking: $f = 1$, $\lambda > 0$, $\xi > 0$, l and p satisfying (8.39)*
- *Backward tracking: $f = -1$, $\lambda > 0$, $\xi > 0$, l and p satisfying (8.40)*

Proof The target performance (8.42) also guarantees that the tracking error $\tilde{z}(t)$ and its derivatives $\dot{\tilde{z}}(t)$ and $\ddot{\tilde{z}}(t)$ are all convergent to zero. Combining all the conditions from Lemmas 8.1, 8.2, and 8.3, we can obtain the similar conditions (8.39) for the look-ahead tracking and (8.40) for the look-behind tracking.

8.3.3 Requirement of Measurements

As stated in the development of the kinematics- and dynamics-based controllers, besides the vehicular state feedbacks such as velocity/acceleration and steering angle, some measurements are required including relative distance between two vehicles d, velocity $\dot{d}$, acceleration $\ddot{d}$, and relative angle ϕ as well as its derivatives $\dot{\phi}$ and $\ddot{\phi}$. These requirements are vital and also implemented in other vehicle-following systems. For example, the inclusion of relative distance, velocity, and even acceleration in the controller have been well known and implemented in longitudinal controls in order to improve the stability of the tracking system [4,5,10]. Likewise, for steering control, the controllers developed based on kinematic models generally need the relative angle and/or its first derivative [2,10] whereas those based on dynamics model [11,12,20] may or may not require the second derivative.

In practice, the relative distance and angle can be measured by a ranging sensor. Relative velocities and particularly relative accelerations are more difficult to obtain. In general, there are two ways of getting those measurements. The first one is to utilize a wireless communication channel to transmit the vehicular measurements such as velocity, acceleration, and yaw rate of the leader vehicle to the follower vehicle [5,10]. The relative velocities and/or accelerations are computed based on the geometric and dynamic relationships of the two vehicles. The second way relies on the high accuracy of the ranging sensor to estimate the derivatives using numerical calculations or derivative filtering [2,12]. This method is less accurate than the first one but more suitable for low-speed applications and does not require a communication channel.

8.4 TRACKING PERFORMANCE EVALUATION

The nonlinear controller developed in the previous section needs verification and the effects of parameter selections are to be evaluated. In this section, we focus on the effects and evaluations of the design parameters l and p for the dynamics-based controller. The closed-loop system's parameters λ and ξ are set as constants of 1 and 0.5, respectively. Different sets of design parameters (l, p) are tested for both look-ahead and look-behind tracking control.

Some limits are chosen based on the real physical limits of our test-bed car-like vehicle. The steering angle of the vehicle is limited as $|\gamma| \leq \gamma_{max} = \pi/9$ rad $(=20°)$. Other limits are chosen as follows: $d_{max} = 8$ m as the reliable range of the sensing; $l_{min} = 1$ m for safety stopping; and $p_{min} = 0.1$ for some minimum sensitivity to steering.

The evaluation is carried out by numerical simulation using the platform integrating ADAMS[®][1] and Simulink[®].[2] The ADAMS is a mechanical proto-typing package and is used to construct two mobile robot vehicles. Simulink is used to model the proposed nonlinear tracking controller. The integration of these two powerful simulation platforms produces a simulation platform for mobile robotics and associated advance control designs. It has the benefits of doing away with the dynamic modeling of vehicles and the motions are

---

[1] Registered trademark of MSC Software Corporation.
[2] Registered trademark of The MathWorks, Inc.

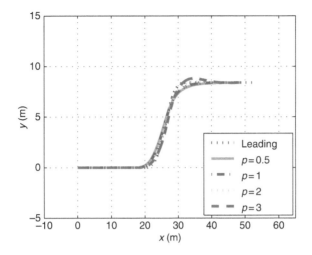

FIGURE 8.3 Vehicle trajectories (x, y) with different values of p.

closer to reality. This platform has served us well and details can be found in Reference 21.

8.4.1 Forward Tracking Control

In this situation, the follower vehicle is initially l meters behind the leader vehicle. The selections of l and p are based on condition (8.39)

$$1 = l_{\min} \ll l \ll d_{\max} = 8 \quad \text{and} \quad 0.1 = p_{\min} \leq p \leq \frac{\pi}{2\gamma_{\max}} = 4.5 \quad (8.58)$$

8.4.1.1 Influence of parameter p

The intervehicular space parameter l is fixed at a desired space of 2.5 m, while different values of p are tested in the range of $(0.1, 4.5)$. Simulation results, in Figure 8.3 to Figure 8.7, show that the follower vehicle successfully tracks the leader vehicle. The tracking errors are small especially along the straight path as shown in Figure 8.4. During turns, the value of parameter p clearly affects the tracking performance. Though starting at the same initial position and orientation, with smaller values of p, for example, $p < 2.5$, the follower vehicle tries to cut corners to catch up with the leader vehicle. In contrast, larger values of p result in overshooting before turning. These maneuvers are not desirable as the follower vehicle might move into a neighboring lane. In practice, a suitable value of p can be obtained from fine tuning with respect to expected performance.

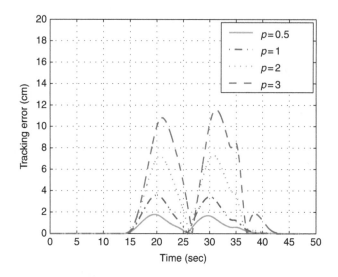

FIGURE 8.4   Tracking errors $\|\tilde{z}\|$ for different values of p.

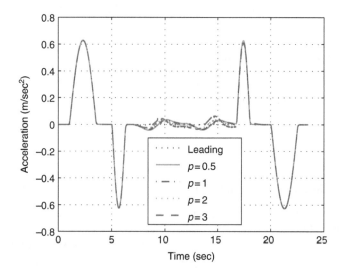

FIGURE 8.5   Vehicle acceleration $\dot{v}$ for different values of p.

Figure 8.5 and Figure 8.6 show that the acceleration and velocity of the follower vehicle are only slightly different from those of the leader vehicle. This indicates that the tracking speed and acceleration are maintained successfully. Larger values of p tend to create more oscillations because the focus point is more sensitive to the steering angle.

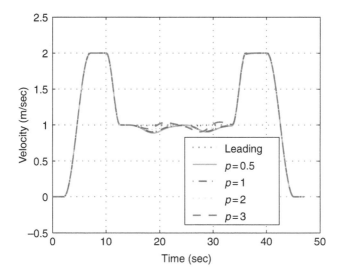

FIGURE 8.6 Vehicle velocity v for different values of p.

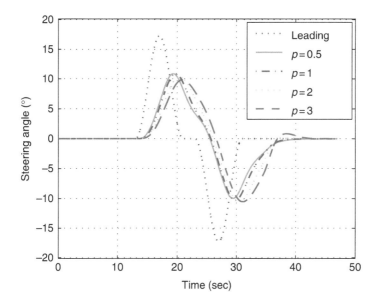

FIGURE 8.7 Steering angle γ for different values of p.

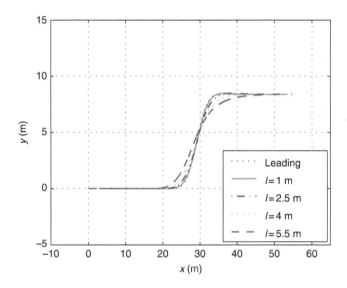

FIGURE 8.8   Vehicle trajectories (x, y) with different values of l.

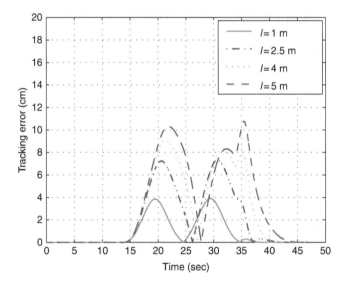

FIGURE 8.9   Tracking errors $\|\tilde{z}\|$ for different values of l.

8.4.1.2   Influence of parameter l

We fix parameter p at value 2. Parameter l varies from 1 to 5.5 m. Simulation results are shown in Figure 8.8 to Figure 8.12. Figure 8.8 and Figure 8.9

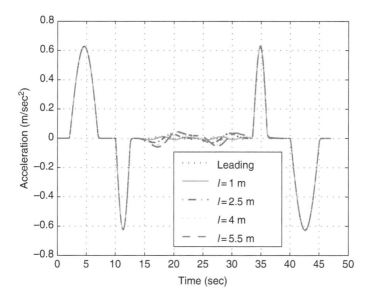

FIGURE 8.10   Vehicle acceleration $\dot{v}$ for different values of l.

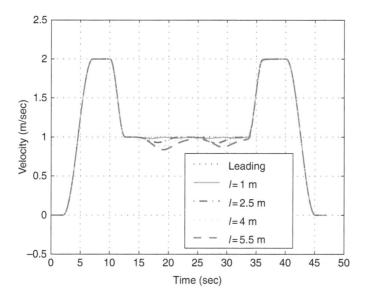

FIGURE 8.11   Vehicle velocity v for different values of l.

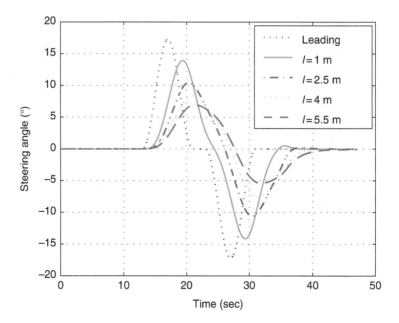

FIGURE 8.12   Steering angle γ for different values of l.

show that the controller is able to drive the vehicle to follow the leader vehicle with the chosen values of l. We note that with shorter desired intervehicular spacing (smaller values of l), the trajectories of the follower vehicle, and that of the leader vehicle are closer. Figure 8.9 shows that although the maximum error increases when l increases, the relative maximum error over the desired spacing l actually decreases. It means that tracking performance is better with larger values of l. The velocity and acceleration track the desired ones. The delay in the steering angle tracking is natural due to the time difference of the vehicles' motions. This delay becomes bigger with a larger value of l. This is because with a larger value of l the relative angle is smaller, assuming that the lateral deviation is the same. As a result, the steering rate command is smaller, and it would take a longer time to converge to that of the leader vehicle.

8.4.2 Backward Tracking Control

In this situation, the leader vehicle is placed l meters behind the follower vehicle and moves backward. Similar to the look-ahead tracking situation, the trajectory of the leader vehicle is generated beforehand and repeated in every test to ease the comparison and analysis.

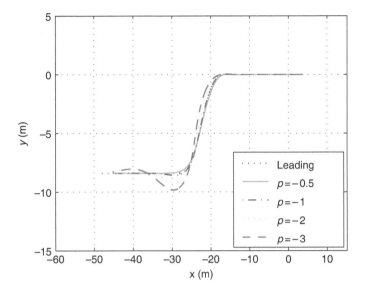

FIGURE 8.13 Vehicle trajectories (x, y) with different values of p.

Condition (8.40) becomes

$$-8 = -d_{\max} \ll l \ll -l_{\min} = -1$$

and   (8.59)

$$-4.5 = -\frac{\pi}{2\gamma_{\max}} \le p \le -p_{\min} = -0.1$$

8.4.2.1 Influence of parameter p

The parameter l is fixed at $l = -2.5$ m and several values of the parameter p in the range of $[-4, -0.1]$ are tested with results shown in Figure 8.13 to Figure 8.17. Along a straight path, the tracking error is very small, and both acceleration and velocity are closely tracked. When the leader vehicle turns, the follower vehicle's steering turns to the opposite side for a while before it turns back to the same direction (Figure 8.17). It is because the tracked point P_d is the front point of the leader vehicle, not the rear point which is considered as the reference point of the leader vehicle. Thus, when the leader vehicle is moving backward and about to turn left, for example, its front point will tend to move to the right side, and vice versa. Similar to the case of forward tracking, the valid range of p can be divided into two parts. Larger values of p make the focus point P_r more sensitive to the steering angle and result in more oscillations and overshoots in its trajectory. The closest tracked trajectory is with $p = -1$.

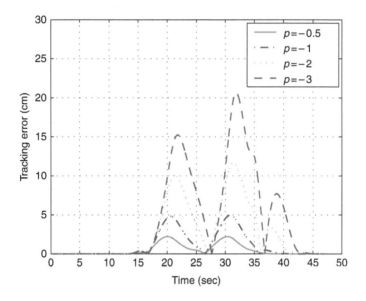

FIGURE 8.14   Tracking errors $\|\tilde{z}\|$ for different values of p.

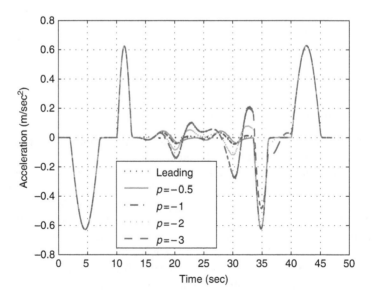

FIGURE 8.15   Vehicle acceleration $\dot{v}$ for different values of p.

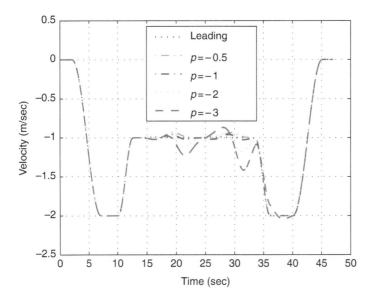

FIGURE 8.16 Vehicle velocity v for different values of p.

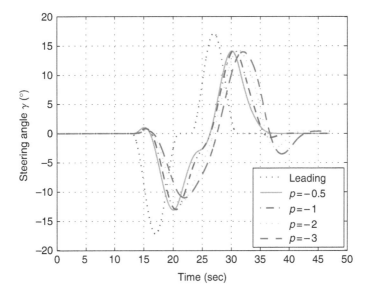

FIGURE 8.17 Steering angle γ for different values of p.

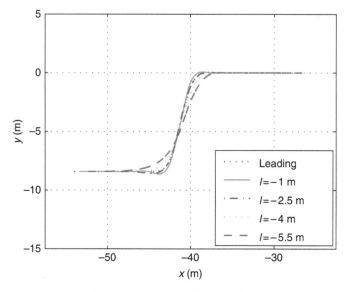

FIGURE 8.18   Vehicle trajectories (x, y) with different values of l.

8.4.2.2   Influence of parameter l

The parameter p is fixed at $p = -1$ and several values of l in the range $[-1 \text{ m}, -8 \text{ m})$ are tested, with results shown in Figure 8.18 to Figure 8.22. The tracking performance in the look-behind case is also influenced by parameter l the same way as that in forward tracking. When l takes larger values, the tracking vehicle takes the "corner-cutting" way to follow the leader vehicle. With smaller absolute values of l, the tracking performance is better in terms of smaller tracking error. However, the steering angle might rise to high values. It is because we are tracking the front point of the leader vehicle which is supposed to be less steady than the back point. Again, the selection of the desired spacing l is subject to the requirement in an application.

8.5   CONCLUSIONS

Many applications such as in outdoor industrial settings and logistics environments require autonomous mobile robot vehicles to carry out frequent and tight turnings, as well as forward and backward maneuvers. The unified nonlinear tracking controller, presented in this chapter, is able to work for both forward and backward maneuvers as well as driving and steering. The design is based on either kinematics or dynamics using an output function as the intervehicular connection. Design parameters of desired intervehicular spacing l and

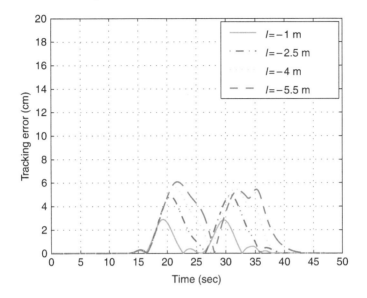

FIGURE 8.19 Tracking errors $\|\tilde{z}\|$ for different values of l.

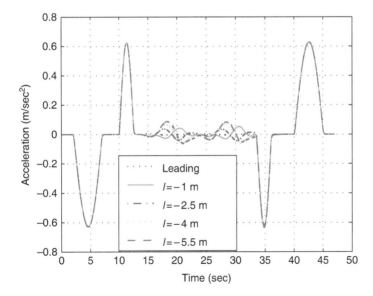

FIGURE 8.20 Vehicle acceleration $\dot{v}$ for different values of l.

Autonomous Mobile Robots

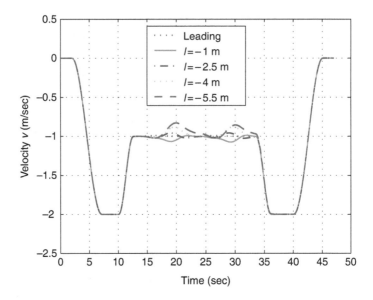

FIGURE 8.21 Vehicle velocity v for different values of l.

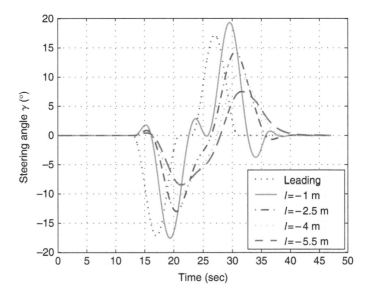

FIGURE 8.22 Steering angle γ for different values of l.

desired multiplier for steering angle p are studied analytically for stable tracking. The derived sufficient conditions of l, p, λ, and ξ can ensure the tracking stability of vehicle following, are simple to choose, and take into account the physical limitations of practical car-like vehicle designs. Extensive numerical simulations also demonstrate the effectiveness of the developed controller and the effects of these design parameters on the tracking performance in various maneuvers.

REFERENCES

1. Hedrick, J. K., Tomizuka, M., and Varaiya, P. Control issues in automated highway systems. *IEEE Control Syst. Mag. 14*, 21, 1994.
2. Daviet, P. and Parent, M. Longitudinal and lateral servoing of vehicles in a platoon. In *Proc. IEEE Intell. Veh. Symp.*, Tokyo, Japan, September 18–20, 1996, p. 41.
3. Yanakiev, D. and Kanellakopoulos, I. Nonlinear spacing policies for automated heavy-duty vehicles. *IEEE Trans. Veh. Technol. 47*, 1365, 1998.
4. Swaroop, D., Hedrick, J. K., and Choi, S. Direct adaptive longitudinal control of vehicle platoons. *IEEE Trans. Veh. Technol. 50*, 150, 2001.
5. No, T. S., To, C. K., and Hwan, R. D. A Lyapunov function approach to longitudinal control of vehicles in a platoon. *IEEE Trans. Veh. Technol. 50*, 116, 2001.
6. Fenton, R. and Selim, I. On the optimal design of an automative lateral controller. *IEEE Trans. Veh. Technol. 37*, 108, 1988.
7. O'Brien, R., Iglesias, P., and Urban, T. Vehicle lateral control for automated highway systems. *IEEE Trans. Contr. Syst. Technol. 4*, 266, 1996.
8. Unyelioğru, K., Hatopoğru, C., and Ozguner, U. Design and stability analysis of a lane following controller. *IEEE Trans. Contr. Syst. Technol. 5*, 127, 1997.
9. Alleyne, A., Williams, B., and DePoorter, M. A lateral position sensing system for automated vehicle following. *IEEE/ASME Trans. Mechatron. 3*, 218, 1998.
10. Fritz, H. Longitudinal and lateral control of heavy duty trucks for automated vehicle following in mixed traffic: experimental results from the CHAUFFEUR project. In *Proc. IEEE Int. Conf. Contr. Applicat.*, Kohala Coast-Island of Hawaii, Hawaii, August 1999, Vol. 2, p. 1348.
11. Haskara, İ., Hatipoğlu, C., and Özgüner, Ü. Combined decentralized longitudinal and lateral controller design for truck convoys. In *Proc. IEEE Conf. Intell. Transport. Syst.*, Boston, Massachusetts, USA, November 9–12, 1997, p. 123.
12. White, R. and Tomizuka, M. Autonomous following lateral control of heavy vehicles using laser scanning radar. In *Proc. American Contr. Conf.*, 2001, Vol. 3, p. 2333.
13. Kato, S., Tsugawa, S., Toduka, K., Matsui, T., and Fujii, H. Vehicle control algorithms for cooperative driving with automated vehicles and intervehicle communications. *IEEE Trans. Intell. Transport Syst. 3*, 155, 2002.

14. Patwardhan, S., Tan, H. S., Guldner, J., and Tomizuka, M. Lane following during backward driving for front wheel steered. In *Proc. American Contr. Conf.*, New Mexico, USA, June 4–6, 1997, Vol. 5, p. 3348.

15. Kim, D. H. and Oh, J. H. Experiments of backward tracking control for trailer system. In *Proc. IEEE Int. Conf. Robot. Automat.*, Detroit, Michigan, USA, May 10–15, 1999, Vol. 1, p. 19.

16. Altafini, C., Speranzon, A., and Wahlberg, B. A feedback control scheme for reversing a truck and trailer vehicle. *IEEE Trans. Robot. Automat. 17*, 915, 2001.

17. Saeki, M. Path following control of articulated vehicle by backward driving. In *Proc. IEEE Int. Conf. Contr. Applicat.*, Glasgow, Scotland, UK, September, 2002, Vol. 1, p. 421.

18. Wang, D., and Xu, G. Full state tracking and internal dynamics of nonholonomic wheeled mobile robots. *IEEE/ASME Trans. Mechatron., Focused Section Adv. Robot Dyn. Control 8*, 203, 2003.

19. Campion, G., Bastin, G., and D'Andréa-Novel, B. Structural properties and classification of kinematic and dynamic models of wheeled mobile robots. *IEEE Trans. Robot. Automat. 12*, 47, 1996.

20. Mammar, S. and Netto, M. Integrated longitudinal and lateral control for vehicle low speed automation. In *Proc. IEEE Int. Conf. Contr. Applicat.*, Taipei, Taiwan, September 1–4, 2004, Vol. 1, p. 350.

21. Wang, D., Pham, M., and Pham, C. T. Simulation study of vehicle platooning maneuvers with full-state tracking control. In *Modeling, Simulation and Optimization of Complex Processes*, H. G. Bock, E. Kostina, H. X. Phu, and R. Rannacher (eds.), Springer-Verlag, Heidelberg, 2004, p. 539.

Biographies

Danwei Wang earned his Ph.D. and M.S.E. degrees from the University of Michigan, Ann Arbor in 1989 and 1984, respectively. He received his B.E. degree from the South China University of Technology, China in 1982. Since 1989, he has been with the School of Electrical and Electronic Engineering, Nanyang Technological University, Singapore. Currently, he is an associate professor and director of the Centre for Intelligent Machines, NTU. He has served as general chairman, technical chairman, and various positions in international conferences, such as *IEEE International Conference on Robotics, Automation and Mechatronics (RAMs), International Conference on Control, Automation, Robotics and Vision (ICARCVs)* and *Asian Conference on Computer Vision (ACCV)*. He is an associate editor of *Conference Editorial Board, IEEE Control Systems Society* and an associate editor of *International Journal of Humanoid Robotics*. He is an active member of IEEE Singapore Section and Robotics and Automation Chapter. He was a recipient of Alexander von Humboldt fellowship, Germany. His research interests include robotics, control theory, and applications. He has published more than 160 technical articles in the areas of iterative learning control, repetitive control, robust control, and adaptive control systems, as well as manipulator/mobile robot dynamics, path planning, and control. (Personal home page: http://www.ntu.edu.sg/home/edwwang)

Minhtuan Pham earned his M.E. degree from Nanyang Technological University, Singapore in 2002. He earned his B.E. degree from Hanoi University of Technology, Vietnam in 1997. He is currently pursuing a Ph.D. degree at School of Electrical and Electronic Engineering, Nanyang Technological University, Singapore. Before joining NTU, he had worked at Department of Automation Technology, Institute of Information Technology, Vietnamese National Centre for Natural Science and Technology, as a research engineer. His research interests include embedded control systems, autonomous vehicle systems, and vehicle platooning.

III

Map Building and
Path Planning

The previous two parts of the book had focused on the sensing and control aspects of autonomous systems. These are fundamental capabilities that any useful mobile robot ought to be equipped with. However, a truly intelligent and autonomous system cannot dispense with the more abstract levels of deliberation, involving planning and a larger amount of information processing and reasoning. Part of these aspects of the autonomous system is dealt with in this portion of the book.

Sensing involves the collection of information, and also involves some preliminary treatment of the collected data, while control makes use of these data for immediate determination of control signals to bring the system's configuration to the desired one. Both these processes lack the sophisticated deliberation that makes a system intelligent. Such intelligent systems should be able to plan their own paths through an unknown environment, make decisions about their goals, and react to the decisions of other robots it senses.

Before the first levels of planning can actually take place, the information obtained from sensors has to be organized into suitable and useful forms. Very often, the success of planning algorithms rely heavily on the accuracy of the robots' estimation of their locations and their internal model (possibly built dynamically) of the world. Map building is the main focus of the first chapter of this part and is given a thorough treatment. In this chapter, the extended Kalman filter (EKF) approach to SLAM is described. For such feature-based approaches, the chapter outlines the essential operations that are required for the successful use of SLAM in uncertain environments. These operations include the ability to extract features from the raw sensor data for inclusion into existing maps, to distinguish between new features and previously detected features which should be associated with known or previously detected features, to be able to determine the robot's location and correct erroneous map information in the presence of ever-increasing uncertainty (i.e., loop closing and relocation techniques). The chapter also describes techniques to handle the required operations for robust performance of SLAM on robots. For the mapping of large environments, the method of *local map joining* is described. The application of the local map joining method to multi-robot mapping, where the relative locations between robots is not known, is also presented.

The success of online robot localization and construction of maps paves the way for more elaborate planning as the robot maneuvers through the unknown environment. The area of path planning has been studied intensively over the years, and is mainly concerned about the generation of a suitable path to the goal, taking into account the obstacles present within the environment. Chapter 10, the second chapter of this part, discusses the incorporation of internal constraints — namely kinematic, dynamic constraints, and visibility constraints — into motion planning. These additional constraints are especially important in systems of embodied mobile robots. Following an overview of conventional classes of approaches to motion planning, the chapter examines the use of randomized sampling techniques for motion planning of robots subjected to kinematic and dynamic constraints. The effect of visibility constraints on motion planning, together with several solution techniques, is investigated through the use of three representative visibility-based planning problems — guarding art galleries, online indoor exploration, and target tracking.

The last chapter of the section examines cooperative motion planning and control in multi-robot systems. This is a natural extension of single robot motion planning, since autonomous systems are seldom made up of a single robot. The planned motions of each robot will no longer be solely to obtain a collision free path, but will also be shaped by the positions of other robots within the team. The control of a robot's path such that it maintains specific relative distances from others, relates to multi-robot formation control, and is treated in detail within the chapter. Specifically, due to the prevalence of nonholonomic robots in real-world applications (Part II), the chapter examines the formation control and stability of teams of nonholonomic robots using formation control graphs, where different formations are achieved through the creation or deletion of edges between robots. Optimization-based control of formations is also investigated, with the focus on an off-line optimization process based on the solution of a mixed integer program, and the use of model predictive control.

The ability to perform map building and path planning operations bring us up the cognitive chain, and sets the stage where still higher levels of planning may take place. These shall be given detailed treatment in the remaining parts of the book.

9 Map Building and SLAM Algorithms

José A. Castellanos, José Neira, and
Juan D. Tardós

CONTENTS

9.1 INTRODUCTION

The concept of autonomy of mobile robots encompasses many areas of knowledge, methods, and ultimately algorithms designed for trajectory control, obstacle avoidance, localization, map building, and so forth. Practically, the success of a path planning and navigation mission of an autonomous vehicle depends on the availability of both a sufficiently reliable estimation of the vehicle location and an accurate representation of the navigation area.

Schematically, the problem of map building consists of the following steps (1) Sensing the environment of the vehicle at time k using onboard sensors (e.g., laser scanner, vision, or sonar); (2) Representation of sensor data (e.g., feature- or raw-data-based approaches); (3) Integration of the recently perceived observations at time k with the previously learned structure of the environment estimated at time $k - 1$.

The simplest approach to map building relies on the vehicle location estimates provided by dead-reckoning. However, as reported in the literature [1], this approach is unreliable for long-term missions due to the time-increasing drift of those estimates (Figure 9.1a). Consequently, a coupling arises between the map building problem and the improvement of dead-reckoning location estimates (Figure 9.1b). Different approaches to the so-called *simultaneous localization and mapping* (SLAM) problem have populated the robotics literature during the last decade.

The most popular approach to SLAM dates back to the seminal work of Smith et al. [2] where the idea of representing the structure of the navigation area in a discrete-time state-space framework was originally presented. They introduced the concept of *stochastic map* and developed a rigorous solution to the SLAM problem using the extended Kalman filter (EKF) perspective. Many successful implementations of this approach have been reported in indoor [1], outdoor [3], underwater [4], and air-borne [5] applications.

The EKF-based approach to SLAM is characterized by the existence of a discrete-time augmented state vector, composed of the location of the vehicle and the location of the map elements, recursively estimated from the available sensor observations gathered at time k, and a model of the vehicle motion, between time steps $k - 1$ and k. Within this framework, uncertainty is represented by probability density functions (pdfs) associated with the state vector, the motion model, and the sensor observations. It is assumed that recursive propagation of the mean and the covariance of those pdfs conveniently approximates the optimal solution of this estimation problem.

The time and memory requirements of the basic EKF–SLAM approach result from the cost of maintaining the full covariance matrix, which is $O(n^2)$ where n is the number of features in the map. Many recent efforts have concentrated on reducing the computational complexity of SLAM in large environments. Several current methods address the computational complexity problem by working on a limited region of the map. Postponement [6]

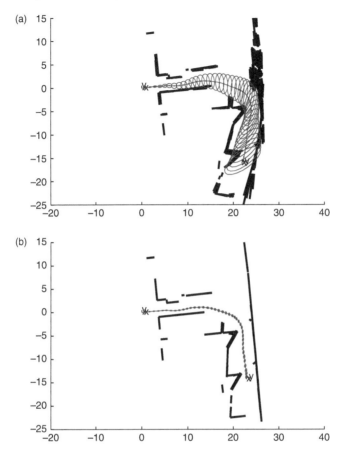

FIGURE 9.1   The need for SLAM: (a) odometric readings and segmented laser walls for 40 m of the trajectory of a vehicle at the Ada Byron building of our campus; (b) map and trajectory resulting from the SLAM algorithm using the same data (95% error ellipses are drawn).

and the compressed filter [3] significantly reduce the computational cost without sacrificing precision, although they require an $O(n^2)$ step on the total number of landmarks to obtain the full map. The split covariance intersection method [7] limits the computational burden but sacrifices precision: it obtains a conservative estimate. The sparse extended information filter [8] is able to obtain an approximate map in constant time per step, except during loop closing. All cited methods work on a single absolute map representation, and confront divergence due to nonlinearities as uncertainty increases when mapping large areas [9]. In contrast, local map joining [10] and the constrained local submap filter [11], propose to build stochastic maps relative to a local reference, guaranteed to be statistically independent. By limiting the size of the local map, this operation is constant time per step. Local maps are joined periodically into a global absolute map, in a $O(n^2)$ step. Given that most of the updates are carried out on a local map, these techniques also reduce the harmful effects of linearization. To avoid the $O(n^2)$ step, the constrained relative submap filter [12] proposes to maintain the independent local map structure. Each map contains links to other neighboring maps, forming a tree structure (where loops cannot be represented). In Atlas [13], network coupled feature maps [14], and constant time SLAM [15] the links between local maps form an adjacency graph. These techniques do not impose loop consistency in the graph, sacrificing the optimality of the resulting global map. Hierarchical SLAM [16] proposes a linear time technique to impose loop consistency, obtaining a close to optimal global map. The FastSLAM technique [17] uses particle filters to estimate the vehicle trajectory and each one has an associated set of independent EKF to estimate the location of each feature in the map. This partition of SLAM into a localization and a mapping problem, allows to obtain a computational complexity $O(\log(n))$ with the number of features in the map. However, its complexity is linear with the number of particles used. The scaling of the number of particles needed with the size and complexity of the environment remains unclear. In particular, closing loops causes dramatic particle extinctions that map result in optimistic (i.e., inconsistent) uncertainty estimations.

Another class of SLAM techniques is based on estimating sequences of robot poses by minimizing the discrepancy between overlapping laser scans. The map representation is the set of robot poses and the corresponding set of laser scans. The work in Reference 18 uses scan matching between close robot poses and global correlation to detect loops. The poses along the loop are estimated using consistent pose estimation [19], whose time complexity is $O(n^3)$ on the number of robot poses, making the method unsuitable for real time execution in large environments. More recently, a similar approach to build consistent maps with many cycles has been proposed in Reference 20. This method obtains correspondences between vehicle poses using the iterative closest point algorithm. Using a quadratic penalty function, correspondences are incorporated into an optimization algorithm that recomputes the whole trajectory. This process is iterated until convergence. Neither computing time nor computational complexity are reported. There are two fundamental limitations in this class of techniques, compared to EKF-based SLAM. First, there is no explicit representation of the uncertainty in the estimated robot poses and the resulting map. As a consequence, their convergence and consistency properties remain unknown. Second, they largely rely on the high precision and density of data provided by laser scanners. They seem hard to extend to sensors that give more imprecise, sparse, or partial information such as sonar or monocular vision.

This chapter describes the basic algorithm to deal with the SLAM problem from the above mentioned EKF-based perspective. We describe techniques that successful SLAM schemes must incorporate: (1) Data association techniques, to relate sensor measurements with features already in the map, as well as to decide those that are spurious or correspond to environment features not previously observed, and (2) Loop closing and relocation techniques, that allow determination of the vehicle location and correct the map when the vehicle uncertainty increases significantly during exploration, or when there is no prior information on the vehicle location. Finally, we point out the main open problem of the current state-of-art SLAM approaches: mapping large-scale areas. Relevant shortcomings of this problem are, on the one hand, the computational burden, which limits the applicability of the EKF-based SLAM in large-scale real time applications and, on the other hand, the use of linearized solutions which jeopardizes the consistency of the estimation process. We point out promising directions of research using nonlinear estimation techniques, and mapping schemes for multivehicle SLAM.

9.2 SLAM Using the Extended Kalman Filter

In feature-based approaches to SLAM, the environment is modeled as a set of geometric features, such as straight line segments corresponding to doors or window frames, planes corresponding to walls, or distinguishable points in outdoor environments. The process of segmentation of raw sensor data to obtain feature parameters depends on the sensor and the feature type. In indoor environments, laser readings can be used to obtain straight wall segments [21,22], or in outdoor environments to obtain two-dimensional (2D) points corresponding to trees and street lamps [3]. Sonar measurement environments can be segmented into corners and walls [10]. Monocular images can provide information about vertical lines [23] or interest points [24]. Even measurements from different sensors can be fused to obtain feature information [25].

In the standard EKF-based approach, the environment information related to a set of elements $\{B, R, F_1, \ldots, F_n\}$ is represented by a map $\mathcal{M}^B = (\hat{\mathbf{x}}^B, \mathbf{P}^B)$, where $\mathbf{x}^B$ is a stochastic state vector with estimated mean $\hat{\mathbf{x}}^B$ and estimated error covariance $\mathbf{P}^B$:

$$\hat{\mathbf{x}}^B = E[\mathbf{x}^B] = \begin{bmatrix} \hat{\mathbf{x}}_R^B \\ \vdots \\ \hat{\mathbf{x}}_{F_n}^B \end{bmatrix}$$

$$(9.1)$$

$$\mathbf{P}^B = E[(\mathbf{x}^B - \hat{\mathbf{x}}^B)(\mathbf{x}^B - \hat{\mathbf{x}}^B)^\mathrm{T}] = \begin{bmatrix} \mathbf{P}_R^B & \cdots & \mathbf{P}_{RF_n}^B \\ \vdots & \ddots & \vdots \\ \mathbf{P}_{F_nR}^B & \cdots & \mathbf{P}_{F_n}^B \end{bmatrix}$$

Vector $\hat{\mathbf{x}}^B$ contains the estimated location of the vehicle R and the environment features $F_1 \ldots F_n$, all with respect to a base reference B. In the case of the vehicle, its location vector $\hat{\mathbf{x}}_R^B = (x, y, \phi)^T$ describes the transformation from B to R. In the case of an environment feature j, the parameters that compose its location vector $\hat{\mathbf{x}}_{F_j}^B$ depend on the feature type, for example, $\hat{\mathbf{x}}_{F_j}^B = (x_j, y_j)^T$ for point features. The diagonal elements of the matrix $\mathbf{P}^B$ represent the estimated error covariance of the different features of the state vector and that of the vehicle location; its off-diagonal elements represent the cross-covariance matrices between the estimated locations of the corresponding features.

Recursive estimation of the first two moments of the probability density function of $\mathbf{x}^B$ is performed following Algorithm 9.1. There, the map is initialized using the current vehicle location as base reference, and thus with perfect knowledge of the vehicle location. Sensing and feature initialization is also performed before the first vehicle motion, to maximize the precision of the resulting map. Prediction of the vehicle motion using odometry and update of the map using onboard sensor measurements are then iteratively carried out.

9.2.1 Initialization

In the creation of a new stochastic map at step 0, a base reference B must be selected. It is common practice to build a map relative to a fixed base reference different from the initial vehicle location. This normally requires the

---

ALGORITHM 9.1
EKF–SLAM

$\mathbf{x}_0^B = \mathbf{0}; \mathbf{P}_0^B = \mathbf{0}$ {*Map initialization*}
$[\mathbf{z}_0, \mathbf{R}_0] = \text{get\_measurements}$
$[\mathbf{x}_0^B, \mathbf{P}_0^B] = \text{add\_new\_features}(\mathbf{x}_0^B, \mathbf{P}_0^B, \mathbf{z}_0, \mathbf{R}_0)$
for $k = 1$ to steps **do**
$\quad [\mathbf{x}_{R_k}^{R_{k-1}}, \mathbf{Q}_k] = \text{get\_odometry}$
$\quad [\mathbf{x}_{k|k-1}^B, \mathbf{P}_{k|k-1}^B] = \text{compute\_motion}(\mathbf{x}_{k-1}^B, \mathbf{P}_{k-1}^B, \mathbf{x}_{R_k}^{R_{k-1}}, \mathbf{Q}_k)$ {*EKF predict.*}
$\quad [\mathbf{z}_k, \mathbf{R}_k] = \text{get\_measurements}$
$\quad \mathcal{H}_k = \text{data\_association}(\mathbf{x}_{k|k-1}^B, \mathbf{P}_{k|k-1}^B, \mathbf{z}_k, \mathbf{R}_k)$
$\quad [\mathbf{x}_k^B, \mathbf{P}_k^B] = \text{update\_map}(\mathbf{x}_{k|k-1}^B, \mathbf{P}_{k|k-1}^B, \mathbf{z}_k, \mathbf{R}_k, \mathcal{H}_k)$ {*EKF update*}
$\quad [\hat{\mathbf{x}}_k^B, \mathbf{P}_k^B] = \text{add\_new\_features}(\mathbf{x}_k^B, \mathbf{P}_k^B, \mathbf{z}_k, \mathbf{R}_k, \mathcal{H}_k)$
end for

---

assignment of an initial level of uncertainty to the estimated vehicle location. In the theoretical linear case [26], the vehicle uncertainty should always remain above this initial level. In practice, due to linearizations, when a nonzero initial uncertainty is used, the estimated vehicle uncertainty rapidly drops below its initial value, making the estimation inconsistent after very few EKF update steps [9].

A good alternative is to use, as base reference, the current vehicle location, that is, $B = R_0$, and thus we initialize the map with perfect knowledge of the vehicle location:

$$\hat{\mathbf{x}}_0^B = \hat{\mathbf{x}}_{R_0}^B = \mathbf{0}; \quad \mathbf{P}_0^B = \mathbf{P}_{R_0}^B = \mathbf{0} \tag{9.2}$$

If at any moment there is a need to compute the location of the vehicle or the map features with respect to any other reference, the appropriate transformations can be applied (see Appendix). At any time, the map can also be transformed to use a feature as base reference, again using the appropriate transformations [10].

9.2.2 Vehicle Motion: The EKF Prediction Step

When the vehicle moves from position $k-1$ to position k, its motion is estimated by odometry:

$$\mathbf{x}_{R_k}^{R_{k-1}} = \hat{\mathbf{x}}_{R_k}^{R_{k-1}} + \mathbf{v}_k \tag{9.3}$$

where $\hat{\mathbf{x}}_{R_k}^{R_{k-1}}$ is the estimated relative transformation between positions $k - 1$ and k, and $\mathbf{v}_k$ (process noise [27]) is assumed to be additive, zero-mean, and white, with covariance $\mathbf{Q}_k$.

Thus, given a map $\mathcal{M}_{k-1}^B = (\hat{\mathbf{x}}_{k-1}^B, \mathbf{P}_{k-1}^B)$ at step $k - 1$, the predicted map $\mathcal{M}_{k|k-1}^B$ at step k after the vehicle motion is obtained as follows:

$$\hat{\mathbf{x}}_{k|k-1}^B = \begin{bmatrix} \hat{\mathbf{x}}_{R_{k-1}}^B \oplus \hat{\mathbf{x}}_{R_k}^{R_{k-1}} \\ \hat{\mathbf{x}}_{F_1,k-1}^B \\ \vdots \\ \hat{\mathbf{x}}_{F_m,k-1}^B \end{bmatrix} \tag{9.4}$$

$$\mathbf{P}_{k|k-1}^B \simeq \mathbf{F}_k \mathbf{P}_{k-1}^B \mathbf{F}_k^T + \mathbf{G}_k \mathbf{Q}_k \mathbf{G}_k^T$$

where $\oplus$ represents the composition of transformations (see Appendix), and:

$$\mathbf{F}_k = \left. \frac{\partial \mathbf{x}_{k|k-1}^B}{\partial \mathbf{x}_{k-1}^B} \right|_{(\hat{\mathbf{x}}_{k-1}^B, \hat{\mathbf{x}}_{R_k}^{R_{k-1}})} = \begin{bmatrix} \mathbf{J}_{1\oplus} \left\{ \hat{\mathbf{x}}_{R_{k-1}}^B, \hat{\mathbf{x}}_{R_k}^{R_{k-1}} \right\} & 0 & \cdots & 0 \\ 0 & \mathbf{I} & & \vdots \\ \vdots & & \ddots & \\ 0 & \cdots & & \mathbf{I} \end{bmatrix}$$

$$\mathbf{G}_k = \left. \frac{\partial \mathbf{x}_{k|k-1}^B}{\partial \mathbf{x}_{R_k}^{R_{k-1}}} \right|_{(\hat{\mathbf{x}}_{k-1}^B, \hat{\mathbf{x}}_{R_k}^{R_{k-1}})} = \begin{bmatrix} \mathbf{J}_{2\oplus} \left\{ \hat{\mathbf{x}}_{R_{k-1}}^B, \hat{\mathbf{x}}_{R_k}^{R_{k-1}} \right\} \\ 0 \\ \vdots \\ 0 \end{bmatrix}$$

where $\mathbf{J}_{1\oplus}$ and $\mathbf{J}_{2\oplus}$ are the Jacobians of transformation composition (see Appendix).

9.2.3 Data Association

At step k, an onboard sensor obtains a set of measurements $\mathbf{z}_{k,i}$ of m environment features E_i ($i = 1, \ldots, m$). Data association consists in determining the origin of each measurement, in terms of the map features $F_j, j = 1, \ldots, n$. The result is a hypothesis:

$$\mathcal{H}_k = [\mathsf{j}_1 \ \mathsf{j}_2 \ \cdots \ \mathsf{j}_m]$$

associating each measurement $\mathbf{z}_{k,i}$ with its corresponding map feature F_{j_i} ($\mathsf{j}_i = 0$ indicates that $\mathbf{z}_{k,i}$ does not come from any feature in the map). The core tools of data association are a prediction of the measurement that each feature would generate, and a measure of the discrepancy between a predicted measurement and an actual sensor measurement.

The measurement of feature F_j can be predicted using a nonlinear measurement function $\mathbf{h}_{k,j}$ of the vehicle and feature location, both contained in the map state vector $\mathbf{x}_{k|k-1}^B$. If observation $\mathbf{z}_{k,i}$ comes from feature F_j, the following relation must hold:

$$\mathbf{z}_{k,i} = \mathbf{h}_{k,j}(\mathbf{x}_{k|k-1}^B) + \mathbf{w}_{k,i} \tag{9.5}$$

where the measurement noise $\mathbf{w}_{k,i}$, with covariance $\mathbf{R}_{k,i}$, is assumed to be additive, zero-mean, white, and independent of the process noise $\mathbf{v}_k$. Linearization of Equation (9.5) around the current estimate yields:

$$\mathbf{h}_{k,j}(\mathbf{x}_{k|k-1}^B) \simeq \mathbf{h}_{k,j}(\hat{\mathbf{x}}_{k|k-1}^B) + \mathbf{H}_{k,j}(\mathbf{x}_k^B - \hat{\mathbf{x}}_{k|k-1}^B) \tag{9.6}$$

with

$$\mathbf{H}_{k,j} = \left. \frac{\partial \mathbf{h}_{k,j}}{\partial \mathbf{x}^B_{k|k-1}} \right|_{(\hat{\mathbf{x}}^B_{k|k-1})} \tag{9.7}$$

The discrepancy between the observation i and the predicted observation of map feature j is measured by the innovation term $v_{k,ij}$, whose value and covariance are:

$$
\begin{aligned}
v_{k,ij} &= \mathbf{z}_{k,i} - \mathbf{h}_{k,j}(\hat{\mathbf{x}}^B_{k|k-1}) \\
\mathbf{S}_{k,ij} &= \mathbf{H}_{k,j} \mathbf{P}^B_k \mathbf{H}^{\mathrm{T}}_{k,j} + \mathbf{R}_{k,i}
\end{aligned}
\tag{9.8}
$$

The measurement can be considered corresponding to the feature if the Mahalanobis distance $D^2_{k,ij}$ [28] satisfies:

$$D^2_{k,ij} = v^{\mathrm{T}}_{k,ij} \mathbf{S}^{-1}_{k,ij} v_{k,ij} < \chi^2_{d,1-\alpha} \tag{9.9}$$

where $d = \dim(\mathbf{h}_{k,j})$ and $1 - \alpha$ is the desired confidence level, usually 95%. This test, denominated individual compatibility (IC), applied to the predicted state, can be used to determine the subset of map features that are compatible with a measurement, and is the basis for some of the most popular data association algorithms discussed later in this chapter.

An often overlooked fact, that will be discussed in more detail in Section 9.3, is that *all* measurements should be jointly compatible with their corresponding features. In order to establish the consistency of a hypothesis $\mathcal{H}_k$, measurements can be jointly predicted using function $\mathbf{h}_{\mathcal{H}_k}$:

$$\mathbf{h}_{\mathcal{H}_k}(\mathbf{x}^B_{k|k-1}) = \begin{bmatrix} \mathbf{h}_{j_1}(\mathbf{x}^B_{k|k-1}) \\ \vdots \\ \mathbf{h}_{j_m}(\mathbf{x}^B_{k|k-1}) \end{bmatrix} \tag{9.10}$$

which can also be linearized around the current estimate to yield:

$$\mathbf{h}_{\mathcal{H}_k}(\mathbf{x}^B_{k|k-1}) \simeq \mathbf{h}_{\mathcal{H}_k}(\hat{\mathbf{x}}^B_{k|k-1}) + \mathbf{H}_{\mathcal{H}_k}(\mathbf{x}^B_k - \hat{\mathbf{x}}^B_{k|k-1}); \quad \mathbf{H}_{\mathcal{H}_k} = \begin{bmatrix} \mathbf{H}_{j_1} \\ \vdots \\ \mathbf{H}_{j_m} \end{bmatrix}$$

$$\tag{9.11}$$

The joint innovation and its covariance are:

$$\nu_{\mathcal{H}_k} = \mathbf{z}_k - \mathbf{h}_{\mathcal{H}_k}(\hat{\mathbf{x}}_{k|k-1}^B)$$
$$\mathbf{S}_{\mathcal{H}_k} = \mathbf{H}_{\mathcal{H}_k}\mathbf{P}_k^B\mathbf{H}_{\mathcal{H}_k}^T + \mathbf{R}_{\mathcal{H}_k} \tag{9.12}$$

Measurements $\mathbf{z}_k$ can be considered compatible with their corresponding features according to $\mathcal{H}_k$ if the Mahalanobis distance satisfies:

$$D_{\mathcal{H}_k}^2 = \nu_{\mathcal{H}_k}^T \mathbf{S}_{\mathcal{H}_k}^{-1} \nu_{\mathcal{H}_k} < \chi_{d,1-\alpha}^2 \tag{9.13}$$

where now $d = \dim(\mathbf{h}_{\mathcal{H}_k})$. This consistency test is denominated joint compatibility (JC).

9.2.4 Map Update: The EKF Estimation Step

Once correspondences for measurements $\mathbf{z}_k$ have been decided, they are used to improve the estimation of the stochastic state vector by using the standard EKF update equations as follows:

$$\hat{\mathbf{x}}_k^B = \hat{\mathbf{x}}_{k|k-1}^B + \mathbf{K}_{\mathcal{H}_k}\nu_{\mathcal{H}_k} \tag{9.14}$$

where the filter gain $\mathbf{K}_{\mathcal{H}_k}$ is obtained from:

$$\mathbf{K}_{\mathcal{H}_k} = \mathbf{P}_{k|k-1}^B\mathbf{H}_{\mathcal{H}_k}^T\mathbf{S}_{\mathcal{H}_k}^{-1} \tag{9.15}$$

Finally, the estimated error covariance of the state vector is:

$$\mathbf{P}_k^B = (\mathbf{I} - \mathbf{K}_{\mathcal{H}_k}\mathbf{H}_{\mathcal{H}_k})\mathbf{P}_{k|k-1}^B$$
$$= (\mathbf{I} - \mathbf{K}_{\mathcal{H}_k}\mathbf{H}_{\mathcal{H}_k})\mathbf{P}_{k|k-1}^B(\mathbf{I} - \mathbf{K}_{\mathcal{H}_k}\mathbf{H}_{\mathcal{H}_k})^T + \mathbf{K}_{\mathcal{H}_k}\mathbf{R}_{\mathcal{H}_k}\mathbf{K}_{\mathcal{H}_k}^T \tag{9.16}$$

9.2.5 Adding Newly Observed Features

Measurements for which correspondences in the map cannot be found by data association can be directly added to the current stochastic state vector as new features by using the relative transformation between the vehicle R_k and the observed feature E. Therefore, updating of $\mathbf{x}_k^B$ takes place as follows:

$$
\mathbf{x}_k^B = \begin{bmatrix} \mathbf{x}_{R_k}^B \\ \vdots \\ \mathbf{x}_{F_{n,k}}^B \end{bmatrix} \Rightarrow \mathbf{x}_{k+}^B = \begin{bmatrix} \mathbf{x}_{R_k}^B \\ \vdots \\ \mathbf{x}_{F_{n,k}}^B \\ \mathbf{x}_{E_k}^B \end{bmatrix} = \begin{bmatrix} \mathbf{x}_{R_k}^B \\ \vdots \\ \mathbf{x}_{F_{n,k}}^B \\ \mathbf{x}_{R_k}^B \oplus \mathbf{x}_E^{R_k} \end{bmatrix}
\tag{9.17}
$$

Additionally, the updated covariance matrix $\mathbf{P}_{k+}^B$ is computed using the linearization of Equation (9.17).

9.2.6 Consistency of EKF–SLAM

A state estimator is called *consistent* if its state estimation error $\mathbf{x}_k^B - \hat{\mathbf{x}}_k^B$ satisfies [29]:

$$
E[\mathbf{x}_k^B - \hat{\mathbf{x}}_k^B] = \mathbf{0}
$$
$$
E[(\mathbf{x}_k^B - \hat{\mathbf{x}}_k^B)(\mathbf{x}_k^B - \hat{\mathbf{x}}_k^B)^{\mathrm{T}}] \le \mathbf{P}_k^B
\tag{9.18}
$$

This means that the estimator is *unbiased* and that the actual mean square error matches the filter-calculated covariances. Given that SLAM is a nonlinear problem, consistency checking is of paramount importance. When the ground truth solution for the state variables is available, a statistical test for filter consistency can be carried out on the normalized estimation error squared (NEES), defined as:

$$
\text{NEES} = (\mathbf{x}_k^B - \hat{\mathbf{x}}_k^B)^{\mathrm{T}} (\mathbf{P}_k^B)^{-1} (\mathbf{x}_k^B - \hat{\mathbf{x}}_k^B)
\tag{9.19}
$$

Consistency is checked using a chi-squared test:

$$
\text{NEES} \le \chi_{d,1-\alpha}^2
\tag{9.20}
$$

where $d = \dim(\mathbf{x}_k^B)$ and $1 - \alpha$ is the desired confidence level. Since in most cases ground truth is not available, the consistency of the estimation is maintained by using only measurements that satisfy the innovation test of Equation (9.13). Because the innovation term depends on the data association hypothesis, this process becomes critical in maintaining a consistent estimation [9] of the environment map.

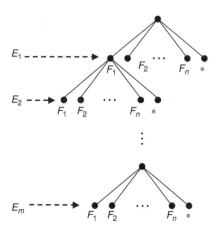

FIGURE 9.2   Interpretation tree of measurements $E_1, \ldots, E_m$ in terms of map features $F_1, \ldots, F_n$.

9.3   DATA ASSOCIATION IN SLAM

Assume that a new set of m measurements $\mathbf{z} = \{\mathbf{z}_1, \ldots, \mathbf{z}_m\}$ of the environment features $\{E_1, \ldots, E_m\}$ have been obtained by a sensor mounted on the vehicle. As mentioned in Section 9.2, the goal of data association is to generate a hypothesis $\mathcal{H} = [\jmath_1 \, \jmath_2 \, \cdots \, \jmath_m]$ associating each measurement E_i with its corresponding map feature $F_{j_i} (\jmath_i = 0$ indicating that $\mathbf{z}_i$ does not correspond to any map feature). The space of measurement-feature correspondences can be represented as an *interpretation tree* of m levels [30] (see Figure 9.2); each node at level i, called an *i-interpretation*, provides an interpretation for the first i measurements. Each node has $n + 1$ branches, corresponding to each of the alternative interpretations for measurement E_i, including the possibility that the measurement be spurious and allowing map feature repetitions in the same hypothesis. Data association algorithms must select in some way one of the $(n + 1)^m$ m-interpretations as the correct hypothesis, carrying out validations to determine the compatibility between sensor measurements and map features.

9.3.1   Individual Compatibility Nearest Neighbor

The simplest criterion to decide a pairing for a given measurement is the *nearest neighbor* (NN), which consists in choosing among the features that satisfy IC of Equation (9.9), the one with the smallest Mahalanobis distance. A popular data association algorithm, the Individual Compatibility Nearest Neighbor (ICNN, Algorithm 9.2), is based on this idea. It is frequently used given its conceptual simplicity and computational efficiency: it performs $m \cdot n$ compatibility tests, making it linear with the size of the map.

ALGORITHM 9.2
ICNN

```
ICNN (E₁...ₘ, F₁...ₙ)
```
for $i = 1$ to m **do** {measurement E_i}
    $D^2_{min} \leftarrow$ `mahalanobis2` (E_i, F_1)
    nearest $\leftarrow 1$
    for $j = 2$ to n **do** {feature F_j}
        $D^2_{ij} \leftarrow$ `mahalanobis2` (E_i, F_j)
        if $D^2_{ij} < D^2_{min}$ **then**
            nearest $\leftarrow j$
            $D^2_{min} \leftarrow D^2_{ij}$
        end if
    end for
    if $D^2_{min} \leq \chi^2_{d_i,1-\alpha}$ **then**
        $\mathcal{H}_i \leftarrow$ nearest
    else
        $\mathcal{H}_i \leftarrow 0$
    end if
end for
return $\mathcal{H}$

The IC considers *individual* compatibility between a measurement and a feature. However, individually compatible pairings are not guaranteed to be *jointly* compatible to form a consistent hypothesis. Thus, with ICNN there is a high risk of obtaining an inconsistent hypothesis and thus updating the state vector with a set of incompatible measurements, which will cause EKF to diverge. As vehicle error grows with respect to sensor error, the discriminant power of IC decreases: the probability that a feature may be compatible with an unrelated (or spurious) sensor measurement increases. ICNN is a greedy algorithm, and thus the decision to pair a measurement with its most compatible feature is never reconsidered. As a result, spurious pairings may be included in the hypothesis and integrated in the state estimation. This will lead to a reduction in the uncertainty computed by the EKF with no reduction in the actual error, that is, inconsistency.

9.3.2 Joint Compatibility

In order to limit the possibility of accepting a spurious pairing, reconsideration of the established pairings is necessary. The probability that a spurious

ALGORITHM 9.3
JCBB

Continuous_JCBB $(E_{1\cdots m}, F_{1\cdots n})$
Best = [ ]
JCBB ([ ], 1)
return Best

procedure JCBB $(\mathcal{H}, i)$: {*find pairings for observationE_i*}
if $i > m$ **then** {*leaf node?*}
   if pairings($\mathcal{H}$) > pairings(Best) **then**
      Best ← $\mathcal{H}$
   end if
else
   for $j = 1$ to n **do**
      if individual_compatibility(i,j) **and then** joint_compatibility$(\mathcal{H}, i, j)$
      then
         JCBB($[\mathcal{H}\ j], i + 1$) {*pairing (E_i, F_j) accepted*}
      end if
   end for
   if pairings($\mathcal{H}$) $+ m - i >$ pairings(Best) **then** {*can do better?*}
      JCBB($[\mathcal{H}\ 0], i + 1$) {*star node, E_i not paired*}
   end if
end if

---

pairing is jointly compatible with all the other pairings of a given hypothesis decreases as the number of pairings in the hypothesis increases. The JC test can be used to establish the consistency of a hypothesis $\mathcal{H}_m$, using Equation (9.13). The JC test is the core of the joint compatibility branch and bound data association algorithm (JCBB, Algorithm 9.3), that traverses the interpretation tree in search for the hypothesis that includes the largest number of *jointly compatible* pairings. The quality of a node at level i, corresponding to a hypothesis $\mathcal{H}_i$, is defined as the number of non-null pairings that can be established from the node. In this way, nodes with quality lower than the best available hypothesis are not explored, bounding the search [30]. The NN rule using the Mahalanobis distance $D^2_{\mathcal{H}_i}$ is used as heuristic for *branching*, so that the nodes corresponding to hypotheses with a higher degree of JC are explored first. The size of both $\mathbf{h}_{\mathcal{H}_i}$ and $\mathbf{S}_{\mathcal{H}_i}$ increase with the size of hypothesis $\mathcal{H}_i$. This makes this test potentially expensive to apply (see References 31 and 32 for techniques for the efficient computation of the Mahalanobis distance).

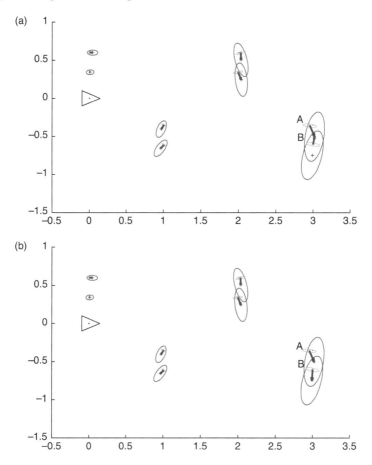

FIGURE 9.3 Predicted feature locations relative to vehicle (large ellipses), measurements (small ellipses), and associations (bold arrows). According to the ICNN algorithm observation B is incorrectly matched with the upper map point (a) and according to the JCBB algorithm (b) all the matches are correct.

During continuous SLAM, data association problems may arise even in very simple scenarios. Consider an environment constituted by 2D points. If at a certain point the vehicle uncertainty is larger than the separation between the features, the predicted feature locations relative to the robot are cluttered, and the NN algorithm is prone to make an incorrect association as illustrated in Figure 9.3a where two measurements are erroneously paired with the same map feature. In these situations, the JCBB algorithm can determine the correct associations (Figure 9.3b), because through correlations it considers the relative location between the features, independent of vehicle error.

The robustness of JCBB is especially important in loop-closing operations (Figure 9.4). Due to the big odometry errors accumulated, simple data association algorithms would incorrectly match the signaled point with a point feature previously observed in the pillar. Accepting an incorrect matching will cause the EKF to diverge, obtaining an inconsistent map. The JC algorithm takes into account the relative location between the point and the segment and has no problem in finding the right associations. The result is a consistent and more precise global map.

Joint compatibility is a highly restrictive criterion, that limits the combinatorial explosion of the search. The computational complexity does not suffer with the increase in vehicle error because the JC of a certain number of measurements fundamentally depends on their *relative error* (which depends on sensor and map precision), more than on their *absolute error* (which depends on robot error). The JC test is based on the linearization of the relation between the measurements and the state (Equation [9.6]). JCBB will remain robust to robot error as long as the linear approximation is reasonable. Thus, the adequacy of using JCBB is determined by the robot orientation error (in practice, we have found the limit to be around 30°). Even if the vehicle motion is unknown (no odometry is available), as long as it is bounded by within this limit, JCBB can perform robustly. In these cases, the predicted vehicle motion can be set to zero ($\hat{\mathbf{x}}_{R_k}^{R_{k-1}} = \mathbf{0}$, Figure 9.5a), with $\mathbf{Q}_k$ sufficiently large to include the largest possible displacement. The algorithm will obtain the associations, and during the estimation stage of the EKF the vehicle motion will be determined and the environment structure can be recovered (Figure 9.5b).

9.3.3 Relocation

Consider now the data association problem known as vehicle relocation, first location, global localization, or "kidnapped" robot problem, which can be stated as follows: *given a vehicle in an unknown location, and a map of the environment, use a set of measurements taken by onboard sensors to determine the vehicle location within the map.* In SLAM, solving this problem is essential to be able to restart the robot in a previously learned environment, to recover from localization errors, or to safely close big loops.

When there is no vehicle location estimation, simple *location independent* geometric constraints can be used to limit the complexity of searching the correspondence space [30]. Given a pairing $p_{ij} = (E_i, F_j)$, the unary geometric constraints that may be used to validate the pairing include length for segments, angle for corners, or radius for circular features. Given two pairings $p_{ij} = (E_i, F_j)$ and $p_{kl} = (E_k, F_l)$, a binary geometric constraint is a geometric relation between measurements E_i and E_k that must also be satisfied between their corresponding map features F_j and F_l (e.g., distance between two points,

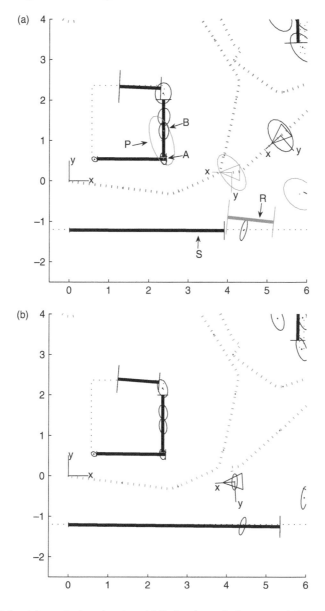

FIGURE 9.4 A loop-closing situation. (a) Before loop closing, potential matches have been found for measurements signaled with an arrow: measurement R is compatible only with feature S, but measurement P is compatible with both features A and B. The NN rule would incorrectly match P with A. (b) The JCBB algorithm has correctly matched both observations with the corner (P with A) and the lower wall (R with B), and the map has been updated.

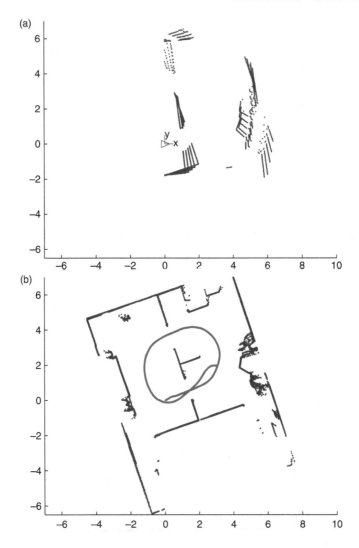

FIGURE 9.5 Data association using JCBB without odometry: (a) laser data in the absolute reference with null vehicle motion; (b) map and vehicle trajectory resulting from the SLAM algorithm.

angle between two segments). For stochastic geometric constraint validation in SLAM, see Reference 33.

Grimson [30] proposed a branch and bound algorithm for model-based geometric object recognition that uses unary and binary geometric constraints. A closely related technique also used in object recognition consists in building a compatibility graph whose nodes are unary compatible matchings and whose arcs represent pairs of binary compatible matchings. Finding the largest hypothesis consistent with unary and binary constraints is equivalent to finding the maximum clique in the compatibility graph (see Reference 30 for a discussion and references). This idea has been applied recently by Bailey et al. [34] to the problem of robot relocation with an a priori map.

Branch and bound algorithms are forced to traverse the whole correspondence space until a good bound is found. In the SLAM relocation problem, when the vehicle is not within the mapped area, a good bound is never found. Since the correspondence space is exponential with the number of measurements, in this worst case the execution times of branch and bound algorithms are very long. To overcome this limitation, the data association process can be done using random sampling (RS) instead of by a full traversal of the interpretation tree. The RS algorithm that we use (Algorithm 9.4) is an adaptation of the RANSAC algorithm [35] for the relocation problem. The fundamental idea is to randomly select p out of the m measurements to try to generate vehicle localization hypotheses using geometric constraints, and verify them with all m measurements using JC. If P_g is the probability that a randomly selected measurement corresponds to a mapped feature (not spurious) and P_{fail} is the acceptable probability of not finding a good solution when it exists, the required number of tries is:

$$t = \left\lceil \frac{\log P_{\text{fail}}}{\log(1 - P_g{}^p)} \right\rceil \tag{9.21}$$

Hypothesis generation–verification schemes such as this one perform better because feature location is a tighter consistency criterion than geometric constraints, and thus branch pruning is more effective. The potential drawback of this approach is that hypothesis verification is *location dependent*, and thus the constraints to be used for validation cannot be precomputed. To limit the amount of location dependent constraints to apply, verification can take place when a hypothesis contains at least *three* consistent pairings. Choosing $P_{\text{fail}} = 0.05$ and considering a priori that only half of the measurements are present in the map $P_g = 0.5$, the maximum number of tries is $t = 23$. If you can consider that at least 90% of the measurements correspond to a map feature, the number of required tries is only three. The RS algorithm randomly permutes the measurements and performs hypothesis generation considering the first three measurements not spurious (without star branch). The number of tries is recalculated to adapt to the current best hypothesis, so that no unnecessary tries are carried out [36].

Notice that the maximum number of tries does *not* depend on the number of measurements. Experiments show that this fact is crucial in reducing the computational complexity of the RS algorithm.

ALGORITHM 9.4
Relocation using RANSAC

```
Relocation_RS (H)
```
$P_{\text{fail}} = 0.05$, p = 3, $P_g = 0.5$
Best = []
$i = 0$
repeat
  $\hat{z}$ = random_permutation($\hat{z}$)
  RS([], 1)
  P_g = max(P_g, pairings(Best) / m)
  t = log P_{fail}/ log $\left(1 - P_g{}^p\right)$
  i = i + 1
until $i \geq t$
return Best

procedure RS (H):
{H : *current hypothesis*}
{i : *observation to be matched*}
if $i > m$ **then**
  if pairings(H) > pairings(Best) **then**
    Best = H
  end if
else if pairings(H) == 3 **then**
  x_R^B = estimate_location_(H)
  if joint_compatibility(H) **then**
    JCBB(H, i) { *hypothesis verification*}
  end if
else {*branch and bound without star node*}
  for j = 1 to n **do**
    if unary(i, j) $\wedge$ binary(i, j, H) **then**
      RS([H j], i + 1)
    end if
  end for
end if

9.3.4 Locality

As explained in Section 9.3.3, the main problem of the interpretation tree approach is the exponential number of possible hypotheses (tree leaves): $N_h = (n + 1)^m$. The use of geometric constraints and branch and bound search dramatically reduce the number of nodes explored, by cutting down entire branches of the tree. However, Grimson [30] has shown that in the general case where spurious measurements can arise, the amount of search needed to find the best interpretation is still exponential. In these conditions, the interpretation tree approach seems impracticable except for very small maps.

To overcome this difficulty we introduce the concept of *locality*: given that the set of measurements has been obtained from a unique vehicle location (or from a set of nearby locations), it is sufficient to try to find matchings with *local* sets of features in the map. Given a map feature F_j, we define its *locality* $L(F_j)$ as the set of map features that are in the vicinity to it, such that they can be seen from the same vehicle location. For a given mapping problem, the maximum cardinality of the locality sets will be a constant c that depends on the sensor range and the maximum density of features in the environment.

During the interpretation tree search, once a matching has been established with a map feature, the search can be restricted to its locality set. For the first measurement, there are n possible feature matchings. Since there are at most c features covisible with the first one, for the remaining $m - 1$ measurements there are only c possible matches, giving a maximum of $n(c + 1)^{m-1}$ hypotheses. If the first measurement is not matched, a similar analysis can be done for the second tree level. Thus, the total number of hypotheses N_h will be:

$$N_h \leq n(c + 1)^{m-1} + \cdots + n + 1 = n\frac{(c + 1)^m - 1}{c} + 1 \qquad (9.22)$$

This implies that, using locality, the complexity of searching the interpretation tree will be *linear* with the size of the map.

There are several ways of implementing locality:

1. SLAM can be implemented by building sequences of independent local maps [10]. If the local maps are stored, the search for matchings can be performed in time linear with the number of local maps. In this case, the locality of a feature is the set of features belonging to the same local map. A drawback of this technique is that global localization may fail around the borders between two local maps.

2. Alternatively, the locality of a feature can be computed as the set of map features within a distance less than the maximum sensor range. There are two drawbacks in this approach: first, this will require $O(n^2)$ distance computations, and second, in some cases features that are close cannot be seen simultaneously (e.g., because they are in different rooms), and thus should not be considered local.

3. The locality of a feature can be defined as the set of features that have been seen simultaneously with it at least once. We choose this last alternative, because it does not suffer from the limitations of the first

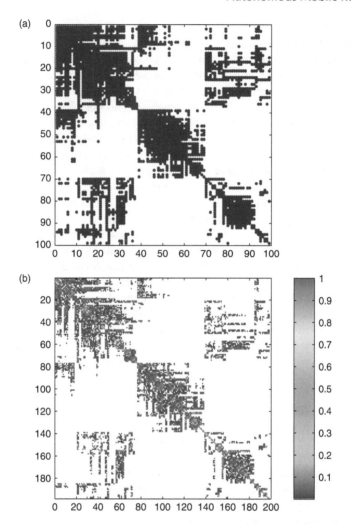

FIGURE 9.6   Covisibility matrix (a) and normalized information matrix (b).

two, and additionally it can be done during map building without extra cost.

Figure 9.6a shows the covisibility matrix obtained during map building for the first 1000 steps of the dataset obtained by Guivant and Nebot [3], gathered with a vehicle equipped with a SICK laser scanner in Victoria Park, Sydney. Wheel encoders give an odometric measure of the vehicle location. The laser scans are processed using Guivant's algorithm to detect tree trunks and estimate their radii (Figure 9.7). As features are added incrementally during map

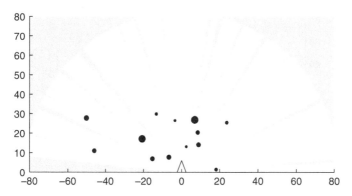

FIGURE 9.7   Segmentation of scan 120, with $m = 13$ tree trunks detected. Radii are magnified $\times 5$.

building, the typical form of the covisibility matrix is band-diagonal. Elements far from the diagonal appear when a loop is closed, because recently added features become covisible with previously mapped features. In any case, the number of elements per row or column only depends on the density of features and the sensor reach. Using a sparse matrix representation, the amount of memory needed to store the covisibility matrix (or any other locality matrix) is $O(n)$.

An important property of the covisibility matrix is its close relation to the information matrix of the map (the inverse of the map covariance matrix). Figure 9.6b shows the normalized information matrix, where each row and column has been divided by the square root of the corresponding diagonal element. It is clear that the information matrix allows the determination of those features that are seen from the vehicle location during map building. The intuitive explanation is that as the uncertainty in the absolute vehicle location grows, the information about the features that are seen from the same location becomes highly coupled.

This gives further insight on the structure of the SLAM problem: while the map covariance matrix is a full matrix with $O(n^2)$ elements, the normalized information matrix tends to be sparse, with $O(n)$ elements. This fact can be used to obtain more efficient SLAM Algorithms [37].

Running continuous SLAM for the first 1000 steps, we obtain a map of $n = 99$ point features (see Figure 9.8). To verify the vehicle locations obtained by our algorithm, we obtained a reference solution running continuous SLAM until step 2500. Figure 9.8 shows the reference vehicle location for steps 1001 to 2500. The RS relocation algorithm was executed on scans 1001 to 2500. This guarantees that we use scans statistically independent from the stochastic map. The radii of the trunks are used as unary constraints, and the distance between the centers as binary constraints.

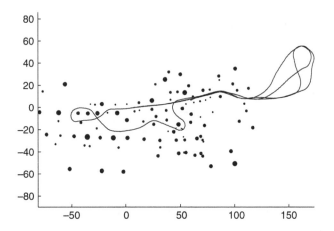

FIGURE 9.8 Stochastic map of 2D points (tree trunks) built until step 1000. There are $n = 99$ features. Reference vehicle trajectory for steps 1001 to 2500. Trunk radii are magnified $\times 5$.

In this experiment, when six or more measurements are paired, the algorithm finds the solution with no false positives. Otherwise, the solution must be discarded as being unreliable. In case that less than six points are segmented from the scan, more sensor information is necessary to reliably determine the vehicle location. When the vehicle is in the map, the RS algorithm finds the solution with a mean execution time of less than 1 sec (in MATLAB®, and executed on a Pentium IV, at 1.7 GHz). When the vehicle is not in the mapped area, for up to 30 measurements, RS runs in less than 2 sec (see Reference 33 for full details).

9.4 MAPPING LARGE ENVIRONMENTS

The EKF–SLAM techniques presented in previous sections have two important limitations when trying to map large environments. First, the computational cost of updating the map grows with $O(n^2)$, where n is the number of features in the map. Second, as the map grows, the estimates obtained by the EKF equations quickly become inconsistent due to linearization errors [9].

An alternative technique that reduces the computational cost and improves consistency is local map joining [10]. Instead of building one global map, this technique builds a set of independent local maps of limited size. Local maps can be joined together into a global map that is equivalent to the map obtained by the standard EKF–SLAM approach, except for linearization errors. As most of the mapping process consists in updating local maps, where errors remain small, the consistency of the global map obtained is greatly improved. In the following sections we present the basics of local map joining.

9.4.1 Building Independent Local Maps

Each local map can be built as follows: at a given instant t_j, a new map is initialized using the current vehicle location as base reference B_j. Then, the vehicle performs a limited motion (say k_j steps) acquiring sensor information about the neighboring environment features $\mathcal{F}_j$. The standard EKF-based techniques presented in previous sections are used to obtain a local map $\mathcal{M}_{\mathcal{F}_j}^{B_j} = (\hat{\mathbf{x}}_{\mathcal{F}_j}^{B_j}, \mathbf{P}_{\mathcal{F}_j}^{B_j})$. This local map is independent of any prior estimation of the vehicle location because it is built relative to the initial vehicle location B_j. The local map depends only on the odometry and sensor data obtained during the k_j steps. This implies that, under the common assumption that process and measurement noise are *white* random sequences, two local maps built with the same robot from *disjoint* sequences of steps are functions of independent stochastic variables. Therefore, the two maps will be *statistically independent* and *uncorrelated*. As there is no need to compute the correlations between features in different local maps and the size of local maps is bounded, the cost of local map building is constant per step, independent from the size of the global map.

The decision to close map $\mathcal{M}_j$ and start a new local map is made once the number of features in the current local map reaches a maximum, or the uncertainty of the vehicle location with respect to the base reference of the current map reaches a limit, or no matchings were found by the data association process for the last sensor measurements (a separate region of the environment is observed). Note that the new local map $\mathcal{M}_{j+1}$ will have the current vehicle position as base reference, which corresponds to the last vehicle position in map $\mathcal{M}_j$. Thus, the relative transformation between the two consecutive maps $\mathbf{x}_{j+1} = \mathbf{x}_{B_{j+1}}^{B_j}$ is part of the state vector of map $\mathcal{M}_j$.

9.4.2 Local Map Joining

Given two uncorrelated local maps:

$$\mathcal{M}_{\mathcal{F}}^{B} = (\hat{\mathbf{x}}_{\mathcal{F}}^{B}, \mathbf{P}_{\mathcal{F}}^{B}); \quad \mathcal{F} = \{B, F_0, F_1, \ldots, F_n\}$$

$$\mathcal{M}_{\mathcal{E}}^{B'} = (\hat{\mathbf{x}}_{\mathcal{E}}^{B'}, \mathbf{P}_{\mathcal{E}}^{B'}); \quad \mathcal{E} = \{B', E_0, E_1, \ldots, E_m\}$$

where a common reference has been identified $F_i = E_j$, the goal of map joining is to obtain one full stochastic map:

$$\mathcal{M}_{\mathcal{F}+\mathcal{E}}^{B} = (\hat{\mathbf{x}}_{\mathcal{F}+\mathcal{E}}^{B}, \mathbf{P}_{\mathcal{F}+\mathcal{E}}^{B})$$

containing the estimates of the features from both maps, relative to a common base reference B, and to compute the correlations appearing in the process. Given that the features from the first map are expressed relative to reference B, to form the joint state vector $\mathbf{x}^B_{\mathcal{F}+\mathcal{E}}$ we only need to transform the features of the second map to reference B using the fact that $F_i = E_j$:

$$\hat{\mathbf{x}}^B_{\mathcal{F}+\mathcal{E}} = \begin{bmatrix} \hat{\mathbf{x}}^B_{\mathcal{F}} \\ \hat{\mathbf{x}}^B_{\mathcal{E}} \end{bmatrix} = \begin{bmatrix} \hat{\mathbf{x}}^B_{\mathcal{F}} \\ \hat{\mathbf{x}}^B_{\mathcal{F}_i} \oplus \hat{\mathbf{x}}^{E_j}_{E_0} \\ \vdots \\ \hat{\mathbf{x}}^B_{\mathcal{F}_i} \oplus \hat{\mathbf{x}}^{E_j}_{E_m} \end{bmatrix} \tag{9.23}$$

The covariance $\mathbf{P}^B_{\mathcal{F}+\mathcal{E}}$ of the joined map is obtained from the linearization of Equation (9.23), and is given by:

$$\begin{aligned} \mathbf{P}^B_{\mathcal{F}+\mathcal{E}} &= \mathbf{J}_{\mathcal{F}} \mathbf{P}^B_{\mathcal{F}} \mathbf{J}^{\mathrm{T}}_{\mathcal{F}} + \mathbf{J}_{\mathcal{E}} \mathbf{P}^{E_j}_{\mathcal{E}} \mathbf{J}^{\mathrm{T}}_{\mathcal{E}} \\ &= \begin{bmatrix} \mathbf{P}^B_{\mathcal{F}} & \mathbf{P}^B_{\mathcal{F}} \mathbf{J}^{\mathrm{T}}_1 \\ \mathbf{J}_1 \mathbf{P}^B_{\mathcal{F}} & \mathbf{J}_1 \mathbf{P}^B_{\mathcal{F}} \mathbf{J}^{\mathrm{T}}_1 \end{bmatrix} + \begin{bmatrix} 0 & 0 \\ 0 & \mathbf{J}_2 \mathbf{P}^{E_j}_{\mathcal{E}} \mathbf{J}^{\mathrm{T}}_2 \end{bmatrix} \end{aligned} \tag{9.24}$$

where

$$\mathbf{J}_{\mathcal{F}} = \frac{\partial \mathbf{x}^B_{\mathcal{F}+\mathcal{E}}}{\partial \mathbf{x}^B_{\mathcal{F}}} \bigg|_{(\hat{\mathbf{x}}^B_{\mathcal{F}}, \hat{\mathbf{x}}^{E_j}_{\mathcal{E}})} = \begin{bmatrix} \mathbf{I} \\ \mathbf{J}_1 \end{bmatrix}$$

$$\mathbf{J}_{\mathcal{E}} = \frac{\partial \mathbf{x}^B_{\mathcal{F}+\mathcal{E}}}{\partial \mathbf{x}^{E_j}_{\mathcal{E}}} \bigg|_{(\hat{\mathbf{x}}^B_{\mathcal{F}}, \hat{\mathbf{x}}^{E_j}_{\mathcal{E}})} = \begin{bmatrix} 0 \\ \mathbf{J}_2 \end{bmatrix}$$

$$\mathbf{J}_1 = \begin{bmatrix} 0 & \cdots & \mathbf{J}_{1\oplus}\left\{\hat{\mathbf{x}}^B_{\mathcal{F}_i}, \hat{\mathbf{x}}^{E_j}_{E_0}\right\} & \cdots & 0 \\ \vdots & & \vdots & & \vdots \\ 0 & \cdots & \mathbf{J}_{1\oplus}\left\{\hat{\mathbf{x}}^B_{\mathcal{F}_i}, \hat{\mathbf{x}}^{E_j}_{E_m}\right\} & \cdots & 0 \end{bmatrix}$$

$$\mathbf{J}_2 = \begin{bmatrix} \mathbf{J}_{2\oplus}\left\{\hat{\mathbf{x}}^B_{\mathcal{F}_i}, \hat{\mathbf{x}}^{E_j}_{E_0}\right\} & \cdots & 0 \\ \vdots & \ddots & \vdots \\ 0 & \cdots & \mathbf{J}_{2\oplus}\left\{\hat{\mathbf{x}}^B_{\mathcal{F}_i}, \hat{\mathbf{x}}^{E_j}_{E_m}\right\} \end{bmatrix}$$

Obtaining vector $\hat{\mathbf{x}}^B_{\mathcal{F}+\mathcal{E}}$ with Equation (9.23) is an $O(m)$ operation. Given that the number of nonzero elements in $\mathbf{J}_1$ and $\mathbf{J}_2$ is $O(m)$, obtaining matrix $\mathbf{P}^B_{\mathcal{F}+\mathcal{E}}$ with Equation (9.24) is an $O(nm + m^2)$ operation. Thus when $n \gg m$, map joining is linear with n.

9.4.3 Matching and Fusion after Map Joining

The map resulting from map joining may contain features that, coming from different local maps, correspond to the same environment feature. To eliminate such duplications and obtain a more precise map we need a data association algorithm to determine correspondences, and a feature fusion mechanism to update the global map. For determining correspondences we use the JCBB algorithm described in Section 9.3.2.

Feature fusion is performed by a modified version of the EKF update equations, which consider a nonlinear measurement equation:

$$\mathbf{z}_{ij} = \mathbf{h}_{ij}(\mathbf{x}) = \mathbf{0} \tag{9.25}$$

with null noise covariance matrix, which constraints the relative location between the duplicates F_i and F_j of an environment feature. Once the matching constraints have been applied, the corresponding matching features become fully correlated, with the same estimation and covariance. Thus, one of them can be eliminated.

The whole process of local map joining, matching, and fusion can be seen in the example of Figure 9.9.

9.4.4 Closing a Large Loop

To compare map joining with full EKF–SLAM we have performed a map building experiment, using a robotized wheelchair equipped with a SICK laser scanner. The vehicle was hand-driven along a loop of about 250 m in a populated indoor/outdoor environment in the Ada Byron building of our campus. The laser scans were segmented to obtain lines using the RANSAC technique. The global map obtained using the classical EKF–SLAM algorithm is shown in Figure 9.10a. At this point, the vehicle was very close to the initial starting position, closing the loop. The figure shows that the vehicle estimated location has some 10 m error and the corresponding 95% uncertainty ellipses are ridiculously small, giving an inconsistent estimation. Due to these small ellipsoids, the JCBB data association algorithm was unable to properly detect the loop closure. This corroborates the results obtained with simulations in Reference 9: in large environments the map obtained by EKF–SLAM quickly becomes inconsistent, due to linearization errors.

The same dataset was processed to obtain independent local maps at fixed intervals of about 10 m. The local maps were joined and fused obtaining the global map shown in Figure 9.10b. In this case the loop was correctly detected by JCBB and the map obtained seems to be consistent. Furthermore, the computational time was about 50 times shorter that the standard EKF approach.

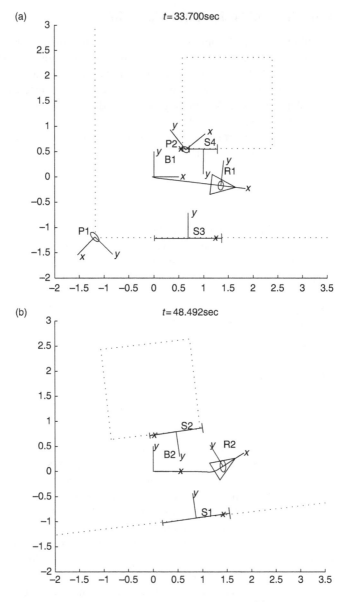

FIGURE 9.9   Example of local map joining. (a) Local map $\mathcal{M}_{\mathcal{F}_1}^{B1}$ with four point features, P1, P2, S3, and a segment S4, with respect to reference $B1$; (b) local map $\mathcal{M}_{\mathcal{F}_2}^{B2}$ with two features, S1 and S2, with respect to reference $B2$; (c) both maps are joined to obtain $\mathcal{M}_{\mathcal{F}_1+\mathcal{F}_2}^{B1}$; (d) map $\mathcal{M}_{\mathcal{F}_{1:2}}^{B1}$ after updating by fusing S3 with S5, and S4 with S6. (Reprinted with permission from Tardós, J. D. et al. *International Journal of Robotics Research*, 21: 311–330, 2002.).

(c)

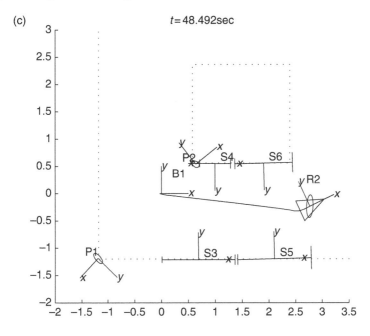

$t = 48.492$ sec

(d)

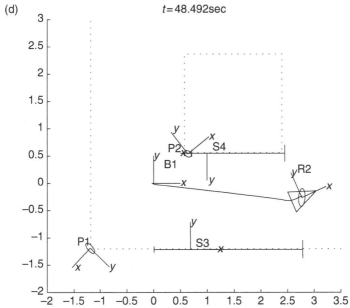

$t = 48.492$ sec

FIGURE 9.9 Continued.

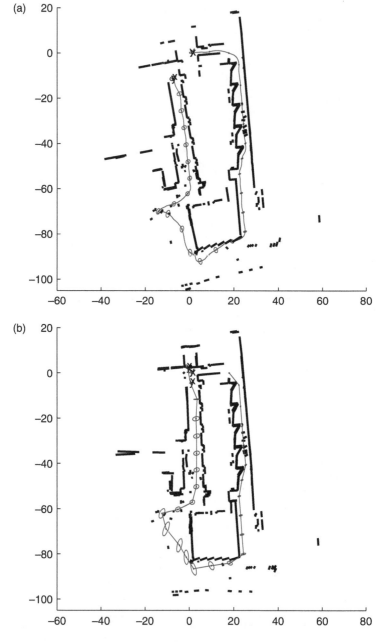

FIGURE 9.10 Global maps obtained using the standard EKF–SLAM algorithm (a) and local map joining (b).

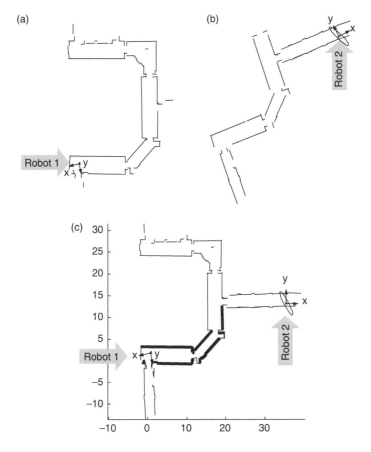

FIGURE 9.11 Maps build by two independent robots (a, b) and global map obtained by joining them (c).

9.4.5 Multi-robot SLAM

The techniques explained above can be applied to obtain global maps of large environments using several mobile robots. In Figure 9.11a and b, we can see the maps built by two independent robots that have traversed a common area. In this case, the relative location between the robots is unknown. The process for obtaining a common global map is as follows:

- Choose at random one feature on the first map, pick its set of covisible features and search for matchings in the second map using the RS relocation algorithm. Repeat the process until a good matching is found, for a fixed maximum number of tries.

- When a match is found, choose a common reference in both maps. In this case the reference is built in the intersection of two nonparallel walls that have been matched by RS. This gives the relative location between both maps. Change the base of the second map to be the common reference using the technique detailed in Reference 10.
- Join both maps (Section 9.4.2), search for more matchings using JCBB, and fuse both maps in a global map (Section 9.4.3) that contains the location of all features and both robots, relative to the base of the first map.

The global map obtained is shown in Figure 9.11c. The bold lines are the covisibility set used to match both maps. After that point, both robots can continue exploring the environment, building new independent local maps that can be joined and fused with the global map.

9.5 CONCLUSIONS

The EKF approach to SLAM dates back to the seminal work reported in Reference 2 where the idea of representing the structure of the navigation area in a discrete-time state-space framework was originally presented. Nowadays the basic properties and limitations of this approach are quite well understood. Three important convergence properties were proven in Reference 26 (1) the determinant of any submatrix of the map covariance matrix decreases monotonically as observations are successively made, (2) in the limit, as the number of observations increases, the landmark estimates become fully correlated, and (3) in the limit, the covariance associated with any single landmark location estimate reaches a lower bound determined only by the initial covariance in the vehicle location estimate at the time of the first sighting of the first landmark.

It is important to note that these theoretical results only refer to the evolution of the covariance matrices computed by the EKF in the ideal linear case. They overlook the fact that, given that SLAM is a nonlinear problem, there is no guarantee that the computed covariances will match the actual estimation errors, which is the true SLAM consistency issue first pointed out in Reference 38. In a recent paper [9], we showed with simulations that linearization errors lead to inconsistent estimates well before the computational problems arise. In Section 9.4 we have presented experimental evidence that methods like map joining, based on building independent local maps, effectively reduce linearization errors, improving the estimator consistency.

The main open challenges in SLAM include efficient mapping of large environments, modeling complex and dynamic environments, multi-vehicle SLAM, and full 3D SLAM. Most of these challenges will require scalable representations, robust data association algorithms, consistent estimation techniques, and different sensor modalities. In particular, solving SLAM with monocular or stereo vision is a crucial research goal for addressing many real life applications.

APPENDIX: TRANSFORMATIONS IN 2D

Two basic operations used in stochastic mapping are transformation inversion and composition, which were represented by Reference 2 using operators $\ominus$ and $\oplus$:

$$\hat{\mathbf{x}}_A^B = \ominus \hat{\mathbf{x}}_B^A$$

$$\hat{\mathbf{x}}_C^A = \hat{\mathbf{x}}_B^A \oplus \hat{\mathbf{x}}_C^B$$

In this chapter, we generalize the $\oplus$ operator to also represent the composition of transformations with feature location vectors, which results in the change of base reference of the feature. The Jacobians of these operations are defined as:

$$\mathbf{J}_{\ominus}\{\hat{\mathbf{x}}_B^A\} = \left. \frac{\partial(\ominus \mathbf{x}_B^A)}{\partial \mathbf{x}_B^A} \right|_{(\hat{\mathbf{x}}_B^A)}$$

$$\mathbf{J}_{1\oplus}\{\hat{\mathbf{x}}_B^A, \hat{\mathbf{x}}_C^B\} = \left. \frac{\partial(\mathbf{x}_B^A \oplus \mathbf{x}_C^B)}{\partial \mathbf{x}_B^A} \right|_{(\hat{\mathbf{x}}_B^A, \hat{\mathbf{x}}_C^B)}$$

$$\mathbf{J}_{2\oplus}\{\hat{\mathbf{x}}_B^A, \hat{\mathbf{x}}_C^B\} = \left. \frac{\partial(\mathbf{x}_B^A \oplus \mathbf{x}_C^B)}{\partial \mathbf{x}_C^B} \right|_{(\hat{\mathbf{x}}_B^A, \hat{\mathbf{x}}_C^B)}$$

In 2D, the location of a reference B relative to a reference A (or transformation from A to B) can be expressed using a vector with three d.o.f.: $\mathbf{x}_B^A = [x_1, y_1, \phi_1]^{\mathrm{T}}$. The location of A relative to B is computed using the inversion operation:

$$\mathbf{x}_A^B = \ominus \mathbf{x}_B^A = \begin{bmatrix} -x_1 \cos \phi_1 - y_1 \sin \phi_1 \\ x_1 \sin \phi_1 - y_1 \cos \phi_1 \\ -\phi_1 \end{bmatrix}$$

The Jacobian of transformation inversion is:

$$\mathbf{J}_{\ominus}\{\mathbf{x}_B^A\} = \begin{bmatrix} -\cos \phi_1 & -\sin \phi_1 & -x_1 \sin \phi_1 - y_1 \cos \phi_1 \\ \sin \phi_1 & -\cos \phi_1 & x_1 \cos \phi_1 + y_1 \sin \phi_1 \\ 0 & 0 & -1 \end{bmatrix}$$

Let $\mathbf{x}_C^B = [x_2, y_2, \phi_2]^T$ be a second transformation. The location of reference C relative to A is obtained by the composition of transformations $\mathbf{x}_B^A$ and $\mathbf{x}_C^B$:

$$\mathbf{x}_C^A = \mathbf{x}_B^A \oplus \mathbf{x}_C^B = \begin{bmatrix} x_1 + x_2 \cos\phi_1 - y_2 \sin\phi_1 \\ y_1 + x_2 \sin\phi_1 + y_2 \cos\phi_1 \\ \phi_1 + \phi_2 \end{bmatrix}$$

The Jacobians of transformation composition are:

$$\mathbf{J}_{1\oplus}\{\mathbf{x}_B^A, \mathbf{x}_C^B\} = \begin{bmatrix} 1 & 0 & -x_2 \sin\phi_1 - y_2 \cos\phi_1 \\ 0 & 1 & x_2 \cos\phi_1 - y_2 \sin\phi_1 \\ 0 & 0 & 1 \end{bmatrix}$$

$$\mathbf{J}_{2\oplus}\{\mathbf{x}_B^A, \mathbf{x}_C^B\} = \begin{bmatrix} \cos\phi_1 & -\sin\phi_1 & 0 \\ \sin\phi_1 & \cos\phi_1 & 0 \\ 0 & 0 & 1 \end{bmatrix}$$

ACKNOWLEDGMENT

This research has been funded in part by the Dirección General de Investigación of Spain under project DPI2003-07986.

REFERENCES

1. J. A. Castellanos, J. M. M. Montiel, J. Neira, and J. D. Tardós. The SPmap: a probabilistic framework for simultaneous localization and map building. *IEEE Transactions on Robotics and Automation*, 15: 948–953, 1999.
2. R. Smith, M. Self, and P. Cheeseman. A stochastic map for uncertain spatial relationships. In Faugeras O. and Giralt G. (eds), *Robotics Research, The Fourth International Symposium*, pp. 467–474. The MIT Press, Cambridge, MA, 1988.
3. J. E. Guivant and E. M. Nebot. Optimization of the simultaneous localization and map-building algorithm for real-time implementation. *IEEE Transactions on Robotics and Automation*, 17: 242–257, 2001.
4. J. J. Leonard, R. Carpenter, and H. J. S. Feder. Stochatic mapping using forward look sonar. *Robotica*, 19, 467–480, 2001.
5. J. H. Kim and S. Sukkarieh. Airborne simultaneous localisation and map building. In *IEEE International Conference on Robotics and Automation*, Taipei, Taiwan, September 2003.
6. J. Knight, A. Davison, and I. Reid. Towards constant time SLAM using postponement. In *IEEE/RSJ International Conference on Intelligent Robots and Systems*, pp. 406–412, Maui, Hawaii, 2001.

7. S. J. Julier and J. K. Uhlmann. Building a million beacon map. In Paul S. Schenker and Gerard T. McKee (eds), *SPIE Int. Soc. Opt. Eng.*, vol. 4571, pp. 10–21, SPIE, Washington DC, 2001.

8. Y. Liu and S. Thrun. Results for outdoor-SLAM using sparse extended information filters. In *IEEE International Conference on Robotics and Automation*, pp. 1227–1233, Taipei, Taiwan, 2003.

9. J. A. Castellanos, J. Neira, and J. D. Tardós. Limits to the consistency of EKF-based SLAM. In *5th IFAC Symposium on Intelligent Autonomous Vehicles*, Lisbon, Portugal, 2004.

10. J. D. Tardós, J. Neira, P. Newman, and J. Leonard. Robust mapping and localization in indoor environments using sonar data. *International Journal of Robotics Research*, 21: 311–330, 2002.

11. S. B. Williams, G. Dissanayake, and H. F. Durrant-Whyte. An efficient approach to the simultaneous localisation and mapping problem. In *IEEE International Conference on Robotics and Automation, ICRA*, vol. 1, pp. 406–411, Washington DC, 2002.

12. S. B. Williams. *Efficient Solutions to Autonomous Mapping and Navigation Problems*. Australian Centre for Field Robotics, University of Sydney, September 2001. http://www.acfr.usyd.edu.au/

13. M. Bosse, P. Newman, J. Leonard, M. Soika, W. Feiten, and S. Teller. An atlas framework for scalable mapping. In *IEEE International Conference on Robotics and Automation*, pp. 1899–1906, Taipei, Taiwan, 2003.

14. T. Bailey. *Mobile Robot Localisation and Mapping in Extensive Outdoor Environments*, Australian Centre for Field Robotics, University of Sydney, August 2002, http://www.acfr.usyd.edu.au/

15. J. J. Leonard and P. M. Newman. Consistent, convergent and constant-time SLAM. In *International Joint Conference on Artificial Intelligence*, Acapulco, Mexico, August 2003.

16. C. Estrada, J. Neira, and J. D. Tardós. Hierarchical SLAM: real-time accurate mapping of large environments. *IEEE Transactions on Robotics and Automation*, 21(4): 588–596, 2005.

17. M. Montemerlo, S. Thrun, D. Koller, and B. Wegbreit. FastSLAM: a factored solution to the simultaneous localization and mapping problem. In *Proceedings of the AAAI National Conference on Artificial Intelligence*, Edmonton, Canada, 2002, AAAI.

18. J. S. Gutmann and K. Konolige. Incremental mapping of large cyclic environments. In *IEEE International Symposium on Computational Intelligence in Robotics and Automation, CIRA*, pp. 318–325, Monterey, California, 1999.

19. F. Lu and E. Milios. Globally consistent range scan alignment for environment mapping. *Autonomous Robots*, 4: 333–349, 1997.

20. S. Thrun, D. Hähnel, D. Ferguson, M. Montemerlo, R. Triebel, W. Burgard, C. Baker, Z. Omohundro, S. Thayer, and W. Whittaker. A system for volumetric robotic mapping of abandoned mines. In *IEEE International Conference on Robotics and Automation*, pp. 4270–4275, Taipei, Taiwan, 2003.

21. J. A. Castellanos and J. D. Tardós. *Mobile Robot Localization and Map Building: A Multisensor Fusion Approach*. Kluwer Academic Publishers, Boston, MA, 1999.

22. P. Newman, J. Leonard, J. D. Tardós, and J. Neira. Explore and return: experimental validation of real-time concurrent mapping and localization. In *IEEE International Conference on Robotics and Automation*, pp. 1802–1809. IEEE, Washington DC, 2002.

23. J.A. Castellanos, J. M. M. Montiel, J. Neira, and J. D. Tardós. Sensor influence in the performance of simultaneous mobile robot localization and map building. In Corke Peter and Trevelyan James (eds), *Experimental Robotics VI. Lecture Notes in Control and Information Sciences*, vol. 250, pp. 287–296. Springer-Verlag, Heidelberg, 2000.

24. A.J. Davison. Real-time simultaneous localisation and mapping with a single camera. In *Proceedings of International Conference on Computer Vision*, Nice, October 2003.

25. J. A. Castellanos, J. Neira, and J. D. Tardós. Multisensor fusion for simultaneous localization and map building. *IEEE Transactions on Robotics and Automation*, 17: 908–914, 2001.

26. M. W. M. G. Dissanayake, P. Newman, S. Clark, H. F. Durrant-Whyte, and M. Csorba. A solution to the simultaneous localization and map building (SLAM) problem. *IEEE Transactions on Robotics and Automation*, 17: 229–241, 2001.

27. A. H. Jazwinski. *Stochastic Processes and Filtering Theory*. Academic Press, New York, 1970.

28. T. Bar-Shalom and T. E. Fortmann. *Tracking and Data Association*. Academic Press, New York, 1988.

29. Y. Bar-Shalom, X. R. Li, and T. Kirubarajan. *Estimation with Applications to Tracking and Navigation*. John Wiley & Sons, New York, 2001.

30. W. E. L. Grimson. *Object Recognition by Computer: The Role of Geometric Constraints*. The MIT Press, Cambridge, MA, 1990.

31. J. M. M. Montiel and L. Montano. Efficient validation of matching hypotheses using Mahalanobis distance. *Engineering Applications of Artificial Ingelligence*, 11: 439–448, 1998.

32. J. Neira and J. D. Tardós. Data association in stochastic mapping using the joint compatibility test. *IEEE Transactions on Robotics and Automation*, 17: 890–897, 2001.

33. J. Neira, J. D. Tardós, and J. A. Castellanos. Linear time vehicle relocation in SLAM. In *IEEE International Conference on Robotics and Automation*, pp. 427–433, Taipei, Taiwan, September 2003.

34. T. Bailey, E. M. Nebot, J. K. Rosenblatt, and H. F. Durrant-Whyte. Data association for mobile robot navigation: a graph theoretic approach. In *IEEE International Conference on Robotics and Automation*, pp. 2512–2517, San Francisco, California, 2000.

35. M. A. Fischler and R. C. Bolles. Random sample consensus: a paradigm for model fitting with applications to image analysis and automated cartography. *Communications of the ACM*, 24: 381–395, 1981.

36. R. Hartley and A. Zisserman. *Multiple View Geometry in Computer Vision*. Cambridge University Press, Cambridge, U.K., 2000.

37. Thrun Sebastian, Liu Yufeng, Koller Daphne, Y. Ng Andrew, Ghahramani Zoubin, and H. F. Durrant-Whyte. Simultaneous localization and

mapping with sparse extended information filters. *The International Journal of Robotics Research*, 23: 693–716, 2004.

38. S. J. Julier and J. K. Uhlmann. A counter example to the theory of simultaneous localization and map building. In *2001 IEEE International Conference on Robotics and Automation*, pp. 4238–4243, Seoul, Korea, 2001.

BIOGRAPHIES

José A. Castellanos was born in Zaragoza, Spain, in 1969. He earned the M.S. and Ph.D. degrees in industrial-electrical engineering from the University of Zaragoza, Spain, in 1994 and 1998, respectively. He is an associate professor with the Departamento de Informática e Ingeniería de Sistemas, University of Zaragoza, where he is in charge of courses in SLAM, automatic control systems, and computer modelling and simulation. His current research interest include multisensor fusion and integration, Bayesian estimation in nonlinear systems, and simultaneous localization and mapping.

José Neira was born in Bogotá, Colombia, in 1963. He earned the M.S. degree in computer science from the Universidad de los Andes, Colombia, in 1986, and the Ph.D. degree in computer science from the University of Zaragoza, Spain, in 1993. He is an associate professor with the Departamento de Informática e Ingeniería de Sistemas, University of Zaragoza, where he is in charge of courses in compiler theory, computer vision, and mobile robotics. His current research interests include autonomous robots, data association, and environment modeling.

Juan D. Tardós was born in Huesca, Spain, in 1961. He earned the M.S. and Ph.D. degrees in industrial-electrical engineering from the University of Zaragoza, Spain, in 1985 and 1991, respectively. He is an associate professor with the Departamento de Informática e Ingeniería de Sistemas, University of Zaragoza, where he is in charge of courses in real time systems, computer vision, and artificial intelligence. His current research interests include perception and mobile robotics.

10 Motion Planning: Recent Developments

Héctor H. González-Baños, David Hsu, and Jean-Claude Latombe

Contents

10.1   INTRODUCTION

A key trait of an autonomous robot is the ability to plan its own motion in order to accomplish specified tasks. Often, the objective of motion planning is to change the state of the world by computing a sequence of admissible motions for the robot. For example, in the path planning problem, we compute a collision-free path for a robot to go from an initial position to a goal position among static obstacles. This is the simplest type of motion planning problems; yet it is provably hard to computational problem [1]. Sometimes, instead of changing the state of the world, our objective is to maintain a set of constraints on the state of the world (e.g., following a target and keeping it in view), or to achieve a certain state of knowledge about the world (e.g., exploring and mapping an unknown environment).

Ideally, the robot achieves its objectives despite the many possible motion constraints, internal or external to the robot. Traditionally, motion planning emphasizes a single *external* constraint: physical obstacles in the environment. This is actually the only constraint considered in path planning. However, real robots have inherent mechanical limitations, such as the nonholonomic constraints that prevent wheeled robots from moving sideways. Robots may also be constrained by sensor limitations, such as obstacles blocking the views of cameras. These internal constraints are important, but taking them into account further complicates motion planning.

In recent years, random sampling has emerged as a powerful approach for motion planning. It is computationally efficient and relatively simple to implement. Its development was originally driven by the need to plan motions for robots with many degrees of freedom (dof), such as cooperating manipulator arms. However, we will downplay this aspect in this chapter. Instead, our main goal is to show how random sampling, combined with geometric and physical insights, can effectively handle motion constraints resulting from robots' mechanical and sensor limitations.

We start with an overview of path planning and proceed to the random-sampling approach to path planning (Section 10.2). Next, we focus on motion planning under two types of *internal* constraints: kinematic, dynamic constraints (Section 10.3) and visibility constraints (Section 10.4). We also briefly touch on the effect of uncertainty on motion planning (Section 10.5).

10.2 PATH PLANNING

In path planning, we are given a complete description of the geometry of a robot and a static environment populated with obstacles. Our goal is to find a collision-free path for the robot to move from an initial position and orientation to a goal position and orientation.

Although path planning algorithms differ greatly in details, most of them follow a common framework (Figure 10.1). The first step is to map a robot, which may have complex geometric shape, to a *point* in a new, abstract space, called the *configuration space* [2]. This mapping transforms the original problem to that of path planning for a moving point. Next we discretize the continuous configuration space and construct a graph that represents the connectivity of the space. Finally, we search this graph to find a path for the robot to reach the goal. If no path is found, we may sometimes repeat the process by refining the discretization and searching for the path again.

An important consideration for path planning algorithms is *completeness*. A path planning algorithm is *complete*, if it finds a path whenever one exists

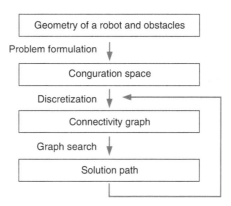

FIGURE 10.1   A common framework for path planning.

and reports none exists otherwise. However, achieving completeness is often computationally intractable. In practice, we have to trade-off some amount of completeness for increased computational efficiency.

In this section, we first present the concept of configuration space (Section 10.2.1). Next, we briefly describe some early approaches to path planning (Section 10.2.2), before focusing on how the random-sampling approach works in this relatively simple setting (Section 10.2.3).

10.2.1 Configuration Space

The *configuration* of a robot is a set of parameters that uniquely determine the position of every point in the robot. For example, the configuration of a mobile robot is usually its position (x, y) and orientation θ for $\theta \in [-\pi, \pi)$. The configuration of an articulated robot manipulator is usually a list of joint angles $(\theta_1, \theta_2, \ldots)$.

Suppose that the configuration of a robot consists of d parameters. It can then be regarded as a point in a d-dimensional space C, called the configuration space. A configuration q is *free*, if the robot placed at q does not collide with the obstacles or with itself. We define the *free space* $\mathcal{F}$ to be the subset of all free configurations in C, and define the obstacle space $\mathcal{B}$ to be the complement of $\mathcal{F}$: $\mathcal{B} = C \backslash \mathcal{F}$. See Figure 10.2b for an illustration.

For a robot that only translates in the plane, we can construct C explicitly by computing the Minkowski difference of the robot and the obstacles. Intuitively, we can think of the computation as "growing" the obstacles by the shape of the robot and shrinking the robot to a point (Figure 10.2). In general, a mobile robot not only translates, but also rotates. In this case, we compute slices of C with the robot in various fixed orientations and then stack and stitch these

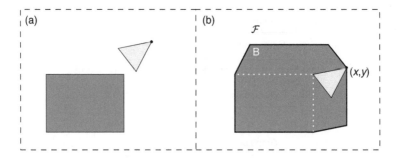

FIGURE 10.2   A robot translating in the plane. (a) The triangular robot moves in an environment with a single rectangular obstacle. (b) The configuration space of the robot. The configuration of the robot is represented by the position (x, y) of a reference point in the robot.

slices together. Computing $\mathcal{C}$ exactly is also possible, though somewhat more complicated [3].

For high-dimensional configuration spaces, explicitly constructing $\mathcal{C}$ is difficult. Instead, we represent $\mathcal{C}$ implicitly by a function CLEARANCE: $\mathcal{C} \mapsto R$, which maps a configuration $q \in \mathcal{C}$ to the distance between a robot at q and the obstacles. If CLEARANCE(q) returns 0, then q is in collision. An efficient implementation of this function can be achieved with hierarchical collision detection or distance computation algorithms [4].

Whether represented explicitly or implicitly, the configuration space encodes the key information of whether a robot at a particular configuration is in collision with obstacles or not. We can thus state the path planning problem formally in the configuration space as follows.

Problem 10.2.1 (path planning) *Given an initial configuration q_{init} and a goal configuration q_{goal}, find a path in the free space $\mathcal{F}$ between q_{init} and q_{goal}.*

In essence, the robot becomes a point in $\mathcal{C}$, and the path planning problem for the robot becomes that of finding a path for a moving point in $\mathcal{F}$. This transformation does not change the problem in any way, but it is often easier to think about the motion of a point than that of a robot with complex geometric shape. It also makes the problem formulation cleaner mathematically, especially when other constraints, in addition to physical obstacles, are considered (see Section 10.3).

10.2.2 Early Approaches

Path planning is fundamentally a question about the connectivity of $\mathcal{F}$: is there a path in $\mathcal{F}$ that connects two given configurations q_{init} and q_{goal}? To answer this question, a path planning algorithm usually discretizes $\mathcal{F}$ and computes a graph that represents its connectivity. It then searches this graph for a suitable path. The first step, constructing the connectivity graph, is the key and is where algorithms differ. The second step, graph search, is accomplished with standard graph-search techniques, such as the Dijkstra's algorithm or the A* algorithm.

There are three general approaches for path planning: roadmap, cell decomposition, and potential field. They differ in the connectivity graphs constructed and their representations. These differences were important a decade ago, when computers were much slower and the differences could affect the computational cost greatly even for path planning in simple 2D configuration spaces. With the advances in computer hardware, these differences are much less important today. All three approaches can solve path planning problems in 2D configuration spaces in a fraction of a second on a modern PC. What is relevant today is whether an approach scales up for configuration spaces of high dimensions (six or more). Unfortunately none of them really does in their original forms. In the following sections, we give selected examples of the three approaches in 2D configuration spaces, for the purpose of comparison with the random-sampling approach to be presented in Section 10.2.3. See Reference 5 for a complete survey of these approaches.

10.2.2.1 Roadmap

The roadmap approach captures the connectivity of $\mathcal{F}$ in a network G of 1D curves, called the *roadmap*. Once G is constructed, the robot is restricted to move along the curves in G. It appears that such a restriction may affect the robot's ability to find a collision-free path to the goal. However, a good roadmap has the property that there is a collision-free path in $\mathcal{C}$ between two configurations if and only if there is a collision-free path using only the curves represented in G. Algorithms that produce such roadmaps are clearly complete.

A classic example of the roadmap approach is the visibility graph algorithm [6], which applies mainly to 2D configuration spaces with polygonal obstacles. It captures the connectivity of $\mathcal{C}$ in a visibility graph G_{vis} (Figure 10.3). The nodes of G_{vis} are the vertices of polygonal obstacles in $\mathcal{C}$, plus q_{init} and q_{goal}. There is an edge between two nodes in G_{vis} if the straight-line path between the two nodes does not intersect the interior of the obstacles. The visibility graph can be computed in $O(n^2 \lg n)$ time using a simple rotational sweep-line algorithm [7], where n is the total number of vertices in the polygonal obstacles. After constructing G_{vis}, we can find the shortest path between q_{init} and q_{goal} by applying the Dijkstra's algorithm to G_{vis}. Furthermore, one can prove that the shortest path in G_{vis} is also the shortest among all possible paths in $\mathcal{F}$ between q_{init} and q_{goal}. This is the main strength of the visibility graph

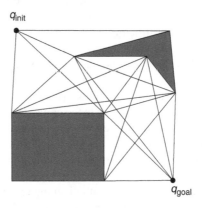

FIGURE 10.3 The visibility graph of a configuration space.

algorithm. However, it produces paths that graze the obstacles and thus bring the robot dangerously close to the obstacles, which is undesirable in practice.

An alternative is the Voronoi diagram algorithm, which captures the connectivity of $\mathcal{F}$ in the Voronoi diagram of $\mathcal{F}$ [8]. By following the curves in the Voronoi diagram, a robot stays as far away from the obstacles as possible, a clear advantage over the visibility graph algorithm. The Voronoi diagram can be computed in $O(n \lg n)$ time, which is also more efficient.

In 2D polygonal configuration spaces, both the visibility graph and the Voronoi diagram capture the connectivity of the space exactly: there is a collision-free path in $\mathcal{C}$ between two given configurations if and only if there is such a path in the corresponding graphs. So both algorithms are complete for 2D polygonal configuration spaces.

10.2.2.2 Cell decomposition

The cell decomposition approach first divides a robot's free space into simple, canonical regions called *cells*. Cells are usually convex so that it takes constant time to compute a path between any two configurations within a cell. We then construct a graph G_{cell} to capture the connectivity of $\mathcal{F}$, just as the roadmap algorithms do. The nodes of G_{cell} are the cells. There is an edge between two nodes if the corresponding cells are adjacent to each other.

The simplest cell decomposition is a grid with a fixed resolution (Figure 10.4a). To find a path between q_{init} and q_{goal}, we locate the two cells containing q_{init} and q_{goal}, respectively, and search for a path in G_{cell} between the two corresponding nodes. The result is a sequence of adjacent free cells that form a channel of free space between q_{init} and q_{goal}. A main advantage of this algorithm is the ease of implementation, giving rise to its great popularity in motion planning of mobile robots. However, its guarantee of completeness is weaker: it finds a path when one exists, only if the resolution of the grid is fine enough. Thus we say that the algorithm is only *resolution-complete*.

A more severe disadvantage of this algorithm is the grid size. If each dimension of a d-dimensional configuration space is discretized into n intervals, we end up with $O(n^d)$ cells in total. This becomes prohibitively expensive to store and process, as d grows. To reduce the total number of cells, one possibility is to start with a coarse grid and refine the grid locally when necessary. This leads to a data structure similar to quad- or oct-tree (see Reference 5 for more details). Another possibility is to analyze the input data carefully and use critical geometric features — such as the vertices or edges of polygonal obstacles — as a basis for discretizing the space, in order to avoid creating unnecessarily small cells. As an example, consider the triangulation algorithm [7], which divides the free space into triangles using the vertices of polygonal obstacles (Figure 10.4b). When there are a small number of simple obstacles, a triangulation contains much fewer cells than a grid with a reasonable resolution.

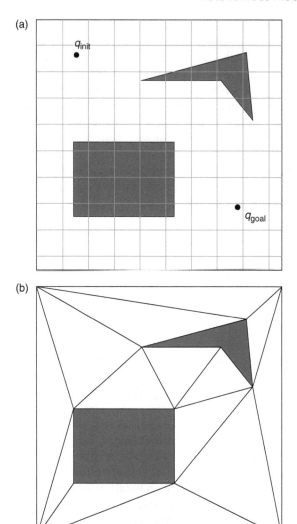

FIGURE 10.4   Cell decomposition with (a) a fixed-resolution grid and (b) a triangulation.

10.2.2.3 Potential field

The potential field approach [9] appears to be of a somewhat different nature from the previous two. It does not build a connectivity graph explicitly. Instead, it constructs an artificial potential function over $\mathcal{F}$ to guide a robot toward the goal. The potential function $U(q)$, which depends on the current configuration q of the robot, consists an attractive component and a repulsive component: $U(q) = U_a(q) + U_r(q)$. The attractive potential $U_a(q)$ pulls the robot toward the goal. The repulsive potential $U_r(q)$ pushes the robot away from the obstacles. The robot moves toward the goal, which is usually the global minimum of $U(q)$, by following the negated gradient of $U(q)$. One important advantage of this approach is that it computes not just a single path, but a feedback control strategy. The potential function $U(q)$ specifies the motion of the robot at any arbitrary configuration $q \in \mathcal{C}$. So the approach is more robust against control and sensing errors. It is also quite efficient. However, the potential field approach, which is based on steepest-descent optimization, suffers from the local minima problem: the robot may be trapped in a local minimum of $U(q)$ without reaching the global minimum, that is, the goal. The problem cannot be eliminated in general, but can be alleviated by constructing better potential functions with few local minima or executing random moves to help the robot escape from the local minima [5].

In some implementations, the potential function is represented on a grid. Such a potential field algorithm is closely related to cell decomposition with a fixed-resolution grid. We can think of the potential function as a heuristic function for graph search on a grid.

10.2.3   Random Sampling

Even for a mobile robot, the dimensionality of its configuration space, $\dim(\mathcal{C})$, sometimes becomes quite high. The position and orientation of a mobile robot operating in the plane can typically be specified by three parameters (x, y, θ), but many mobile robots are wheeled differential-drive systems subject to nonholonomic or dynamic constraints. To represent these constraints, we may need to consider the velocities $(\dot{x}, \dot{y}, \dot{\theta})$ in addition to (x, y, θ), resulting in a 6D space. If there are multiple robots cooperating in the same environment, $\dim(\mathcal{C})$ becomes even higher. As one expects, path planning becomes increasingly difficult as $\dim(\mathcal{C})$ grows.

During the past decade, random sampling has emerged as a powerful tool for path planning in high-dimensional configuration spaces. Algorithms based on random sampling, for example, the probabilistic roadmap (PRM) planners, are both efficient and simple to implement. They have solved path planning problems for multiple robots with dozens of dof [10]. Although these algorithms are originally intended for robot manipulators with many dof, the configuration space framework allows us to use them for mobile robots equally well.

As the name suggests, a PRM planner uses the roadmap approach. It tries to build a network of 1D curves that captures the connectivity of $\mathcal{F}$. Compared with the classic roadmap algorithms presented in Section 10.2.2.1, the main difference is that the nodes of a probabilistic roadmap are free configurations, sampled randomly according to a suitable probability distribution.

ALGORITHM 10.1
Roadmap construction for multi-query PRM planning

1: **loop**
2:    Pick q from $\mathcal{C}$ at random with probability $\pi(q)$.
3:    **if** CLEARANCE$(q) > 0$ **then**
4:       Insert q into the roadmap G as a milestone.
5:       **for** every milestone $q' \in G$ such that $q' \neq q$ **do**
6:          **if** LINK(q, q') returns TRUE **then**
7:             Insert an edge into G between q and q'.
8:          **end if**
9:       **end for**
10:    **end if**
11: **end loop**

There are two main classes of random-sampling algorithms. The first class precomputes a roadmap so that multiple planning queries in the same static environment can then be processed quickly. The second class performs no precomputation and builds a small roadmap on the fly in order to process a single query as fast as possible. The latter scenario occurs if environments change frequently and precomputation is not feasible. We refer to the first class as *multi-query* planning, and the second class as *single-query* planning.

10.2.3.1 Multi-query planning

In multi-query planning, we proceed in two stages. The first stage is precomputation, whose objective is to compute a roadmap G that captures the connectivity of $\mathcal{F}$ as accurately as possible in a reasonable amount of time. We sample $\mathcal{C}$ at random according to a suitable probability distribution π and retain the free configurations, called *milestones*, as nodes in G. Let LINK(q, q') denote a function that returns true if two milestones q and q' can be connected by a collision-free, straight-line path. We insert an edge in G between two milestones q and q' if LINK(q, q') returns true. Algorithm 10.1 shows the main steps of this stage. The second stage is query processing. Each query asks for a collision-free path connecting q_{init} and q_{goal}. We first find two milestones q'_{init} and q'_{goal} in G such that q_{init} (q_{goal}, respectively) and q'_{init} (q'_{goal}, respectively) can be connected by a collision-free path. We then search for a path in G between q'_{init} and q'_{goal}.

The key issue in constructing probabilistic roadmaps is the sampling distribution for generating milestones. The first PRM planner uses a straightforward uniform distribution, followed by an enhancement step to increase sampling

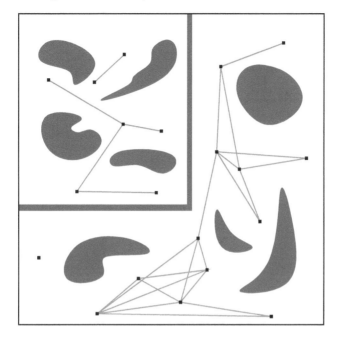

FIGURE 10.5 A probabilistic roadmap generated by the uniform sampling strategy for multi-query planning in a 2D configuration space.

density in critical regions [11]. See Figure 10.5 for an example. The success of the first PRM planner led to intensive research. Many different sampling strategies for PRM planning have been proposed [12–20] (see Chapter 7 of Reference 78, for a survey). Most of them try to increase the sampling density inside narrow passages, which are small regions critical for capturing the connectivity of $\mathcal{F}$ well.

Another important issue for PRM planners is the representation of $\mathcal{C}$. The configuration space $\mathcal{C}$ is generally represented implicitly in PRM planning. In Algorithm 10.1, CLEARANCE(q) determines whether q is collision-free, and LINK(q, q') determines whether there is a collision-free, straight-line path between q and q'. Both can be implemented efficiently using hierarchical bounding volume representation [4,21].

10.2.3.2 Single-query planning

In contrast to multi-query planning, there is no precomputation in the single-query setting. Instead, we construct a small roadmap on the fly to answer a single query. We sample only the connected components of $\mathcal{F}$ that contain either q_{init}

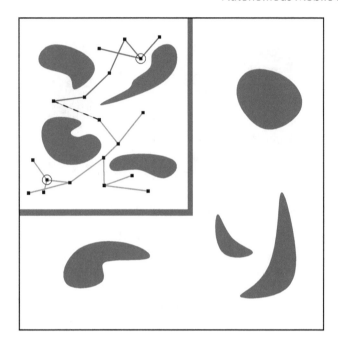

FIGURE 10.6 A roadmap for single-query planning in a 2D configuration space. The two circles mark q_{init} and q_{goal}.

or q_{goal} [22,23]. The reason is that although $\mathcal{F}$ may contain several connected components at most *two* of them, which contain q_{init} or q_{goal}, are relevant to the query being processed. It is clearly undesirable to construct a roadmap for the entire space. The roadmap for the single-query setting typically consists of two trees rooted at q_{init} and q_{goal}, respectively (Figure 10.6). We expand the two trees by sampling new milestones at random from $\mathcal{C}$ and inserting them into the trees as milestones, until the two trees "meet," that is, a milestone in one tree is connected to a milestone in the other.

The two trees are expanded in an identical way. To add a new milestone to a tree T, we pick at random an existing milestone q in T with probability $\pi_T(q)$ and sample a new free configuration q' at random from the neighborhood of q with probability $\pi_q(q')$. If there is a straight-line path between q and q', then q' is inserted into T as a milestone along with an edge between q and q'. In contrast to Algorithm 10.1, a new configuration is inserted into T only if it can be connected to some existing milestone in T. So by construction, there is a path between the root of T and every milestone in T. The pseudocode in Algorithm 10.2 sketches out the algorithm for building a tree rooted at a given configuration.

ALGORITHM 10.2
Building a tree T rooted at configuration q_0

1: Insert q_0 into T.
2: **loop**
3:    Pick an existing milestone q from T with probability $\pi_T(q)$.
4:    Sample a new configuration q' at random from the neighborhood of q
      with probability $\pi_q(q')$.
5:    **if** CLEARANCE$(q) > 0$ and LINK(q, q') returns TRUE **then**
6:       Insert q into T along with an edge between q and q'.
7:    **end if**
8: **end loop**

In Algorithm 10.2, we must avoid oversampling any region of $\mathcal{F}$, especially around q_{init} and q_{goal}. Ideally we would like the milestones to eventually distribute rather uniformly over the connected components containing q_{init} or q_{goal}. Two common ways to achieve this are the *expansive space tree* (EST) [22] and the *rapidly exploring random tree* (RRT) [23]. EST assigns every milestone q in T a weight that measures how densely the neighborhood of q has already been sampled. We then pick an existing milestone q with a suitable distribution $\pi_T(q)$ (line 3) so that low-density neighborhoods are more likely to be sampled. RRT uses a target distribution, for example, the uniform distribution, and pick q so that the final distribution of milestones are close to the target distribution.

Another interesting idea for single-query planning is to delay executing LINK, an expensive operation, until it becomes necessary [4,10].

10.2.3.3 Probabilistic completeness

In general, path planning algorithms based on random sampling cannot detect whether any path exists. We must explicitly set the maximum number of milestones to be sampled. We may also try to estimate how well $\mathcal{C}$ has been sampled and terminate the algorithm if $\mathcal{C}$ has been sampled adequately and no path has been found. Because of this, these algorithms are not complete. Instead they can only guarantee *probabilistic completeness*: a path planning algorithm is *probabilistically complete* if it finds a path with high probability when one exists. Probabilistic completeness provides a guarantee of performance only if a solution path exists. No assurance is implied, if there is no path. It can be shown that under reasonable geometric assumptions on the configuration space, both the multi-query and the single-query algorithms with suitable sampling distributions are probabilistically complete with exponentially fast convergence rate [22,25–27].

10.2.3.4   Advantages of random sampling

The success of random sampling in path planning results from several factors:

- It can handle high-dimensional configuration spaces efficiently.
- It is easy to implement, partly due to the availability of good programming libraries for collision checking and pseudorandom number generation.
- It benefits from a probabilistic framework, which provides powerful tools for designing new sampling strategies and analysis techniques.
- It is difficult for an adversary to construct worst-case input, because of the random decisions made by the algorithm, thus improving the robustness of the algorithm on the average.

10.3   MOTION PLANNING UNDER KINEMATIC AND DYNAMIC CONSTRAINTS

Path planning is a purely geometric problem. It ignores some key aspects of real robots: inherent limits on mechanical systems restrict the range of possible motion. For example, a car cannot move sidewise. These limits cause certain configurations to be invalid, even if a robot does not collide with obstacles at those configurations. In this section, we consider two important classes of constraints, kinematic constraints and dynamic constraints, together referred to as *kinodynamic constraints*. Unlike the physical obstacles, kinodynamic constraints cannot always be represented in the configuration space. They involve not only the configuration, but also the velocity and possibly the acceleration of the robot.

To address this issue, we use *state space*, a straightforward generalization of configuration space. Every point in the state space contains information on both the configuration and the velocity of a robot. Our objective is to find, in the state space, an *admissible* path that is both collision-free and satisfies kinodynamic constraints. This class of problems is called *kinodynamic motion planning* [28].

10.3.1   Kinematic and Dynamic Constraints

Kinematic constraints impose a relationship between the configuration q of a robot and its velocity $\dot{q}$. They can be written mathematically as

$$F(q, \dot{q}) = 0 \qquad (10.1)$$

Kinematic constraints can be further classified into holonomic and nonholonomic ones.

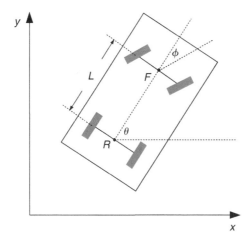

FIGURE 10.7 A simplified model for a car-like robot.

Holonomic constraints do not involve the velocity of a robot; they have the special form $F(q) = 0$. A set of holonomic constraints can be used to eliminate some of the configuration parameters and reduce the dimensionality of C. By choosing a suitable parameterization of C, we may be able to convert a problem with holonomic constraints into one with no constraints and apply the algorithms from Section 10.2.

Nonholonomic constraints are fundamentally different. They are not integrable, meaning that we cannot eliminate $\dot{q}$ via integration and convert them to the form $F(q) = 0$. A classic example is the constraints on the motion of car-like mobile robots (Figure 10.7). Let (x, y) be the position of the midpoint R between the rear wheels of the robot and θ be the orientation of the rear wheels with respect to the x-axis. Assuming that the wheels do not skid, the robot cannot move sidewise. This constraint can be written as $\tan \theta = \dot{y}/\dot{x}$, which clearly has the form $F(q, \dot{q}) = 0$. What is less obvious is that the constraint is not integrable. We will not get into the details here. It suffices to say that the mathematical conditions for integrability is known, but for a given set of constraints, checking these conditions is a nontrivial task. (See Reference 5, pp. 403–451, for details.)

Although most of the work on nonholonomic motion planning focuses on car-like or tractor-trailer robots, many results are applicable to other problems, including object pushing [29] and dextrous manipulation [30].

Dynamic constraints are closely related to nonholonomic constraints, but they involve not only the configuration and the velocity of a robot, but also the acceleration. Consider the Lagrange's equations of motion, which have the form $G(q, \dot{q}, \ddot{q}) = 0$, where q, $\dot{q}$, and $\ddot{q}$ are the robot's configuration, velocity, and acceleration. Defining $s = (q, \dot{q})$, we can rewrite the equation as $F(s, \dot{s}) = 0$ which is the same as (10.1).

The motion of a robot may also be constrained by inequalities of the form $F(q, \dot{q}) \leq 0$ or $G(q, \dot{q}, \ddot{q}) \leq 0$. Such constraints restrict the set of admissible states to a subset of the state space.

The presence of kinodynamic constraints implies that not all collision-free path are admissible, because they may violate the constraints. For some robots, we can represent motion constraints explicitly by constructing a class Γ of admissible path segments. Ideally Γ has the property that if there is an admissible path between two states, then one can construct another admissible path as a sequence of segments from Γ. This property is necessary for algorithms using Γ to be complete. Examples of such path segments include jump curves [31] or Reeds and Shepp curves [32] for car-like robots. In general, one can prove such a class of path segments can be constructed for any locally controllable system using tools from nonlinear control theory [33–35]. Unfortunately, the path segments generated by the proof are often inefficient in practice, because they may contain many unnecessary maneuvers.

An alternative representation of motion constraints is a *control system*

$$\dot{s} = f(s, u) \tag{10.2}$$

which constitutes the robot's equations of motion under suitable control. In Equation (10.2), $s \in S$ is the robot's state, which encodes the robot's configuration and optionally velocity as well; $\dot{s}$ is the derivative of s with respect to time; $u \in \Omega$ is the control input. The set S and Ω are called the *state space* and *control space*, respectively. We assume that S and Ω are bounded manifolds of dimensions n and m, with $m \leq n$. By defining appropriate charts on these manifolds, we can treat S as a subset of R^n, and Ω a subset of R^m.

Equation (10.2) can represent both kinematic and dynamic constraints described earlier. Suppose that we have ℓ kinodynamic constraints $G_i(s, \dot{s}) = 0$ for $i = 1, 2, \ldots, \ell$. We can solve these ℓ equations for $\dot{s}$. In general, if ℓ is less than n, the solution is not unique, but we can parameterize the set of solutions by $u \in R^{n-\ell}$ and write them down, at least formally, as $\dot{s} = f(s, u)$ for some suitable function f. More precisely, it can be shown that under suitable conditions, the set of constraints $G_i(s, \dot{s}) = 0$ for $i = 1, 2, \ldots, \ell$ is equivalent to (10.2), in which u is a point in $R^m = R^{n-\ell}$ [33].

To deal with inequality constraints of the form $G(s, \dot{s}) \leq 0$, we typically restrict the state space S and control space Ω to suitable subsets of R^n and R^m, respectively.

Let us now look at an example to illustrate the above notions.

Example 10.3.1 (simplified nonholonomic car navigation) *Consider the car example in Figure 10.7. The state of the car is specified by $(x, y, \theta) \in R^3$.*

The nonholonomic constraint $\tan\theta = \dot{y}/\dot{x}$ *is equivalent to the system*

$$\dot{x} = v\cos\theta$$
$$\dot{y} = v\sin\theta$$
$$\dot{\theta} = (v/L)\tan\phi$$

This reformulation corresponds to defining the car's state to be its configuration (x, y, θ) *and choosing the control input to be the vector* (v, ϕ), *where* v *and* ϕ *are the car's speed and steering angle, respectively. Bounds on* (x, y, θ) *and* (v, ϕ) *can be used to restrict* $\mathcal{S}$ *and* Ω *to subsets of* R^3 *and* R^2, *respectively. For instance, if the maximum speed of the car is 1, we require* $|v| \leq 1$.

10.3.2 General Approaches

Sometimes, the path planning approaches described in Section 10.2 can be applied to kinodynamic motion planning after some modifications. To construct a roadmap for car-like robots, we may discretize the boundaries of polygonal obstacles and connect pairs of points on the boundaries with jump curves composed of circular and straight-line segments [31]. To apply this idea to other robots would require a suitable class of admissible path segments to be constructed. Alternatively, we may consider the cell decomposition approach by placing a regular grid over the state space [28,33]. We represent the motion constraints as a control system and search for an admissible path in the discretized state space. As we have mentioned before, the cell-decomposition approach works only for robot with few dof, because the grid size increases exponentially with $\dim(\mathcal{C})$. We may also use the potential field approach by projecting the potential forces onto the surface defined by the motion constraints and applying the projected forces on the robot.

One approach unique to kinodynamic motion planning is path transformation. It proceeds in three steps [36]. First, we generate a collision-free path γ that disregards the motion constraints. We then discretize γ into a sequence of short path segments and replace each segment with one from a class Γ of admissible path segments, thus transforming γ into an admissible path γ'. Finally we smooth γ' to remove the unnecessary maneuvers and obtain a more efficient admissible path. This algorithm can be extended in various ways, which are all based on the idea of successive path transformation, but differ in what transformations to use and how to perform the transformations [37–39]. A natural question to ask about these path transformation algorithms is whether it is always possible to transform a collision-free path into an admissible path that obeys the motion constraints. In theory, the answer is yes, if the robot is locally controllable [36], for example, car-like robots. However, the approach is only practical for robots for which a class Γ of efficient admissible path segments can be easily constructed. It is not applicable to robots that are not locally controllable, for example, car-like robots that can only go forward.

10.3.3 Random Sampling

Random sampling has also been successful for kinodynamic motion planning, including robots that are *not* locally controllable. In this section, we give two representative examples.

The first one follows the multi-query approach [27], described in Section 10.2.3. It applies to car-like robots and assumes the existence of a class Γ of admissible path segments. It proceeds in almost the same way as Algorithm 10.1, with one major difference. When connecting two milestones in the roadmap, the algorithm uses path segments from Γ instead of straight-line paths. Thus every path in the roadmap is not only collision-free, but also admissible.

The second example follows the single-query approach. It represents the motion constraints as a control system. The main steps of the algorithm are similar to Algorithm 10.2. The difference occurs in lines 3 and 5. In Algorithm 10.2, we sample a new configuration and connect it to an existing milestone with a straight-line path. However, straight-line paths often violate the motion constraints. So instead, we choose a random control function and integrate the robot's equations of motion forward under this control function for a small period of time. The motion constraints are enforced automatically during the integration. If the resulting path is admissible, we then insert the endpoint of the path into the tree being constructed as a new milestone. Intuitively, we map a random sample in the control space Ω to a random sample in the state space S by integrating the equations of motion. Of course, we must still avoid oversampling. We can use the same methods described in Section 10.2.3, but they work less effectively here, because the motion constraints skew the density estimate and the target distribution.

It may appear somewhat surprising that, in the random-sampling approach, algorithms for path planning and kinodynamic motion planning are very similar. This is in fact one major advantage of the approach: it applies to a wide class of problems with relatively small, local changes related to the specifics of robots. This greatly eases implementation.

10.3.4 Case Studies on Real Robotic Systems

Having seen a number of motion planning algorithms, we now look into some important practical issues in the context of two real robotic systems.

FIGURE 10.8 A trailer-truck carrying aircraft components on a narrow road with many obstacles nearby.

10.3.4.1 Motion planning of trailer-trucks for transporting Airbus A380 components

Airbus A380 is the largest commercial aircraft that has ever been built. The main components — wings, fuselage sections, and the tail plane — are produced in different European cities and transported by trailer-trucks to a central location for assembly (see Figure 10.8). The transport itinerary must go through small towns and villages with sometimes very narrow roads. The enormous size of the cargo, the length of the itinerary, and the narrow roads along the way pose unique challenges. It is highly desirable to have an automated system to help validate the itinerary in advance and guide the truck driver [40].

Trailer-trucks have been studied extensively in nonholonomic motion planning. In this case, a path transformation algorithm is used for motion planning [41]. An initial admissible path is computed and then iteratively improved to make it more efficient. Obstacle avoidance is achieved with a potential field method.

The automated system is used to validate the itinerary for the trailer-trucks and determine which parts of the itinerary must be adapted to fit the vehicle size. The system also optimizes the trajectories to maximize the distance between the truck and the surrounding obstacles, such as buildings and trees. The validated

FIGURE 10.9   An air-cushioned robot among moving obstacles.

trajectory is then fed into a computer-aided driving system to help the driver
follow the trajectory.

10.3.4.2   A space robotics test-bed

A variant of the single-query random-sampling planner described in
Section 10.3.3 has been implemented on a real robot in an environment
with moving obstacles [42]. The robot system was developed in the Stanford
Aerospace Robotics Laboratory for testing space robotics technology. The air-
cushioned robot moves frictionlessly on a flat granite table (Figure 10.9). It has
eight air thrusters providing omni-directional motion capability, but the force
is small compared to the robot's mass, resulting in tight acceleration limits.

We model the robot as a disc in the plane for planning purposes. To deal
with moving obstacles, the planner augments the state space with a time axis
and computes a trajectory for the robot in the state-time space instead of the
usual state space. An overhead vision system estimates the motion of moving
obstacles in the environment and sends the information to the planner, which
runs on an off-board computer. The planner is then allocated a short, predefined
amount of time to compute a trajectory, as required by the real-time nature of
the system,

The success of random sampling for motion planning in real-time system indicates its effectiveness despite many adversarial conditions, including

(i) severe dynamic constraints on the robot's motion, (ii) moving obstacles, and (iii) various time delays and uncertainties inherent to an integrated system operating in a physical (as opposed to a simulated) environment.

10.4 MOTION PLANNING UNDER VISIBILITY CONSTRAINTS

Often, we picture robots as intelligent machines maneuvering autonomously through a cluttered environment, transporting parts, or assembling products. These tasks fall strictly within the domain of classic motion planning. However, acquiring information about environments through *sensing* is another important task: surveillance and mapping unknown environments are all examples of tasks in which observing the world is the main objective. It may not be immediately obvious, but motion planning plays a key role in these problems.

The goal of sensing is to extract an understanding of the world from sensor data. The basic act of sensing is passive. It becomes active when an algorithm directs the robot to move in order to make sensing more effective. The motion may help the robot keep a target within the sensor range or gain new information about an unknown environment. More generally, motion is executed to maintain a set of constraints on the state of the world or achieve a certain state of knowledge about the world. Here, the term "state" reflects not only the robot's physical configuration, as in the previous sections, but also the robot's observations and knowledge. The admissible paths for the robot are constrained not only by the robot's geometry and mechanics, but also by a set of *visibility constraints* due to the robot's sensors.

To understand the role of visibility constraints, consider the example of a robot following a target. Suppose that at its initial location, the robot has the target in view. As the target moves, it may get out of the robot's sensor range. The robot must move to a new location to keep the target in view. The path that the robot takes must, of course, be collision-free. In addition, at every point along the path, the robot must maintain target visibility. The visibility constraints reduce the set of admissible paths available to the robot, just as the kinodynamic constraints do. To deal with visibility constraints effectively, we must now leave the realm of classic motion planning and enter the realm of *motion planning under visibility constraints*.

This section presents three motion planning problems under visibility constraints: sensor placement (Section 10.4.1), indoor exploration (Section 10.4.2), and target tracking (Section 10.4.3). In the first problem, we compute a set of robot sensing locations to build a model of an environment effectively. This is the simplest scenario, because we ignore the cost of robot motion. The second problem, often called the *next best view*, is an extension of the first, when the environment is not known in advance. Motion planning becomes important, because the robot may inadvertently collide with unknown obstacles in the environment. The last problem is that of computing the motion of a robot observer following a target. This is probably the most complex problem of the three, because it involves both visibility and kinodynamic constraints. Moreover, the robot is sometimes expected to track an unpredictable target in real time.

10.4.1 Sensor Placement

Nowadays, robots equipped with laser range sensors are often used to build 3D models of the environment [43–46]. Acquiring high-quality 3D information is a costly operation, and it is desirable to minimize the number of sensing operations. To do this, we use an initial 2D map of the environment and compute a set of locations from which a range sensor (e.g., laser) scans the environment. We call this problem *sensor placement*.

Sensor placement is related to the classic art gallery problem [47], which asks for the minimum number of guards whose joint visibility region covers the interior of an art gallery. In its simplest form, the problem considers the art gallery to be a polygonal environment. It also assumes a simple line-of-sight visibility model, where two points are visible to each other if the line segment between them is unobstructed. The problem seems deceptively simple, but finding the minimum number of guards is actually NP-hard. In robotics, the visibility model is rarely as clean as that assumed in the art gallery problem. So the art gallery results are usually not directly applicable.

To derive a practical sensor placement algorithm, the visibility model must take into account the limitations of laser range sensors. The visibility definition below lists three constraints, which, we believe, are most relevant (Figure 10.10).

Definition 10.4.1 (constrained visibility) *Let the bounded and open set $W \subset R^2$ denote the robot's free space, and ∂W denote the boundary of W. A point $w \in \partial W$ is visible from a point $q \in W$ if the following conditions hold:*

- Line-of-sight constraint: *The open segment $S(q, w)$ joining q and w does not intersect ∂W.*
- Range constraint: $d_{\min} \leq d(q, w) \leq d_{\max}$, *where $d(q, w)$ is the Euclidean distance between q and w, and $d_{\min} \geq 0$ and $d_{\max} > d_{\min}$ are constants.*
- Incidence constraint: $\angle(\boldsymbol{n}, \boldsymbol{v}) \leq \tau$, *where $\boldsymbol{n}$ is the vector perpendicular to ∂W at w, $\boldsymbol{v}$ is the vector oriented from w to q, and $\tau \in [0, \pi/2]$ is a constant.*

We are interested in finding a minimal set of sensor locations that cover ∂W.

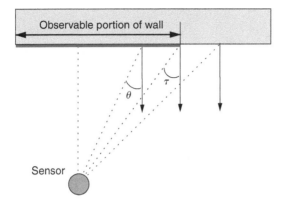

FIGURE 10.10   The incidence constraint of laser range sensors: wall sections are seen reliably, only if $|\theta| \leq \tau$.

Problem 10.4.1 (sensor placement)   *Given a bounded, open set $\mathcal{W} \subset R^2$, compute the minimal set of sensor locations $\mathcal{G}$ in $\mathcal{W}$, such that every point $w \in \partial\mathcal{W}$ is visible from at least one point in $\mathcal{G}$ under the visibility model given in Definition 10.4.1.*

Like the art gallery problem, Problem 10.4.1 is NP-hard, and we have to settle for an approximate solution, one that covers most of, but not the entire boundary, $\partial\mathcal{W}$. We use random sampling to transform the sensor placement problem into a *set cover problem* [48].

10.4.1.1   Sampling

Sample at random a set of m points from $\mathcal{W}$. Denote the set by $\mathcal{G}_{\text{sam}}$. For every edge $e \in \partial\mathcal{W}$, compute the fraction seen by each point in $\mathcal{G}_{\text{sam}}$. The arrangement of all covered fractions decomposes each edge into cells such that all points within the same cell are visible to the same subset of $\mathcal{G}_{\text{sam}}$ (see Figure 10.11a for an example). Now enumerate all the cells in the decomposition of $\partial\mathcal{W}$ and group them under the ground set $X = \{1, 2, \ldots, l\}$, where l is the number of cells. This ground set represents the decomposition of $\partial\mathcal{W}$.

Let R_i be the subset of X that is visible to a sample point $g_i \in \mathcal{G}_{\text{sam}}$. The set family $\mathcal{R} = \{R_1, R_2, \ldots, R_m\}$ is thus a collection of subsets of X. The set system $\Sigma = (X, \mathcal{R})$ can be regarded as an encoding of the sampled or discretized version of Problem 10.4.1, and the original problem is reduced to that of computing the optimal set cover of the set system Σ: find the smallest subcollection $\hat{\mathcal{R}} \subseteq \mathcal{R}$, such that the union of all the R_i's in $\hat{\mathcal{R}}$ equals X.

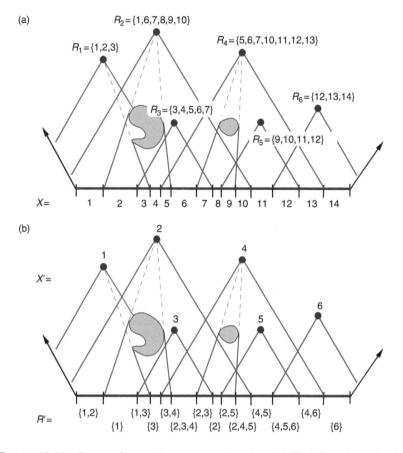

FIGURE 10.11   Sensor placement seen as a set system. (a) Each boundary edge is decomposed into cells. All points within the same cell are visible to the same subset of $\mathcal{G}_{\text{sam}}$. Each cell is then labeled with an integer and grouped under X. A subset $R_i \subseteq X$ is the set of cells visible from the sample point g_i. (b) In the *dual* representation, candidate sensor locations are grouped and labeled under X'. Each set $R'_i \in \mathcal{R}'$ is the set of locations covering cell i in the boundary decomposition.

The sampled problem is clearly not the same as the original. Finding the optimal set cover of Σ may not lead to an optimal set of sensor locations: $\mathcal{G}_{\text{sam}}$ may contain incorrectly distributed points, or $\mathcal{W}$ admits no finite solution due to its geometry. Sampling, however, often produces a satisfactory solution at a small cost, because the probability that $\mathcal{G}_{\text{sam}}$ contains the optimal set of guards quickly approaches 1 in most practical scenarios. Even when no finite solution exists, sampling produces reasonable solutions, as Σ encodes a "portion" of the original problem that actually admits a finite solution for realistic sensor models.

10.4.1.2 Near-optimal set covers

After sampling, we ask the question: has the problem become easier? Unfortunately, the set cover problem is also NP-hard. However, finding optimal set covers is a well-studied problem, and efficient algorithms that produce near-optimal solutions are available. More interestingly, the set cover problem has a dual, the *hitting set problem*.

Every set system has a dual. Consider $\Sigma = (X, \mathcal{R})$. Its *dual* $\Sigma' = (X', \mathcal{R}')$ is defined by $X' = \mathcal{R}$ and $\mathcal{R}' = \{\mathcal{R}_x | x \in X\}$, where $\mathcal{R}_x$ consists of all the sets $R \in \mathcal{R}$ that contain x. Figure 10.11b illustrates the dual set system for our sensor placement problem. Note that the set of candidate sensor locations now becomes the ground set X'. A *hitting set* for $\Sigma' = (X', \mathcal{R}')$ is a subset $H' \subseteq X'$ such that $H' \cap R' \neq \emptyset$ for every set R' in $\mathcal{R}'$. In other words, the hitting set H' contains members from all the sets in $\mathcal{R}'$. The problem of finding the smallest set cover for Σ is *equivalent* to that of finding the smallest hitting set for Σ'. For a set system with finite VC-dimensions,[1] an efficient algorithm exists for finding near-optimal hitting sets [49].

Assume that $\mathcal{W}$ is represented as a polygon with holes caused by obstacles. The VC-dimension of the set system for the sampled version of Problem 10.4.1 is then bounded by $O(\log(n + h))$, where n is the number of vertices describing $\partial \mathcal{W}$ and h is the number of holes [48]. Using the algorithm in Reference 49, we can find a set of sensor locations that is within a factor $O(\log(n + h) \cdot \log(c \log(n + h)))$ of the optimal size c. In other words, we can compute a near-optimal set of sensor locations within a logarithmic factor of the optimal.

Sensor placement is a set cover problem in nature, and the same is true for art gallery problems in general. A key development in recent years is to transform such problems into set systems, which may have finite VC-dimensions and lead to efficient approximation algorithms. For example, it has been shown that for a polygon with h holes, the VC-dimension of the set system for the classic art gallery problem is $O(h)$ [50] under the simple line-of-sight visibility model. This fact is exploited to produce a polynomial-time algorithm that finds a solution within a factor $O(\log(h) \cdot \log(c \log(h)))$ of the optimal size c [51].

10.4.1.3 Extensions

A straightforward extension of the sensor placement problem is to generate routes instead of locations for sensing tasks involving mobile robots. If the cost of sensing is very expensive compared to that of motion, then motion costs can be ignored. The problem remains the same as that defined in Problem 10.4.1. If the converse is true, then the cost of sensing can be ignored, and motion

---

[1] VC-dimension stands for the Vapnik–Červonenkis dimension. It is a measure of the complexity of a set system.

incurs the dominant cost. The problem becomes the *watchman route problem* [47]: find the shortest closed path from which the entire environment is visible. Developing sampling techniques to compute watchman routes is an interesting topic for future research.

A more difficult problem requires both the cost of sensing and the cost of motion to be considered. This topic remains largely unexplored, but some limited work exists [52,53].

10.4.2 Indoor Exploration

Automatic map building is an important problem in robotics. Research in this area has traditionally focused on developing techniques to extract environmental features, such as edges and corners, from sensor data and integrating these features into a consistent map. The former is a computer vision problem, and the latter is the *simultaneous localization and mapping* (SLAM) problem [54].

The SLAM algorithms seek the best way to integrate sensor data acquired by a robot during navigation. It, however, does not answer the following question: Given the map known so far, where should the robot move next to observe the unexplored regions? From the point of view of motion planning, this is the most interesting question in automatic map building. It involves the computation of successive sensing locations by iteratively solving the *next best view* (NBV) problem. At each location, the robot must not only observe large unexplored areas of the environment, but also a portion of the known environment to allow for image registration [55]. NBV is complementary to SLAM [54]. A SLAM algorithm builds a map by making the best use of the available sensor data, whereas an NBV algorithm guides the robot through locations that provide the best possible sensor data. In addition to robotics and computer vision, NBV arises in computer graphics [56] and many other areas.

The NBV is an *on-line* version of the sensor placement problem, where the 2D map of the environment is unknown initially and only revealed incrementally as new sensor data are acquired.

10.4.2.1 Constraints on the NBV

In mobile robotics, two important constraints must be considered by NBV algorithms. First, a mapping robot must not collide with obstacles, whether they are known or unknown in advance. The second constraint results from imperfect robot localization. Due to errors in inertial navigation (e.g., wheel slippage), a mobile robot must constantly relocalize itself as the map is built. New laser scan images must be aligned with the current map, a problem called *image registration*. Image registration requires an overlap between each new image and previously seen portions of the environment. An NBV algorithm must take this requirement into account.

A NBV can thus be viewed as an optimization problem where the best sensing position is computed subject to safe-navigation and image-registration constraints. As is often the case in optimization, the problem can be solved more effectively if the search domain is characterized explicitly. In motion planning terms, the NBV is a position in the free space, where the free space is collision-free with respect to both the known and unknown obstacles. Is it possible to characterize this free space explicitly?

It seems odd to define a free space that depends on obstacles yet to be discovered, for if they are not discovered, how can we use them to build the free space? The key is to view free space from the sensor's perspective, and not from the environment's perspective. That is, construct the largest region guaranteed to be free of obstacles, mapped or not, given the history of sensor data. Such a region is called the *safe region* to distinguish it from the usual notion of free space.

10.4.2.2 Safe regions

Consider a 2D range sensor that obeys the visibility model in Definition 10.4.1, with $d_{min} = 0$. Figure 10.12a shows a sample sensor reading. Here, the sensor detects the obstacle contour shown in bold black. From this reading, we want to construct a closed region that is obstacle-free. One possibility is to join the detected contour to the range limit of the sensor using radial line segments. This region is shown in light color in Figure 10.12b. Unfortunately, such a region is guaranteed to be free of obstacles only in the absence of incidence constraints. Consider Figure 10.12c, which shows the actual environment. Notice how the region from Figure 10.12b overlaps with walls oriented at a grazing angle (roughly 70°) with respect to the sensor position. In contrast, the region in Figure 10.12d, which takes into account the incidence constraint, is indeed safe.

Assume that the sensor output is an ordered list Π of piecewise continuous curves. The *local safe region* $s_l(q)$ is the largest closed region guaranteed to be free of obstacles given an observation $\Pi(q)$ made at location q. Such a region is bounded by the curves in $\Pi(q)$, representing the visible sections of the free space boundary ∂W, plus additional curves joining the disjoint visible sections and calculated from the information in $\Pi(q)$ [57] (see Figure 10.12d for an example). The safe region $s_l(q)$ is topologically equivalent to a classic visibility region. In fact, when the visibility constraints in Definition 10.4.1 are relaxed, the safe region becomes exactly the visibility region. Several properties and algorithms that apply to visibility regions also apply to safe regions. For example, $s_l(q)$ is a *star-shaped* set, a set that is entirely visible from at least one interior point.

A global safe region is constructed iteratively from local safe regions. First, a local safe region $s_l(q_0)$ is constructed from the sensor reading $\Pi(q_0)$ made

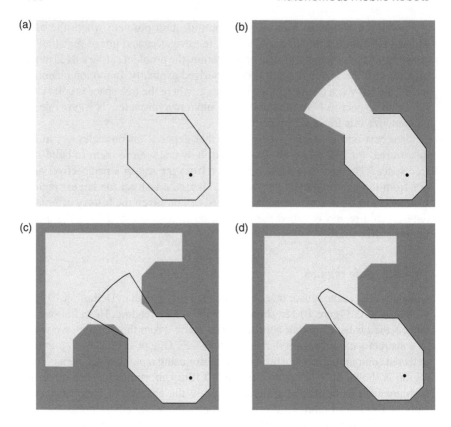

FIGURE 10.12   The effect of incidence constraints on safe regions.

at the robot's initial position q_0. The global safe region $S_g(q_0)$ is initially equal to $s_l(q_0)$. Next, the robot moves to a position q_1 and gets a new sensor reading $\Pi(q_1)$, yielding a new local safe region $s_l(q_1)$. Now, $S_g(q_1) = S_g(q_0) \cup s_l(q_1)$. The robot again moves, now to q_2. A new reading $\Pi(q_2)$ is made, yielding $s_l(q_2)$, and $S_g(q_2) = S_g(q_1) \cup s_l(q_2)$, and so on. The region $S_g(q_t)$ represents both a map of the environment at time t and the search domain for computing the next best view for $t + 1$.

10.4.2.3 Image registration

Robots cannot localize with perfect precision. An algorithm ALIGN is used to compute the transform T that aligns $s_l(q_{t+1})$ with $S_g(q_t)$ before the union operation. Image registration has been studied widely, and many techniques exist [45]. The details of ALIGN are inconsequential to the NBV computation, but it is important to note that most image registration algorithms are based on feature matching. It is thus essential that the NBV for $t + 1$ ensures a minimum overlap between the current $S_g(q_t)$ and the anticipated $s_l(q_{t+1})$.

10.4.2.4 Evaluating next views

Suppose that at time t, the robot is positioned at q_t and the global safe region is $S_g(q_t)$. The goal is to compute the future position of the robot, given $S_g(q_t)$. The unexplored areas of the environment can only be revealed through the free boundary of $S_g(q_t)$, that is, the portions of $S_g(q_t)$ not blocked by obstacles. Therefore, a potential candidate q is good, if it sees large unexplored areas outside of $S_g(q_t)$ through the free boundary of $S_g(q_t)$. We say that such q has high *potential visibility gain*, measured by a function $V_g(q, t)$.

Several definitions of $V_g(q, t)$ are possible. One way is to first compute the visibility region from q assuming that the free boundary is transparent, and intersect this region with the complement of $S_g(q_t)$ [57]. The gain $V_g(q, t)$ is the area of the resulting intersection. This definition works well for office environments, even in cluttered conditions. As an alternative, the next view can be chosen to maximize entropy reduction, and the gain $V_g(q, t)$ becomes a measure of the expected entropy reduction at position q [58].

The computation of NBV must also factor in the cost of motion, which is weighed against the potential visibility gain. Again, this can be done in several ways. One way is to define the overall merit of q, factoring in both visibility gain and motion cost, as

$$g(q, t) = V_g(q, t) \exp(-\lambda L(q, t)) \qquad (10.3)$$

where $L(q, t)$ is the length of the collision-free path computed by a path planner between position q and the current robot position at time t. The constant $\lambda \geq 0$ is used to weigh the cost of motion against the visibility gain. A small λ gives priority to the gain of information. Conversely, a large λ gives priority to motion economy, favoring locations near q_t that potentially produce marginal information gain.

10.4.2.5 Computing the NBV

At this point, the only remaining issue is to search for the NBV. This is simple, as the global safe region $S_g(q_t)$ completely characterizes the search domain. Following a random-sampling approach akin to those described in Section 10.2.3, a set $\mathcal{N}$ of NBV candidate positions is generated along the free boundary of $S_g(q_t)$. This set is processed in three steps. First, for each $q \in \mathcal{N}$, we determine the extent to which $s_l(q)$ and $S_g(q_t)$ overlap. The overlap $\zeta(q)$ is measured by the length of the visible part of $S_g(q_t)$'s boundary abutting obstacles. If $\zeta(q)$ is

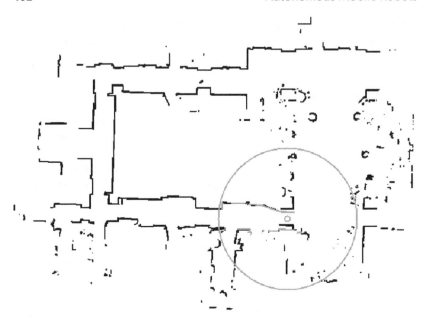

FIGURE 10.13 A partial map of a wing of the computer science building at Stanford University. The total length of the circuit is approximately 40 m. The circled region corresponds to the last local measurement.

smaller than a threshold imposed by ALIGN, then q is removed from $\mathcal{N}$. Second, a path planner computes a collision-free path between q_t and each remaining candidate q in $\mathcal{N}$. Those candidates that yield no feasible paths are removed from $\mathcal{N}$. Finally, the merit of each remaining candidate in $\mathcal{N}$ is evaluated according to Equation (10.3), and the best candidate is selected.

Figure 10.13 shows a sample map constructed using the NBV algorithm in Reference 57. The figure shows the partial map of a wing of the Computer Science Building at Stanford University after 14 iterations. Note the final mismatch after the robot completed a circuit around the lab (about 40 m). The discrepancy appears, because every image alignment transform was computed locally. To reduce the discrepancy, the NBV algorithm should be combined with a SLAM algorithm.

10.4.2.6 Extensions

We have so far ignored any error-recovery capabilities in the NBV computation. Any serious errors in sensing or image registration lead to unacceptable maps. An experimental system must be designed conservatively to avoid this, perhaps forcing the robot to take more measurements or travel longer paths to produce the final map. A better solution is to combine the NBV computation with SLAM algorithms and exploit their complementary strengths.

Another extension is to have multiple robots building a map cooperatively. Centralized approaches are acceptable, if the relative positions of all the robots are known. A single map can be generated from all the sensor readings, and a centralized NBV algorithm then computes the aggregate NBV for the entire team. The problem becomes far more difficult, if the relative positions of the robots are not known. In this case, the robots act independently, perhaps communicating their positions and findings only sporadically. A distributed approach is then needed.

10.4.3 Target Tracking

Tracking in the sense of detecting targets in images is studied widely in computer vision. In contrast, target tracking in motion planning is concerned with computing the motion of a robotized camera in order to keep a target in view [59]. Variations of this problem arise in different applications, for example, visual servoing [60,61] and computer-assisted surgery [62]. Target tracking is also called *target following* to distinguish it from the tracking problem in computer vision.

Target tracking is a motion planning problem that combines visibility constraints with kinodynamic constraints. It takes into account the actions of an external agent — the target — acting as a potential opponent. Thus target tracking can be treated as a problem in game theory [63]. The game-theoretic view provides a clean mathematical formulation of the problem.

10.4.3.1 State transition equations

Suppose that both the robot observer and the target are rigid bodies moving in the plane. The free configuration space for the observer is a subset of R^2 and denoted by $\mathcal{F}^o$, while that for the target is denoted by $\mathcal{F}^t$. Define $s^o(t)$ as the observer's *state* at time t. Suppose that the state transition equation for the observer is given by $\dot{s}^o = f^o(s^o, u)$, where $u(t)$ is the action selected from an action set $\mathcal{U}$ at time t. The function f^o models the observer's dynamics and may encode nonholonomic constraints or other types of kinodynamic constraints. Similarly, the transition equation for the target is given by $\dot{s}^t = f^t(s^t, \theta)$, with the action $\theta(t)$ selected from a target action set Θ. The state of the observer–target system is given by $s = (s^o, s^t)$. Let $\mathcal{X}$ be the joint state space, which is the Cartesian product of the individual state spaces of both the observer and the target. A state may encode both the configuration of a robot and its velocity. So, in general, $\mathcal{X}$ is *not* equal to $\mathcal{F}^o \times \mathcal{F}^t$, the Cartesian product of individual configuration spaces.

10.4.3.2   Visibility constraints

The distinction between state space and configuration space is important. The state space, along with the associated transition equation, focuses on the kinodynamic constraints. The configuration space, on the other hand, focuses on where the robot observer can see the target. Now let us identify those configurations where the target is visible.

Let $V(q^o)$ be the visibility region at the observer position q^o, that is, the set of all locations from which the target is visible to an observer located at q^o. Usually, the target is said to be visible if the line-of-sight to the observer is unobstructed, but this model can be extended. For example, the field of view can be restricted to some fixed visibility cone or limited by lower- and upper-bounds on the distance range. Incidence constraints such as those in Definition 10.4.1 can also be added.

Tracking algorithms usually compute the visibility region from a synthetic model or reconstruct it from sensor data. In the former case, a sweep-line algorithm can be used [7]. In the latter case, laser range sensors or similar sensors are installed on the robot observer (see Figure 10.14a), but some sensors cannot provide reliable measurements, thus complicating the reconstruction of the visibility region. For example, stereo vision systems often produce unreliable range measurements if the object's surface is textureless.

An important concept in target tracking is that of the *visibility sweeping line* $\ell(t)$, defined as the line passing through the target position at time t and a reflex vertex of the free space (Figure 10.15). At any time t, the observer must stay on the side of $\ell(t)$ which allows it to see the target. The observer's path is influenced by the behavior of these sweeping lines, and some tracking algorithms exploit them explicitly [64,65].

10.4.3.3   Tracking strategies

Target tracking consists of computing a function $u(t)$, called a *strategy*, so that the target remains in view for all $t \in [0, T]$, where T is the target's *stopping time*, also known as the *horizon* of the problem. It may also be important to optimize secondary criteria such as the total distance traversed by the observer and the final distance to the target. Various tracking strategies are known, and they can be compared from different angles.

Predictable vs. unpredictable targets. The target is predictable if the target action $\theta(t) \in \Theta$ is known in advance for all $t \le T$. Thus the location of the target is known for all t, and its state transition equation simplifies to $\dot{s}^t = f^t(s^t)$. The target is unpredictable if its actions are not known in advance, though the action set Θ may be known.

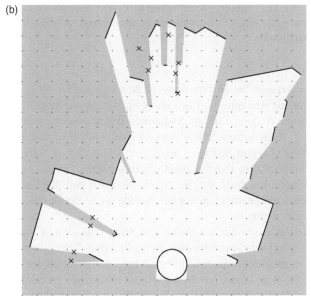

FIGURE 10.14 Measuring the visibility region with a laser range sensor.

Off-line vs. online. Off-line tracking strategies have access to future states, while online strategies do not. In other words, online algorithms are *causal*, whereas off-line ones are *noncausal*. Causality is a characteristic of the algorithm, not a logical requirement of target predictability. Obviously, an off-line strategy that relies on the target's future positions implies that the

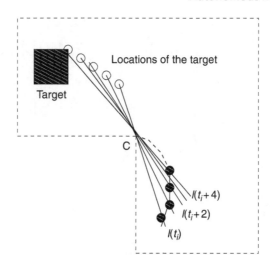

FIGURE 10.15 The visibility sweeping line $\ell(t)$ going through the target position at time t and the reflex vertex C. In this example, the observer must remain above $\ell(t)$ to keep the target visible.

target is predictable, but an algorithm can be noncausal for other reasons. Also note that on-line strategies may or may not run in closed loop.

Critical vs. average tracking. Sometimes it is impossible to track the target for all $t \leq T$. Thus some strategies maximize the target's *escape time* t_{esc}, the time when the observer first loses the target. An alternative is to maximize the *exposure*, the total time that the target remains visible. The former choice, critical tracking, implies that losing the target effectively ends the task, whereas the latter choice, average tracking, implies that the observer can possibly reacquire the target after losing it.

Expected vs. worst-case analysis. A tracking strategy may maximize either worst-case or expected performance. In the first case, a tracking strategy maximizes the minimum escape time given all the adversarial choices for $\theta(t) \in \Theta$ during the problem's horizon. This approach is suitable for tracking antagonistic targets. In the second case, the expected escape time is maximized given a probability distribution over the target's actions. In both cases, the problem is intractable, and we have to settle for approximate solutions. A typical one is to solve the problem for a time horizon much smaller than the target's stopping time T.

Open vs. closed loop. A strategy operates in closed loop, if the strategy u is computed as a function of the state $s(t)$. Otherwise, the strategy runs in open loop, and u depends explicitly on t. Closed-loop strategies are preferred over open-loop ones even when the target is predictable, unless it is guaranteed that the state transition models and observations are exact, for example, the case in Reference 66. Open-loop strategies are often used in theoretical studies, but they rarely work well in practice.

10.4.3.4 Backchaining and dynamic programming

One way to compute an observer trajectory for a predictable target is through *preimage backchaining*. Suppose that both the observer and the target are modeled as points in the plane. Let $\bar{\mathcal{V}}(q^t) \subset \mathcal{F}^o$ be the set of observer configurations from which a target at q^t is visible. Let $\mathcal{A}(t) \subset \mathcal{F}^o$ be the set of all configurations at time t from which the observer could move into $\bar{\mathcal{V}}(q^t(t+1))$ at time $t+1$. Since the observer must see the target at time t and move to a configuration that sees the target at $t+1$, its configuration at t must be contained in $\bar{\mathcal{V}}(q^t(t)) \cap \mathcal{A}(t)$, which can often be computed easily for the 2D case. Thus, the observer's trajectory can be obtained by backchaining from the final stage, guaranteeing visibility at each step, until a set of possible initial states is obtained or the problem is shown to have no solution.

Backchaining can be generalized into higher dimensions using dynamic programming (DP) [63]. Kinodynamic constraints and secondary optimization criteria can also be added. However, DP is computationally intensive. A brute-force implementation of DP leads to a grid whose size grows exponentially with the dimensionality of the state space. Random sampling may ease the computational burden, but to achieve real-time performance, approximate local strategies are needed.

10.4.3.5 Escape-time approximations

The time horizon is often reduced in practice to handle unpredictable targets. In the extreme case, only one step into the future is considered. If there are no kinodynamic constraints, maximizing the minimum escape time is equivalent to maximizing the shortest distance to escape. The observer's action for the next step can be selected to maximize this distance. This is sometimes achieved through randomized techniques [67]. The shortest distance to escape is easy to compute, but it could be a poor approximation of the escape time for longer time horizons or under kinodynamic constraints.

Alternatively, the escape time can be approximated with a quantity called the *escape risk* [65]. The negative gradient of the escape risk is composed of a *reactive* component and a *look-ahead* component. The reactive component drives the observer to swing around corners as a target is about to be occluded, while the look-ahead component drives the observer towards a corner in order to make future tracking easier. The algorithm relies on an *escape-path tree*, a data structure encoding all the locally worst-case paths that a target may use to escape the observer's visibility region (Figure 10.16). This data structure can be

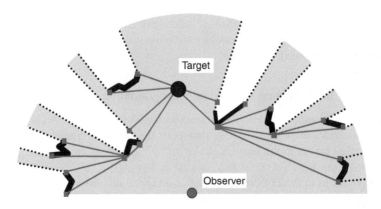

FIGURE 10.16   An example of the escape-path tree. The area in gray is the observer's visibility region, while obstacle boundaries are shown in bold. The squares indicate the nodes of the tree.

computed in $O(n)$ time for 2D environments, where n is the number of polygon vertices describing the observer's visibility region.

10.4.3.6   Robot localization

If a tracking strategy uses a global map of the environment to determine the observer's actions, tracking is tied to robot localization. This connection potentially leads to a conflict between the goals of tracking and localization. Suppose, for example, that the observer relocalizes whenever a ceiling landmark is visible. The target may force an observer trajectory without any landmarks, resulting in the localization error becoming so large that tracking fails.

A simple solution to this problem is to increase the number of landmarks, or to use more robust localization techniques based on (hopefully) abundant natural features. A better solution is to explicitly add the relocalization constraint into the tracking problem. For example, the observer actions maximize the sum of two utility functions: one based on the probability of observing the target and the other based on localization precision [68].

An entirely different approach is to abandon the global map and avoid the localization problem altogether. For example, the observer's actions could depend only on the gradient of the escape risk, which can be computed from purely local sensor information [65].

10.4.3.7   Other results and extensions

We often ignore the kinodynamic constraints on the observer and the target in order to simplify the tracking problem. However, it is important to assume bounded target velocity; otherwise, the target's escape time may become zero. The effect of velocity bounds on tracking has been studied [69]. Assuming bounded target velocity, an optimal strategy can be computed efficiently for polygonal environments and predictable targets [64].

An interesting extension of the tracking problem is that of *stealth tracking*: the observer tracks the target while remaining hidden from it. The work in Reference 70 extends the linear-time algorithm in Reference 65 to account for the additional stealth constraint. This involves computing the subset of the target's visibility region contained inside the observer's visibility region. The computation can be done efficiently so that the total cost of the strategy remains linear per step.

A more difficult problem is to track multiple targets with multiple observers. If a centralized strategy is used, the problem is not fundamentally different from tracking a single target with a single observer. However, the dimensionality of the state space gets higher, and visibility regions may become disconnected [67]. Distributed strategies, on the other hand, require a coordination scheme among observers.

10.5 OTHER IMPORTANT ISSUES

Uncertainty is an important issue in motion planning, but we will only touch on it very briefly here (see Chapter 13 for more details). Except for Section 10.3.4 and Section 10.4.2, we have mostly assumed that a planning algorithm knows exactly the geometry of the robot, the shapes and locations of obstacles in the environment, and when and how the environment changes. We have also assumed that the robot can exactly execute the path computed by a planning algorithm. These assumptions are satisfied to various degrees in real robotic systems.

Depending on the degree of uncertainty present and the amount of prior knowledge available, there are different ways to deal with uncertainty. If the uncertainties are small, we can largely ignore them during planning and use closed-loop control during path execution to reduce its effects (e.g., the air-cushioned robot in Section 10.3.4). If uncertainty is bounded or modeled by a probability distribution, we can incorporate it into planning using methods such as preimage backchaining [71] or partially observable Markov decision processes (POMDP) [72]. In this case, path planning and execution together form a closed-loop process. However, the computational cost of incorporating uncertainty into planning is often high and sometimes intractable. Also, uncertainty is difficult to model effectively for lack of prior knowledge, and we must rely on a worst-case analysis of various possible scenarios (e.g., in the target tracking problem of Section 10.4.3). In the extreme case, no prior knowledge of the environment is available. Planning is then of little use, and the robot must rely on sensor-based reaction.

Another important topic is multi-robot motion planning. Conceptually, we can take the cross product of the state spaces of all the robots involved and plan in this composite space. This is called centralized planning, which is computationally expensive due to the high dimensionality of the composite space. Alternatively, we may plan the motion for each individual robot separately and then coordinate their motion afterwards. This is called distributed planning, which is computationally more efficient, but sacrifices completeness and optimality. Chapter 11 provides a more in-depth discussion of this topic.

In recent years, bipedal humanoid robots have become more prevalent, for example, Honda's ASIMO and Sony's QRIO. A bipedal robot has the ability to navigate on uneven surfaces and step over obstacles along its path, but efficient footstep planning algorithms that take into account the robot's dynamics are needed to realize this potential [73]. Motion planning for humanoid robots is an important area of research, but is outside the scope of this chapter.

10.6 Conclusion

Motion planning has moved far beyond its original form of computing a collision-free path for a mobile robot to move from an initial to a final goal position. We have seen in this chapter how kinodynamic constraints and visibility constraints come into play. Nowadays motion planners compute footsteps for humanoid robots [73], paths for inserting a probe into an airplane engine with hundred of parts [74], and motion trajectories for minimal-invasive procedures in robot-assisted surgery [75]. Motion planning also continues to grow into unexpected domains, for example, exploring molecular energy landscapes [76,77]. In all these disparate problems, our objective remains the same: find a sequence of admissible motions, to transform the world from an initial to a final state, or to maintain a set of constraints on the state. The notion of what constitutes a state has certainly expanded to cover an increasing number of applications; yet, motion remains the crux of the problem.

In recent years, we have also witnessed a trend towards the unification of principles. In essence, motion planning is a collection of common principles for analyzing motion combinatorially. "Motion" refers to the continuous process of state changes, and "combinatorial" refers to the partition of the continuous process into discrete elements. Motion planning studies those problems where the rearrangement of these elements is the result of motion — problems that cannot be reduced to pure instances of computational geometry or control theory. As we have seen in this chapter, random sampling plays a critical role in solving these problems and has shown great success.

REFERENCES

1. J. H. Reif. Complexity of the mover's problem and generalizations. In *Proceedings of IEEE Symposium on Foundations of Computer Science*, pp. 421–427, 1979.
2. T. Lozano-Pérez and M. A. Wesley. An algorithm for planning collision-free paths among polyhedral obstacles. *Communications of the ACM*, 22: 560–570, 1979.
3. F. Avnaim and J.-D. Boissonnat. Polygon placement under translation and rotation. In *Proceedings of Annual Symposium on Theoretical Aspects of Computer Science*, vol. 294 of *LNCS*, pp. 322–333. Springer-Verlag, Heidelberg, 1988.
4. M. Lin and D. Manocha. Collision and proximity queries. In J. E. Goodman and J. O'Rourke (eds), *Handbook of Discrete and Computational Geometry*, Chapter 35. CRC Press, Boca Raton, FL, 2004.
5. J. C. Latombe. *Robot Motion Planning*. Kluwer Academic Publishers, Boston, MA, 1991.
6. N. J. Nilsson. A mobile automation: an application of intelligence techniques. In *Proceedings of International Conference on Artificial Intelligence*, pp. 509–520, 1969.
7. M. de Berg, M. van Kreveld, M. Overmars, and O. Schwarzkopf. *Computational Geometry: Algorithms and Applications*. Springer-Verlag, Berlin, 2nd ed., 2000.
8. C. Ó'Dúnlaing and C. K. Yap. A retraction method for planning the motion of a disc. *Journal of Algorithms*, 6: 104–111, 1982.
9. O. Khatib. Real-time obstacle avoidance for manipulators and mobile robots. *International Journal of Robotics Research*, 5: 90–98, 1986.
10. G. Sánchez and J. C. Latombe. On delaying collision checking in PRM planning — application to multi-robot coordination. *International Journal of Robotics Research*, 21: 5–26, 2002.
11. L. E. Kavraki, P. Švestka, J. C. Latombe, and M. H. Overmars. Probabilistic roadmaps for path planning in high-dimensional configuration space. *IEEE Transactions on Robotics and Automation*, 12: 566–580, 1996.
12. N. M. Amato, O. B. Bayazit, L. K. Dale, C. Jones, and D. Vallejo. OBPRM: an obstacle-based PRM for 3D workspaces. In P. K. Agarwal et al. (eds), *Robotics: The Algorithmic Perspective: 1998 Workshop on the Algorithmic Foundations of Robotics*, pp. 155–168. A. K. Peters, Wellesley, MA, 1998.
13. V. Boor, M. H. Overmars, and F. van der Stappen. The Gaussian sampling strategy for probabilistic roadmap planners. In *Proceedings of IEEE International Conference on Robotics and Automation*, pp. 1018–1023, 1999.
14. M. Foskey, M. Garber, M. C. Lin, and D. Manocha. A Voronoi-based hybrid motion planner. In *Proceedings of IEEE/RSJ International Conference on Intelligent Robots and Systems*, pp. 55–60, 2001.
15. C. Holleman and L. E. Kavraki. A framework for using the workspace medial axis in PRM planners. In *Proceedings of IEEE International Conference on Robotics and Automation*, pp. 1408–1413, 2000.

16. D. Hsu, T. Jiang, J. Reif, and Z. Sun. The bridge test for sampling narrow passages with probabilistic roadmap planners. In *Proceedings of IEEE International Conference on Robotics and Automation*, pp. 4420–4426, 2003.

17. D. Hsu, L. E. Kavraki, J. C. Latombe, R. Motwani, and S. Sorkin. On finding narrow passages with probabilistic roadmap planners. In P. K. Agarwal et al. (eds), *Robotics: The Algorithmic Perspective: 1998 Workshop on the Algorithmic Foundations of Robotics*, pp. 141–154. A. K. Peters, Wellesley, MA, 1998.

18. J.-M. Lien, S. L. Thomas, and N. M. Amato. A general framework for sampling on the medial axis of the free space. In *Proceedings of IEEE International Conference on Robotics and Automation*, pp. 4439–4444, 2003.

19. T. Simeon, J. P. Laumond, and C. Nissoux. Visibility-based probabilistic roadmaps for motion planning. *Journal of Advanced Robotics*, 14: 477–494, 2000.

20. Y. Yang and O. Brock. Adapting the sampling distribution in PRM planners based on an approximated medial axis. In *Proceedings of IEEE International Conference on Robotics and Automation*, 2004.

21. F. Schwarzer, M. Saha, and J. C. Latombe. Exact collision checking of robot paths. In J. D. Boissonnat et al. (eds), *Algorithmic Foundations of Robotics V*, pp. 25–41. Springer-Verlag, 2002.

22. D. Hsu, J. C. Latombe, and R. Motwani. Path planning in expansive configuration spaces. In *Proceedings of IEEE International Conference on Robotics and Automation*, pp. 2719–2726, 1997.

23. S. M. LaValle and J. J. Kuffner. Randomized kinodynamic planning. In *Proceedings of IEEE International Conference on Robotics and Automation*, pp. 473–479, 1999.

24. R. Bohlin and L. E. Kavraki. Path planning using lazy PRM. In *Proceedings of IEEE International Conference on Robotics and Automation*, pp. 521–528, 2000.

25. L. E. Kavraki, J. C. Latombe, R. Motwani, and P. Raghavan. Randomized query processing in robot path planning. In *Proceedings of ACM Symposium on Theory of Computing*, pp. 353–362, 1995.

26. A. M. Ladd and L. E. Kavraki. Theoretic analysis of probabilistic path planning. *IEEE Transactions on Robotics and Automation*, 20: 229–242, 2004.

27. Švestka and M. H. Overmars. Motion planning for car-like robots, a probabilistic learning approach. *International Journal of Robotics Research*, 16: 119–143, 1997.

28. B. R. Donald, P. Xavier, J. Canny, and J. Reif. Kinodynamic motion planning. *Journal of the ACM*, 40: 1048–1066, 1993.

29. K. M. Lynch and M. T. Mason. Stable pushing: mechanics, controllability, and planning. *International Journal of Robotics Research*, 15: 533–556, 1996.

30. P. Hsu, Z. Li, and S. Sastry. On grasping and coordinated manipulation by a multifingered robot hand. In *Proceedings of IEEE International Conference on Robotics and Automation*, pp. 384–389, 1988.

31. P. Jacobs and J. Canny. Planning smooth paths for mobile robots. In *Proceedings of IEEE International Conference on Robotics and Automation*, pp. 2–7, 1989.

32. J. A. Reeds and L. A. Shepp. Optimal paths for a car that goes forwards and backwards. *Pacific Journal of Mathematics*, 145: 367–393, 1990.

33. J. Barraquand and J. C. Latombe. Nonholonomic multibody mobile robots: controllability and motion planning in the presence of obstacles. *Algorithmica*, 10: 121–155, 1993.

34. J.-P. Laumond, S. Sekhavat, and F. Lamiraux. Guidelines in nonholonomic motion planning for mobile robots. In J. P. Laumond (ed.), *Robot Motion Planning and Control*, Lectures Notes in Control and Information Sciences vol. 229, pp. 1–53. Springer-Verlag, Heidelberg, 1998.

35. Z. Li, J. F. Canny, and G. Heinzinger. Robot motion planning with nonholonomic constraints. In H. Miura et al. (eds), *Robotics Research: The Fifth International Symposium*, pp. 309–316. The MIT Press, Cambridge, MA, 1989.

36. J.-P. Laumond. Feasible trajectories for mobile robots with kinematic and environmental constraints. In *Proceedings of International Conference on Intelligent Autonomous Systems*, pp. 346–354, 1986.

37. J. E. Bobrow, S. Dubowsky, and J. S. Gibson. Time-optimal control of robotic manipulators along specified paths. *International Journal of Robotics Research*, 4: 3–17, 1985.

38. P. Ferbach. A method of progressive constraints for nonholonomic motion planning. *IEEE Transactions on Robotics and Automation*, 14: 172–179, 1998.

39. S. Sekhavat, P. Švestka, J.-P. Laumond, and M. H. Overmars. Multi-level path planning for nonholonomic robots using semi-holonomic subsystems. In J.-P. Laumond et al. (eds), *Algorithms for Robotic Motion and Manipulation: 1996 Workshop on the Algorithmic Foundations of Robotics*, pp. 79–96. A. K. Peters, Wellesley, MA, 1996.

40. F. Lamiraux, J.-P. Laumond, C. Van Geem, D. Boutonnet, and G. Raust. Trailer-truck trajectory optimization for airbus A380 component transportation. *IEEE Robotics and Automation Magazine*, 20: 14–71, 2005.

41. F. Lamiraux, D. Bonnafous, and C. Van Geem. Path optimization for nonholonomic systems: application to reactive obstacle avoidance and path planning. In *Control Problems in Robotics*. Springer-Verlag, Heidelberg, 2002.

42. D. Hsu, R. Kindel, J. C. Latombe, and S. Rock. Randomized kinodynamic motion planning with moving obstacles. *International Journal of Robotics Research*, 21: 233–255, 2002.

43. C. I. Conolly. The determination of next best views. In *Proceedings of IEEE International Confernce on Robotics and Automation*, pp. 432–435, 1985.

44. J. Maver and R. Bajcsy. Occlusions as a guide for planning the next view. *IEEE Transactions on Pattern Analysis and Machine Intelligence*, 15: 417–433, 1993.

45. R. Pito. A solution to the next best view problem for automated CAD model acquisition of free-form objects using range cameras. Technical Report 95-23, GRASP Lab, University of Pennsylvania, May 1995.

46. L. Wixson. Viewpoint selection for visual search. In *Proceedings of IEEE Conference on Computer Vision and Pattern Recognition*, pp. 800–805, 1994.

47. T. Shermer. Recent results in art galleries. *Proceedings of IEEE*, 80: 1384–1399, 1992.

48. H. H. González-Baños and J. C. Latombe. A randomized art-gallery algorithm for sensor placement. In *Proceedings of ACM Symposium on Computational Geometry*, pp. 232–240, 2001.

49. H. Brönnimann and M. T. Goodrich. Almost optimal set covers in finite vc-dimension. *Discrete and Computational Geometry*, 14: 463–479, 1995.

50. P. Valtr. Guarding galleries where no point sees a small area. *Israel Journal of Mathematics*, 104: 1–16, 1998.

51. A. Efrat and S. Har-Peled. Locating guards in art galleries. In *Proceedings of IFIP International Conference on Theoretical Computer Science*, pp. 181–192, 2002.

52. T. Danner and L. E. Kavraki. Randomized planning for short inspection paths. In *Proceedings of IEEE International Conference on Robotics and Automation*, pp. 971–976, 2000.

53. S. P. Fekete, R. Klein, and A. Nüchter. Online searching with an autonomous robot. In *Proceedings of The Sixth International Workshop on the Algorithmic Foundations of Robotics*, 2004.

54. S. Thrun. Robotic mapping: a survey. In G. Lakemeyer and B. Nebel (eds), *Exploring Artificial Intelligence in the New Millenium*. Morgan Kaufmann, 2002.

55. R. Pito. A sensor based solution to the next best view problem. In *Proceedings of IEEE International Conference on Pattern Recognition*, vol. 1, pp. 941–945, 1996.

56. B. Curless and M. Levoy. A volumetric method for building complex models from range images. *SIGGRAPH 96 Conference Proceedings*, pp. 303–312, 1996.

57. H. H. González-Baños and J.-C. Latombe. Navigation strategies for exploring indoor environments. *International Journal of Robotics Research*, 21: 829–848, 2002.

58. Y. Yu and K. Gupta. C-space entropy: a measure for view planning and exploration for general robot-sensor systems in unknown environments. *International Journal of Robotics Research*, 23: 1197, 2004.

59. C. Becker, H. H. González-Baños, J.-C. Latombe, and C. Tomasi. An intelligent observer. In *Proceedings of International Symposium on Experimental Robotics*, pp. 153–160, 1995.

60. S. Hutchinson, G. D. Hager, and P. I. Corke. A tutorial on visual servo control. *IEEE Transactions on Robotics and Automation*, 12: 651–670, 1996.

61. N. P. Papanikolopoulos, P. K. Khosla, and T. Kanade. Visual tracking of a moving target by a camera mounted on a robot: a combination of control and vision. *IEEE Transactions on Robotics and Automation*, 9: 14–35, 1993.

62. S. M. Lavallée, J. Troccaz, L. Gaborit, A. L. Benabid, P. Cinquin, and D. Hoffmann. Image guided operating robot: a clinical application in stereotactic neurosurgery. In R. H. Taylor et al. (eds), *Computer Integrated*

Surgery: Technology and Clinical Applications, pp. 342–351. The MIT Press, Cambridge, MA, 1995.

63. S. M. LaValle, H. H. González-Baños, C. Becker, and J. C. Latombe. Motion strategies for maintaining visibility of a moving target. In *Proceedings of IEEE International Conference on Robotics and Automation*, pp. 731–736, 1997.

64. A. Efrat, H. H. González-Baños, S. Kobourov, and L. Palaniappan. Optimal strategies to track and capture a predictable target. In *Proceedings of International Conference on Robotics and Automation*, pp. 3789–3796, 2003.

65. H. H. González-Baños, C.-Y. Lee, and J.-C. Latombe. Real-time combinatorial tracking of a target moving unpredictably among obstacles. In *Proceedings of International Conference on Robotics and Automation*, pp. 1683–1690, 2002.

66. T. Y. Li, J. M. Lien, S. Y. Chiu, and T. H. Yu. Automatically generating virtual guided tours. In *Proceedings of Computer Animation*, pp. 99–106, 1999.

67. R. Murrieta-Cid, H. H. González-Baños, and B. Tovar. A reactive motion planner to maintain visibility of unpredictable targets. In *Proceedings of IEEE International Conference on Robotics and Automation*, pp. 4242–4248, 2002.

68. P. Fabiani and J. C. Latombe. Dealing with geometric constraints in game-theoretic planning. In *Proceedings of International Joint Conference on Artificial Intelligence*, pp. 942–947, 1999.

69. R. Murrieta-Cid, A. Sarmiento, S. Bhattacharya, and S. Hutchinson. Maintaining visibility of a moving target at a fixed distance: the case of observer bounded speed. In *Proceedings of IEEE International Conference on Robotics and Automation*, pp. 479–484, 2004.

70. T. Bandyopadhyay, Y. P. Li, M. H. Ang Jr., and D. Hsu. Stealth tracking of an unpredictable target among obstacles. In *Proceedings of The Sixth International Workshop on the Algorithmic Foundations of Robotics*. Springer-Verlag, Heidelberg, 2004.

71. T. Lozano-Pérez, M. T. Mason, and R. H. Taylor. Automatic synthesis of find-motion strategies for robot. *International Journal of Robotics Research* 3: 3–24, 1984.

72. L. P. Kaelbling, M. L. Littman, and A. R. Cassandra. Planning and acting in partially observable stochastic domains. *Artificial Intelligence*, 101: 99–134, 1998.

73. J. J. Kuffner, K. Nishiwaki, S. Kagami, Y. Kuniyoshi, M. Inaba, and H. Inoue. Online footstep planning for humanoid robots. In *Proceedings of IEEE International Conference on Robotics and Automation*, 2003.

74. H. Chang and T.-Y. Li. Assembly maintainability study with motion planning. In *Proceedings of IEEE International Conference on Robotics and Automation*, pp. 1012–1019, 1995.

75. A. Schweikard, J. R. Adler, and J.-C. Latombe. Motion planning in stereotaxic radiosurgery. *IEEE Transactions on Robotics and Automation*, 9: 764–774, 1993.

76. A. P. Singh, J.-C. Latombe, and D. L. Brutlag. A motion planning approach to flexible ligand binding. In *Proceedings of International Conference on Intelligent Systems for Molecular Biology*, pp. 252–261, 1999.

77. G. Song and N. M. Amato. Using motion planning to study protein folding pathways. In *Proceedings of ACM International Conference on Computational Biology (RECOMB)*, pp. 287–296, 2001.
78. H. Choset, K. M. Lynch, S. Hutchinson, G. Kantor, W. Burgard, L. E. Kavraki, and S. Thrun. *Principles of Robot Motion: Theory, Algorithms, and Implementations (Intelligent Robotics and Autonomous Agents)*. The MIT Press, Cambridge, MA, 2005.

BIOGRAPHIES

Héctor H. González-Baños is a senior research scientist at Honda Research Institute USA Inc. He received his Ph.D. in electrical engineering from Stanford University in 2001. His research interests are robotics and computer vision.

David Hsu is the Sung Kah Kay assistant professor of computer science at the National University of Singapore and a fellow of the Singapore–MIT Alliance. He received B.Sc. in computer science and mathematics from the University of British Columbia in 1995 and Ph.D. in computer science from Stanford University in 2000. From 2000 to 2001, he was a member of the research staff at Compaq Computer Corporation's Cambridge Research Lab. He moved to Singapore in 2002. His main research interests include robotics, motion planning, computational biology, and geometric algorithms.

Jean-Claude Latombe is the Kumagai professor of computer science at Stanford University. He received his Ph.D. from the National Polytechnic Institute of Grenoble (INPG) in 1977. He was on the faculty of INPG from 1980 to 1984, when he joined Industry and Technology for Machine Intelligence (ITMI), a company that he had cofounded in 1982. He moved to Stanford in 1987. At Stanford, he served as the chairman of the Computer Science Department from 1997 to 2001, and on the BioX Leadership Council from 2002 to 2004. His main research interests are in artificial intelligence, robotics, motion planning, computational biology, computer-aided surgery, and graphic animation.

11 Multi-Robot Cooperation

*Rafael Fierro, Luiz Chaimowicz, and
Vijay Kumar*

Contents

11.1 Introduction

Cooperative multi-robot systems have recently received a great deal of attention, motivated by recent technological advances in communication, computation, and sensing. Another major factor behind this interest is that there are many tasks that single robots cannot efficiently accomplish. Multi-robot applications include environmental monitoring, search and rescue, cooperative manipulation, collaborative mapping and exploration, battlefield assessment, and health monitoring of civil infrastructure. In these applications, a system composed of multiple cooperative robots is desirable because of its size, cost, flexibility, and fault tolerance.

The research challenges encountered in cooperative multi-robot systems require the integration of different disciplines including control systems, artificial intelligence (AI), biology, optimization, and robotics. Therefore, it is not surprising that the related literature enjoys the flavor of a broad spectrum of approaches which have been utilized in an attempt to come up with a solution for cooperative control problems [1]. To name just a few, in *behavior-based* approaches [2,3] the main idea is to compose primitive behaviors (i.e., controllers) in order to produce a useful *emergent* behavior. Closely related methods originating from the field of distributed artificial intelligence (DAI) consider a cooperative multi-robot system as interacting software agents [4]. Yet another perspective comes from the research in biological systems. Here, the notions of *swarm intelligence* [5], and *flocking* and *schooling* [6] constitute a basis to investigate behavior of multi-robot systems composed of large number of agents. Also, *game theory* provides a rigorous framework to understand complex behaviors of multiple robots engaged in competitive or cooperative tasks [7].

A fundamental problem in cooperative multi-robot systems is designing a mechanism of *cooperation* between agents so that the overall performance of the system improves. This design can include control, communication, computation, and sensing aspects. For example, multiple robots which are supposed to push an object within the workspace without grasping it [8,9]. These robots may need to map the environment and find the object of interest. Once the object has been localized, robots approach the object maintaining some formation and sensing constraints. Additionally, they need to communicate and perform common computations utilizing their sensing readings in order to push the object toward a desired direction with a minimum deviation from it.

The rest of the chapter is organized as follows: Cooperative multi-robot systems and some tools for their analysis are introduced in Section 11.2. Section 11.3 is devoted to formation control of multi-robot systems. Section 11.4 describes a method for incorporating optimization-based tools into systems composed of multiple robots. Here, notions of model predictive control are explained. Two real-world cooperative multi-robot applications

FIGURE 11.1   Robots performing a cooperative task (Courtesy: NASA/JPL-Caltech).

are discussed in Section 11.5. Finally, Section 11.6 gives concluding remarks and a brief discussion of open problems and future research opportunities in multi-robot cooperation.

11.2   COOPERATIVE MULTI-ROBOT SYSTEMS

Due to recent substantial developments in electronics and computing, it is now possible to find onboard embedded computers which have more computing power than the super computers available a few years ago. Exchanging inform-ation between robots distributed over an area is now possible by means of off-the-shelf ad hoc wireless network devices. Furthermore, there are various small size, light weight sensing devices on the market ranging from laser range sensors to color CCD cameras. As a result, by exploiting current technology, one can build a group of relatively small robots each having satisfactory capabilities within a reasonable budget. In adversary and dangerous missions, it might be desirable to have multiple cost-effective robots because even if some of the team members are lost due to some failure, the others can continue to operate. This leads to fault tolerance and robustness. For tasks such as obtaining sensory measurements over a wide area, multiple robots are desirable because they can accomplish the task more efficiently than a single robot. Figure 11.1 depicts NASA's vision for cooperating robots.

A good survey of cooperative robots can be found in Reference 10. This survey reviewed about 200 papers published before 1997. This chapter is not intended to be a survey of the state-of-the-art in cooperative systems. Rather, it focuses on cooperating multi-robot systems performing tasks that require or involve (i) concurrent, coordinated operation and execution; (ii) formation of spatial patterns; (iii) mobility; and (iv) distributed sensing, computation, and actuation. However, task decomposition and planning [11], multi-robot learning, communication protocols, architectures for multi-robot cooperation [12], and human–robot interaction are outside the scope of this chapter.

Although a team of robots has certain benefits as stated above, if the individual members do not cooperate with their teammates for a common task, the whole group may perform poorly. Therefore, cooperation, in general, is the key aspect of a multi-robot team strategy. At this point, the following definition may be helpful to get more insight into what cooperative behavior means.

Definition 11.2.1 *[10] Given some task specified by the designer, a multiple-robot system displays cooperative behavior if, due to some underlying mechanism (i.e., the mechanism of cooperation), there is an increase in the total utility of the system.*

As can be inferred from the definition, the main goal of a cooperative system design should be to build a mechanism of cooperation so that the overall performance of the system improves for a given task.

In many cooperative control problems, robots move in a coordinated fashion to achieve some common goal and seek to maintain some geometrical relationships among themselves. Often movement is dictated by measurement of gradients of some actual sensor measurements, or some artificial potential field. Solutions defined with inter-robot distance relationships are explored in Reference 13, where methods to measure and project gradient information are discussed. The applications for these methods are in, for example, data acquisition in large areas such as oceans where the most advantageous arrangement of sensors may not be to distribute them evenly, but to have them adapt to concentrate more sensors in areas where the measured variable has steeper gradients.

A different application is in the control strategies of robotic games such as RoboCup Soccer [14]. RoboCup provides an ideal platform for the development of cooperative robotic systems. Successful control strategies are often behavior-based, with robots operating in certain modes with heuristic rules determining how each robot reacts to best benefit its team. Yet within this class, good control strategies may require more than simply a set of "if–then–else" rules defining the mode switching.

We are mainly concerned with cooperative tasks that require robots to plan their trajectories and maintain a certain *formation* shape. Before discussing

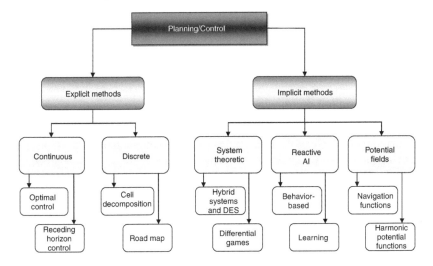

FIGURE 11.2 Motion planning and control approaches.

formation control in more detail, we give an outline of motion planning and control methodologies (based on Reference 15) and graph theory. These are fundamental tools for the formal analysis and design of cooperative multi-robot systems.

11.2.1 Motion Planning and Control

Motion planning approaches can be grouped into two main categories as illustrated in Figure 11.2.

Explicit approaches provide open-loop control policies to motion planning. On the other hand, *implicit* or *reactive* approaches provide closed-loop plans by composing low-level feedback controllers. In this chapter, we focus on implicit system theoretic methodologies.

11.2.1.1 Explicit approaches

An explicit motion planning approach produces a path or trajectory through the configuration space from an initial configuration to the goal configuration. Explicit methods can be further classified into *continuous* and *discrete*. Optimal motion planning/control (see Reference 16) and receding horizon control (RHC) methods are typically formulated in continuous-time. Recently RHC methods have received considerable attention due to their ability to incorporate motion constraints and changes in mission objectives. The inherent ability of RHC to handle nonlinear constrained systems makes it a natural technique for multi-robot cooperative tasks. *Discrete* explicit methods (e.g., road maps and cell decomposition) [17] use computational geometric tools to determine a solution to the planning problem. Due to their computational complexity, discrete methods are appropriate for off-line planning in static environments.

11.2.1.2 Implicit approaches

Implicit approaches produce on-line solutions to the motion planning problem by mapping the state of the robot and state of the environment (e.g., obstacles, moving targets) to a set of inputs for controlling the robot. Implicit approaches include (1) system theoretic methods, (2) reactive AI approaches, and (3) artificial potential fields, as illustrated in Figure 11.2.

System theoretic methods. These methods use control theory to develop closed-loop control laws for the robots. Hybrid systems (i.e., the combination of continuous dynamics and discrete dynamics) offer a suitable framework to represent robots performing cooperative tasks [18]. More specifically, a robot can be modeled by a hybrid automaton whose locations or states determine robot behaviors. Thus, complex behaviors can emerge by parallel or sequential composition of basic behaviors [9].

Reactive AI approaches. AI techniques address the problem of combining simple controllers (i.e., behaviors) into an aggregate system that exhibits an emergent useful behavior. If behaviors are composed sequentially, then the control input is due to the behavior that is currently active. In contrast, in parallel composition the system input is computed as the weighted sum of the outputs from all active behaviors or controllers.

Potential field approaches. The main idea behind potential field approaches is to define a scalar field V (called *potential function*) over the robot's free space. This artificial field produces a force $-\nabla V$ acting on the robot. Obstacles and goals produce repulsive and attractive potentials, respectively. The resultant force is mapped to contoller/actuator commands. Thus, the robot, at least in theory, would navigate toward its goal destination while avoiding collisions. Several researchers have extended potential field methods to make them suitable for multi-robot systems [19].

The main drawback of potential field techniques is that the robot might get stuck in a local minimum before reaching the goal. Several variants have been proposed to overcome this limitation.

11.2.2 Graph Theory Preliminaries

We present a brief review of some definitions and results from algebraic graph theory. A detailed treatment can be found in Reference 20. A *directed graph* $\mathcal{X} = (V, E)$ consists of a vertex set $V(\mathcal{X})$ and a directed *edge* or *arc* set $E(\mathcal{X})$, where an arc is an ordered pair of distinct vertices. An arc (v_i, v_j) between two vertices v_i, v_j in a directed graph is said to be *incoming* with respect to v_j and *outcoming* with respect to v_i. The *in(out)-degree* of a vertex in a directed graph is defined as the number of incoming (outcoming) edges at this vertex. A *path* of length r in a directed graph is a sequence $v_0, \ldots, v_r$ of distinct vertices such that for every $i \in [1, r]$, $(v_{i-1}, v_i) \in E$. The *distance* between two vertices v_i and v_j in a graph $\mathcal{X}$ is the length of the shortest path from v_i to v_j. The *diameter* of a graph is the maximum distance between two distinct vertices. A (directed) *cycle* is a connected graph where every vertex has one incoming and one outcoming edge. An *acyclic graph* is a graph with no cycles. A connected acyclic graph is also called a *tree*. A *subgraph* of a graph $\mathcal{X}$ is a graph $\mathcal{Y}$ such that $V(\mathcal{X}) \subseteq V(\mathcal{Y})$, $E(\mathcal{X}) \subseteq E(\mathcal{Y})$. When $V(\mathcal{X}) = V(\mathcal{Y})$, $\mathcal{Y}$ is called a *spanning subgraph*. A subgraph $\mathcal{Y}$ *of* $\mathcal{X}$ *is an* induced subgraph if vertices $v_i, v_j \in V(\mathcal{Y})$ are adjacent in $\mathcal{Y}$ if and only if they are adjacent in $\mathcal{X}$. A spanning subgraph with no cycles is called a *spanning tree*. If (v_i, v_j) is an edge, v_i, v_j are said to be *adjacent* or v_j is a *neighbor* of v_i ($v_j \sim v_i$). The *adjacency matrix* of a directed graph $\mathcal{X}$ with n vertices is an $n \times n$ matrix $A(\mathcal{X})$, the *ij*-element of which (denoted by A_{ij} henceforth) is the number of arcs from v_i to v_j. In most cases this is a binary matrix with $A_{ij} = 1$ when arc (v_i, v_j) exists and 0 otherwise. It is easy to see that the adjacency matrix for an undirected graph is symmetric. If there are no loops (cycles of length zero) the diagonal entries are zero.

11.3 FORMATION CONTROL

Many systems in nature exhibit stable formation behaviors, for example, swarms, schools, and flocks [21,22]. In these highly robust systems, individuals follow distant leaders without colliding with neighbors. Thus, a coordinated grouping behavior emerges by composing individual control actions and sensor information in a distributed control architecture. One possibility to realize such a grouping behavior is using artificial potential functions as a coordination mechanism [23]. In some application domains, the group of vehicles are to move as a rigid structure. In this case, centralized strategies are used to control the team [24]. However, in many practical situations (e.g., cooperative manipulation), a target formation needs to be established for a given task or environment. In these cases, reconfiguration of robots in formation is required [25]. Additionally, formation control has applications where rigorous coverage of an area is required, such as in collaborative exploration [26] or mine sweeping [27].Conceptually, fighter jets flying in a delta formation at an air show are an example of manually controlled vehicles tracking a trajectory in formation.

This section presents a methodology to coordinate a team of N robots in formation. The problem of stabilizing a group of mobile robots in formation has received considerable attention in the last few years [25,28–31]. We assume

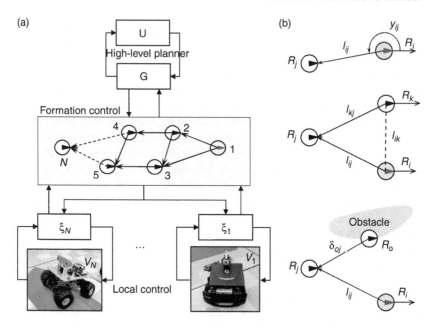

Figure 11.3 (a) Modular architecture for formation control. (b) Basic formation control algorithms.

that formation control is required to accomplish a given cooperative task. Furthermore, the formation shape is represented by a *control graph*[1] that changes over time to accommodate kinematic, sensor, and communication constraints [32]. The desired formation shape is achieved by deleting or creating links between robot neighbors. This process implies switching between controllers and leaders in a stable fashion [33].

Specifically, a group of robots is required to follow a prescribed trajectory, while achieving and maintaining a desired formation shape. We address the development of complex formations by composing simple building blocks in a *bottom-up approach*. The building blocks consist of controllers (see Figure 11.3b) and estimators, and the framework for composition allows for tightly coupled perception-action loops. While this philosophy is similar in spirit to a behavior-based control paradigm [2], it differs in the more formal, control-theoretic approach in developing the basic components and their composition.

We are particularly interested in applications like *cooperative manipulation*, where a semi-rigid formation may be necessary to transport a grasped object

---

[1] In a control graph, vertices and edges correspond to robots and control laws/communication links, respectively.

to a prescribed location, and *cooperative exploration*, where the formation may be defined by a set of sensor constraints. In our approach, each node (i.e., robot) has a definite identity that can be determined by visual observations as well as by communication. All nodes can *hear* each other up to a finite distance. Nodes that cannot *talk* or cannot *listen* are left out of the group. Also, it is assumed that a planned trajectory $g(t)$ and a desired formation shape vector r^d are specified either by a human operator or by the task specification at a higher level. Figure 11.3a shows a hierarchical modular architecture for formation control.

At the *formation control* layer, a network of N vehicles is built on three different networks: a *physical network* that captures the physical constraints on the dynamics, control, and sensing of each robot; a *communication network* that describes the information flow between the robots; and a *computational network* that describes the computational resources available to each robot. Each network is modeled by a graph with N nodes. $\mathcal{R}$ is a finite set of nodes $\{R_1, R_2, \ldots, R_N\}$. The physical network G_p is a directed graph representing the flow of sensory information (i.e., relative state). The communication network G_c is an undirected graph where edges represent communication channels. The topology of these networks are determined by constraints of the hardware, the physical distribution of the robots, and the characteristics of the environment.

The design of the computational network involves the assignment of (kinematic) control policies for each robot. The ability of a node R_j to sense another node R_i allows R_j to use a state feedback controller that regulates relative position and/or orientation of R_j with respect to R_i. The ability of R_j to listen to R_i allows R_i to broadcast feed-forward information and for R_j to use feed-forward control.

In our previous work [25,29], we developed three basic controllers depicted in Figure 11.3b using input–output feedback linearization. By composing these basic controllers, complex formation shapes can be built. In Section 11.3.1, we present a controller that uses *dynamic feedback linearization* [34] and implements a basic leader–follower algorithm.

11.3.1 Full-State Linearization via Dynamic Feedback

We describe a control algorithm that allows a follower robot R_j to maintain a separation l_{ij} and relative bearing ψ_{ij} with respect to a leader robot R_i. Robots are nonholonomic platforms modeled with the unicycle model:

$$
\begin{aligned}
\dot{x}_i &= v_i \cos \theta_i \\
\dot{y}_i &= v_i \sin \theta_i \\
\dot{\theta}_i &= \omega_i
\end{aligned}
\tag{11.1}
$$

where x_i, y_i, θ_i, v_i, and ω_i are the x-position, y-position, orientation angle, linear velocity, and angular velocity of robot i, respectively. The leader–follower kinematic model is given by

$$
\begin{aligned}
\dot{l}_{ij} &= v_j \cos \gamma - v_i \cos \psi_{ij} \\
\dot{\psi}_{ij} &= \frac{v_i \sin \psi_{ij} - v_j \sin \gamma - \omega_i l_{ij}}{l_{ij}} \\
\dot{\theta}_{ij} &= \omega_i - \omega_j \\
\gamma &= \psi_{ij} + \theta_i - \theta_j
\end{aligned}
\tag{11.2}
$$

The output vector of interest is

$$
z = \begin{bmatrix} l_{ij} \\ \psi_{ij} \end{bmatrix}
\tag{11.3}
$$

Taking the derisssvative of (11.3) with respect to time, we have

$$
\dot{z} = \begin{bmatrix} \dot{l}_{ij} \\ \dot{\psi}_{ij} \end{bmatrix} = \begin{bmatrix} \cos \gamma & 0 \\ -\dfrac{\sin \gamma}{l_{ij}} & 0 \end{bmatrix} \begin{bmatrix} v_j \\ \omega_j \end{bmatrix} + \begin{bmatrix} -v_i \cos \psi_{ij} \\ \dfrac{v_i \sin \psi_{ij}}{l_{ij}} - \omega_i \end{bmatrix}
\tag{11.4}
$$

Since ω_j does not appear, the decoupling matrix is singular. In order to overcome the singularity of decoupling matrix an integrator is added before the first input

$$
\begin{aligned}
v_j &= \zeta_1 \\
\dot{\zeta}_1 &= a_j
\end{aligned}
\tag{11.5}
$$

where a_j is the new auxiliary input which is the linear acceleration of the follower robot. By using (11.5), Equation (11.4) is rewritten as

$$
\dot{z} = \begin{bmatrix} \cos \gamma & 0 \\ -\dfrac{\sin \gamma}{l_{ij}} & 0 \end{bmatrix} \begin{bmatrix} \zeta_1 \\ \omega_j \end{bmatrix} + \begin{bmatrix} -v_i \cos \psi_{ij} \\ \dfrac{v_i \sin \psi_{ij}}{l_{ij}} - \omega_i \end{bmatrix}
\tag{11.6}
$$

Differentiating (11.6) with respect to time results in

$$
\ddot{z} = \begin{bmatrix} \cos \gamma & \zeta_1 \sin \gamma \\ -\dfrac{\sin \gamma}{l_{ij}} & \dfrac{\zeta_1}{l_{ij}} \cos \gamma \end{bmatrix} \begin{bmatrix} a_j \\ \omega_j \end{bmatrix} + \begin{bmatrix} s_1 \\ s_2 \end{bmatrix}
\tag{11.7}
$$

where

$$s_1 = \frac{\zeta_1^2}{2l_{ij}}(1 - \cos 2\gamma) - a_i \cos \psi_{ij} - 2\frac{v_i\zeta_i}{l_{ij}} \sin \gamma \sin \psi_{ij}$$

$$- v_i\omega_i \sin \psi_{ij} + \frac{v_i^2}{2l_{ij}}(1 - \cos \psi_{ij})$$

$$s_2 = \frac{a_i \sin \psi_{ij}}{l_{ij}} + \frac{v_i^2}{l_{ij}^2} \sin 2\psi_{ij} - 2\frac{v_i\zeta_1}{l_{ij}^2} \sin(\gamma + \psi_{ij})$$

$$+ \frac{\zeta_1^2}{l_{ij}^2} \sin 2\gamma - \frac{v_i\omega_i}{l_{ij}} \cos \psi_{ij} - \alpha_i$$

where a_i and α_i represent linear and angular acceleration of the leader robot, respectively. The decoupling matrix is nonsingular if $\zeta_1/l_{ij} \neq 0$. Using this condition, the system can be written as

$$\ddot{z} = Au + B \tag{11.8}$$

where $u = [a_j \quad \omega_j]^T$ is computed as

$$u = A^{-1}(P - B) \tag{11.9}$$

where $P = [p_1 \quad p_2]^T$, and $B = [s_1 \quad s_2]^T$. From (11.7), we can derive

$$A^{-1} = \frac{l_{ij}}{\zeta_1} \begin{bmatrix} \dfrac{\zeta_1 \cos \gamma}{l_{ij}} & -\zeta_1 \sin \gamma \\ \dfrac{\sin \gamma}{l_{ij}} & \cos \gamma \end{bmatrix} \tag{11.10}$$

After some algebraic manipulations, the auxiliary inputs become

$$a_j = p_1 \cos \gamma - s_1 \cos \gamma - p_2 l_{ij} \sin \gamma + s_2 l_{ij} \sin \gamma$$

$$\omega_j = \frac{1}{\zeta_1}[p_1 \sin \gamma - s_1 \sin \gamma + p_2 l_{ij} \cos \gamma - s_2 l_{ij} \cos \gamma] \tag{11.11}$$

Thus (11.5) and (11.11) transform the original leader–follower system into two decoupled chains of two integrators.

$$\ddot{z} = \begin{bmatrix} p_1 \\ p_2 \end{bmatrix} \tag{11.12}$$

The closed-loop system becomes

$$p_1 = \ddot{l}^d_{ij} + k_1(\dot{l}^d_{ij} - \dot{l}_{ij}) + k_2(l^d_{ij} - l_{ij})$$
$$p_2 = \ddot{\psi}^d_{ij} + k_3(\dot{\psi}^d_{ij} - \dot{\psi}_{ij}) + k_4(\psi^d_{ij} - \psi_{ij})$$

(11.13)

where l^d_{ij} and ψ^d_{ij} define the desired formation geometry, and k_i, $i = 1, \ldots, 4$ are positive feedback gains whose values can be found using well-known linear control techniques. Figure 11.4 shows simulation results verifying the validity of the control algorithm. The desired initial formation variables are $l^d_{ij} = 5$ m and $\psi^d_{ij} = \pi/2$ rad. After 140 sec, ψ^d_{ij} is changed to $4\pi/3$ rad while l^d_{ij} remains constant.

11.3.2 Formation Reconfiguration

We would like to build formations of N robots in a modular fashion by composing basic formation controllers. *Formation control graphs* provide the tool to achieve this objective. In order to proceed we need to define the shape of a formation.

Definition 11.3.1 *The shape S of a formation of N robots moving in $\mathbb{R}^2$ with one robot identified as the lead robot is a point in a $2(N-1)$-dimensional submanifold of $\mathbb{R}^{2N}$ with coordinates $[(r_2 - r_1) \quad (r_3 - r_1) \cdots (r_N - r_1)]^T$.*

In the case where the *desired* shape S^d can be locally parameterized as a vector $r \in \mathbb{R}^{2(N-1)}$ then the shape error is simply given by the Euclidean distance:

$$\tilde{S} = \|r^d - r\|$$

(11.14)

The robot interconnections can then be designed to implement the desired shape. The formation will be identified by a directed graph that represents a parametrization of the formation shape S and the control specifications that realize it.

Definition 11.3.2 *A control graph $\mathcal{H} = (V, E)$ is a directed graph with:*

- *A finite set $V = \{R_1, \ldots, R_N\}$ of N vertices and a map assigning to each vertex R_i a control system $\dot{q}_i = f(q_i, u_i)$ where $q_i \in \mathbb{R}^n$ and $u_i \in \mathbb{R}^m$.*
- *An edge set $E \subset V \times V$ encoding leader–follower relationships and controller assignments between robots. The ordered pair $(R_i, R_j) \triangleq e_{ij}$ belongs to E if u_j depends on the state of robot i, q_i.*

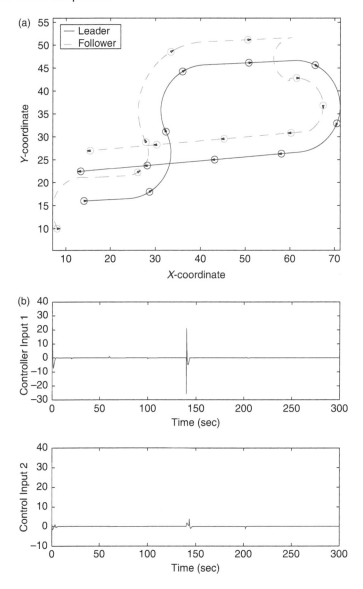

FIGURE 11.4 (a) Leader–follower trajectory. (b) Controller inputs.

According to Definition (11.3.2), each vertex represents the dynamics of a particular robot. From the adjacency matrix H for the control graph $\mathcal{H}$, column j corresponds to robot R_j, and a 1 in row i ($H(i,j) = 1$) denotes an incoming edge originating at a neighboring leader R_i. In Figure 11.3a for example, the adjacency matrix of the first five vertices in the control graph shown becomes

$$
H = \begin{bmatrix}
0 & 1 & 1 & 0 & 0 \\
0 & 0 & 1 & 1 & 0 \\
0 & 0 & 0 & 0 & 1 \\
0 & 0 & 0 & 0 & 1 \\
0 & 0 & 0 & 0 & 0
\end{bmatrix}
$$

Robot R_j has to control its shape r_j relative to all such R_i (for our specific controllers the number of leaders and hence the indegree at each vertex is bounded by $m = 2$). Similarly, we can identify the neighboring followers of robot j by destination of the outcoming edges. Vertices of indegree zero, represent formation leaders. We restrict our formation to have one lead robot. For the formation leader no control specification is prescribed with respect to other robots. Instead, the formation leader aims at achieving group objectives such as following a reference trajectory $g(t)$ or navigating within an obstacle populated environment. Also, the control graphs are restricted to be acyclic.

The structure of the control graph will affect the stability of a multi-robot formation system. Stability means that $\tilde{S} \rightarrow 0$ as $t \rightarrow \infty$ where $\tilde{S}$ is given in (11.14). It is not difficult to show that for a team of fully actuated planar robots, acyclicity of $\mathcal{H}$ guarantees stability. The interested reader is referred to References 30 and 32 for a detailed treatment of formation stability issues.

The formation shape $\mathcal{S}$ can be specified with respect to some common reference frames, which can be assumed to be the local frames of the formation leaders. Formation specifications are defined in terms of a desired shape vector r^d. If r_j^d is the desired shape component corresponding to robot j, then the desired state for j can be expressed as $q_j^d = q_i + r_j^d$, where q_i can be the state of any formation leader. We assume each robot derives relative localization information by some sensing modality (see for instance Reference 28).

A fundamental formation control problem can be posed as follows.

Problem 11.3.3 *Given a distribution of N robots in the plane and a desired planar shape $\mathcal{S}^d$ parameterized by r^d, find an optimal control graph $\mathcal{H}^*$ that assigns a controller for each robot subject to the following two constraints (a) kinematic constraints that must be satisfied by the relative position and orientation between neighboring robots; and (b) sensor constraints based on the limits on range and field of view of sensor and communication device(s) that prevent a robot from obtaining complete information about its neighbors.*

ALGORITHM 11.1
Control graph assignment algorithm

initialize adjacency matrix $H(i,j) := 0$;
for all robot $k \in \{1, 2, \ldots, n\}$, $k \neq leader$ **do**
   $H(i,k):=1$ for $SB_{ik}C$, edges(i,k) $\in$spanning tree of G_p;
   $d_k :=$ depth of node k in G_c;
   find set P_k of robots visible to k with depths $d_k, d_k - 1$;
   if $P_k = \emptyset$(disconnected) **then**
      report failure at k, break;
   end if
   $S_k := P_k$ sorted by ascending Δt_k^i with k;
   if $numOfElements(S_k) \geq 2$ **then**
      pick last two elements $i,j \in P_k$;
      if $\epsilon_{ijk} = (l_{ik} + l_{jk} - l_{ij}) \neq 0$ **then**
         $H(i,k):= 1, H(j,k):= 1$ for $SS_{ijk}C$;
      else
         repeat above check for remaining $j \in S_k$ in order;
      end if
   end if
end for

An algorithmic approach to addressing this problem is developed in Reference 32. Algorithm 11.1 assigns control policies to different robots, based on sensor and actuator constraints. Among the feasible control graphs that satisfy the constraints, it selects those control graphs that globally minimize the tracking error in formation shape.

The approach follows a two-step procedure (a) assign an initial acyclic leader–follower graph $\mathcal{H}_0$ with single-leader-based control links (this is a tree); and (b) refine (add/delete edges) control graph based on local optimality measures. Once the leader is identified, $\mathcal{H}_0$ is derived via communication by having each robot identify its neighbors in the physical network. If each robot communicates the identities of its neighbors in a prescribed order, a breadth-first search can be used to establish a spanning tree $\mathcal{H}_0$. If there are robots with no neighbors in the physical network (i.e., with no visible neighbors), we have a disconnected graph. Obstacles are treated as virtual robots, in this way a group can navigate within a dynamic environment and exhibit formation reconfiguration as shown in Figure 11.5a. An algorithm that allows formation splitting/rejoining, Figure 11.5b, is an area of current research.

Section 11.4 describes an optimization-based approach to address coordination and formation control of cooperative multi-robot systems.

(a)

Final formation

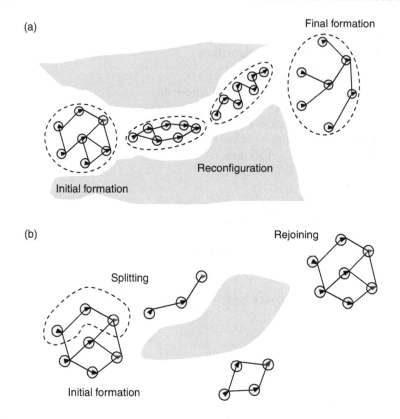

Reconfiguration

Initial formation

(b)

Rejoining

Splitting

Initial formation

FIGURE 11.5   Examples of formation reconfiguration.

11.4 OPTIMIZATION-BASED COOPERATIVE CONTROL

Many researchers are addressing multi-robot cooperation problems using optimization techniques. Contributions in this area include the work in Reference 35, where the focus is on autonomous vehicles performing distributed sensing tasks. Optimal motion planning is considered in Reference 24. In Reference 36, the task of repositioning a formation of robots to a new shape while minimizing either the maximum distance that any robot travels or the total distance traveled by the formation is considered. One of the goals of this work is to extend the mission lives of robot formations and mobile ad hoc networks (MANETs). More recently, the use of model predictive control (MPC) is becoming popular in the multi-robot system literature [37,38]. In Reference 37, a *distributed* MPC algorithm for stabilizing multi-vehicle formations is developed. This section describes the optimization-based solution of a formation control problem using receding-horizon model predictive control MPC methods. Also, we discuss some difficulties in optimization, and point out some potential drawbacks to model predictive control that we will need to address in the development.

Generally, MPC algorithms rely on an optimization of a predicted model response with respect to plant input to determine the best input changes for a given state. The MPC algorithms can in general handle nonlinear models, and it may be reconfigurable. Either hard constraints (that cannot be violated) or soft constraints (that can be violated but with some penalty) can be incorporated into the optimization, giving MPC a potential advantage over passive state feedback control laws. However, MPC has some possible disadvantages when applied to mobile robot formation control, the foremost being the computational cost.

In this section, we describe a different approach based on *terminal constraints* to ensure stability of the MPC algorithm. From the terminal constraint region, we use a local stabilizing controller (e.g., the leader–follower controller derived in Section 11.3.1) in the manner described in References 39 and 40, resulting in a *dual-mode* MPC algorithm. This may have the advantage of conceptual simplicity, but may suffer the disadvantage of infeasible optimization problems outside a necessarily limited region of convergence for the algorithm.

Now consider a dual-mode MPC algorithm for a single leader–follower pair. The unicycle model will represent the kinematics of the leader robot R_i and the controlled robot R_j. For the follower robot, the state vector is $x(k) = [x_j(k)\ y_j(k)\ \theta_j(k)]^T$, and the input vector is $u(k) = [v_j(k)\ \omega_j(k)]^T$. The allowable inputs are limited to the set $\{u(k) \mid u_{\min} \le u(k) \le u_{\max}\}$ so that the kinematic inputs are limited to some reasonable magnitudes.

Let the discrete-time dynamic model of the system be $x(k + 1) = f(x(k), u(k))$, so that $f(\cdot)$ is time-invariant and nonlinear and has an equilibrium point at $u(k) = 0$. An MPC algorithm uses a model to predict the trajectory based on the current state and some input series, and generates a control signal that results in a trajectory that is optimum in some quantifiable way. At the discrete time k, the state of the system $x(k)$ is measured, and the plant model is used to predict the system trajectory from $k + 1$ to $k + H_p$, where H_p is called the *prediction horizon*. The optimality of a trajectory is evaluated by calculating some *objective function* having the general form

$$V(k) = \sum_{m=1}^{H_p} \rho(x_e(k + m|k), u_e(k - 1 + m|k)) \qquad (11.15)$$

where $\rho(\cdot)$ is an incremental cost, $x_e(k + m|k) = [x(k + m|k) - x_d(k + m|k)]$, and $x_d(k + m|k)$ is the desired value of $x(k + m|k)$. The notation $(k + m|k)$ indicates a value at time $(k+m)$ calculated at time k. Each $x_e(\cdot)$ is an element of a convex, closed set $\mathcal{X} \subset \mathbb{R}^3$ containing the origin. The vector $u_e(k-1+m|k) = [u(k-1+m|k) - u_d(k-1+m|k)]$, with $u_d(k-1+m|k)$ as the input necessary to achieve $x_d(k + m|k)$ from $x_d(k - 1 + m|k)$. Each $u_e(\cdot)$ is an element of a convex, compact set $\mathcal{U} \subset \mathbb{R}^2$ containing the origin.

Let any control input series that stabilizes the system (to zero error) in H_p steps or fewer while not violating any constraints be denoted $\tilde{u}(k) = (u_e(k|k), u_e(k + 1|k), \ldots, u_e(k + H_p - 1|k))$, and let an optimum control input series that stabilizes the system in the best way according to the value of the objective function while not violating any constraints be denoted $\tilde{u}^o(k)$, with an associated objective function value $V^o(k)$. An optimization routine varies the control input over some *control horizon* H_u, where $H_u \leq H_p$, to minimize the objective function, and the inputs for $(k + H_u + 1, \ldots, k + H_p - 1)$ are fixed at $u(H_u)$ in the optimization. For simplicity, let $H_u = H_p$ in the remainder of this section.

The optimization will keep the input and state vectors within any existing constraints that may apply, and generates an open-loop prediction of system behavior. However, the MPC algorithm is fundamentally a feedback controller, so only the first element of $\tilde{u}^o(k)$ is applied in practice. At the next time interval, the state is remeasured, providing feedback that closes the loop, and the process is repeated.

For our discrete-time nonlinear system, we assume the following.

Assumption 11.4.1   *The objective function $V(k)$ is C^2, with $V \geq 0$ and $V(k) = 0$ only when the system error $x_e(k)$ is zero.*

Assumption 11.4.2   *At every calculation interval k, the optimizing input series $\tilde{u}^o(k)$ is determined in negligible time.*

Stability can be achieved if we can require the constraint that $\tilde{u}^o(k)$ stabilizes the system to zero error in H_p steps or fewer via some feasible path. However, the optimization problem may be difficult or impossible to solve in real systems. Instead, relax the definition of $\tilde{u}(k)$ so that it drives the system to some *terminal constraint set* $\mathcal{X}_f$. Here we allow $x_e(k + H_p|k) \in \mathcal{X}_f \subset \mathcal{X}$. The terminal constraint must be closed and compact, and must contain $x_e = \mathbf{0}$. The controller that uses $\mathcal{X}_f$ instead of the origin as a terminal constraint may certainly impose a much smaller computational burden at each time interval, but it is not stabilizing because it only drives the system to a region around the origin. Whenever $x_e(k) \in \mathcal{X}_f$, control switches to a local stabilizing controller $\mathcal{K}(x)$ to drive the system to zero error. Thus, this is a *dual-mode model predictive controller*.

Assumption 11.4.3   *No collisions will occur for any $x_e \in \mathcal{X}_f$. The leader velocity $v_i > 0$ is required for the local controller to be asymptotically stable.*

In addition to providing a region in which the local controller is asymptotically stable, the terminal constraint must be small enough to effectively prevent collisions for any $x_e \in \mathcal{X}_f$.

The objective function at discrete time k is

$$V(k) = V_{\text{pos}}(k) + V_{\text{input}}(k) + V_{\text{col}}(k) \tag{11.16}$$

where $V_{\text{pos}}(k)$ is a position (or state) error cost term, $V_{\text{input}}(k)$ is an input power cost term, and $V_{\text{col}}(k)$ is a collision cost term. The position error cost term is defined by

$$V_{\text{pos}}(k) = \sum_{m=1}^{H_p} x_e^{\text{T}}(k + m|k)Qx_e(k + m|k) \tag{11.17}$$

and here we will use $Q = \text{diag}[q_1 \; q_2 \; q_3]$ with scalar weights $q_i > 0$. Some cost is associated with the input effort according to the term $V_{\text{input}}(k)$, where

$$V_{\text{input}}(k) = \sum_{m=1}^{H_p} u_e^{\text{T}}(k - 1 + m|k)Ru_e(k - 1 + m|k) \tag{11.18}$$

where $R = \text{diag}[r_1 \; r_2]$ with scalar weights $r_i > 0$.

A collision cost term will penalize trajectories which result in inter-robot collisions. Given some minimum inter-robot separation r_{min}, separation distance is defined as

$$c_{i,j}(k + m|k) = \|x_i(k + m|k) - x_j(k + m|k)\|_2 - r_{\text{min}} \tag{11.19}$$

where $c_{i,j}(k + m|k) \geq 0$ must be maintained to prevent a collision between robots i and j. Ideally, the objective function should penalize any trajectory having some $c_{i,j}(k + m|k) < 0$ quite heavily, while having virtually no cost for $c_{i,j}(k + m|k) \geq 0$. Thus, the nonlinear cost term

$$V_{\text{col}}(k) = \sum_{m=1}^{H_p} e^{-c_{i,j}(k+m|k)/\tau} \tag{11.20}$$

is used, with $0 < \tau \ll 1$. This cost term can correspond to an undefined (infinite in magnitude) artificial potential for $c_{i,j}(k + m|k) < 0$ (cf. [23]). It can result in an objective function that is not strictly increasing with $\|x_e(k)\|_2$ and $\|u_e(k)\|_2$ if trajectories pass through areas where $c_{i,j}(k + m|k) < 0$. However, for trajectories that start in the feasible region and controllers that have sufficient power available for stabilization, it effectively prevents collisions.

A constant prediction horizon is chosen to be long enough to ensure that the trajectory can enter $\mathcal{X}_f$, and a local stabilizing control law is applied in a manner similar to that of Reference 40. Let $\mathcal{X}_p \subset \mathcal{X}$ be the closed, compact set of states of x_e that can be driven to some terminal constraint set $\mathcal{X}_f$ in H_p steps or fewer. Thus, $\mathcal{X}_p$ is a kind of region of convergence of x_e to $\mathcal{X}_f$, and it is defined so that $\mathcal{X}_p \cap \mathcal{X}_f = \varnothing$. The determination of the required H_p depends on input limits; in the kinematic sense, the motion of the robot is limited by the available maximum absolute values of v_j and ω_j. More available input power means that a shorter H_p can be used for a given $\mathcal{X}_p$.

Assumption 11.4.4 *For any $x_e(k) \in \mathcal{X}_p$, the prediction horizon will be long enough so that there exists a feasible trajectory that does not cause a collision. This is equivalent to having at least one set $\tilde{u}^o(k)$ such that $V_{\text{col}}(k) \to 0$.*

When the collision cost term is thus defined, the cost term $V_{\text{col}}(k)$ is not finite for every $x_e \in \mathcal{X}_p$, and trajectories that pass through the region for which $V_{\text{col}}(k) \to \infty$ have no feasible solution of the optimization problem which is intended to determine $\tilde{u}^o(k)$. If the only trajectories which meet the terminal constraint $\mathcal{X}_f$ pass through such a region, then no feasible $\tilde{u}^o(k)$ exists and the optimization is not tractable.

Assumption 11.4.5 *For each increment there is some function $\lambda(\cdot)$ such that $\rho(x_e(k+m|k), u_e(k+m-1)) \geq \lambda(\|[x_e^T(k+m|k) \ u_e^T(k-1+m)]^T\|_2) > 0$ $\forall x_e \in \mathcal{X}_p$. Furthermore, it is required that $\rho(x_e(k), u_e(k-1)) = 0 \ \forall x_e(k) \in \mathcal{X}_f$.*

The MPC problem is constrained by the kinematics of the robot and any bounds on x or u that may exist, particularly an upper bound on the magnitude of u. With these constraints the dual-mode MPC problem is then

$$\text{minimize} \quad V(k)$$
$$\text{with respect to} \quad \tilde{u}(k)$$
$$\text{such that}$$
$$x(k+1+m|k) = f(x(k+m|k), u(k+m|k)) \quad (11.21)$$
$$u_{\min} \leq u(k+m|k) \leq u_{\max}$$
$$\text{for } m \in \{0, 1, \ldots, (H_p - 1)\}$$
$$x_e(k+H_p|k) \in \mathcal{X}_f$$

Given the above definitions and assumptions, one can show that any state within $\mathcal{X}_p$ can be driven to $\mathcal{X}_f$ in finite time, from which the system is stable.

For a control task with a positive-velocity leader, the proper choice of H_p allows the controlled robot to find the proper orientation under dual-mode MPC

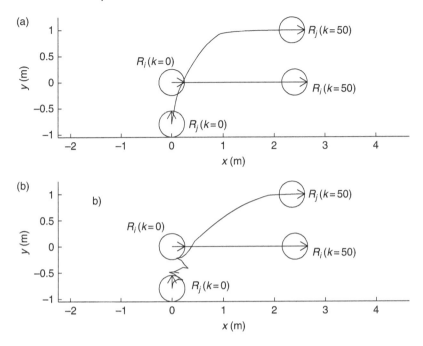

FIGURE 11.6 The dual-mode MPC algorithm with $V_{col} = 0$ results in a collision in (a), whereas with V_{col} as in Equation (11.20), collision is avoided in (b).

while avoiding collisions, as shown in Figure 11.6. In Figure 11.6a, the collision cost term has been disabled, whereas in Figure 11.6b it effectively prevents a collision of the two robots (while delaying recovery). Steady-state operation is under the local leader–follower controller. Note that from the initial condition shown, the recovery of the system under local control would not result in a collision with the leader robot. However, under local control collisions with obstacles and robots other than the leader can only be prevented by redefining pseudo-leaders as shown in Reference 33. The MPC algorithm offers a less cumbersome method to ensure that collisions are avoided.

11.4.1 Control of a Chain of Robots

Here the general characteristics of a formation of N robots are considered, where robot R_1 moves independently and robots R_i, $i \in 2, \ldots, N$ each follow R_{i-1}. This uses $N - 1$ dual-mode MPC algorithms as described earlier, assuming the robots sense one another within some radius but do not share information (i.e., internal calculations) explicitly. This form of architecture represents one possible method for control of teams of robots, where global information is not available to a centralized controller, but onboard sensors with a reasonable degree of precision and accuracy are presumed. This is similar to the distributed MPC problem described by Reference 41, where the overall formation control is formulated as a series of subproblems; in this case the subproblem for R_i is much smaller than the overall formation control problem and coupled only to subproblems for R_{i-1} and R_{i+1}, for $1 < i < N$. Information, in the form of sensor readings giving the current state and kinematic input for R_{i-1}, is exchanged only once per time interval, and each dual-mode MPC controller assumes behavior (specifically, constant kinematic input for the duration of the prediction horizon) for R_{i-1}. Eventually, each controller may be programmed to react to obstacles as well as other robots within its sensing radius.

For a chain of five robots $\{R_1, \dots, R_5\}$, with $\{R_2, \dots, R_5\}$ under dual-mode MPC, the performance of this system depends in part upon how much information is available to the controller. For example, if an infinite sensing radius for every robot is assumed, then they could all follow R_1 directly, but this scenario is unlikely in practice. Instead, the assumption is that each robot R_i can "see" only far enough to sense robot R_{i-1}; this distance is assumed to be large enough so that any other potentially dangerous robots or obstacles that may collide with R_i will also be sensed, allowing future implementation of some obstacle avoidance. From a practical standpoint, it could be difficult for R_i to properly sense the kinematic velocities of R_{i-1}. The response of R_i to information about R_{i-1} does not occur until the next time segment, so there is a one time unit delay for every follower robot in the chain. Figure 11.7 depicts a five-robot system moving in formation. In this case the entire chain is asymptotically stable, in the sense that $\|x_s(k)\|_2 \to 0$ asymptotically as $k \to \infty$, where $x_s(k)$ is the total error vector. Here $x_s(k) = [x_{e1}^T(k)\, x_{e2}^T(k) \cdots x_{eN}^T(k)]^T$, $x_{ej} = x_j(k) - x_{gj}(k)$ for $j = 1, \dots, N$, and $x_{gj}(k)$ is the desired position of robot j if there were no error anywhere in the formation.

Although a basic MPC formulation may seem quite intuitive, using a seemingly commonsense formulation without analyzing stability can lead to divergent responses. It may not be difficult to prove stability for some infinite-horizon control laws, or for MPC algorithms that require a zero-state constraint at the end of the horizon at each step, but these approaches often do not lead to tractable problems. With shorter horizons that can be handled computationally, a stabilizing solution may not exist. Much of the recent research in MPC has centered in analyzing stability using Lyapunov analysis centered on appropriate terminal costs or constraints (see, for example, Reference 42 for a good survey of recent MPC research).

Stability of an MPC algorithm may be ensured by setting requirements on the prediction horizon, the terminal constraint set, the terminal cost, or some combination of these factors. The first and easiest way to ensure stability is to require some minimum prediction horizon length; in fact it is easy to prove that if $H_p \to \infty$ and V has the properties of a Lyapunov function then the algorithm

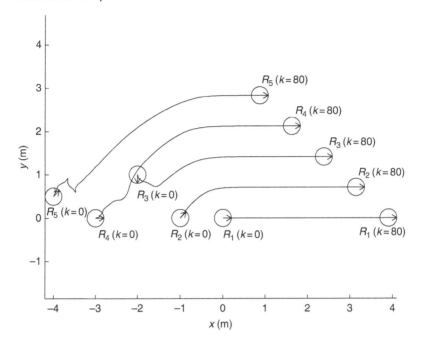

FIGURE 11.7 Stability is achieved when R_1 has $v_1 > 0$, and the local stabilizing controller can be implemented for each controller in the chain.

is stable for an unconstrained plant. However, this approach is as unsuitable as it is trivial for real applications. Stability of the system can be assured in some sense if we use a prediction horizon that is long enough to drive the system to a region small enough that from it the system can be driven to equilibrium using some local feedback controller [39].

Some key limitations of the input–output feedback linearization controller include limited regions of stability, lack of robustness to unmodeled kinematics and dynamics, and the inability to incorporate higher level formation-control heuristics. These are all problems that can be solved, or at least moderated, by using some form of model predictive control, provided some computational difficulties do not preclude feasible solutions. Using chains of dual-mode model predictive controllers may show promise for larger formations of mobile robots, provided certain assumptions can be satisfied.

The previous sections present a framework for formation control. Formation control can be seen as a tool that a higher-level mission planner (see Figure 11.3b) might use in order to perform a complex task. In Section 11.5, two real-world cooperative multi-robot systems that require formation control, inter-robot communication, decision-logic, and optimization are described.

11.5 Applications

In this section, we present two cooperative multi-robot applications. First, a team of robots is supposed to transport an object and cooperate through a dynamic role assignment mechanism. Second, a network of cooperative mobile sensors is employed to detect and track the perimeter defined by a certain substance (e.g., chemical spill). We show that both cooperative tasks can be modeled as hierarchical hybrid systems where robots need to switch between different behaviors/modes/controllers in order to accomplish the task.

11.5.1 Cooperative Manipulation

In a cooperative manipulation task as depicted in Figure 11.1, robots cooperate to carry a large object in an environment containing static and dynamic obstacles. Cooperative manipulation is a classical example of a tightly coupled task because it cannot be performed by a single robot working alone and requires a tight coordination to grasp and transport objects without dropping them.

Here, we describe a paradigm for coordinating multiple robots in the execution of cooperative manipulation tasks. The basic idea is to assign to each robot in the team a *role* that determines its actions during the cooperation. By dynamically assuming and exchanging roles in a synchronized manner, the robots are able to perform cooperative tasks, adapting to unexpected events in the environment and improving their individual performance in benefit of the team. Basically, each role can be viewed as a behavior or a reactive controller. However, more generally, roles may define more elaborate functions of the robot state and on information about the environment and other robots including the history of these variables, and may encapsulate several behaviors or controllers. It not only dictates what controllers are used and how the state of a robot changes, but also how information flows between robots. The reader is referred to Reference 43 for a more detailed discussion.

A hybrid system framework is used to model dynamic role assignment between multiple robots in this application. Hybrid systems explicitly capture the discrete and continuous dynamics in a unified framework, allowing us to model the interaction of these two types of dynamics.

11.5.1.1 Dynamic role assignment

In general, to execute cooperative tasks a team of robots must be coordinated: they have to synchronize their actions and exchange information. In this approach, each robot performs a role that determines its actions during the cooperative task. According to its internal state and information about the other robots and the task received through communication, a robot can dynamically change its role, adapting itself to changes and unexpected events in the environment. The mechanism for coordination is completely decentralized. Each robot has its own controllers and takes its own decisions based on local and global information. In general, each team member has to explicitly communicate with other robots to gather information but they normally need not construct a complete global state of the system for the cooperative execution. We consider that each team member has a specification of the possible actions that should be performed during each phase of the cooperation in order to complete the task. These actions must be specified and synchronized considering several aspects, such as robot properties, task requirements, and characteristics of the environment. The dynamic role assignment will be responsible for allocating the correct actions to each robot and synchronizing the cooperative execution.

Before describing in detail the role assignment mechanism, it is necessary to define what a role in a cooperative task is. Webster's Dictionary defines it as follows:

Definition 11.5.1 *(a) Role is a function or part performed especially in a particular operation or process and (b) role is a socially expected behavior pattern usually determined by an individual's status in a particular society.*

Here, a role is defined as a function that one or more robots perform during the execution of a cooperative task. Each robot will be performing a role while certain internal and external conditions are satisfied, and will assume another role otherwise. The role will define the behavior of the robot in that moment, including the set of controllers used by the robot, the information it sends and receives, and how it will react in the presence of dynamical and unexpected events.

The role assignment mechanism allows the robots to change their roles dynamically during the execution of the task, adapting their actions according to the information they have about the system and the task. Basically, there are three ways of changing roles during the execution of a cooperative task: the simplest way is the *Allocation*, in which a robot assumes a new role after finishing the execution of another role. In the *Reallocation* process, a robot interrupts the performance of one role and starts or continues the performance of another role. Finally, robots can *Exchange* their roles. In this case, two or more robots synchronize themselves and exchange their roles, each one assuming the role of one of the others.

The role assignment mechanism depends directly on the information the robots have about the task, the environment, and about their teammates. Part of this information, mainly the information concerning the task, is obtained a priori, before the start of the execution. The control software for each robot includes programs for each role it can assume. However, the definition of the task includes an a priori specification of the roles it can assume during the execution of the task and the conditions under which the role is reassigned or exchanged. The rest of the information used by the robots is obtained dynamically during the task execution and is composed by local and global parts. The local information consists of the robot's internal state and its perception about the environment. Global information contains data about the other robots and their view of the system and is normally received through explicit communication (message passing). A key issue is to determine the amount of global and local information necessary for the role assignment. This depends on the type of task being performed. Tightly coupled tasks require a higher level of coordination and consequently a greater amount of information exchange. On the other hand, robots executing loosely coupled tasks normally do not need much global information because they can act more independently from each other.

This approach allows for two types of explicit communication. In synchronous communication, the messages are sent and received at a constant rate, while in asynchronous communication an interruption is generated when a message is received. Asynchronous communication is used to broadcast unexpected events such as the presence of obstacles.

It is important to define when a robot should change its role. In the role allocation process, the robot detects that it has finished its role and assumes another available role. The possible role transitions are defined a priori and are modeled using a hybrid automaton as will be explained later. In the reallocation process, the robots should know when to relinquish the current role and assume another. A possible way to do that is to use a function that measures the utility of performing a given role. A robot performing a role r has a utility given by μ_r. When a new role r' is available, the robot computes the utility of executing the new role $\mu_{r'}$. If the difference between the utilities is greater than a threshold τ ($\mu_{r'} - \mu_r > \tau$) the robot changes its role. The function μ can be computed based on local and global information and may be different for distinct robots, tasks, and roles. Also, the value τ must be chosen such that the possible overhead of changing roles will be compensated by a substantial gain on the utility and consequently a better overall performance. It is also possible for two robots to exchange their roles. In this case, one robot assumes the role of the other. For this, the robots must agree to exchange roles and should synchronize the process, which is done using communication.

11.5.1.2 Modeling

The dynamic role assignment can be described and modeled in a more formal framework. In general, a cooperative multi-robot system can be described by its state (X), which is a concatenation of the states of the individual robots:

$$X = [x_1, x_2, \ldots, x_N]^{\mathrm{T}} \tag{11.22}$$

Considering a simple control system, the state of each robot varies as a function of its continuous state (x_i) and the input vector (u_i). Also, each robot may receive information about the rest of the system ($\hat{z}_i$) that can be used in the controller. This information consists of estimates of the state of the other robots that are received mainly through communication. We use the hat ($\hat{}$) notation to emphasize that this information is an estimate because the communication can suffer delays, failures, etc. Using the role assignment mechanism, in each moment each robot will be controlled by a different continuous equation according to its current role in the task. Therefore, we use the subscript q, $q = 1, \ldots, S$, to indicate the current role of the robot. Following this description, the state equation of each robot i, $i = 1, \ldots, N$, during the execution of the task can be defined as:

$$\dot{x}_i = f_{i,q}(x_i, u_i, \hat{z}_i) \tag{11.23}$$

Since each robot is associated with a control policy,

$$u_i = g_{i,q}(x_i, \hat{z}_i) \tag{11.24}$$

and since $\hat{z}_i$ is a function of the state X, we can rewrite the state equation:

$$\dot{x}_i = f_{i,q}(X) \tag{11.25}$$

or, for the whole team,

$$\dot{X} = F_\Sigma(X), \quad \text{where } F_\Sigma = [f_{1,q_1}, \ldots, f_{N,q_N}]^{\mathrm{T}}, \quad q_i \in \{1, \ldots, S\} \tag{11.26}$$

The equations shown above model the continuous behavior of each robot and consequently the continuous behavior of the team during the execution of a cooperative task. These equations, together with the roles, role assignments, variables, communication, and synchronization can be better understood and formally modeled using a hybrid automaton.

A hybrid automaton is a finite automaton augmented with a finite number of real-valued variables that change continuously, as specified by differential equations and inequalities, or discretely, according to specific assignments. It is used to describe hybrid systems, that is, systems that are composed by discrete and continuous states. A hybrid automaton H can be defined as: $H = \{Q, V, E, f, \text{Inv}, G, \text{Init}, R\}$. $Q = \{1, 2, \ldots, S\}$ is the set of discrete states, also called *control modes*. The set V represents the variables of the system and can be composed by discrete (V_d) and continuous (V_c) variables: $V = V_d \cup V_c$. Each variable $v \in V$ has a value that is given by a function $v(v)$. This is called *valuation* of the variables. Thus, the state of the system is given by a pair (q, v), composed by the discrete state $q \in Q$ and the valuation of the variables. The dynamics of the continuous variables are determined by the flows f, generally described as differential equations inside each control mode (f_q). Discrete transitions between pairs of control modes (p, q) are specified by the control switches E (also called *edges*). Invariants (Inv) and guards (G) are predicates related to the control modes and control switches respectively. The system can stay in a certain control mode while its invariant is satisfied, and can take a control switch when its guard (jump condition) is satisfied. The initial states of the system are given by Init, and each control switch can also have a reset statement R associated, to change the value of some variable during a discrete transition.

In this model, each role is a control mode of the hybrid automaton. Internal states and sensory information within each mode can be specified by continuous and discrete variables of the automaton. The variables are updated according to the equations inside each control mode (flows) and reset statements of each discrete transition. The role assignment is represented by discrete transitions and the invariants and guards define when each robot will assume a new role. Finally, we can model the cooperative task execution using a parallel composition of several automata as described in Reference 44.

Communication (e.g., message passing) among robots can also be modeled in this framework. To model this message passing in a hybrid automaton, we consider that there are communication channels between agents and use the basic operations send and receive to manipulate messages. In the hybrid automaton, messages are sent and received in discrete transitions. These actions are modeled in the same way as assignments of values to variables (reset statements). It is very common to use a self-transition, that is, a transition that does not change the discrete state, to receive and send messages.

The execution of the cooperative manipulation uses a leader–follower architecture [45]. One robot is identified as a leader, while the others are designated as followers. The assigned leader has a planner and broadcasts its estimated position and velocity to all the followers using asynchronous messages. Each follower has its own trajectory controller that acts in order to cooperate with the leader. The planner and the trajectory controllers send set points to the low level controllers that are responsible for the actuators. All robots have a coordination module that controls the cooperative execution of the task. This module receives information from the sensors and exchanges synchronous messages with the other robots. It is responsible for the role assignment and for other decisions that directly affect the planners and trajectory controllers.

In this cooperative manipulation task, the robots can be executing one of the following roles: *Dock*, where they must coordinate themselves to approach and pick up the box; and *Transport*, where they march in a coordinated fashion. The Transport role is obtained by composing the roles Lead and Follow using *sequential composition*. Thus, a robot transporting a box will be performing either a leader or a follower role. The control modes of the robots' automata

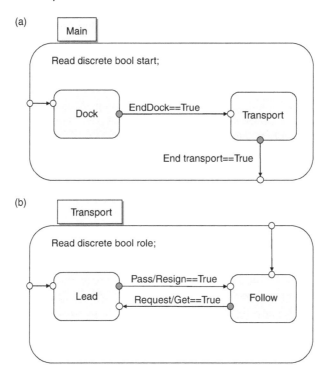

FIGURE 11.8 Robot's roles in the execution of a cooperative manipulation task.

are shown in Figure 11.8. The role assignment is used here mainly to exchange the leadership responsibilities among the robots: at any moment during transportation, the robot performing the leading role can become a follower, and any follower can take over the leadership of the team. One reason for possibly exchanging leadership is when one of the followers is better suited to be the leader during the execution of the task. For example, when the leader's sensors are occluded or the follower is better positioned to avoid an obstacle. The role assignment is also used for synchronizing actions, in such a way that the robots are able to go from the dock role to the transport role in a coordinated manner.

Different controllers and planners are used by each robot depending on its role in the task. In the *Dock* mode, robots use a proportional feedback controller based on the distance to the object to move in order to grasp the object. In the *Transport* mode, the robots have different behaviors when leading or following. In the *Lead* mode, they are controlled by planners that send set-points to the actuators. In the *following* mode, the controllers are designed to enable the robots to follow a trajectory that is compatible with the leader's in order to follow and cooperate with the leader [45,46].

As mentioned, the main purpose of the leadership exchange mechanism used here is to allow the robots to react and adapt easily to unexpected events such as obstacle detection and sensor failures. It is also important to assign the leadership to the appropriate robot in such a way that, in each phase of the cooperation, the robot that is best suited in terms of sensory power and manipulation capabilities will be leading the group. A method for executing the leadership exchange under the role assignment paradigm is as follows. One of the follower robots sends a message requesting the leadership. This normally happens when one of the robots is not able to follow the leader's plan or knows a better way to lead the group in that moment or both. For example, if one of the followers detects an obstacle, it can request the leadership, avoid the obstacle, and then return the leadership to the previous leader.

The above approach is illustrated in simulation using four holonomic robots that cooperate in order to carry an object from an initial position to the goal. In this experiment, one of the robots requests the leadership when it senses that it will not be able to follow the path determined by the leader. Figure 11.9 shows snapshots of the simulator during the task execution.

Snapshot (a) in Figure 11.9 shows the robots performing the Dock role. The robots are represented by circles and the object to be carried is the square in the middle of them. Each robot has a sensing area represented by a dashed circle around it. The other rectangles on the environment are obstacles and the goal is marked with a small x on the right of the figure. When they finish docking, they are allocated to the Transport role and there is a leadership exchange to avoid the obstacle in the top (snapshot [b]). As it can be seen the robots are able to transport the object to the goal position (snapshot [c]) while avoiding collisions.

11.5.2 Multi-Robot Perimeter Detection and Tracking

Many applications of multi-robot cooperation have been studied including area coverage, search and rescue, manipulation, exploration and mapping, and perimeter detection [47–49]. A perimeter is an area enclosing some type of substance. We consider two types of perimeters (1) static and (2) dynamic. A static perimeter (e.g., a minefield) does not change over time. On the other hand, dynamic perimeters (e.g., a radiation leak) are time-varying and expand/contract over time. Perimeter detection has a wide range of uses in several areas, including (1) Military (e.g., locating minefields or surrounding a target), (2) Nuclear/Chemical industries (e.g., tracking radiation/chemical spills), (3) Oceans (e.g., tracking oil spills), and (4) Space (e.g., planetary exploration). In many cases, humans are used to perform these dull and/or dangerous tasks, but if robotic swarms could replace humans, it could be beneficial.

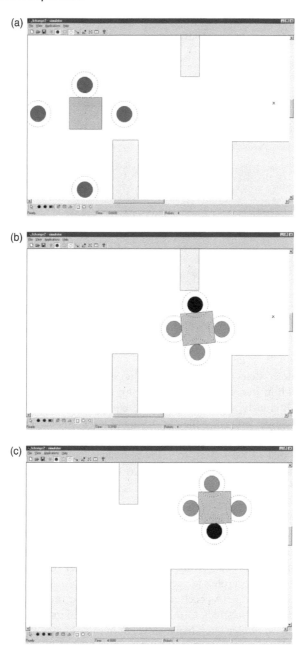

FIGURE 11.9 Cooperative manipulation with four holonomic robots.

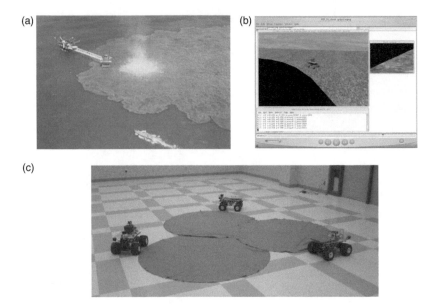

FIGURE 11.10 (a) Example perimeter: oil spill, (b) simulation environment, and (c) experimental testbed.

In perimeter detection tasks, a robotic swarm locates and surrounds a substance, while dynamically reconfiguring as additional robots locate the perimeter. Obviously, the robots must be equipped with sensors capable of detecting whatever substance they are trying to track. Substances could be airborne, ground-based, or underwater. If the perimeter moves with a velocity greater than that of the robots, then the perimeter cannot be tracked. Abrupt perimeter changes requiring sharp turns may be difficult to track because of the robots' limited turning radius. See Figure 11.10a for an example of a perimeter, an oil spill.[2]

In this section, a decentralized, cooperative hybrid system is presented utilizing biologically inspired emergent behavior [50]. Each controller is composed of finite state machines and it is assumed that the robots have a suite of sensors and can communicate only within a certain range. A relay communication scheme is used. Once a robot locates the perimeter, it broadcasts the location to any robots within range. As each robot receives the perimeter location, it also begins broadcasting, in effect, forming a relay. Other groups have used the terms *perimeter* and *boundary* interchangeably, but in this chapter, there is a distinct difference. The perimeter is the *chemical substance* being tracked, while the *boundary* is the limit of the exploration area.

---

[2] Courtesy of the NOAA Office of Response and Restoration.

11.5.2.1 Cooperative hybrid controller

In one of our previous work reported in Reference 9, we developed an object-oriented software architecture that supports hierarchical composition of robot agents and behaviors or modes. Key features of the software architecture are summarized below:

- *Architectural hierarchy.* The building block for describing the system architecture is an *agent* that communicates with its environment via shared variables and also communication channels. In this application, the team of mobile sensors defines the *group* agent. The group agent receives information about the area, that is, boundary where the perimeter is located.
- *Behavioral hierarchy.* The building block for describing a flow of control inside an agent is a *mode*. A mode is basically a hierarchical state machine, that is, a mode can have submodes and transitions connecting them. Modes can be connected to each other through entry and exit points. We allow the instantiation of modes so that the same mode definition can be reused in multiple contexts.
- *Discrete and continuous variable updates.* Discrete updates are specified by *guards* labeling transitions connecting the modes. Such updates correspond to mode-switching, and are allowed to modify variables through assignment statements.

The cooperative hybrid systems is modeled by (11.25) and (11.26) as in the manipulation task. Furthermore, the overall finite automaton consists of three states (1) *Random Coverage Controller* (RCC), (2) *Potential Field Controller* (PFC), and (3) *Tracking Controller* (TC). These three controllers are composed such that the sensor/robot network is able to locate and track a perimeter. See Figure 11.11 for hierarchical automata of the cooperative hybrid system described herein.

In Section 11.5.2.2, details of the controller agents are presented.

11.5.2.2 Random coverage controller

The goal of the RCC is to efficiently cover as large an area as possible while searching for the perimeter and avoiding collisions. The robots move fast in this state to quickly locate the perimeter. The RCC consists of three states (1) *spiral search*, (2) *boundary avoidance*, and (3) *collision avoidance*. The spiral search is a random search for effectively covering the area. The boundary and collisions are avoided by adjusting the angular velocity.

The logarithmic spiral, seen in many instances in nature, is used for the search pattern. In Reference 51, a spiral search pattern such as that used by

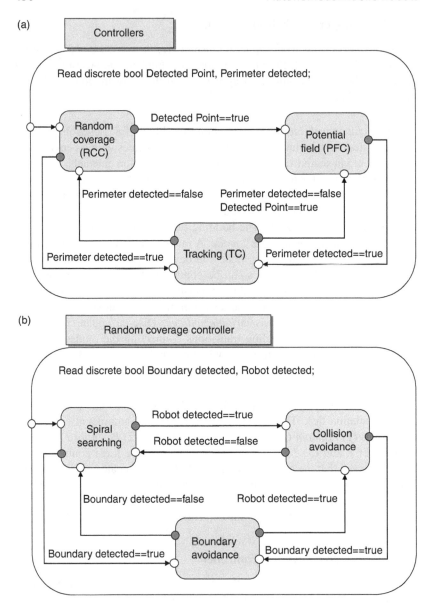

FIGURE 11.11   Hierarchical finite automata for perimeter detection and tracking.

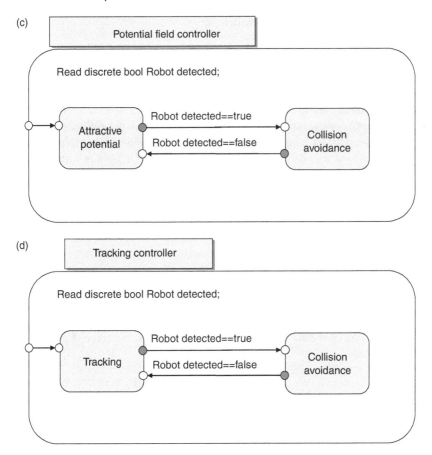

(c) Potential field controller

Read discrete bool Robot detected;

Attractive potential

Robot detected==true

Robot detected==false

Collision avoidance

(d) Tracking controller

Read discrete bool Robot detected;

Tracking

Robot detected==true

Robot detected==false

Collision avoidance

FIGURE 11.11   Continued.

moths is utilized for searching an area. It has been shown that the spiral search is not optimal, but effective [52]. Some examples are hawks approaching prey, insects moving toward a light source, sea shells, spider webs, and so forth.

$$v_i = v_s(1 - e^{-t}) \tag{11.27}$$

$$\omega_i = ae^{b\theta_i} \tag{11.28}$$

where v_s is a positive constant, a is a constant, and $b > 0$. If $a > 0$ (<0), then the robots move counterclockwise (clockwise). Collision and *boundary* (limit of the exploration area here) avoidance are handled in simulation by sharply turning, while in experiments, the robots back up and turn, then go forward.

11.5.2.3 Potential field controller

Potential fields have been used by a number of groups for controlling a swarm [47,49,53]. In Reference 53, a method using artificial potentials and virtual bodies is shown in which the robot network forms regular polygons upon uniformly surrounding a target. In Reference 49, virtual potential fields and graph theory are used for area coverage.

The PFC uses an attractive potential which allows the robots to quickly move to the perimeter once it has been detected. The first robot to detect the perimeter *broadcasts* its location to the other robots. If a robot is within range, then the PFC is used to quickly move to the perimeter. Otherwise, the robot will continue to use the RCC unless it comes within range, at which point it will switch to the PFC. As a robot moves towards the goal, if it detects the perimeter before it reaches the goal, it will switch to the TC.

The PFC has two states (1) *attractive potential* and (2) *collision avoidance*. The attractive potential, $\mathbf{P_a}(x_i, y_i)$, is:

$$\mathbf{P_a}(x_i, y_i) = \tfrac{1}{2}\epsilon[(x_i - x_g)^2 + (y_i - y_g)^2] \tag{11.29}$$

where (x_i, y_i) is the position of robot i, ϵ is a positive constant, and (x_g, y_g) is the position of the attractive point (goal). The attractive force, $\mathbf{F_a}(x_i, y_i)$, is derived below:

$$\mathbf{F_a}(x_i, y_i) = -\nabla \mathbf{P_a}(x_i, y_i) = - \begin{bmatrix} \dfrac{\partial \mathbf{P_a}}{\partial x_i} \\ \dfrac{\partial \mathbf{P_a}}{\partial y_i} \end{bmatrix}$$

$$\mathbf{F_a}(x_i, y_i) = \epsilon \begin{bmatrix} x_g - x_i \\ y_g - y_i \end{bmatrix} = \begin{bmatrix} F_{a,x_i} \\ F_{a,y_i} \end{bmatrix} \tag{11.30}$$

Equation (11.30) is used to get the desired orientation angle, $\theta_{i,d}$, of robot i:

$$\theta_{i,d} = \arctan 2(F_{a,y_i}, F_{a,x_i}) \tag{11.31}$$

Depending on θ_i and $\theta_{i,d}$, the robot will turn the optimal direction to quickly line up with the goal using the following proportional angular velocity controller:

$$\omega_i = \pm k \,(\theta_{i,d} - \theta_i) \tag{11.32}$$

where $k = \omega_{\max}/2\pi$ and $\omega_{\max} = 0.3$ rad/sec and θ_i is the orientation angle of robot i. Collisions are avoided in the same manner as in the RCC.

11.5.2.4 Tracking controller

The TC changes ω and v in order to track the perimeter and avoid collisions, respectively. Cyclic behavior *emerges* as multiple robots track the perimeter. The robots' goal in this state is to accurately track the perimeter counterclockwise. The TC consists of two states (1) *tracking*, and (2) *collision avoidance*. Tracking is accomplished by adjusting the robots' angular velocity. On the other hand, collisions are avoided by changing the linear velocity.

A vision sensor (blobfinder algorithm) is being used to detect the perimeter. Smooth tracking is accomplished with the following proportional angular velocity controller:

$$\omega_i = k_P \left(\gamma_o - \gamma_i \right) \qquad (11.33)$$

where $k_P > 0$, and γ_o and γ_i are the areas outside and inside the perimeter seen by the blobfinder, respectively. Counterclockwise tracking is assumed, which implies that the robot will turn left (right) if the robot is too far outside (inside) the perimeter.

An experiment is shown in Figure 11.10c in which three robots search for, locate, and track a perimeter while avoiding collisions. Refer to Figure 11.12a and b for trajectory and state transitions plots, respectively. Collision avoidance is accomplished through the use of IR sensors while position/orientation information comes from the encoders. Notice in Figure 11.12a that the perimeter is not exactly like the perimeter in Figure 11.10c, but it is fairly accurate and allows the user to estimate the location of the substance.

11.6 CONCLUSIONS

Recent advances in communication, computation, and embedded technologies are enabling a growing interest in developing cooperative multi-robot systems. In the near future, small, affordable mobile robots equipped with embedded sensors and processors will be able to cooperatively execute tasks within unknown, dynamic environments with limited human intervention.

In this chapter, we have presented a set of tools and methodologies that are suitable for the analysis and design of multi-robot systems engaged in cooperative tasks that require coordinated operation and execution, formation of spatial patterns, mobility, and distributed sensing, computation, and actuation. Specifically, these tools include graph theory, distributed optimization, formation control, and hierarchical hybrid systems.

Additionally, two real-world cooperative examples are given (i) A role assignment paradigm for coordinating multiple robots in the execution of manipulation tasks, and (ii) a cooperative mobile sensor network for perimeter

(a)

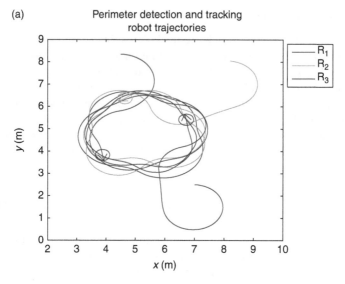

(b)

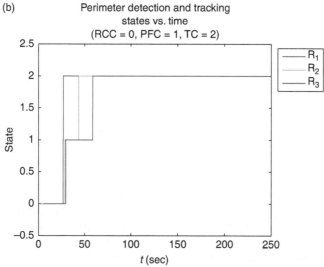

FIGURE 11.12 (a) Three robots defining a perimeter and (b) discrete state transition plot.

detection and tracking. We show that both systems can be modeled as hierarchical hybrid systems.

We anticipate that, given the increasing interest in applications of multi-robot systems, we will witness significant developments in this field. Advances in distributed and hierarchical optimization-based algorithms, highly reconfigurable hardware for example, FPGA's, communication protocols, multi-robot learning, and mobile sensor networks will positively impact the development of cooperative robotic systems.

An important topic to be addressed is the definition of suitable performance metrics to evaluate the efficiency and performance of mechanisms of cooperation. Optimization-based approaches hold the promise to provide efficient solutions to some cooperative robotic system problems. Although, some progress has been made in constrained robust optimal control, the design of computationally efficient *online* optimization algorithms for multi-robot systems remains a challenging task.

ACKNOWLEDGMENTS

The work of R. Fierro was partially supported by NSF grants #0311460 and #0348637 and by the U.S. Army Research Office under grant DAAD19-03-1-0142 (through the University of Oklahoma). The work of L. Chaimowicz was partially supported by the National Council for Scientific and Technological Development (CNPq-Brazil). The work of V. Kumar was partially supported by NSF grants CCR02-05336, IIS02-22927, and CNS-0410514.

REFERENCES

1. V. Kumar, N. E. Leonard, and A. S. Morse (eds). *Cooperative Control*, volume 309 of *LNCIS*. Springer-Verlag, Berlin, 2005.
2. T. Balch and R. Arkin. Behavior-based formation control for multi-robotic teams. *IEEE Transactions on Robotics and Automation*, 14: 926–934, 1998.
3. R. Arkin. *Behavior-Based Robotics*. MIT Press, Cambridge, MA, 1998.
4. A. H. Bond and L. Gasser. *Readings in Distributed Artificial Intelligence*. Morgan Kaufmann Publishers, San Francisco, CA, 1988.
5. Y. Liu and K. M. Passino. Biomimicry of social foraging bacteria for distributed optimization: models, principles, and emergent behaviors. *Journal of Optimization Theory and Applications*, 115: 603–628, 2002.
6. A. Jadbabaie, J. Lin, and A. S. Morse. Coordination of groups of mobile autonomous agents using nearest neighbor rules. *IEEE Transactions on Automatic Control*, 48: 988–1001, 2003.
7. L. R. M. Johansson, N. Xiong, and H. I. Christensen. A game theoretic model for management of mobile sensors. In *The Sixth International Conference on Information Fusion*, Queensland, Australia, July 2003.
8. L. Parker. ALLIANCE: an architecture for fault tolerant multirobot cooperation. *IEEE Transactions on Robotics and Automation*, 14: 220–240, 1998.
9. R. Fierro, A. Das, J. Spletzer, J. Esposito, V. Kumar, J. P. Ostrowski, G. Pappas, C. J. Taylor, Y. Hur, R. Alur, I. Lee, G. Grudic, and J. Southall. A framework and architecture for multi-robot coordination. *International Journal of Robotics Research*, 21: 977–995, 2002.

10. Y. U. Cao, A. S. Fukunga, and A. B. Kahng. Cooperative mobile robotics: antecedents and directions. *Autonomous Robots*, 4: 1–23, 1997.
11. B. P. Gerkey and M. J. Matarić. A formal analysis and taxonomy of task allocation in multi-robot systems. *International Journal of Robotics Research*, 23: 939–954, 2004.
12. R. Alami, S. Fleury, M. Herrb, F. Ingrand, and F. Robert. Multi-robot cooperation in the MARTHA project. *IEEE Robotics and Automation Magazine*, 36–47, March 1998.
13. R. Bachmayer and N. E. Leonard. Vehicle networks for gradient descent in a sampled environment. In *Proceedings of IEEE Conference on Decision and Control*, pp. 112–117, Las Vegas, NV, December 10–13, 2002.
14. P. Lima, T. Balch, M. Fujita, R. Rojas, M. Veloso, and H. A. Yanco. Robocup 2001: a report on research issues that surfaced during the competitions and conference. *IEEE Robotics and Automation Magazine*, 9: 20–30, 2002.
15. J. M. Esposito. *Simulation and Control of Hybrid Systems with Applications to Mobile Robotics*. PhD thesis, University of Pennsylvania, Philadelphia, PA, 2002. Mechanical Engineering and Applied Mechanics.
16. M. Žefran, V. Kumar, and C. Croke. On the generation of smooth three-dimensional rigid body motions. *IEEE Transactions on Robotics and Automation*, 14: 576–589, 1998.
17. J. F. Canny. *The Complexity of Robot Motion Planning*. MIT Press, Cambridge, MA, 1988.
18. R. Alur, T. Dang, J. Esposito, R. Fierro, Y. Hur, F. Ivancic, V. Kumar, I. Lee, P. Mishra, G. Pappas, and O. Sokolsky. Hierarchical hybrid modeling of embedded systems. In T. A. Henzinger and C. M. Kirsch (eds), *EMSOFT 2001*, vol. 2211 of *LNCS*, pp. 14–31. Springer-Verlag, Berlin, Heidelberg, 2001.
19. J. S. Baras, X. Tan, and P. Hovareshti. Decentralized control of autonomous vehicles. In *Proceedings of IEEE Conference on Decision and Control*, pp. 1532–1537, Maui, Hawaii, December 2003.
20. C. D. Godsil and G. Royle. *Algebraic Graph Theory*. Graduate Texts in Mathematics. Springer-Verlag, New York, 2001.
21. A. Okubo. Dynamical aspects of animal grouping: swarms, schools, flocks and herds. *Advances in Biophysics*, 22: 1–94, 1986.
22. J. Toner and Y. Tu. Flocks, herds and schools: a quantitative theory of flocking. *Physical Review E*, 58: 4828–4858, 1998.
23. N. E. Leonard and E. Fiorelli. Virtual leaders, artificial potentials and coordinated control of groups. In *Proceedings of IEEE Conference on Decision and Control*, vol. 3, pp. 2968–2973, Orlando, FL, December 2001.
24. C. Belta and V. Kumar. Optimal motion generation for groups of robots: a geometric approach. *ASME Journal of Mechanical Design*, 126: 63–70, 2004.
25. A. K. Das, R. Fierro, V. Kumar, J. P. Ostrowski, J. Spletzer, and C. J. Taylor. A vision-based formation control framework. *IEEE Transactions on Robotics and Automation*, 18: 813–825, 2002.
26. W. Burgard, M. Moors, D. Fox, R. Simmons, and S. Thrun. Collaborative multi-robot exploration. In *Proceedings of IEEE International Conference on Robotics and Automation*, pp. 476–481, San Francisco, CA, April 2000.

27. A. J. Healey. Application of formation control for multi-vehicle robotic minesweeping. In *Proceedings of the IEEE Conference on Decision and Control*, pp. 1497–1502, Orlando, FL, December 2001.

28. J. Spletzer, A. Das, R. Fierro, C. J. Taylor, V. Kumar, and J. P. Ostrowski. Cooperative localization and control for multi-robot manipulation. In *IEEE/RSJ International Conference on Intelligent Robots and Systems*, pp. 631–636, Maui, Hawaii, October 2001.

29. J. P. Desai, J. P. Ostrowski, and V. Kumar. Modeling and control of formations of nonholonomic mobile robots. *IEEE Transactions on Robotics and Automation*, 17: 905–908, 2001.

30. H. Tanner, V. Kumar, and G. Pappas. Leader-to-formation stability. *IEEE Transactions on Robotics and Automation*, 20: 443–455, 2004.

31. G. Lafferriere, A. Williams, J. Caughman, and J. J. P. Veerman. Decentralized control of vehicle formations. *Systems & Control Letters*, 54, 899–910, 2005.

32. A. K. Das, R. Fierro, and V. Kumar. Control graphs for robot networks. In S. Butenko, R. Murphey, and P. Pardalos (eds), *Cooperative Control: Models, Applications and Algorithms*, vol. 1 of *Cooperative Systems*, chapter 4, pp. 55–73. Kluwer Academic Publishers, Dordrecht, 2003.

33. R. Fierro, A. Das, V. Kumar, and J. P. Ostrowski. Hybrid control of formations of robots. In *Proceedings of the IEEE International Conference on Robotics and Automation*, pp. 157–162, Seoul, Korea, May 2001.

34. A. Isidori. *Nonlinear Control Systems*. Springer-Verlag, London, 3rd ed., 1995.

35. J. Cortés, S. Martínez, T. Karatas, and F. Bullo. Coverage control for mobile sensing networks. *IEEE Transactions on Robotics and Automation*, 20: 243–255, 2004.

36. J. Spletzer and R. Fierro. Optimal position strategies for shape changes in robot teams. In *Proceedings of the IEEE International Conference on Robotics and Automation*, pp. 754–759, Barcelona, Spain, April 18–22, 2005.

37. W. B. Dunbar and R. M. Murray. Receding horizon control of multi-vehicle formations: A distributed implementation. In *43rd IEEE Conference on Decision and Control*, vol. 2, pp. 1995–2002, 14–17 Dec.

38. T. Keviczky, F. Borrelli, and G. J. Balas. A study on decentralized receding horizon control for decoupled systems. In *Proceedings of American Control Conference*, vol. 6, pp. 4921–4926, Boston, MA, June 2004.

39. H. Michalska and D. Q. Mayne. Robust receding horizon control of constrained nonlinear systems. *IEEE Transactions on Automatic Control*, 38: 1623–1633, 1993.

40. P. O. M. Scokaert, D. Q. Mayne, and J. B. Rawlings. Suboptimal model predictive control (feasibility implies stability). *IEEE Transactions on Automatic Control*, 44: 648–654, March 1999.

41. E. Camponogara, D. Jia, B. H. Krogh, and S. Talukdar. Distributed model predictive control. *IEEE Control Systems Magazine*, 22: 44–52, 2002.

42. D. Q. Mayne, J. B. Rawings, C. V. Rao, and P. O. M. Scokaert. Constrained model predictive control: stability and optimality. *Automatica*, 36: 789–814, 2000.

43. L. Chaimowicz, V. Kumar, and M. Campos. A paradigm for dynamic coordination of multiple robots. *Autonomous Robots*, 17: 7–21, 2004.
44. L. Chaimowicz. *Dynamic Coordination of Cooperative Robots: A Hybrid System Approach*. PhD thesis, Universidade Federal de Minas Gerais, Brazil, June 2002.
45. L. Chaimowicz, T. Sugar, V. Kumar, and M. Campos. An architecture for tightly coupled multi-robot cooperation. In *Proceedings of the IEEE International Conference on Robotics and Automation*, pp. 2292–2297, 2001.
46. T. Sugar and V. Kumar. Control of cooperating mobile manipulators. *IEEE Transactions on Robotics and Automation*, 18: 94–103, 2002.
47. D. J. Bruemmer, D. D. Dudenhoeffer, M. D. McKay, and M. O. Anderson. A robotic swarm for spill finding and perimeter formation. In *Spectrum 2002*, Reno, NV, USA, August 2002.
48. J. T. Feddema, C. Lewis, and D. A. Schoenwald. Decentralized control of cooperative robotic vehicles: theory and application. *IEEE Transactions on Robotics and Automation*, 18: 852–864, 2002.
49. J. Tan and N. Xi. Peer-to-peer model for the area coverage and cooperative control of mobile sensor networks. In *SPIE Symposium on Defense and Security*, Orlando, FL, April 12–16, 2004.
50. R. Fierro, J. Clark, D. Hougen, and S. Commuri. A multi-robot testbed for bilogically inspired cooperative control. In L. E. Parker, F. E. Schneider, and A. C. Schultz (eds), *Multi-Robot Systems. From Swarms to Intelligent Automata*, vol. III, pp. 171–182, Naval Research Laboratory, Springer, Washington DC, March 14–16, 2005.
51. A. T. Hayes, A. Martinoli, and R. M. Goodman. Distributed odor source localization. *IEEE Sensors*, 2: 260–271, 2002.
52. D. W. Gage. Randomized search strategies with imperfect sensing. In *Proceedings of SPIE Mobile Robots VIII*, vol. 2058, pp. 270–279, Boston, MA, September 1993.
53. P. Ögren, E. Fiorelli, and N. E. Leonard. Cooperative control of mobile sensor networks: adaptive gradient climbing in a distributed environment. *IEEE Transactions on Automatic Control*, 49: 1292–1302, 2004.

BIOGRAPHIES

Rafael Fierro received his M.Sc. degree in control engineering from the University of Bradford, England, and his Ph.D. degree in electrical engineering from the University of Texas at Arlington in 1990 and 1997, respectively. From 1999 to 2001, he held a postdoctoral research appointment with the GRASP Lab, University of Pennsylvania. He is currently an assistant professor in the School of Electrical and Computer Engineering at Oklahoma State University. His research interests include hierarchical hybrid and embedded systems, optimization-based cooperative control, and robotics. Dr. Fierro was the recipient of a Fulbright Scholarship. He was also a finalist in the Best

Paper Conference Competition at the 2001 *IEEE International Conference on Robotics and Automation (ICRA)*. Dr. Fierro is the recipient of a 2004 National Science Foundation CAREER Award.

Luiz Chaimowicz received his M.Sc. and Ph.D. in computer science from the Federal University of Minas Gerais, Brazil in 1996 and 2002 respectively. From 2003 to 2004 he held a postdoctoral research appointment with the GRASP laboratory at University of Pennsylvania where he worked in several NSF and DARPA. He is currently an Assistant Professor in the Computer Science Department at the Federal University of Minas Gerais, Brazil. Dr. Chaimowicz's research interests include cooperative robotics, swarming behaviors, multi-robot simulation, and robot programming.

Vijay Kumar received his M.Sc. and Ph.D. in mechanical engineering from The Ohio State University in 1985 and 1987 respectively. He has been on the faculty in the Department of Mechanical Engineering and Applied Mechanics with a secondary appointment in the Department of Computer and Information Science at the University of Pennsylvania since 1987. He is currently the UPS Foundation Professor and chair of Mechanical Engineering and Applied Mechanics. He was the director of GRASP Laboratory, a multidisciplinary robotics and perception laboratory from 1998–2005, and the Deputy Dean for the School of Engineering and Applied Science from 2000–2004. He is a Fellow of the American Society of Mechanical Engineers and a Fellow of the Institute of Electrical and Electronic Engineers. He has served on the editorial board of the *IEEE Transactions on Robotics and Automation, Journal of the Franklin Institute*, and the *ASME Journal of Mechanical Design*. He is the recipient of the 1991 National Science Foundation Presidential Young Investigator award and the 1997 Freudenstein Award for significant accomplishments in mechanisms and robotics. His research interests include robotics, dynamics, control, design, and biomechanics.

IV

Decision Making and Autonomy

At one of the highest rungs of the cognitive ladder of autonomous robots is the ability to manipulate, organize, and reason about available information, and from there, make relevant operational plans that conform to the abstract object-ives of complex missions demanded of the robots. This sets the present portion of the book apart from the subject matter of the previous portion (Part III), where the type of planning involved is usually more explicit and of a more immediate nature (for instance, motion planning involves deliberation over the immediate actions of the robots and is less involved with strategizing about long-term mission goals and objectives).

Central to the cognitive capabilities of robots, is their ability to manipulate information (known a priori or gained during runtime) into forms amenable to analysis and reasoning (by the robots themselves). Knowledge representa-tion and decision making is discussed in detail in Chapter 12, the first chapter of this section. The chapter aims to provide readers with a broad overview of the various techniques and paradigms in knowledge representation and decision making, as well as the inter-relationships between the representa-tion and decision-making systems. It begins with an introduction to the more commonly used knowledge representation approaches, with increasing levels of abstraction, in mobile robotics — namely, the spatial, topological and symbolic, andontological approaches. These different forms of representation techniques facilitate understanding, reasoning, and knowledge creation by the robots, on

461

different levels. With the many forms of knowledge representation, each with its own advantages, researchers have also investigated the use of multiple representations within systems, in order to facilitate planning within each level. Given the knowledge it possesses, an intelligent robotic system must utilize the knowledge efficiently for deciding its actions. Decision-making mechanisms are tightly coupled to knowledge representations, and the second part of the chapter describes the commonly used techniques of computation-based closed loop control, cost-based search strategies, finite state machines (FSM), and rule-based systems. The chapter concludes with a detailed case study of the knowledge representation and decision-making aspects within the 4D/RCS architecture.

Despite the efficiency within knowledge representation systems, there is still the presence of uncertainties within the real world that cannot be modeled adequately and accounted in its entirety within representations. As such, any abstract planning algorithm has to be able to cope robustly with any inaccuracies of available knowledge to ensure good performance of the autonomous robots within a real environment. The second chapter of this part therefore deals with the planning capabilities of robots in the presence of uncertainties, in particular prediction and sensing uncertainties. Prediction uncertainties occur when the robot is unable to have an accurate prediction of the future effects of actions, while sensing uncertainty deals with inaccuracies in the current knowledge base (due perhaps to sensing imperfections). This chapter deals with planning under uncertainties by explicitly accounting for and modeling uncertainties as a "*game against nature*." The chapter first discusses planning under only prediction uncertainty with a discussion of both optimal (through value iteration and policy iteration) and approximate solution methods. Techniques for planning (with methods designed in discrete spaces) in continuous spaces are also examined. When sensing imperfections exist, the information space is introduced for use in planning, in place of the usual state space. In order to reduce the complexity of information space representations, the chapter introduces the mapping (collapse) of the information into a smaller, collapsed, information space for easier manipulation.

The presence of multiple interacting robots greatly complicates decision making, and this leads to the need for effective coordination amongst robots. Efficient coordination mechanisms can potentially bring about greater team autonomy. Several coordination mechanisms exist, and one of the most popular framework, is that of behavior-based control, which is presented in Chapter 14 of this book. Interaction between robots is indispensable if any form of coordination is to be possible. Thus, the chapter examines the use of different interaction mechanisms (through the environment, through sensing, and through communications) for coordination and illustrates the effectiveness through an analysis of various case studies. The chapter also describes the microscopic and macroscopic models of multi-robot systems, and investigates the ways in which multi-robot systems may be systematically synthesized. Such studies will greatly benefit the design of effective multi-robot systems that are both efficient and autonomous.

This part of the book covers the abstract mechanisms behind knowledge representation, decision making and planning, and coordination between intelligent mobile robots. These mechanisms form the last general module, which, through a cohesive integration with the other modules discussed in the earlier parts of the book, make up an intelligent and autonomous robotic system.

12 Knowledge Representation and Decision Making for Mobile Robots

Elena Messina and Stephen Balakirsky

CONTENTS

12.1 INTRODUCTION

Knowledge is central to a mobile robot's ability to carry out its missions and adapt to changes in the environment. The knowledge subsystem must support acquisition of information from external sources, maintain prior knowledge, infer new knowledge from the knowledge that has been captured, and provide appropriate input to the planning subsystem. In order to carry out these responsibilities, there are different categories of knowledge required, such as task (also known as functional or procedural), and declarative, which includes spatial (or metrical). Representation schemes for the various types of knowledge must be chosen so as to provide the best performance and reliability. Many design decisions must be made, taking into account the real-time requirements of the robot control system, the resolution of the sensors, as well as the onboard processing and memory.

Decision making must be tightly coupled with knowledge representation because the decisions must be based on the knowledge available to the robot. Roboticists have drawn from fields as varied as symbolic artificial intelligence, operations research, and control theory as well as creating many ad hoc methods such as behavior-fusion.

In this chapter, we introduce several commonly used approaches to both knowledge representation and decision making in mobile robot systems. We discuss the inter-relationship between the representation format and content and the decision-making systems. The chapter delves more deeply into systems that accommodate multiple representation types and decision algorithms. Design considerations are presented to provide the reader a brief introduction to the many complexities of selecting knowledge representation types and decision-making approaches. The chapter concludes with a high-level implementation example.

12.2 INTRODUCTION AND A BRIEF SURVEY OF REPRESENTATION APPROACHES

12.2.1 Grounding Representation

Mobile robots have especially challenging knowledge requirements in order to negotiate within and interact with uncertain and dynamic environments. The internal representation of the world has many implications for the effectiveness, reliability, efficiency, validity, and robustness of the mobile robot. The symbol grounding problem is one perspective of this challenge and it is defined in Reference 1 as "the problem of how to causally connect an artificial agent with its environment such that the agent's behaviour, as well as the mechanisms, representations, etc. underlying it, can be intrinsic and meaningful to itself, rather than dependent on an external designer or observer." There are two principal approaches as described by Ziemke, each with a focus on a different aspect of the robot's interaction with the world. Cognitivism takes a computational view and divides the world into input systems (i.e., sensing) and central systems (where the problem-solving takes place). A predefined fixed representation resides in the central system and is populated by the input systems. Therefore *cognitivism* emphasizes the sensing interaction. Contrast this with the *enaction* paradigm, which emphasizes the actuation by the robot. In the enaction view, cognition is considered the outcome of the interaction between the robot and its environment. "Consequently, cognition is no longer seen as problem solving on the basis of representations; instead, cognition in its most encompassing sense consists in the enactment or bringing forth of a world by a viable history of structural coupling" [2]. A common interpretation of the enactive paradigm is that no explicit world model is required — rather that the combination of the robot and the world itself are adequate, for they capture the "real thing" [3].

But knowledge representation need not hew to one extreme or the other. Putting aside the more philosophical considerations of knowledge representation and focusing on requirements for enabling a mobile robot to perform its mission, we now look at the classes of knowledge and the different representation paradigms.

12.2.2 Representation Approaches

This section describes the most common single-representation approaches used for mobile robots, but is not meant to be exhaustive. Note that this chapter does not address the issues of simultaneous localization and map building, which are covered elsewhere in the book.

12.2.2.1 Spatial representations

A large number of the mobile robot systems implemented have relied on spatial representations. Decomposing the space that the robot has to travel within into uniform or nonuniform regions (a geometric space) is one approach. Two commonly used geometric spaces are world space and configuration space (Cspace). World space is defined as the physical space that the robot, obstacles, and goals exist in Reference 4. A particular location in world space can typically be represented by two to four parameters, where planar worlds with static environments require two parameters (x and y location) and 3D worlds with dynamic environments require four (x, y, z, and *time*). An example of a robot arm with world space obstacles may be seen in Figure 12.1a. A configuration of an object may be defined as the independent set of parameters that completely specify the location of every point (or the pose) of the object [4]. The set of all possible configurations is known as the configuration space, and represents all of the possible poses of an object. The number of parameters necessary to specify the Cspace (or the dimensionality of the space) is also known as the degrees of freedom of the object. For a point object, the Cspace and world space are identical. The Cspace representation of the robot arm from Figure 12.1a may be seen in Figure 12.1b. World space has the advantage of having objects in the world directly integrated into the space as opposed to having to compute potential object configuration interactions in Cspace. However, for nonholonomic robots, any path found in the Cspace is guaranteed to be collision free and realizable whereas a path found in world space may cause collisions with parts of the robot [5]. Identical spatial structures may be used to represent both spaces. For simplicity, the examples in this section will concentrate on world space representations.

Grid-based structures [6,7] are a convenient means of capturing input from the robot's sensors, especially if multiple readings from one or more sensors are to be fused. They have the advantage of being easy to implement and maintain, due to their uniform, array-like structure. A probability or certainty measure can be assigned to each grid cell indicating the degree of confidence that the cell is really occupied as opposed to purely open space, resulting in each location being marked as probably occupied, probably empty, and unknown. This type of representation is also referred to as an evidence grid [8]. Figure 12.2a shows an example of a grid representation.

Furthermore, it is fairly straightforward to implement path planning and obstacle avoidance algorithms that use a regular structure, which can be readily translated into a graph that is searched (for instance using a Dijkstra [9] or A* [10]) to find the lowest-cost path — typically based on shortest distance. A node is placed at the center of each grid location and it can be connected to adjacent cells via arcs that are assigned costs. The costs may be based on distance traveled between cells and usually reflect a penalty if the motion would take the robot into an occupied area. In the most simplistic approach, each cell can be connected to its nearest four or eight neighbors (see Figure 12.2b). More efficient approaches build the graph connecting only empty cells or by using other techniques such as visibility graphs [11] or Voronoi borders [12]. The grid itself can be represented more compactly by using adaptive tesselation approaches. These include quadtrees which are efficient if the environment is not uniformly cluttered or when additional spatial information such as depth must be captured. References 13 and 14 describes the use of quadtrees as a two-and-a-half-dimensional (2.5D) approach to capturing the geometry of a lake bed for underwater autonomous vehicles. Multi-resolutional approaches,

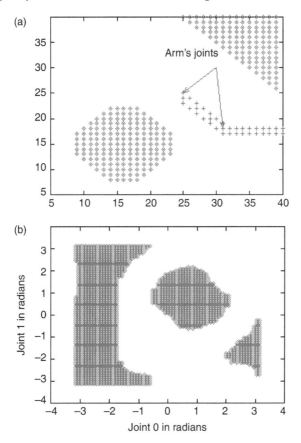

Figure 12.1   Representation of a robot arm and obstacles in both world and configuration space. (a) The 2D world space contains a two-jointed robot arm (represented by the "+" signs) and obstacles (represented by filled boxes). (b) The world space is also 2D with the axis representing the joint angles of the robot arm. The clear regions in this figure represent joint angle combinations that are collision free.

such as in Reference 15 also improve efficiency by giving the cells closer to the robot higher resolution than those further away.

Approaches that tessellate space may need to represent more than two (or two and a half) dimensions. For instance, in cases where a 2D spatial representation is inadequate, the evidence grid approach has been extended to 3D [16]. In many applications, it is insufficient to have the robot plan a path that only avoids obstacles. Additional constraints, such as nonplanar terrain and the robot's own kinematics and dynamics often need to be taken into consideration. Considering velocity and acceleration while generating the

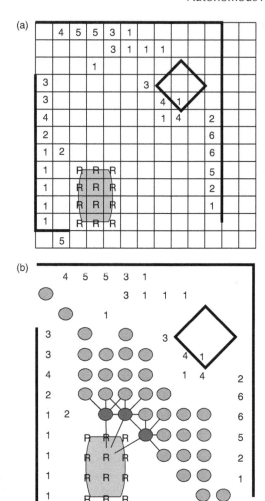

Figure 12.2 Robot in a room with a table. The ground truth, indicating where the walls, robot, and table are located, is shown by solid lines. (a) Evidence grid showing grid cells where sensor has detected an obstacle. The robot's location is indicated by R's. The higher the number in a cell, the higher the "confidence" that there is an object in that space. Note the spurious detections in some cells, which often happens due to noisy sensors and inaccurate localization. (b) Nodes, indicated by shaded circles are placed at traversable locations in the grid space and are connected potentially to up to 8 neighbors (8-connected). Nodes are placed everywhere except where there is an obstacle detected or directly (4-connected) adjacent to a cell marked as an obstacle. The darker nodes are the first level of accessible nodes from the robot's current location. Only the arcs connecting these nodes to the next layer of potential locations are shown.

robot's path significantly increases the state space for planning, so it tends to be done in two stages. Generally, the path planning process produces a coarse set of waypoints, which are then smoothed by another process that takes into account the robot's dynamic constraints. However, for systems with complex dynamics (e.g., legged robots, two-wheeled vehicles [17], soccer playing robots, or hovercraft [18]), it may be inadequate to ignore dynamics during the obstacle-avoidance planning process, therefore explicitly modeling the systems' dynamics may be necessary to guarantee collision-free trajectories.

Some researchers have successfully demonstrated mobile robot systems that use only the sensor image ("windshield view"), also known as the iconic representation, to plan within. From Reference 19: "According to the model being proposed here, our ability to discriminate inputs depends on our forming 'iconic representations' of them ... These are internal analog transforms of the projections of distal objects on our sensory surfaces." This may be 2D spatially, as is the case for Charged Coupled Device (CCD) cameras, or 3D, in the case of range sensors, such as Laser Radars (LADARs). Some mobile robots successfully accomplish their goals by planning based on purely the sensor image view. This is particularly true for road-following systems, such as those by Dickmanns [20] and Jochem [21], where road edges are extracted by sensor processing algorithms and used to plan the vehicle's steering command in the image frame.

Grid-based and other spatial representations vary in choice of coordinate systems and in the relationship to the robot itself. Some implementations use polar coordinates because the sensor data is returned in the form of distance (to object) and angle, reducing the number of calculations in constructing the map and in planning motion. The robot is always at the origin of the coordinate system in this case. However, it is more difficult to maintain a global map as the robot traverses the environment. The majority of implementations use a Cartesian coordinate system. In some approaches, the map is centered on the robot's current location and oriented with respect to the robot. Sensor information is easily placed within the map, but the entire map must be transformed when the robot changes location or orientation (assuming that previous information is kept). Some systems maintain the maps in an absolute global reference frame (for e.g., based on magnetic north). This facilitates localization with respect to global positioning systems, registration with a priori maps, and landmark-based navigation, but requires the transformation from the local sensor frames to the global one.

Other spatial representations are based on the geometric boundaries within the environment [22], such as planar surfaces [23]. These representations may augment the iconic or grid-based ones and often provide efficiencies by providing more compact descriptions of an environment, especially for indoor applications or highly structured environments. Describing a wall as a plane or a line is more efficient storagewise vs. a set of grid cells. However, additional

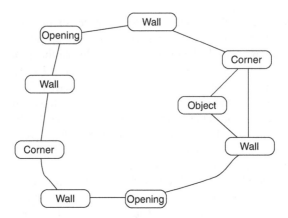

FIGURE 12.3   Topological map capturing significant features in environment. The map is a simple representation of the same room shown in the example in Figure 12.

computations by grouping algorithms that process adjacent occupied cells or convert individual pixels into higher-level geometric entities are required to achieve this reduction in memory or disk requirements.

12.2.2.2 Topological representations

Some systems represent the world via topological information (e.g., [24–27]). This enables them to reduce the amount of data stored and relate individual local maps together into a more global one. Topological maps provide qualitative information, noting significant entities in the environment, such as landmarks, and the connectivities and adjacencies amongst them but do not provide exact coordinates or relative distances. Typically, topological information is implemented via graph structures, where the features are the nodes. The resulting maps are much more sparse and provide computational advantages in planning. They can also provide more natural interfaces for humans by referring to places by name, rather than coordinates. Figure 12.3 conceptually shows a topological map of the same room used in Figure 12.2a and b. For a full discussion and comparison of the grid-based and topological paradigms, see Reference 28.

12.2.2.3 Symbolic representations

Symbolic representations provide ways of expressing knowledge and relationships, and of manipulating knowledge, including the ability to address objects by property. Much early work in robotics was carried out in the context of artificial intelligence (AI) research using symbolic representations [29–31]. This had the result of uncoupling robotics from the geometry and dynamics of the real world, and focusing on purely symbolic approaches to perception, planning, and reasoning [32]. Probably the best-known symbolic representation developed in classical AI is frame-based [33]. A frame defines a stereotypical situation, which is instantiated when appropriate. There are slots to be filled out for the particular instantiation, and actions to be carried out when conditions defined by the frame are met. For example, there would be a series of frames related to a building, essentially defining what the robot may be expected to encounter as it travels inside the building. A frame for a room may have concepts for "floor," "ceiling," "right wall," "left wall," "far wall," and so on. The robot would try to find entities using its vision system to fill in the slots for these concepts. One of the difficulties is that the robot has to be able to realize when a particular scene does not match any of the existing stereotypical situations defined within its frame system.

After struggling for the better part of two decades, the AI community turned away from robotics and focused on expert systems, knowledge representations, and problem solving in the symbolic domain. Little of this early work ever found practical application in mobile robots, although work which couples higher-level planners or agents to real systems has found new advocates, for example, in space applications [34,35].

Tying symbolic knowledge back into the spatial representation provides symbol grounding, thereby solving the previously noted problem inherent to purely symbolic knowledge representations, It also provides the valuable ability to identify objects from partial observations and then extrapolate facts or future behaviors from the symbolic knowledge.

A common type of symbolic representation for representing rules is ontological. Ontologies are definitions and organizations of classes of facts and formal rules for accessing and manipulating (and possibly extending) those facts. There are two main approaches to creating ontologies, one emphasizing the organizational framework, with data entered into that framework, and the other emphasizing large-scale data creation with relationships defined as needed to relate and use that data. Cyc [36] is an example of the latter, an effort to create a system capable of common sense, natural language understanding, and machine learning.

Ontologies provide mechanisms for reasoning over information. This includes being able to infer information that may not be explicitly represented, as well as the ability to pose questions to the knowledge base and receive answers in return. One way of enabling this functionality is to represent the symbolic information in the world model in a logic-based, computer-interpretable format, such as in the Knowledge Interface Format (KIF) representation [37] or description logics such as OWL (Web Ontology Language) [38]. Tools are starting to be developed to make this information entry process easier, primarily by hiding the intricacies of the syntax of the underlying language. Protégé is an example of such a tool [39,40].

12.2.2.4 No representation

Some have argued that representations such as those described earlier are too expensive to maintain and not valid due to the uncertainty inherent in trying to model the world [41]. This mindset paved the way for the robot architecture known as subsumption or behavior-based [42]. Brooks argues that "When we examine very simple level intelligence we find that explicit representations and models of the world simply get in the way. In turns out to be better to use the world as its own model" [43].

12.2.3 Multi-Representational Systems

Perhaps because of the overall complexity and difficulty of implementing a mobile robot, most implementations have relied entirely on a single represent-ation approach as discussed in the preceding sections. There are researchers who have chosen to expand the types of knowledge representations within their robotic systems to incorporate more than one type. For instance, the Spatial Semantic Hierarchy (SSH) is comprised of several distinct but interacting rep-resentations, each with its own ontology [44]. The SSH is based on properties of the human cognitive map and incorporates both quantitative and qualitat-ive representations organized within a hierarchy. *Large-scale space*, which is defined as space whose structure is at a much larger scale than the sensory horizon of the agent, poses additional challenges for constructing maps and facilitating exploration by robots. As the robot traverses space, it collects sets of information (maps) in a local frame, which must then be "stitched" together into a global frame. Qualitative knowledge includes names of objects, control laws, views, causal schemas, and topological information, such as places, paths, connectivity, and order. Quantitative knowledge includes sensor values, local and global 2D geometry, distances and angles/headings. Sensor and control level information is based on various types of control laws leading to locally distinctive states. Local geometric maps with their individual frames of refer-ence are constructed at the control level. Above this is a causal level, which derives discrete models of action from the control level. A topological level contains an ontology of places, paths, and regions, which connects the various local metrical maps into a patchwork, which can be merged into a single global frame of reference.

The Polybot architecture [45] is designed to enable various modes of reason-ing based on multiple types of data representations. Polybot is built upon a series of specialist modules that use any algorithm or data structure in order to perform inferences or actions. Since specialists may need to share knowledge, which they internally represent in different manners, a common propositional language for communicating information is part of the Polybot system. Examples of spe-cialists implemented in Polybot include perception, a reactive motion planner, spatial location (using a cognitive map), causation (which uses production rules), and object identifier (using neural networks).

A third example of a multi-representational architecture for mobile robots is 4D/RCS (4D Real-Time Control System) [46]. This architecture, with its hierarchical and heterogeneous world model, has been used in numerous types of implementations, ranging from underwater robots to autonomous scout vehicles. Several U.S. Department of Defense programs have selected 4D/RCS. These include the Army Research Laboratory's Demo III eXperimental Unmanned Vehicle (XUV) [47] and the Army Future Combat Systems Autonomous Navigation Systems. The Army XUV has successfully navigated many kilometers of off-road terrain, including fields, woods, streams, and hilly terrain, given only a few waypoints on a low-resolution map by an Army scout. The XUV used its onboard sensors to create high-definition multi-resolution maps of its environment and then navigated successfully through very difficult terrain [48]. The following sections describe the many dimensions of knowledge in 4D/RCS.

12.2.4 Decision Making

Any intelligent system has a limited vocabulary of actions that it may take in order to accomplish its goals. The agent must *decide* which of these actions to perform, and when to perform them. The responsibility for making this decision is shared by the process that creates the knowledge representation and the process that constructs a plan of action based on this knowledge representation. The choice of which representation is used and what knowledge is stored helps to decide the division of this responsibility. As an example of one extreme, the knowledge representation may be formulated as a grid-based structure that contains the cost/benefit of the agent being in a particular state. Very complex reasoning may be required to condense all of the available information into this single measure. The planning process then becomes the optimization problem of finding the lowest-cost path through a graph.

As an example at the other extreme, all knowledge may be stored in a raw form. An example would be the storing of a priori map data directly in the autonomous vehicle's database without any further processing of the map data in order to make it more readily usable by the decision-making system. In this case, no decisions are made in creating the knowledge representation, but complex reasoning or decision making must occur to determine a plan of action.

There are many different forms of decision making that exist in the current literature. Popular techniques include computation-based closed-loop control, cost-based search strategies, finite state machines (FSM), and rule-based systems. Computation-based closed-loop controllers put most of the decision burden on the planning task. They attempt to maintain stability in an operating system by taking corrective action anytime that there is a deviation in the system from a desired value (the system "setpoint"). What action to take may be determined by techniques such as fuzzy logic, neural networks, Petri nets, and proportional-integral-derivative (PID) control strategies. PID control is the most common control methodology in process control. In PID control, the state of the world is observed (either directly from sensors or from the stored knowledge representation) and matched against the system setpoint. If an error exists, corrective action is sent to the actuators based on a computation that takes into account the error (proportional), the sum of all previous errors (integral), and the rate of change of the error (derivative) [49].

In cost-based search strategies, the decision burden is mostly placed on the knowledge representation. The knowledge representation must contain a discrete representation of a reduced system state space (e.g., a mobile robot state space may be represented by 2D occupancy grid that ignores time) along with a mapping that maps a single cost/benefit value to each state transition. In most cases, this state space is completely instantiated to the planning horizon of the agent, however some systems do exist that incrementally build the representation as the search progresses [50]. While the formulation of this cost/benefit value may be as simple as the average or maximum value of some attribute over the region encompassed by the discrete area of state space, more complex assignment techniques may be applied. For example, the planning systems described in Reference 50 implement a knowledge representation that is composed of a combination of "knowledge layers." In this approach, some areas of knowledge are represented in traditional grid-based structures (e.g., the traversability of the terrain), while others are computed by the repository on-the-fly (e.g., whether it violates any road driving laws to transition from one state to another). All of the information from the various layers is fed into a value judgment process that combines the knowledge with the goals and objectives of the system to formulate a single cost/benefit number. These cost/benefit numbers are then used to build a graph structure that the planning process may search using a number of different search algorithms.

While cost-based search controllers work solely of a representation of the system state space, FSM-based controllers enhance their knowledge by building a representation of the system event space. These controllers operate off of a preconfigured state-graph structure where each state represents the internal state of the agent and decisions on state transitions are made based on periodic event input from the knowledge representation. As such, the decision making is shared between the sensory processing that decides that a particular event has happened, and the a priori planning process that decides what to do in response to that event. It should be noted that in the FSM approach, all of the planning decisions are made by a domain expert before the first operation of the system. To be complete, the system designer must anticipate every combination of events that may occur during the system operation. An example of a simple

FSM may be a system for controlling a vehicle's right turn signal. This system would have the two states of "turn signal on" and "turn signal off" and the transition events would be to transition from "turn signal off" to "turn signal on" when a "prepare for right turn" event is detected and transition from "turn signal on" to "turn signal off" when a "right turn complete" event is detected. In this case, it may be the responsibility of the knowledge representation to examine the system and world knowledge and decide that a turn is imminent or has just been completed. A comprehensive look at hierarchical FSMs may be found in Reference 51.

Rule-based systems may be used to construct decisions for both the knowledge representation as well as for the planning process. As noted earlier, a rule-based system may be used to make decisions that feed into the cost/benefit value of a cost-based planning engine. Planning systems such as deduction systems combine the application of rules with a graph search to compute a set of actions that will achieve the goal set. Systems such as Graphplan [52] examine preconditions that are necessary for the application of rule, and then the postconditions that will apply after the rule has fired. A plan is formulated by searching for a combination of rule firings that accomplishes the agent's goals. Systems such as this have been very successful in solving planning problems in the domain independent planning arena [53].

There is no single correct answer as to where the decision making should occur, or what form of decision making should take place. In fact, many robotic systems combine multiple strategies into a single system. For example, the lower levels of the processing in the Demo III XUV program utilize a graph search on a cost map for formulating steering and acceleration decisions. The arcs represent the cost of moving from one node location to another using a specific steering angle and acceleration profile. In this case, the majority of the decision making may be said to lie in processing that determines the cost values associated with each arc within this cost map. In addition to using several cost maps, the higher-level planning system also examines a priori data and a knowledge base of constraints in order to apply a rule-based planning system. This system takes advantage of decision making in both the process that creates the knowledge representation and the process that decides a course of action.

12.3 CASE STUDY: KNOWLEDGE REPRESENTATION AND DECISION MAKING WITHIN A 4D/RCS

A 4D/RCS is designed to accommodate multiple types of representation formalisms. It is a hierarchical control structure, composed of nodes, with different range and resolution in time and space at each level. Each level of the hierarchy is a control loop unto itself, but very different types of entities are tracked and controlled. Each of the control nodes receives input commands from its supervisor node, performs sensory perception, behavior generation (decision making), world modeling, and other supporting functions, to produce a set of commands to provide its subordinate nodes. The functionality of each level in the 4D/RCS hierarchy is defined by the functionality, characteristic timing, bandwidth, and algorithms chosen by Behavior Generation processes for decomposing tasks and goals at each level. Hierarchical layering enables optimal use of memory and computational resources in the representation of time and space. At each level, state variables, images, and maps are maintained at the resolution in space and time that is appropriate to that level. At each successively lower level in the hierarchy, as detail is geometrically increased, the range of computation is geometrically decreased. Also, as temporal resolution is increased, the span of interest decreases. This produces a ratio that remains relatively constant throughout the hierarchy, yet enables the overall control system to attain sophisticated behaviors within complex environments. Although the overall capabilities of the autonomous mobile robot are enhanced through the implementation of a multi-level, multi-representational knowledge base, there are design and engineering complexities that must be dealt with. These are discussed in Section 12.3.2.

The lower levels of the hierarchy are concerned with controlling servo motors and other actuation devices. The world models at the lowest levels primarily contain state variables such as actuator positions, velocities, and forces, pressure sensor readings, position of switches, gearshift settings, and inertial sensors for detecting gravitational and locomotion acceleration and rotary motion. Decisions at this level usually occur through the behavior generation process operating on raw data to close PID control loops for servo control or operate finite state machines to determine the appropriate time to change switches or gearshift settings. The time horizons are very short, the representation is typically not multi-dimensional but rather is single-valued parameters. Further up the hierarchy, a combination of map-based representations and object knowledge bases are used, which contain names and attributes of environmental features such as road edges, holes, obstacles, ditches, and targets. In order to form these higher-level knowledge bases, decision making must be part of the knowledge representation construction process. These maps represent the shape and location of terrain features and obstacle boundaries and are used to perform obstacle avoidance (reactive) and path or mission planning (deliberative). The higher levels of the hierarchy may be concerned with controlling the tactical behaviors of one or several vehicles. Knowledge is primarily symbolic, although it may be tied to global locations on a map, and it represents concepts such as targets, landmarks, and features such as buildings, roads, woods, fences, intersections, etc. The symbolic nature of the knowledge requires complex reasoning for the behavior generation in order to formulate a course of action. However, it should also be noted that the creation of the representation may have also involved several levels of reasoning and decision making. In this case, the spatial extents over which the system plans and functions are large scale, but the resolution is low and the temporal horizons for closing the control loops are longer.

A 4D/RCS defines an explicit knowledge database (KD), although it is not a single monolithic structure, but rather is heterogeneous and distributed across the hierarchy in order to most efficiently and effectively serve the processes that populate, update, and access it. The KD consists of data structures that contain the static and dynamic information that collectively form a model of the world. The KD contains the information needed by the world model to support the behavior generation, sensory processing, and value judgment processes within each node. Knowledge in the KD includes the system's best estimate of the current state of the world plus parameters that define how the world state can be expected to evolve in the future under a variety of circumstances. An important feature of knowledge representation within 4D/RCS is the concept of continually updating knowledge throughout the hierarchy, which supports continual replanning, albeit at different update rates for each level. Figure 12.4 shows the many different types of knowledge representation formalisms that

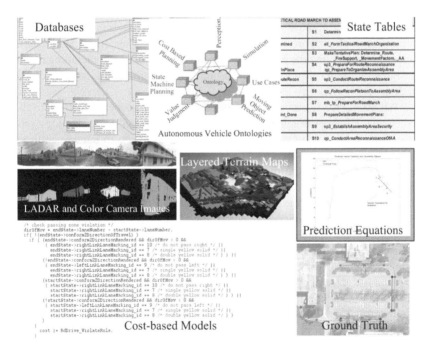

FIGURE 12.4 Different knowledge structures in 4D/RCS. These various types of knowledge representations capture different aspects of the information that the decision-making systems must use within the control system.

are currently being implemented within the 4D/RCS architecture as applied to autonomous driving. These formalisms range from iconic and grid-based to symbolic and from procedural to declarative [54].

12.3.1 Procedural vs. Declarative Knowledge in 4D/RCS

There are many different ways of classifying knowledge. In addition to the classification shown earlier (Spatial, Topological, and Symbolic), 4D/RCS classifies knowledge as either procedural or declarative, as described in the following sections.

12.3.1.1 Procedural knowledge

Procedural knowledge is the knowledge of how to perform tasks. Procedural knowledge is different from other kinds of knowledge, such as declarative knowledge, in that it can be directly applied to a task. Within 4D/RCS, procedural knowledge is primarily used for decision making and control purposes.

Two primary planning approaches are implemented in 4D/RCS, each representing procedural knowledge differently: FSM and cost-based paradigms. In both cases, the application and domain-specific tasks and commands are first defined through a rigorous domain analysis process. The control hierarchy is designed by detailing the responsibilities of each control node, including inputs from the higher-level supervisor and outputs (as commands) to its subordinate nodes.

Within 4D/RCS, procedural knowledge may be encoded directly into the executable FSMs or graph structures, or it may be stored in an ontology representation. An ontology is being created, for instance, to capture military behaviors for autonomous ground vehicles [39]. Focusing initially on a route reconnaissance to be performed by a scout platoon, this effort details in a hierarchical fashion the activities necessary in order to perform this activity. The troop commander is at the top level and decides the priority items on the route, defines the march column organization, specifies the formation and movement technique, and dispatches a scout platoon to conduct the reconnaissance. The scout platoon leader will do finer level decision making, organizing the platoon's sections of vehicles and assigning commands to each section leader to do reconnaissance of different areas along the route while maintaining security. Each section leader will evaluate the environment to provide detailed tactical goal paths for each of his vehicles, coordinating their movement by the use of detailed motion commands to control points along with security overwatch commands. The decision-making responsibilities are thus refined and narrowed at each subsequent level, down to that of individual vehicles and subsystems. This ontology is based upon the OWL-S specification (Web Ontology Language-Services) [55]. In this context, behaviors are actions that an autonomous vehicle is expected to perform when confronted with a predefined situation. The ontology is stored within the 4D/RCS knowledge database, and the behaviors are spawned when situations in the world are determined to be true, as judged by sensor information and the value judgment components.

In the FSM approach, each of these command decompositions at each node will be represented in the form of a state-table of ordered production rules. The sequence of simpler output commands required to accomplish the input command and the named situations (branching conditions) that transition the state-table to the next output command are the primary knowledge represented in this approach. Each node therefore contains labeled representations of the states and transitions, which is beneficial in terms of making the reasoning of the system explicit [56] (see Figure 12.5 for an example). FSM's have the advantage of making the decision criteria and logic obvious to a human reading the code. However, they require the programmers to consider and handle all possible situations ahead of time, which is often not realistic for robots operating in complex situations and environments.

The cost-based approach combines a graph-based search technique with a set of knowledge modules that simulate the effects of alternative actions and provide input to a unified cost model [50,57]. Different feature layers are discretized. Examples of feature layers are elevation, road networks, and vegetation. The planner at a given level sends candidate trajectories to simulators that compute the cost of state transitions for each of the relevant feature layers. For instance, a proposed path may take the vehicle from an on-road location to off-road. The cost associated with this is dependent on the context of the situation — if going off-road avoids a pedestrian on the road (which would be noted by another feature layer, possibly the obstacle one) this is an acceptable cost. Similarly, the cost/benefit of running a red light would be substantially different for a casual driver than it would be for a police vehicle responding to an emergency. Ontologies and other knowledge bases support the generation of cost models during execution. Whereas this cost-based approach is more general than the FSM, it is also more challenging in terms of defining the appropriate costs for each action, especially since they will be combined. This is a good candidate for the application of learning to develop the cost models. In general, graph-based representations can result in an explosion of data (nodes and arcs connecting the nodes) and hence can have very poor performance characteristics when the graph is being searched. This is a concern especially for real-time systems, such as mobile robots. When a robot plans its motions, it must be able to react within an appropriate amount of time to obstacles or events. However, there are several techniques to mitigate these concerns, some of which were noted earlier in this chapter. Other means of mitigating performance issues include reusing parts of the already-processed graph (e.g., Dynamic A*) [58] and using sparse representations that include only relevant features, such as the extrema of an obstacle instead of a uniform grid of the environment [50].

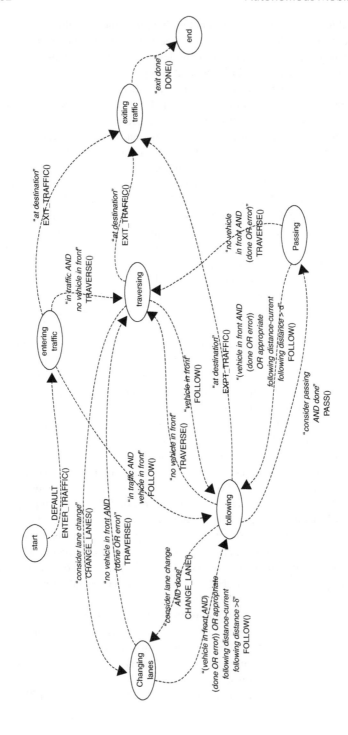

FIGURE 12.5   Example of a FSM.

12.3.1.2 Declarative knowledge

Unlike procedural knowledge, declarative knowledge does not describe how to perform a given task. Instead, it provides the ability to use knowledge in ways that the system designer did not foresee. The declarative knowledge is used by the decision-making processes and is updated by the sensor processing subsystems. The following sections describe declarative knowledge within 4D/RCS, although several similar concepts are also found in other robot architectures.

12.3.1.2.1 Parametric level knowledge

The lowest levels of any control system, whether for an autonomous robot, a machine tool, or a refinery, are at the servo level, where knowledge of the value of system parameters is needed to provide position and velocity and torque control of each degree of freedom by appropriate voltages sent to a motor or a hydraulic servo valve. The control loops at this level can generally be analyzed with classical techniques and the "knowledge" embedded in the world model is the specification of the system functional blocks, the set of gains and filters that define the servo controls for a specific actuator, and the current value of relevant state variables. These are generally called the system parameters, so we refer to knowledge at this level as parametric knowledge.

Learning or adaptive control systems (e.g., [59,60]) may allow changes in the system parameters, autonomous identification of the system parameters, or even behavioral parameters, but the topology of the control loops is basically invariant and set by the control designer. We would not expect a robot to invent a torque loop for itself in the field, although it could well change the gain or phase of a position or velocity loop as it learns to optimize a task.

12.3.1.2.2 Spatial level knowledge

Above the lowest servo level are a series of control loops that coordinate the individual servos and that require what can be generally called "geometric knowledge," "iconic knowledge" (in the case it represents the sensor view), "metrical maps," or "patterns." This knowledge is spatial in nature and is either in 2D or 3D grids and higher-level geometric constructs, such as edges and surfaces. The value of each grid cell may be Boolean data (e.g., indicating whether the cell is occupied or not) or real number data representing a physical property such as light intensity, color, altitude, range, or density. Each cell may also contain spatial or temporal gradients of intensity, color, range, or rate of motion. Cells may also point to specific geometric entities (such as an edge, vertex, surface, or object) to which its contents belong.

Digital maps are a natural way of modeling the environment for path planning and obstacle avoidance. Digital terrain maps are referenced to some coordinate frame tied to the ground or Earth and hence also facilitate data fusion, be it from multiple sensors or from a priori data. Although commercial

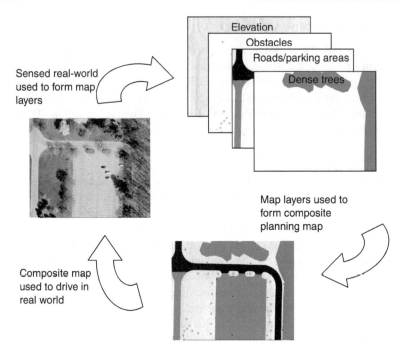

Sensed real-world used to form map layers

Elevation

Obstacles

Roads/parking areas

Dense trees

Map layers used to form composite planning map

Composite map used to drive in real world

FIGURE 12.6   Multiple feature layers in 4D/RCS.

digital terrain map often have a grid-based implementation (especially for the elevation layer), features are typically represented as vectors. The underlying database implementation facilitates spatial queries even for features that are represented by polygons or polylines. In many mobile robots, as previously discussed, a grid-based approach is easier to implement and maintain in real-time. In this case, a map may have multiple layers that represent different "themes" or attributes at each grid element. For instance, there may be an elevation layer, a road layer, a dense tree layer, and an obstacle layer as shown in Figure 12.6. The software can query if there is a road at grid location $[x, y]$ and similarly query for other attributes at the same $[x, y]$ coordinates. If the system being implemented is truly 3D, then the queries can be made according to $[x, y, z]$. This feature is important for accurately capturing features such as road overpasses and subterranean tunnels.

12.3.1.2.3 Symbolic knowledge

Within 4D/RCS, mainly two types of symbolic representations have been implemented thus far: ontologies and relational databases. As noted earlier, ontologies are also used for procedural knowledge. On the declarative side, an ontology for driving determines if objects in the environment are potential obstacles to the autonomous vehicle [61,62]. The system is composed of an ontology of objects representing "things" that may be encountered in our current environment, in conjunction with rules for estimating the damage that would be incurred by collisions with the different objects as a function of the characteristics of the autonomous vehicle, including the type of vehicle, speed, etc. Automated reasoning is used to estimate collision damage, and this information is fed to the route planner to help it decide whether to avoid the object.

Relational databases have also been developed to house symbolic information. Among these is a Road Network Database [63] that includes detailed information about the roadway, such as where the road lies, rules dictating the traversal of intersections, lane markings, road barriers, road surface characteristics, etc. The purpose of the Road Network Database is to provide the data structures needed to capture all of the necessary information about road networks to allow a planner or control system on an autonomous vehicle to plan routes along the roadway at any level of abstraction. At one extreme, the database provides structures to represent information so that a low-level planner can develop detailed trajectories to navigate a vehicle over the span of a few meters. At the other extreme, the database provides structures to represent information so that a high-level planner can plan a course across a country. Each level of planning requires data at different levels of abstraction and, as such, the Road Network Database accommodates these requirements.

12.3.2 Additional Considerations

Thus far, we have described several considerations when implementing the knowledge representation for a mobile robot system. The types of knowledge and the structures for capturing the knowledge have been discussed. Expressivity of the representation, the real-time requirements of the robot control system, the resolution of the sensors, and the onboard processing and memory were among the issues to be considered. Further discussion on the types of performance considerations for knowledge representation in real-time control systems can be found in Reference 64. Additional aspects that affect the design decisions in multi-representational systems such as 4D/RCS, are briefly presented in this section.

12.3.2.1 Integration considerations

Representing multiple classes of knowledge within an intelligent control system introduces the challenge of integrating fundamentally different representations into a single, unified knowledge base. This knowledge base must behave as a single, cohesive entity, and as such, there must be seamless information exchange and interoperability between all knowledge sources. In the case of autonomous mobility, as alluded to previously, parametric knowledge may be stored as a set of numbers in a computer program representing the values of the state variables, the spatial knowledge may be a set of digital terrain maps in 2D grid structures tied to feature information stored as vector fields, and the symbolic knowledge may be a set of entities with pertinent attributes stored in a database.

12.3.2.2 Integration within a single representation

There are integration challenges within a single representation, as well as among disparate representations. For an autonomous vehicle, within solely the symbolic level, one must integrate a priori information about the types of entities that one expects to see in the environment with instances of the entities that are encountered. When both types of information are represented in database format, association of database keys is often sufficient to provide the necessary integration.

Within solely the spatial representation, one must integrate processed sensed data about the environment with a priori terrain maps. This is a difficult challenge due to the noise associated with sensed data as well as the varying level of resolution between a priori maps and the sensed data. In addition, one must integrate two or more sensed images, which may be taken by two different sensors, or by the same sensor at different times. Often described as "data registration," researchers are actively addressing this challenge, for example, see References 65 and 66. The multi-resolutional approach to grid-based representation requires methods for integrating higher-resolution knowledge into lower resolution. For instance, in a 4D/RCS hierarchy, the information contained in the autonomous mobility level, which is typically at a 40 cm resolution, must be abstracted into coarser 4 m cells for use at the higher, vehicle level. This means that the "quilt" of individual cells having different attributes must be consolidated into a single representative large cell. A mixture of cells containing roads and fields could be merged and classified as having both roads and fields at the higher level, perhaps with an indication of percentage of "roadiness."

12.3.2.3 Integration among disparate representations

Similar challenges exist when integrating knowledge captured in different representations. Although the representations differ, there will undoubtedly be direct correlations between the data in each representation. In the case of object recognition [67], information that can be inferred by analyzing the data stored in a grid structure (obtained from a sensor) must be compared to the class attributes stored in the symbolic knowledge base to determine if there is a correspondence. For example, if a cluster of occupied cells in a spatial representation can be grouped into a single object, one can create an object frame and link all the pixels in the spatial representation to the object frame. This object frame contains a list of object attributes that are measured properties of the cluster of pixels in the spatial representation. Depending on the information that is stored in the spatial representation, one may be able to tell the object's dimensions, average color, velocity, location, etc. Based on this information, one can compare the attributes of an observed object to attributes of a class prototype of objects that are expected to be seen in the environment. If a correspondence is found (within a desired threshold), links are established between the object frame and the class prototype in the database. This is the process of classification. Links established through the classification process are bi-directional pointers. Thus, class names and class attributes can be linked back to the object frame, and from there back to the cells in the spatial representation.

Figure 12.7 shows an example of integrating a spatial representation with a symbolic representation. In Figure 12.7a, the number in the cells represent the probability that the cell is occupied, with 10 being the greatest. Other information that is stored in each cell that is not shown in Figure 12.7a, such as the color and the height of the object that is occupying that cell. In Figure 12.7b, the information in the spatial representation is processed and stored as a list of attributes in an object frame. This involves clustering cells that appear to be part of the same object, and determining overall characteristics of that object. The cluster of cells have an overall X-dimension of between 9.5 and 10 m, an overall Y-dimension of 3.5 to 4 m, an average height of 2.8 to 3.0 m, and an average color of green. The perceived attributes are then compared to a priori attributes stored in a list of class prototypes as shown in Figure 12.7c to determine if there is correspondence. In this case, there appears to be a clear match between the observed attributes measured from the sensed data and the attributes in the

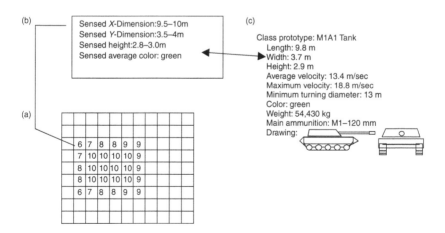

FIGURE 12.7 Integrating spatial with symbolic knowledge.

class prototype of a M1A1 tank. Therefore, links are created between the class prototype and the cells in the spatial representation.

Although the scenario described above is an oversimplified example, it shows the steps that need to be accomplished to establish a link between stored class prototypes and objects observed in the world. These links would ground the symbolic representations in the world model to the objects in the world.

12.3.2.4 Integration of decision systems

When functioning in a hierarchical system, both the knowledge representation and behavior generation must be integrated between levels. Two commonly used techniques for behavior generation integration are plan refinement and cooperative planning. In plan refinement, the highest-level planning system creates a coarse set of decisions that will accomplish the goals of the agent. A subset of these decisions that covers the planning horizon of the next lower level is passed down to that level. This level then takes the decisions of the upper level as its goals and refines the system actions to achieve these goals. This process is repeated throughout the hierarchy until the lowest level behavior generation system is reached. This plan refinement procedure is continuously repeated to account for dynamics and noise in the knowledge representation. An example of this may be a finite state machine that drives a mobile robot. The high level may create a course of action that consists of several driving commands. One of these commands; "turn right at next intersection" is passed down to the next lower level. This level will then decompose this command into a series of actions, the first of which may be "turn on right turn signal." This action will be passed down to the next lower level where it may be further decomposed, and so on until an action by the platform is performed.

In cooperative planning, each planning level is responsible for creating a course of action that covers the area from its subordinate's planning horizon to its own planning horizon. For a two level mobility planning system using cost-based planning, this may be viewed as a doughnut where the lower level creates plans for the doughnut hole and the upper level planner plans from the hole to the edge of the doughnut. In this case, the low-level planner would compute the cost from the vehicle location to each position along the circumference of the doughnut hole. The high-level planner would then use these costs as the starting point for its planning, and find a single jointly optimal path to the goal. This jointly optimal path would then be executed. Once again, continuous replanning is utilized to account for dynamics and noise in the knowledge representation.

12.3.2.5 Implications for system maintainability

Beyond the obvious considerations of the suitability of particular knowledge classes and representations for a given robot's missions/job, there are additional concerns when selecting the knowledge representation(s) for a system. The 4D/RCS architecture is supported by a knowledge engineering methodology, whereby knowledge from subject matter experts is mined, analyzed, and transformed into appropriate data structures within the resulting control hierarchy. Reference 68 describes the methodology. One key aspect of this method is the goal of maintaining correspondence between the human's terminology and semantics within the implemented code to facilitate validation, maintenance, and reuse.

12.3.2.6 Implications for perception design

An additional design and engineering advantage of the task-based approach is that it can be used to derive not only the knowledge requirements, but the sensor and perception ones as well. Given the behavior requirements of the robot, which in turn drive the knowledge representation requirements, one can determine the performance necessary from the sensors and sensory processing algorithms.

Barbera et al. [69] describe the process for defining the sensor-processing requirements for a mobile robot that drives on roads in traffic. They discuss how the sensing requirements of different driving tasks have significantly different resolutions, identification, and classification requirements which suggests that performance metrics should be defined on a task-by-task basis. For example, the task of driving the vehicle along a highway requires the sensor system to identify large objects moving nearby, their direction, speed, acceleration, positions in the lanes (which means the sensory processing system must identify road lanes), and state of the brake and turn signal indicator lights on these objects. There is little requirement for detailed recognition of object types or the need to see them at a distance or to read signs alongside or overhead of the road. However, if the autonomous vehicle decides to pass a vehicle on an undivided two lane road, then an extraordinarily detailed world representation must be sensed that identifies additional entities (e.g., upcoming intersections, rail road crossings, vehicles in the oncoming lane out to very large distances, lane marking types, and roadside signs). The goal of Barbera et al. is to first develop a list of required driving tasks, and then to identify the detailed world model entities, features, attributes, resolutions, recognition distances, minimum data update times, and timing for task stability for each of these decomposed subtask activities.

12.4 AN IMPLEMENTATION EXAMPLE

A reference model architecture such as 4D/RCS is essential for guiding the design and engineering of complex real-time control systems, including the knowledge. The previous sections have introduced general knowledge representation approaches within 4D/RCS. This section briefly describes an instantiation of 4D/RCS that was created during the development and enhancement of the Demo III XUV. The overall autonomous mobile robot's control system is assembled from a basic software component, which is referred to as an RCS Node and is described later. Figure 12.8 shows a simplified 4D/RCS hierarchy similar to the one implemented for Demo III. Only the locomotion portion of the overall hierarchy is discussed.

Each RCS Node contains the same functional elements, yet is tailored for that level of the hierarchy and the node's particular responsibilities. An RCS Node contains Sensory Processing (SP), Behavior Generation (BG), World Modeling (WM), and Value Judgment (VJ). At every level of the control hierarchy there are the same basic elements:

- Deliberative planning processes receive goals and priorities from superiors and decompose them into subgoals for subordinates at levels below.
- Reactive loops respond quickly to feedback to modify planned actions so that goals are accomplished despite unexpected events.
- Sensory processing filters and processes information derived from observations by subordinate levels. Events are detected, objects recognized, situations analyzed, and status reported to superiors at the next higher level. The sensory processing results are stored in the world model for that particular level.
- Sensory processing and behavior generation processes have access to a model of the world that is resident in a knowledge database. This world model enables the intelligent system to analyze the past, plan for the future, and perceive sensory information in the context of expectations.
- Cost functions enable value judgments and determine priorities that support intelligent decision making, planning, and situation analysis. The cost functions can be dynamic and are determined by current commands, priorities, user preferences, past experiences, and other sources.

Therefore, the design of the knowledge requirements at each level is driven by the responsibilities of that level. What commands will an RCS Node be able to execute and what decisions will it be required to make? What is its required control loop response time? What spatial scope does it need to understand? What types of entities does it have to deal with? These questions are addressed below.

At the servo level, an RCS Node receives commands to adjust set points for vehicle steering, velocity, and acceleration or for pointing sensors. It must convert these commands to motion or torque commands for each actuator and

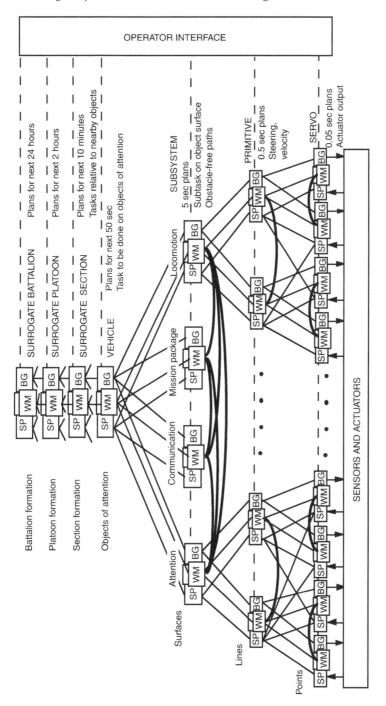

FIGURE 12.8   A simplified example of the Demo III 4D/RCS control hierarchy.

issue them at high frequencies (e.g., every 5 msec). The planning horizon is about 50 msec. The knowledge used at the servo level is primarily single-valued state variables: actuator positions, velocities, and forces, pressure sensor readings, position of switches, and gear shift settings. Decisions that need to be made include the choice of acceleration or torque profile to follow and whether or not it is safe to change switch settings. These decisions are typically made in the behavior generation module.

At the Primitive level, each RCS Node received commands with goal points about 500 msec in the future. The primitive level computes dynamic trajectories expressed in terms of vehicle heading, speed, and acceleration and sends commands to the servo level about every 50 msec.

At the Subsystem level, an Autonomous Mobility node generates a schedule of waypoints that are sent to the subordinate Primitive controller. Commands that the Autonomous Mobility RCS Node accepts include directives to follow a schedule of waypoints to avoid obstacles, maintain position relative to nearby vehicles, and achieve desired vehicle heading and speed along the desired path. Knowledge used at this level supports planning movement through 3D terrain, hence digital terrain maps with multiple registered attribute layers are appropriate. Planning for mobility at this level is concerned with obstacles (both positive and negative, i.e., significant depressions or holes in the ground), elevation, roads, and observability, if it is to perform stealthy movements. A cost-based search through a graph whose nodes are derived from elements of the regular terrain grid is used to find the lowest-cost path that achieves the specified objectives. The map-based format also provides a convenient "receptacle" for registering and fusing information from multiple sensors with each other and with a priori information, such as from digital terrain maps. At this level of the hierarchy, cooperative decision making between sensor processing and behavior generation begins to become apparent. The sensor processing is responsible for interpreting the output of the sensors combined with predictions from the world model to decide what features exist in the world. Without this processed information, planning actions such as road following would be impossible. Depending on the level of sophistication, behavior generation may perform a simple feature-based graph search or sophisticated reasoning. For example, a simple road follower may perform trajectory generation through a graph search where cells with roads are cheap and off-road cells are expensive. A more sophisticated road follower may employ a reasoner that takes into account symbolic information such as lane markings, traffic, road signs, etc. The subsystem level of the hierarchy outputs a new plan about every 500 msec, and the planning horizon at this level is about 5 sec into the future. The spatial scope is roughly 100 m, with a resolution of about 40 cm. The extents of the space considered are based on the planning horizon and vehicle velocity. The grid resolution is based on engineering considerations, like computational resources available and what resolution the onboard sensors can provide.

At the vehicle level, all subsystems on an individual vehicle are coordinated. These may include mobility, communication, weapons, and reconnaissance subsystems. Maps extend to 500 m, with resolution of about 4 m. Plans extend to a time horizon of about 1 min into the future, and may be recomputed every 5 sec. At this level of resolution and coordination, sensor processing decisions become critical for all aspects of behavior generation. Even simple grid-based map structures now require complex decisions. For example, how should two adjacent 4-m areas be represented if each is not traversable when taken alone but traversable when taken in conjunction with its neighbor?

Higher still in the hierarchy is the section level. This is the controller for a group (two or more) of individual vehicles. The section level is responsible for assigning duties to the individual vehicles and coordinating their actions. Orders coming into the section level are tactical maneuvers, including mission goals, timing, and coordination requirements. The planning horizon is 10 min into the future, and new plans are sent to subordinates approximately every minute. Knowledge at the section level includes digital terrain maps, typically covering about 2 to 5 km, at low resolution (30 m), with multiple attribute layers, such as roads (of various types), vegetation, fences, buildings, as well as enemy locations, and militarily significant attributes. Enemy locations may be noted within the map, but more extensive symbolic information about the situation is associated with the grid locations. The symbolic information could include details about the enemy force such as number of soldiers, weapons, and estimated travel direction. This type of information is largely symbolic in nature and may be amenable to rule or case-based reasoning tools. At the section level, a Value Judgment function may convert the knowledge that "a band of 23 soldiers and 1 tank is moving toward location x, y with 60% probability at velocity of 16 km/day" into a set of costs that can be tied to the map grid and utilized by the graph-based search to generate the vehicle plans. The autonomous vehicle then plans a route toward the enemy locations while remaining concealed by terrain features marked in its maps.

At most of the levels, there is some combination of a priori knowledge and *in situ* knowledge. At lower levels concerned with mobility, the maps are primarily sensor-generated, however, there may be precomputed, kinematically correct steering curves that are used to prune the planning graph by eliminating nonfeasible trajectories [70]. At higher levels, more a priori knowledge is used, for example, as digital terrain maps and descriptions of enemy vehicles and capabilities.

12.5 CONCLUSION

No one type of knowledge representation is adequate for all purposes. Davis [71] argues that representation and reasoning at the symbolic level are inextricably intertwined, and that different reasoning mechanisms, such as rules and frames, have different natural representations that must be integrated in a representation architecture to achieve the advantages of multiple approaches to reasoning.

The introduction of spatial data — often grid-based — integrated with symbolic data and parametric data in a multi-resolution hierarchical world model enables the real-time control of complex systems interacting with the real world. The knowledge representation and decision-making systems must form a cohesive unit within autonomous mobile robots in order to attain intelligent behavior.

REFERENCES

1. Ziemke, Rethinking Grounding, in Riegler, Peschl, and von Stein (eds) *Understanding Representation in the Cognitive Sciences*, pp. 177–190, Plenum Press, New York, 1999.
2. Varela, F. J., Thompson, E., and Rosch, E., *The Embodied Mind*, MIT Press, Cambridge, MA, 1991.
3. Brooks, R., "Intelligence Without Representation," *Artificial Intelligence Journal*, 47, pp. 139–159, 1991.
4. Hwang, Y. K. and Ahuja, N., "Gross Motion Planning — A Survey," *ACM Computing Surveys*, 24, pp. 219–291, 1992.
5. Russell, S. and Norvig, P., *Artificial Intelligence: A Modern Approach*, Prentice-Hall, New York, 1995.
6. Borenstein, J. and Koren, Y., "Real Time Obstacle Avoidance for Fast Mobile Robots in Cluttered Environments," *Proceedings, 1990 IEEE ICRA*, Cincinnati, OH, 1990.
7. Moravec, H. and Elfes, A., "High Resolution Maps from Wide Angle Sonar," *Proceedings of the 1985 IEEE ICRA*, St. Louis, MO, 1985.
8. Martin, M. and Moravec, H., Robot Evidence Grids, Technical Report CMU-RI-TR-96-06, Robotics Institute, Carnegie Mellon University, March 1996.
9. Dijkstra, E. W., "A Note on Two Problems in Connexion with Graphs," *Numerische Mathematik*, 1, pp. 269–271, 1959.
10. Nilsson, N., *Artificial Intelligence: A New Synthesis*, Morgan Kaufmann Publishers, Inc., San Francisco, 1998.
11. Nilsson, N., "A Mobile Automaton: An Application of Artificial Intelligence Techniques," *Proceedings of the 1st International Joint Conference on Artificial Intelligence*, Washington, DC, 1969.
12. Dunlaing, C. and Yap, C., "A 'Retraction' Method for Planning the Motion of a Disc," *Journal of Algorithms*, 6, pp. 104–111, 1986.
13. Oskard, D., Hong, T., and Shaffer, C., "Real-Time Algorithms and Data Structures for Underwater Mapping," *SPIE Advances in Intelligent Robotics Systems Conference*, Boston, MA, November 1988.
14. Yahja, A., Stentz, A., Singh, S., and Brummit, B., "Framed-Quadtree Path Planning for Mobile Robots Operating in Sparse Environments," *Proceedings of the IEEE Conference on Robotics and Automation (ICRA)*, Leuven, Belgium, May 1998.

15. Behnke, S., "Local Multiresolution Path Planning," in Browning, B., Polani, D., Bonarini, A., and Yoshida, K. (eds) *RoboCup-2003: Robot Soccer World Cup VII, Lecture Notes in Computer Science*, pp. 332–343. Springer-Verlag, 2004.

16. Moravec, H. and Martin, M., Robot Navigation by 3D Spatial Evidence Grids. Mobile Robot Laboratory, CMU, Internal Report, 1994.

17. Kobilarov, M. and Sukhatme, G., Time Optimal Path Planning on Outdoor Terrains for Mobile Robots under Dynamic Constraints, Technical Report CRES-04-009, USC, 2004.

18. LaValle, S. M., and Kuffner, J. J., Jr., "Randomized Kinodynamic Planning," *1999 IEEE International Conference on Robotics and Automation*, 1, pp. 473–479, 1999.

19. Harnad, S., "The Symbol Grounding Problem," *Physica D*, 42, pp. 335–346, 1990.

20. Dickmanns, E. D., "A General Dynamic Vision Architecture for UGV and UAV," *Journal of Applied Intelligence*, vol 2, pp. 251–270, Kluwer Academic Publishers, Boston, The Netherlands, 1992.

21. Jochem, T. and Pomerlau, D., "Vision Based Neural Network Road and Intersection Detection," in Herbert, M. H., Thorpe, C., and Stentz, A. (eds) *Intelligent Unmanned Ground Vehicles*, pp. 73–86, Kluwer Academic Publishers, Boston, MA, 1997.

22. Schwartz, J. T. and Sharir, M., "A Survey of Motion Planning and Related Geometric Algorithms," *Artificial Intelligence*, 37, pp. 157–169, 1988.

23. Liu, Y., Emery, R., Chakrabarthi, D., Burgard, W., and Thrun, S., "Using EM to Learn 3D Models of Indoor Environments with Mobile Robots," in Brodley, O. E. and Danyluk, A. P. (eds) *Proceedings of the Eighteenth International Conference on Machine Learning (ICML), June 28–July 01, 2001*, pp. 329–336, Morgan Kaufmann Publishers, San Francisco, CA, 2001.

24. Fabrizi, E. and Saffiotti, A., "Augmenting Topology-Based Maps with Geometric Information," *Robotics and Autonomous Systems*, 40, pp. 91–97, 2002.

25. Kortenkamp, D. and Weymouth, T., "Topological Mapping for Mobile Robots using a Combination of Sonar and Vision Sensing," *Proceedings of the Twelfth National Conference on Artificial Intelligence*, pp. 979–984, Menlo Park, AAAI, AAAI Press/MIT Press, July 1994.

26. Kuipers, B. and Byun, Y.-T., "A Robot Exploration and Mapping Strategy Based on a Semantic Hierarchy of Spatial Representations," *Journal of Robotics and Autonomous Systems*, 8, pp. 47–63, 1991.

27. Matarić, M., A distributed model for mobile robot environment-learning and navigation. Master's thesis, MIT, Cambridge, MA, January 1990. Also available as MIT AI Lab Technical Report AITR-1228.

28. Thrun, S., "Learning Metric-Topological Maps for Indoor Mobile Robot Navigation," *Artificial Intelligence*, 99, pp. 21–71, 1998.

29. Laird, J. E., Newell, A., and Rosenbloom, P. S., "Soar: An Architecture for General Intelligence," *Artificial Intelligence*, 33, pp. 1–64, 1987.

30. Newell, A. and Simon, H., GPS, *A Program that Simulates Human Thought*, McGraw-Hill, New York, 1963.

31. Pearson, J. D., Huffman, S. B., Willis, M. B., Laird, J. E., and Jones, R. M., "A Symbolic Solution to Intelligent Real-Time Control," *Robotics and Autonomous Systems*, 11, pp. 279–291, 1993.

32. Etherington, D., "What Does Knowledge Representation Have to Say to Artificial Intelligence?" *Proceedings of the Fourteenth National Conference on Artificial Intelligence*, p. 762, Menlo Park, Calif., AAAI Press, 1997.

33. Minsky, M., "A Framework for Representing Knowledge," in Winston, P. (ed.), *The Psychology of Computer Vision*, pp. 211–277, McGrow Hill, New York, 1975.

34. Volpe, R., Estlin, T., Laubach, S., Olson, C., and Balaram, J., "Enhanced Mars Rover Navigation Techniques," *Proceedings of the IEEE International Conference on Robotics and Automation (ICRA)*, San Francisco, CA, 2000.

35. Wasson, G., Kortenkamp, D., and Huber, E., "Integrating Active Perception with an Autonomous Robot Architecture," *Robotics and Automation Journal*, 29, pp. 175–186, 1999.

36. Lenat, D., Guha, R., Pittman, K., Pratt, D., and Shephard, M., "CYC: Toward Programs with Common Sense," *Communications of the ACM*, 33, pp. 30–49, 1990.

37. Genesereth, M. and Fikes, R., "Knowledge Interchange Format," *Stanford Logic Report Logic-92-1*, Stanford University, 1992.

38. Harmelen, F. and McGuiness, D., "OWL Web Ontology Language Overview," W3C web site: http://www.w3.org/TR/2004/REC-owl-features-20040210/, 2004.

39. Schlenoff, C., Washington, R., and Barbera, T., "Experiences in Developing an Intelligent Ground Vehicle (IGV) Ontology in Protege," *Proceedings of the 7th International Protege Conference*, Bethesda, MD, 2004.

40. Stanford Medical Informatics, "The Protege Homepage," http://protege. stanford.edu/, 2005.

41. Kortenkamp, D., Bonasso, P., and Murphy, R. (eds), *Articial Intelligence and Mobile Robots: Case Studies of Successful Robot Systems*. MIT Press, Cambridge, MA, 1998.

42. Brooks, R. "A Robust Layered Control System for a Mobile Robot," *IEEE Journal of Robotics and Automation*, 2, 14–23, 1986.

43. Brooks, R., *Cambrian Intelligence: The Early History of the New AI*, MIT Press, Cambridge, MA, 1999.

44. Kuipers, B., "The Spatial Semantic Hierarchy," *Artificial Intelligence*, 119, pp. 191–233, 2000.

45. Cassimatis, N., Trafton, G., Bugajska, M., and Schultz, A., "Integrating Cognition, Perception and Action through Mental Simulation in Robots," *Robotics and Autonomous Systems*, 49, 13–23, 2004.

46. Albus, J. et al., "4D/RCS Version 2.0: A Reference Model Architecture for Unmanned Vehicle Systems," *NISTIR 6910*, August 2002.

47. Shoemaker, C. and Bornstein, J. A., "Overview of the Demo III UGV Program," *Proceedings of the SPIE Robotic and Semi-Robotic Ground Vehicle Technology Conference*, 3366, pp. 202–211, 1988.

48. Lacaze, A., Murphy, K., and Delgiorno, M., "Autonomous Mobility for the Demo III Experimental Unmanned Vehicles," *Proceedings of the AUVSI 2002 Conference*, Orlando, FL, July 8–12, 2002.

49. Levine, W. (ed.), *The Control Handbook*. CRC Press, Boca Raton, FL, 1996.

50. Balakirsky, S., *A Framework for Planning with Incrementally Created Graphs in Attributed Problem Spaces*, IOS Press, Berlin, Germany, 2003.

51. Albus, J., *Brain, Behavior, and Robotics*, McGraw-Hill, New York, 1981.

52. Blum, A. L. and Furst, M. L., "Fast Planning Through Planning Graph Analysis," *Artificial Intelligence*, 90, pp. 281–300, 1997.

53. Bacchus, F., "The AIPS '00 Planning Competition," *AI Magazine*, 22, pp. 47–56, 2001.

54. Schlenoff, C., Madhavan, R., Albus, J., Messina, E., Barbera, T., and Balakirsky, S., "Fusing Disparate Information within the 4D/RCS Architecture," *Proceedings of the 8th International Conference on Information Fusion*, Philadelphia, PA, July 2005.

55. The OWL Services Coalition, "OWL-S 1.0 Release," http://www.daml.org/services/owls/ 1.0/owl-s.pdf, 2003.

56. Barbera, A., Messina, E., Huang, H., Schlenoff, C., and Balakirsky, S., "Software Engineering for Intelligent Control Systems," *Special issue on Software Engineering for Knowledge-Intensive Systems of Künstliche Intelligenz*, 3, pp. 22–26, 2004.

57. Hong, T., Balakirsky, S., Messina, E., Chang, T., and Shneier, M., "A Hierarchical World Model for an Autonomous Scout Vehicle," *Proceedings of the SPIE 16th Annual International Symposium on Aerospace/Defense Sensing, Simulation, and Controls*, Orlando, FL, April 1–5, 2002.

58. Stentz, A., "Optimal and Efficient Path Planning for Partially-Known Environments," *Proceedings of the IEEE International Conference on Robotics and Automation (ICRA)*, vol. 4, pp. 3310–3317, 8–13 May 1994.

59. Astrom, K. and Wittenmark, B., *Adaptive Control*, Addison-Wesley, Reading, MA, 1995.

60. Lee, J., Likhachev, M., and Arkin, R., "Selection of Behavioral Parameters: Integration of Discontinuous Switching via Case-Based Reasoning with Continuous Adaptation via Learning Momentum," *Proceedings of the 2002 IEEE International Conference on Robotics and Automation (ICRA)*, vol. 2, pp. 1275–1281, 2002.

61. Province, R., Uschold, M., Smith, S., Balakirsky, S., and Schlenoff, C., "Ontology-based Methods for Enhancing Autonomous Vehicle Path Planning," *Robotics and Autonomous Systems Journal: Special Issue on the 2004 AAAI knowledge Representation and Ontologies for Autonomous Systems Spring Symposium*, vol. 49, pp. 123–133, 2004.

62. Schlenoff, C., Balakirsky, S., Uschold, M., Provine, R., and Smith, S., "Using Ontologies to Aid in Navigation Planning in Autonomous Vehicles," *Knowledge Engineering Review*, 18, pp. 243–255, 2004.

63. Schlenoff, C., Balakirsky, S., Barbera, T., Scrapper, C., Ajot, J., Hui, E., and Paredes, M., "The NIST Road Network Database: Version 1.0," *National Institute of Standards and Technology (NIST)*, NISTIR 7136, 2004.
64. Messina, E., Evans, J., and Albus, J., "Evaluating Knowledge and Representation for Intelligent Control," *Proceedings of the 2001 Performance Metrics for Intelligent Systems (PerMIS) Workshop*, in association with IEEE CCA and ISIC, Mexico City, Mexico, 2001.
65. Madhavan, R. et al., "Issues in Autonomous Navigation of Underground Vehicles," *Journal of Mineral Resources Engineering*, 8, pp. 313–324, 1999.
66. Madhavan, R. and Messina, E., "Performance Evaluation of Temporal Range Registration for Unmanned Vehicle Navigation," *Proceedings of the 2004 Performance Metrics for Intelligent Systems Workshop*, NIST Special publication, 1037, August 2004.
67. Schlenoff, C., "Linking Sensed Images to an Ontology of Obstacles to Aid in Autonomous Driving," Ontologies and the Semantic web. *Papers from the 2002 AAAI Workshop WS-02-11*, pp. 56–62, AAAI Press, Menlo Park, CA, 1998.
68. Barbera, A., Albus, J., Messina, E., Schlenoff, C., and Horst, J., "How Task Analysis Can Be Used to Derive and Organize the Knowledge For the Control of Autonomous Vehicles," *Robotics and Autonomous Systems*, 49, 67–78, 2004.
69. Barbera, A., Horst, J., Schlenoff, C., Wallace, E., and Aha, D. W., "Developing World Model Data Specifications as Metrics for Sensory Processing for On-Road Driving Tasks," *Proceedings of the 2003 Performance Metrics for Intelligent Systems, NIST Special Publication 1014*.
70. Lacaze, A., "Hierarchical Planning Algorithms," *Proceedings of the Sixteenth SPIE International Symposium on Aerospace/Defense Sensing, Simulation, and Controls*, Orlando, FL, April 1–5, 2002.
71. Davis, R., Shrobe, H., and Szolovits, P., "What is a Knowledge Representation?" *AI Magazine*, 14, 17–33, Spring 1993.

BIOGRAPHIES

Elena Messina is leader of the Knowledge Systems Group in the Intelligent Systems Division at the National Institute of Standards and Technology. She has worked extensively on various aspects pertaining to autonomous mobile robots, ranging from knowledge representation and software architectures to development tools and simulation environments. She plays a leadership role in the performance evaluation of intelligent systems. She has organized and chaired the annual Performance Metrics for Intelligent Systems workshop series since 2000. Under her direction, several projects have been established to define the performance requirements for mobile robots and develop test methods to evaluate their effectiveness. These include the development and world-wide replication of the NIST-developed reference test arenas used in theinternational RoboCup Rescue and AAAI Mobile robot competitions. Ms. Messina has recently begun work on performance standards for bomb-disposal robots and for urban search and rescue robots.

Stephen Balakirsky received the Ph.D. degree from the University of Bremen, in Bremen, Germany. He is currently a researcher in the Knowledge Systems Group of the Intelligent Systems Division at the National Institute of Standards and Technology. He has over 15 years of experience in multiple areas of robotic systems that include simulation, autonomous plan and behavior generation, human-computer interfaces, automatic target acquisition, and image stabilization. He has been co-organizer of several workshops on knowledge representations for autonomous systems and is extensively involved in the RoboCup Rescue virtual competition and real/virtual robotic operation. Dr. Balakrisky's research interests include planning systems, simulation development environments, knowledge representations, world modeling, and architectures for autonomous systems.

13 Algorithms for Planning under Uncertainty in Prediction and Sensing

Jason M. O'Kane, Benjamín Tovar,
Peng Cheng, and Steven M. LaValle

Contents

13.1   INTRODUCTION AND PRELIMINARIES

For mobile robots, uncertainty is everywhere. Wheels slip. Sensors are affected by noise. Obstacles move unpredictably. Truly autonomous robots (and decision makers or agents in general) must act in ways that are robust to these sorts of failures and unexpected events which we may think of in general as *uncertainty*. In this chapter, we attempt to meet uncertainty head-on by explicitly modeling it and reasoning about it. We use the term *decision theoretic planning* to refer to this broad class of planning methods characterized by explicit accounting for uncertainty. We will consider a number of formulations for the problem of planning under uncertainty and present algorithms for planning under these formulations.

Uncertainty can take many forms, but for brevity and clarity we will restrict our attention to only two important types:

- *Prediction uncertainty* occurs when the effects of actions are not fully predictable. This can be thought of as an uncertainty in *future* states.
- *Sensing uncertainty* is uncertainty in the *current* state. This occurs, for example, in robots that have limited or imperfect sensing. We also admit the case where robots have no sensing at all.

Some systems can be adequately modeled without either form of uncertainty. Problems in this category can still be quite challenging and are the subject of many earlier chapters in this book. Problems with only prediction uncertainty are addressed in Section 13.2. This manner of formulation is appropriate for robots in environments in which the effects of an action are not fully predictable, but with sufficient sensing capability to fully determine the effects of each action a posteriori. When a robot's sensors are no longer adequate to fully determine the current state, the problem moves from the familiar *state space* to a richer space called an *information space*. Formulations with sensing uncertainty — with or without prediction uncertainty — are the topic of Section 13.3.

In the remainder of this section, we discuss some preliminary ideas that are relevant no matter what sort of uncertainty is present.

Uncertainty as a game against nature A unifying theme will be the idea of uncertainty as a "game against nature." Imagine an external decision maker called *nature* whose decisions determine the values of all uncertain parameters. Executing a plan becomes an interaction with nature as well as with the environment. Both our robot and nature make decisions and the outcome is fully determined given both of these decisions. In a sense, we are pushing all of the uncertainty in a system off to nature. Then, if we can develop some model for how nature will make its decisions, we can build plans to react accordingly. We use the term *uncertainty model* for this description of how nature will make its decisions.

The uncertainty model we select will directly influence the solution concepts we use. That is, an uncertainty model determines the answer to the question "What does 'optimal' mean?" As a result, the mechanics of each planning algorithm will also change. In this chapter, we will consider two distinct types of uncertainty models:

- Under *nondeterministic* models [1, 2], uncertainty is expressed as a set of possible outcomes. This model is also sometimes called the "possibilistic," "worst case," or "set membership" model. Domains in which firm guarantees are required or that involve interaction with a strong antagonist are good candidates for nondeterministic uncertainty models.
- Under *probabilistic uncertainty* [3] we express uncertain events in terms of a conditional probability distribution over possible outcomes, given certain current conditions. This model is particularly well-suited for cases where uncertainty arises from precision errors in sensing or actuation, or from random exogenous events.

The reader should note that legitimate criticisms can be leveled against both of these uncertainty models, some of which are elaborated in Section 13.2.1.2. Consequently, selecting an uncertainty model can sometimes be more of an art than a science. Most of the algorithms we will present are essentially independent of uncertainty model in the sense that they can be adapted to the type of uncertainty we select. Generally, we will derive similar but distinct versions for these two uncertainty models.

What is a plan? The concept of a solution for a planning problem in the absence of uncertainty is well understood: We seek a sequence of actions that transforms the system of interest from an initial state into a goal region, possibly optimizing some cost functional along the way. Uncertainty will force us to reconsider this notion of what a solution is.

Certainly the idea of a solution as a sequence of actions is made inadequate by the introduction of prediction uncertainty. Since state transitions are not fully predictable, we must prepare our agent to act in any state it may reach, rather than only those along a single path we have intended for it. Sensing uncertainty complicates the matter further because the agent will no longer even know its current state with certainty and instead must be able to react to any sensor/action history it encounters. These ideas will be made more formal in subsequent sections. The important idea here is that by allowing uncertainty we are forced to revise our notion of what constitutes a plan; for each new formulation we study, we will ask "What is a plan?"

Discrete vs. continuous spaces Many decision-theoretic planning algorithms are easiest to understand and implement under the assumption the spaces of states, actions, and observations are finite, or at least countable. Indeed, we will adopt this assumption in our initial presentations of most techniques. However, in robotics, the most natural models often involve continuous spaces. For this reason, we must pay careful attention to how these methods can be used to deal with continuous-space problems. Any algorithm designed for a digital computer must have discrete versions of these spaces in some way. Such discrete spaces will generally fall into one of the two broad categories given below:

- *Critical events:* For some problems, there is a natural, finite partition of the state or action space into equivalence classes in such a way that the planning problem can be solved by considering only these equivalence classes, rather than individual states or actions.
- *Sampling:* When no critical event decomposition is available, we can resort to techniques that approximate continuous state or action spaces by a finite selection of *samples*.

13.2 PLANNING UNDER PREDICTION UNCERTAINTY

We now begin with algorithms for planning with uncertainty in prediction. Our primary concern here is the need for feedback. Since we cannot plan an explicit sequence of states, we must instead prepare our decision maker for any state it may encounter. Thus we replace the usual action sequences with functions called *policies* that map from state space to action space. To simplify the presentation, we begin with a certain class of degenerate planning problems, namely those in which only a single decision needs to be made. The appropriate extensions to allow multi-stage decision making (i.e., planning) will be made in Section 13.2.2.

13.2.1 Making a Single Decision

Let us first consider the problem of making a single decision in the face of uncertainty in the outcome. We will model this uncertainty as decision to be made by another decision maker called *nature*. To formalize, a *single-stage decision problem* is defined by:

- A nonempty action set U that represents the set of choices available to our robot.
- A nonempty parameter set Θ that represents set of choices available to nature. This set should encode all of the uncertainty in the outcome of our agent's decision. In other words, given u and θ, the outcome is fully determined. The value of θ is hidden from the robot.
- A cost (or loss) function $L : U \times \Theta \rightarrow \mathbb{R}$ encoding the relative undesirability of each possible outcome. This is the quantity we will want to minimize. Equivalently, we may define a *reward function* we attempt to maximize.
- An uncertainty model for Θ. Under probabilistic uncertainty, this is the distribution $P(\theta)$. Under nondeterministic uncertainty, we need only a set of possibilities for θ. We may assume that any $\theta \in \Theta$ is allowed, hence, no additional information needs to be specified. (Nondeterministic uncertainty will not be so simple for later formulations.)

The objective is to choose a u that will result in the smallest possible $L(u, \theta)$. However, the outcome of any particular trial is unpredictable. Instead, we will use the uncertainty model for θ to describe an anticipated outcome. Under nondeterministic uncertainty, the best we can do is to consider *worst case cost*. The worst case optimal decision u^* is

$$u^* = \operatorname*{argmin}_{u \in U} \max_{\theta \in \Theta} L(u, \theta) \tag{13.1}$$

With the probabilistic uncertainty model, the choice of θ is random, so the relevant measure is the *expected cost*. The decision u^* that minimizes expected cost is

$$u^* = \operatorname*{argmin}_{u \in U} E_\theta[L(u, \theta)] \tag{13.2}$$

$$= \operatorname*{argmin}_{u \in U} \sum_{\theta \in \Theta} P(\theta) L(u, \theta) \tag{13.3}$$

In either case, a plan is simply a choice of some $u \in U$ and the problem can be solved with ordinary optimization techniques.

13.2.1.1   Including an observation

The previous formulation gave the decision maker no special information about what selection would be made for θ on a particular trial. We may extend the model by including an *observation space* Y. Each $y \in Y$ will correspond to a measurement or reading that we can think of as giving the decision maker a "hint" about the θ that will be selected. The decision maker is given some $y \in Y$ and can use this value when selecting a $u \in U$. Thus, a plan is a *decision rule* (or *strategy* or *policy*) $\gamma : Y \rightarrow U$. The presence of observations will change our uncertainty models to be conditioned on the value of y:

- *Nondeterministic:* We assume that y restricts the set of choices available for θ. This can be expressed as a function $F : Y \rightarrow 2^{\Theta}$ so that $F(y) \subseteq \Theta$ represents possible choices for θ given y. Now the optimal decision rule γ^* is simply the one that makes the best worst-case decision for each y:

$$\gamma^*(y) = \operatorname*{argmin}_{u \in U} \ \max_{\theta \in F(y)} L(u, \theta) \qquad (13.4)$$

  Notice that the only change from (13.1) is that the max operation is over only $F(y)$, rather than all of Θ as before.
- *Probabilistic:* The distribution for θ is now conditioned on y. That is, for each $y \in Y$ and $\theta \in \Theta$, we assume that the conditional probability $P(\theta|y)$ is known.

  Given y and u, we can write the expected cost (also called *conditional Bayes risk* in this context) as

$$E^{\theta}[L(u, \theta)] = \sum_{\theta \in \Theta} P(\theta|y)L(u, \theta) \qquad (13.5)$$

The decision rule to minimize this is

$$\gamma^*(y) = \operatorname*{argmin}_{u \in U} \sum_{\theta \in \Theta} P(\theta|y)L(u, \theta) \qquad (13.6)$$

Two prominent examples of single-stage decision-making with observations are *parameter estimation* [4,5] and *classification* [6–8]. In both, we have $U = \Theta$ and $L(x, \theta) = 0$ if and only if $u = \theta$. The observation y will give some information about θ, perhaps as a feature vector or a noise-tainted estimate of θ.

13.2.1.2 Criticisms of decision theory

This is an appropriate point to scrutinize the assumptions implicit in the the use of decision-theoretic methods.

Generating cost functions. First, most decision-theoretic methods depend on a cost function L which must be selected by hand for each problem. Choosing an appropriate cost function may be difficult. *Utility theory* [5,9,10] deals with the existence and, to a lesser degree, construction of these cost functions under the assumption that the decision maker is, in a precisely defined way, reasonably rational. Note also that some formulations can be reworked to eliminate the need for quantification of costs. For example, the minimax formulation of (13.1) really only requires a total ordering on $U \times \Theta$, rather than a real-valued cost function, to make sense. More generally, many decision-theoretic methods can be augmented with *sensitivity analysis*, which is a way of quantifying the amount of disturbance in L needed to make some change in the optimal policy. The idea is that if the policy is fairly robust to changes in L, then a poorly crafted cost function will not have much effect on the decisions made.

Pessimism and nondeterministic uncertainty. Nondeterministic models for uncertainty are often criticized for being overly pessimistic. In fact, using non-deterministic uncertainty with worst-case analysis can cause serious limitations on the planning problems that can be solved. Section 13.2.3.1 will highlight this problem in the context of the convergence of value iteration. Of course, the fact that we express uncertainty as a set of possible outcomes does not constrain us to worst case analysis. One can easily imagine an optimistic "best-case" version of (13.4):

$$\gamma^*(y) = \operatorname*{argmin}_{u \in U} \ \min_{\theta \in F(y)} L(u, \theta) \tag{13.7}$$

This is still unsatisfying because we have simply traded excessive pessimism for an equal measure of optimism. A compromise approach called *Hurwicz weighting* involves selecting a parameter $\alpha \in [0, 1]$ that is in some sense a "coefficient of optimism." We can use α to blend (13.4) with (13.7):

$$\gamma^*(y) = \operatorname*{argmin}_{u \in U} \left\{ \alpha \left[\min_{\theta \in F(y)} L(u, \theta) \right] + (1 - \alpha) \left[\max_{\theta \in F(y)} L(u, \theta) \right] \right\} \tag{13.8}$$

What is probability? There is also debate about the proper understanding of probabilities. The *Bayesian* interpretation views probability as a belief about a single trial. This is essentially the interpretation we have used so far. Given y, a Bayesian thinks of $P(\theta|y)$ as a degree of belief that nature will select θ. In contrast, the *frequentist* interpretation believes that probability is only properly understood in the limit as the number of trials goes to infinity; a probability value says nothing to a frequentist about the next trial, but only about the limit of an infinite sequence of trials. Frequentist interpretations of probability have led to a different, more conservative form of decision theory [10].

13.2.2 Making a Sequence of Decisions

In the previous section, we considered the problem of making a single decision in the face of some uncertainty in the outcome. We may think of planning under prediction uncertainty as a generalization of this idea by introducing a state space X and allowing a sequence of successive decisions to influence the system's transitions between states in X.

We divide time into *stages* and number them starting with one. Both the robot and nature make a decision at each stage. For the moment, suppose that the number of stages is limited to K. We will relax this restriction momentarily. Let $\tilde{u} = (u_1, u_2, \ldots, u_K)$ and $\tilde{\theta} = (\theta_1, \theta_2, \ldots, \theta_K)$ denote the sequences of decisions made by the robot and nature respectively. Given an initial state x_1, we can define a state sequence $\tilde{x} = (x_1, x_2, \ldots, x_{K+1})$ according to a deterministic transition function: $x_{k+1} = f(x_k, u_k, \theta_k)$. Figure 13.1 summarizes this situation for a single stage. To state the problem more formally, we need:

- A nonempty state set X.
- A nonempty action set U. Alternatively, the set of available actions may depend on the current state, that is, we have a set $U(x)$ for each $x \in X$. Since this variation only clutters the notation without making the problem more interesting, we assume that the same actions are available from each state. One possible realization of this action set is that lower-level techniques like motion planning, map building, and manipulation are implementations of the abstract actions we consider. This sort of layered approach has been used in a number of successful robotic systems [11–16].
- A nature action set Θ. As with U, nature's available actions may depend on x.
- A deterministic state transition function $f : X \times U \times \Theta \rightarrow X$.

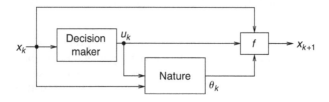

FIGURE 13.1   Planning with prediction uncertainty.

- An initial state x_1.
- A *stage-additive* cost functional

$$L(\tilde{x}, \tilde{u}, \tilde{\theta}) = \sum_{k=1}^{K} l(x_k, u_k, \theta_k) + l_F(x_{K+1}) \qquad (13.9)$$

The cost functional L is defined in terms of *single-stage cost function* $l : X \times U \times \theta \rightarrow \mathbb{R} \cup \{\infty\}$ that gives the cost for each possible transition, and a *termination cost function* $l_F : X \rightarrow \mathbb{R} \cup \{\infty\}$ that gives a cost for being in each state when execution ends after K stages. Sometimes it will be convenient to discuss the special case where the single-stage cost depends only on x_k and u_k. In such cases we write simply $l(x_k, u_k)$.
- A goal region X_G. For each $x_g \in X_G$, we require $l_F(x_g) = 0$.
- An uncertainty model for Θ. As usual, we allow either probabilistic or nondeterministic uncertainty. For nondeterministic uncertainty we need for each x and u a set of possibilities $\Theta(x, u)$. In the probabilistic case, we need a distribution $P(\theta | x, u)$.

Feasible planning. As an example, suppose we are not interested in optimizing any cost measure but only in reaching X_G. To accomplish this, we can set $l(x, u) = 0$ for all $x \in X$ and $u \in U$ and set

$$l_F(x) = \begin{cases} 0 & \text{if } x \in X_G \\ \infty & \text{otherwise} \end{cases} \qquad (13.10)$$

With this cost functional, any plan execution that terminates in X_G will have cost 0; any execution that terminates outside X_G will have infinite cost.

Allowing executions of indefinite length. Now we relax the assumption that our decision maker will act for a predetermined number of stages. Introduce into U a fictitious *termination action* u_F which indicates the decision maker's intention to end the execution. Create a fictitious state x_F, to which selecting u_F always leads. Select $U(x_F) = \{u_F\}$ so that once the agent has terminated, it cannot restart. Lastly, assign $l(x_F, u_F, \theta) = 0$ for all θ.

Now we imagine that stages continue infinitely, so that $\tilde{x}$, $\tilde{u}$, and $\tilde{\theta}$ become infinite sequences and the accumulated cost for an execution is

$$L(\tilde{x}, \tilde{u}, \tilde{\theta}) = \sum_{k=1}^{\infty} l(x_k, u_k, \theta_k) \qquad (13.11)$$

Now we can *define* K in terms of the actions selected, instead of assuming it is known ahead of time:

$$K = \min\{k | u_k = u_F\} \qquad (13.12)$$

If the robot eventually selects u_F, then K is well defined. We neglect cases in which the robot never chooses u_F because the cost of such an execution will generally increase without bound. By defining $l_F(x) = l(x, u_F, \theta)$ for all x and θ, we ensure that (13.9) still holds.

Defining an optimal policy. A solution to this type of problem is a *policy* γ : $X \to U$ that produces an action for each state. In the sequel, the terms plan and policy are interchangeable.

Consider nondeterministic uncertainty. Just as we did in the single-stage case, we want to select a policy that minimizes the worst-case cost. For a single decision, that maximization was over nature's choices for θ. Now the cost of a single execution of a plan depends on the entire sequence of choices made by both the robot and nature, namely $\tilde{u}$ and $\tilde{\theta}$, as well as $\tilde{x}$, which they determine. For a policy π, let $\mathcal{H}(\pi, x_1)$ denote the set of all such histories that can result from executing π starting at x_1. This is the set over which we must consider the worst case. Let $G_\pi(x_1)$ denote the worst-case cost of executing the policy π starting from state x_1:

$$G_\pi(x_1) = \max_{(\tilde{x}, \tilde{u}, \tilde{\theta}) \in \mathcal{H}(\pi, x_1)} L(\tilde{x}, \tilde{u}, \tilde{\theta}) \qquad (13.13)$$

The probabilistic case is similar, using expectation instead of worst-case analysis:

$$G_\pi(x_1) = E_{\mathcal{H}(\pi, x_1)}[L(\tilde{x}, \tilde{u}, \tilde{\theta})] \qquad (13.14)$$

For either sort of uncertainty, an *optimal policy* π^* is one that minimizes G:

$$\pi^* = \operatorname*{argmin}_{\pi} G_\pi(x_1) \qquad (13.15)$$

where the minimum is over all possible policies. Some readers may have noticed that this definition depends on the initial state x_1. Fortunately, there will exist a single policy that is optimal regardless of initial state. Suppose a policy π^* achieves the minimum in (13.15) for a fixed x_1 and let x denote a state reachable from x_1. If π^* were not optimal from x, then the goal could be reached from x_1 via x with lower cost than by executing π^*, contradicting the optimality of π^*. Consequently there will exist a single policy that is optimal regardless of initial state.

13.2.3 Methods for Finding Optimal Solutions

We have defined a general type of planning problem that includes uncertainty in state transitions and defined a notion of a solution to such a problem. Now we turn our attention to general-purpose solution methods for these problems. As one might expect, we must carefully weigh the trade-offs between generality, optimality, and tractability. Since optimality will come only at a high computational cost, we consider approximate solution methods in Section 13.2.4. In one sense, computing an optimal plan is just an optimization problem over the extremely large space of all policies. Fortunately, our optimality criterion G exhibits enough structure to make several different kinds dynamic programming possible.

13.2.3.1 Value iteration

Value iteration [17] is so named because it gradually develops a *value function* or *cost-to-go function* from which an optimal policy can be extracted. We will derive a recursive expression for this value function; this recurrence will lead directly to a planning algorithm. The derivation proceeds slightly differently depending on the uncertainty model.

Nondeterministic uncertainty. Fix a stage k and let $G_k^*(x_k)$ denote the worst-case cost that could accumulate if the robot executes π^* starting at x_k. We can write $G_k^*(x_k)$ as an alternation of minimum (from the optimality of π^*) and maximum (from the use of worst-case analysis) operations:

$$G_k^*(x_k) = \min_{u_k} \max_{\theta_k} \min_{u_{k+1}} \max_{\theta_{k+1}} \cdots \min_{u_K} \max_{\theta_K} \left\{ \sum_{i=k}^{K} l(x_i, u_i, \theta_i) + l_F(x_{K+1}) \right\}$$

(13.16)

Suppose we separate the first term $l(x_k, u_k, \theta_k)$ from the summation. Since this term only affects the outermost minimum and maximum operations, we can extract it from all of the others to get

$$G_k^*(x_k) = \min_{u_k} \max_{\theta_k} \left\{ l(x_k, u_k, \theta_k) + \min_{u_{k+1}} \max_{\theta_{k+1}} \cdots \min_{u_K} \max_{\theta_K} \right.$$
$$\left. \left[\sum_{i=k+1}^{K} l(x_i, u_i, \theta_i) + l_F(x_{K+1}) \right] \right\}$$

(13.17)

Notice that the innermost portion of (13.17) is simply $G_{k+1}(x_{k+1})$, leaving a simple recurrence:

$$G_k^*(x_k) = \min_{u_k} \max_{\theta_k} \{l(x_k, u_k, \theta_k) + G_{k+1}^*(x_{k+1})\} \qquad (13.18)$$

We also have a simple base case:

$$G_{K+1}^*(x_{K+1}) = l_F(x_F) \qquad (13.19)$$

The value iteration algorithm is a direct implementation of this recurrence. In iteration i of the algorithm, we use the values of G_{K-i+1}^* from the previous iteration (or, when $i = 0$, from the base case) to compute G_{K-i}^* according to (13.18). Of course, K, the number of actions taken by the robot before terminating, is not known ahead of time. One way to think of this is that the algorithm starts with the stage in which the robot terminates and move backward in time, considering progressively longer executions that lead to termination. The value of K never becomes relevant to the execution of the algorithm.

An implementation might be based on two tables, each with one entry for each state. At iteration i, one table holds the values of G_{K-i+1}^* while the other is filled in with G_{K-i}^*. After an iteration finishes, the roles of these tables can be swapped in preparation for the next iteration.

We want to terminate the value iteration algorithm when we reach an iteration in which no change occurs, that is, when an iteration i is reached in which $G_{K-i}^* = G_{K-i+1}^*$. If this occurs, then we will have reached a *stationary value function* $G^* = G_{K-i}^*$ that gives the worst-case cost that will result from executing an optimal policy starting from each state. This convergence will occur for all states from which there exists some policy that can guarantee reaching X_G. If no policy can guarantee reaching X_G, then no stationary value function exists and value iteration will not converge. Figure 13.2 shows a simple example of each case.

Finally, given a stationary value function G^*, we can extract an optimal policy π^* in a straightforward way. When the robot is in state x_k, we want to choose the u_k that achieves the minimum in (13.18):

$$\pi^*(x) = \operatorname*{argmin}_u \max_\theta \{l(x, u, \theta) + G^*(f(x, u, \theta))\} \qquad (13.20)$$

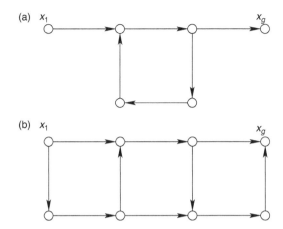

FIGURE 13.2 Two simple nondeterministic planning problems. In (a), nature can always prevent the decision maker from reaching the goal. In (b), all flows lead to the goal.

Probabilistic uncertainty. Under probabilistic uncertainty, a very similar approach will work, because of the linearity of expectation:

$$G_k^*(x_k) = \min_{u_k,\ldots,u_K} E_{\theta_k,\ldots,\theta_K} \left[\sum_{i=k}^{K} l(x_i, u_i, \theta_i) + l_F(x_{K+1}) \right] \quad (13.21)$$

$$= \min_{u_k} E_{\theta_k}[l(x_k, u_k, \theta_k) + G_{k+1}^*(x_{k+1})] \quad (13.22)$$

The base case is the same:

$$G_{K+1}^*(x_{K+1}) = l_F(x_F) \quad (13.23)$$

Equation (13.19) and Equation (13.22) provide the base case and recursive case for value iteration, which works in just the same way as in the nondeterministic case. Convergence, however, is an even thornier question than in the nondeterministic case, because of the possibility that the costs-to-go will converge only in the limit. Figure 13.3 shows an extremely simple example in which this is the case. This phenomenon will occur any time there is nonzero probability of being forced by nature to traverse cycles in the state space. For many applications, the costs-to-go will converge quickly to good approximations of the optimal values. More importantly, recall that we are not directly interested in G^*, but in the policy π^* we extract from it. Thus, we only need the cost-to-go to converge to a point where we are reasonably certain of which action is the correct choice from each state.

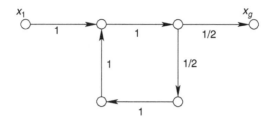

FIGURE 13.3 Probabilistic uncertainty can cause value iteration to converge only in the limit. Edges are labeled with transition probabilities. In this example, nature can cause executions of arbitrary length. However, executions that traverse the cycle in the graph many times are unlikely. Assuming the cost of each transition is 1, the cost-to-go for x_1 converges $2 + 4 \sum_{i=0}^{\infty} i(\frac{1}{2})^{i+1} + 1 = 7$.

Finally, when the dynamic programming iterations finish, we can use the resulting G^* to extract an optimal policy. The probabilistic analog to (13.20) is

$$\pi^*(x) = \operatorname*{argmin}_u E_\theta[l(x, u, \theta) + G^*(f(x, u, \theta))] \qquad (13.24)$$

13.2.3.2 Policy iteration

Value iteration was a dynamic programming technique in the space of states. Only after the stationary cost-to-go function (or an approximation of it) is reached can a policy be extracted. In contrast, *policy iteration* [17, 18] performs dynamic programming directly in the space of policies. At each iteration, a fully-formed policy is generated.

Each step of policy iteration has two parts: *policy evaluation*, in which the expected cost of executing the current policy is computed and *policy improvement*, in which this information is used to construct a policy better than the current one. To simplify notation, assume that the cost of each transition depends on only x and u, so that we can write $l(x, u)$ rather than $l(x, u, \theta)$.

Policy evaluation. First, how can we evaluate a fixed policy π? Recall that $G_\pi(x)$ denotes the expected cost of executing π starting at x. The values of $G_\pi(x)$ will serve as our criteria for evaluating π. We can derive an expression for $G_\pi(x)$ in a similar manner to the derivation of (13.22), but in which we restrict the available actions in each state to the single action suggested by π:

$$G_\pi(x) = E_\theta[l(x, \pi(x)) + G_\pi(f(x, u, \theta))] \qquad (13.25)$$

$$= l(x, \pi(x)) + \sum_{x' \in X} G_\pi(x')P(x'|x, u) \qquad (13.26)$$

The transition probability $P(x'|x, u)$ can be obtained by marginalizing over θ:

$$P(x'|x, u) = \sum_{\{\theta | f(x,u,\theta)=x'\}} P(\theta|x, u) \qquad (13.27)$$

Define $n = |X|$. Equation (13.26) is a linear equation with n unknowns, namely $G_\pi(x)$ for each x. If we make n copies of (13.26), one for each $x \in X$, we get a linear system with n variables and n equations. Solving this system with standard linear algebra methods (e.g., singular value decomposition [19]) gives values for $G_\pi(x)$.

Policy improvement. Now we will show how to use G_π to generate a new policy π' that is an improvement over π in the sense that $G_{\pi'}(x) \leq G_\pi(x)$ for all x. We can construct π' in a relatively direct way. For each x define $\pi'(x)$ according to

$$\pi'(x) = \operatorname*{argmin}_{u \in U} \left\{ l(x, u) + \sum_{x' \in X} G_\pi(x')P(x'|x, u) \right\} \qquad (13.28)$$

This is probably best understood in relation to (13.24). The real difference is that during execution of policy iteration, G^* is unknown. Instead, we use G_π as an estimate for G^*. Since π' will take the best action from x under the assumption that $G_\pi(x')$ is the cost-to-go after this step, we can conclude that π' is at least as good as π. If $\pi' = \pi$, then the algorithm has converged to π^* and can terminate.

  One important property of policy iteration is that, unlike value iteration, it is guaranteed to terminate in finite time. Since an improvement is made on each iteration, no policy will occur more than once. But there are only $|U|^{|X|}$ different policies to consider. Therefore, this algorithm will terminate with the optimal policy in at most $|U|^{|X|}$ iterations, but generally much faster than this.

13.2.3.3 Other methods

We have focused on only two optimal algorithms in order to provide some amount of depth to the subject, and because many other algorithms can be seen as variants of either policy iteration or value iteration. The versions we describe are a form of backward dynamic programming in the sense that they begin with termination and progress backward in time. Forward versions are also possible [17], but slightly more complex conceptually. We described value iteration as a series of sweeps across the state space performing updates but the ordering of updates allows more flexibility. In some cases this property can be exploited to find good policies faster using so-called *asynchronous methods* [20–22]. Under certain restrictions, Dijkstra's shortest-path algorithm can be adapted to account for uncertainty [23].

13.2.4 Methods for Finding Approximate Solutions

Now let us turn our attention to algorithms for planning that is only approximately optimal. Suboptimal planning is important for problems that are too complex to solve optimally and for situations in which resources for computation are limited. For example, if an autonomous robot suddenly discovers an error in its model of the world (say, an unexpected obstacle in its path), it must quickly *replan* under its new world model. In such a circumstance, a plan must be generated quickly and computing an optimal plan may not be possible. We have already seen one suboptimal planning algorithm — the prematurely terminated version of value iteration that arose for probabilistic problems that converge only in the limit. There are also a number of more specialized algorithms.

13.2.4.1 Certainty equivalent control

Allowing even only prediction uncertainty makes planning much more difficult. What happens if we ignore the uncertainty when generating a plan? This is the idea behind *certainty equivalent control*. More precisely, we create an uncertainty-free planning problem by assuming that uncertain parameters will take on "typical" values. So in our formulation, we might form a deterministic planning problem by defining a deterministic state transition function $\bar{f}$ according to the most likely successor:

$$\bar{f}(x_k, u_k) = f(x_k, u_k, \arg\max_{\theta_k} P(\theta_k | x_k, u_k)) \tag{13.29}$$

In the special case where states are numbers, a "typical" result might be the expected one:

$$\bar{f}(x_k, u_k) = f(x_k, u_k, E_{\theta_k}[x_k]) \tag{13.30}$$

By solving the planning problem with transition function $\bar{f}$, we get a plan for the original problem under f. Remarkably, for a certain classes of systems (e.g., linear systems with quadratic cost), this method has been shown to generate optimal plans [24].

13.2.4.2 Limited lookahead

Limited lookahead (or *rolling horizon approximation*) is an approximation technique that aims to reduce the computation required in value iteration. Suppose we have some estimate of the optimal cost-to-go $\hat{G} \approx G^*$. We use this as the base case for value iteration, replacing (13.19). (Some readers may recognize this as essentially the same method that drives computer game-playing, in which $\hat{G}$

is called an *evaluation function*.) If we run i rounds of value iteration with $\hat{G}$ as the base case, the resulting policy will be optimal for the simplified problem in which the decision maker acts for i stages before terminating with cost $\hat{G}(x_k)$. The similarity of this policy to π^* depends directly on the similarity of $\hat{G}$ to G^*.

One question that remains is how to select $\hat{G}$. One possibility is to use some heuristic method to generate a *base policy* $\hat{\pi}$ and use its cost-to-go as $\hat{G}$:

$$\hat{G} = G_{\hat{\pi}} \tag{13.31}$$

One-step lookahead algorithms built on a base policy in this way are called *rollout algorithms*. This rollout can be viewed as a single step of policy iteration in the sense that the cost-to-go function of one policy, $\hat{\pi}$, is used to create a new, improved policy.

13.2.5 Conquering Continuous Spaces

Until now, we have talked about methods which handle problems with finite state spaces and finite action sets. In many situations, especially in robotics, continuous state spaces and action sets are more natural. By continuous, we mean that either the state space, the action set, or both have an uncountably infinite number of elements. In this section, we will extend the methods mentioned above to problems with continuous spaces. The main difficulty is that the techniques we have presented depend on iterating over the elements of X and U. The key idea of the extension is to find a suitable finite representation of the original problem. Then the new problem can be solved with methods similar to those we have already developed.

The first step of the transformation process is to approximate the continuous state space with a finite sampling point set. These sampling points could be obtained by any of the several methods, such as random sampling [25], quasi-random sampling [26], grid sampling [27], or lattice point sampling [28–30]. In selecting one of these sampling methods, one must consider several issues, including the *uniformity* of the points (How well are the points spread out?) and *neighborhood structure* (Given a point in the underlying space, how easy it it to locate its neighbors in the sample set?). Some example sets of sampling points in the unit square are shown in Figure 13.4. A more thorough characterization and comparison of sampling techniques appears in Section 5.2 of Reference 23.

After the continuous state space is represented by a finite set, standard value iteration as described in Section 13.2.3.1 can be applied with the following modification. As the algorithm proceeds, we maintain the value function $G_k^*(x_k)$ only for states in the sampling set. Recall that the update equation (either [13.18]

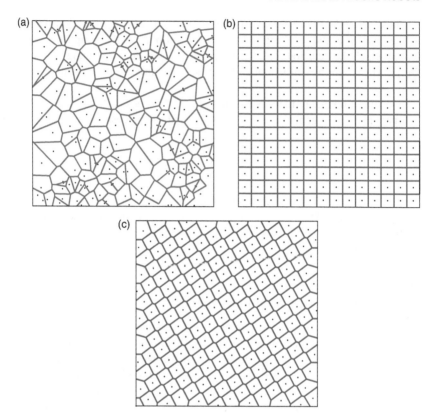

FIGURE 13.4   Three ways to select samples in the unit square. Dots represent samples, the lines show their respective Voronoi cells. Larger Voronoi cells indicate poor uniformity. (a) Pseudorandom samples. (b) Grid samples. (c) Lattice samples.

or [13.22]) depends on knowing $G^*_{k+1}(x_{k+1})$ for each choice of u_k and θ_k.[1] If x_{k+1} is not in the sample set, then $G^*_{k+1}(x_{k+1})$ will not be available, as illustrated in Figure 13.5.

   To make value iteration work, the value function at x_{k+1} needs to be approximated with values of states in the sampled set [31]. In References 32 and 33, a neural network is used to approximate the value function. A more conventional way is by *interpolation*. Interpolation involves designating some set of samples as "neighbors" of x_{k+1} and using as the value function at x_{k+1} some weighted combination of the value function at each of these neighbors. Interpolation has been most thoroughly studied for the case in which the sample set is a grid. In this case, the interpolation can be performed in at least two

---

[1] If U and Θ are themselves continuous, it may be necessary to sample them as well.

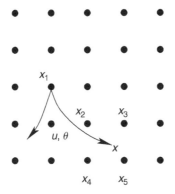

FIGURE 13.5   Dots denote sample states. Applying an action u and nature action θ on a sampled state x_1 moves the system to state x which is not in the sampled set. Some form of interpolation is needed to estimate the value function at x.

different ways:

- *Multi-linear interpolation.* In the 1D case, multi-linear interpolation is just linear interpolation between data points. In higher dimensions, the procedure is recursive:

  1. Choose any axis and project the point onto two faces that are perpendicular to the chosen axis.
  2. Use $(n-1)$-dimensional multi-linear interpolation to calculate the value function at these two points.
  3. Linearly interpolate to calculate the value of the given point according to the value of two points on the two faces.

  Multi-linear interpolation will process 2^n data points for one interpolation in n-dimensional state space. It can be very time-consuming for high-dimensional problems.

- *Simplex-based interpolation [34–37].* This method uses Kuhn triangulation to decompose the n-dimensional hypercube into $n!$ simplices, each of which has $n+1$ vertices. Then the simplex-based interpolation is to calculate the value according to the $n+1$ vertices of the simplex containing the given point. Since only $n+1$ data points and $O(n\log n)$ time will be needed in one interpolation, it is much more efficient than multi-linear interpolation method.

Finally, with the finite sampled set and a chosen interpolation method, we could run value iteration as in the discrete case, but interpolating to estimate the value function for states that are not in the sample set. Similarly, the optimal policy can be extracted from either (13.20) or (13.24), again using interpolation to estimate G^*. More details on sampling-based dynamic programming in continuous state spaces appear in References 17 and 38–41.

13.2.6 Variations

We conclude this section with a survey of variations to the problem formulated in Section 13.2.2.

13.2.6.1 Infinite horizon models

The model presented in Section 13.2.2 deals with planning problems with a well-defined goal set. The decision maker interacts with the environment for a finite period of time before selecting a termination action. What happens if we eliminate the termination actions and allow the robot to continue executing for an infinite number of stages? Problems of this type are called *infinite horizon* problems and have been studied extensively in artificial intelligence and stochastic control theory.

Since the process is infinite, we can omit x_1, X_G, and l_F from the model. However, the most striking change is in the cost functional $L(\tilde{x}, \tilde{u}, \tilde{\theta})$. Excluding the case in which there are cycles with zero or negative cost in which the robot can linger, allowing K to approach infinity in (13.9) will cause L to diverge. As a result, (13.9) is no longer a suitable optimality criterion. We must find a way of keeping the cost finite for an infinite sequence of actions. Two possibilities are:

- *Average cost per stage* (or *gain-optimal* cost). One way to keep the cost finite is to divide by the number of stages:

$$L(\tilde{x}, \tilde{u}, \tilde{\theta}) = \lim_{K \to \infty} \frac{1}{K} \sum_{k=1}^{K} l(x_k, u_k, \theta_k) \qquad (13.32)$$

  If l is bounded by some constant, then it is clear that L must remain less than this constant. One major problem with this model is that costs over any initial prefix are overshadowed by long-run performance [42].
- *Discounted cost*. Pick a parameter $\alpha \in (0, 1)$ called a *discount factor* and use α to define L in the following way:

$$L(\tilde{x}, \tilde{u}, \tilde{\theta}) = \sum_{k=1}^{\infty} \alpha^{k-1} l(x_k, u_k, \theta_k) \qquad (13.33)$$

The intuition is to place less weight on costs that occur further into the future. We may think of α as a measure of "far-sightedness." If α is increased, loss from later stages has greater influence on the value of L. The average cost model can be seen as a limiting case as α approaches 1 [43].

It is important to understand that α is a part of the definition of the optimal policy. A change in α can result in a change in which actions are considered optimal from each state. For this reason, the average cost model is sometimes preferred because it does not introduce any new parameters to tune. Regardless, discounted cost is the dominant model because it is a simple and mathematically manageable way to keep finite the cost of an infinite length execution.

The dynamic programming methods of Sections 13.2.3 can be adapted to find optimal policies for discounted costs, but care must be taken to ensure the stability and convergence properties of these algorithms. A detailed treatment of these methods and others for infinite horizon problems is given in Reference 43. In Section 13.2.6.2, we discuss reinforcement learning, which generally uses a discounted cost model, but assumes that the uncertainty model for θ is unknown.

13.2.6.2 Reinforcement learning

One of the major problems with decision-theoretic planning as we have presented it is that there is a heavy modeling burden associated with the assumption that an uncertainty model for Θ is given. In the probabilistic case, this is the assumption that the distribution $P(x'|x, u)$ is known. When X and U are finite, there are still $|X| \, |U|$ separate values needed to describe this distribution. In a physical system, each of these would require many trials to estimate accurately. For many nontrivial environments, this can be quite impractical.

An alternative is to force the robot to learn these probabilities along the way instead of specifying them upfront. The family of methods that takes this approach is generally called *reinforcement learning* (RL), and occasionally *neuro-dynamic programming* [32] (although that term has a somewhat more specific meaning) or *simulation-based methods* [17]. The primary difference from methods which assume that transition probabilities are known is that reinforcement learning is an *online* model. This means that there is no separation between planning and execution. Rather, as the robot interacts with the environment, it gradually refines its plan. We will very briefly describe an algorithm that, for reasons that will soon be obvious, is called *Q-learning* [44]. *Q*-learning is quite simple, but worth understanding because nearly all other RL algorithms can be seen as variations on the same basic themes.

We can think of Q-learning as a type of value iteration in which, instead of using G^* (which gives the value of each state), use a function $Q : X \times U \to \mathbb{R}$ that gives values for state–action pairs. More precisely, we define $Q(x, u)$ to be the expected cost of starting from state x, taking action u, and acting optimally (i.e., according to π^*) thereafter.

The algorithm works by maintaining a table that lists, for each state–action pair, an estimate $\hat{Q}(x, u)$ of the real $Q(x, u)$. We initialize $\hat{Q}(x, u)$ arbitrarily. After each action, the agent is informed of the new state x' and the cost l of the corresponding transition and an update to the table is performed:

$$\hat{Q}(x, u) \leftarrow (1 - \rho)\hat{Q}(x, u) + \rho \left(l + \min_{u' \in U(x')} \hat{Q}(x', u') \right) \qquad (13.34)$$

If we use the discounted-cost infinite horizon model (as is the custom in the reinforcement learning literature), we must include the discount factor:

$$\hat{Q}(x, u) \leftarrow (1 - \rho)\hat{Q}(x, u) + \rho \left(l + \alpha \min_{u' \in U(x')} \hat{Q}(x', u') \right) \qquad (13.35)$$

In either update rule ρ is a designer-specified *convergence rate* or *learning rate*. It has been shown (e.g., [8]) that with certain assumptions about the sequence of actions chosen, $\hat{Q}$ will converge to Q under this update rule. It may seem conspicuous that the update rule never mentions the transition probability $P(x'|x, u)$. In fact, Q-learning is an example of the so-called *model-free* algorithms that never build an explicit model of the transition probabilities. Instead, the probabilities are hidden by the fact that the update in (13.34) or (13.35) is performed repeatedly, with the distribution of resulting states chosen according to $P(x'|x, u)$. Thus, successor states that are more likely will have greater influence over $\hat{Q}(x, u)$.

Exploration vs. exploitation. To this point, we have shown how Q-learning maintains an estimate for the value of each state-action pair without saying anything about which actions to choose. Suppose $Q(x, u)$ is known for all x and u. Then the best action to choose (cf. [13.20]) is

$$\pi^*(x) = \operatorname*{argmin}_{u} Q(x, u) \qquad (13.36)$$

Unfortunately, the decision maker must choose actions without knowledge of the real Q. To obtain the best cost over a limited period of time, there is a tension between *exploring* the space in order to make $\hat{Q}$ a better estimate of Q and *exploiting* actions that have been effective so far — that is, actions for which $\hat{Q}$ is currently small.

The exploration–exploitation dilemma is an important enough problem in reinforcement learning that many different methods have been suggested to deal with it, including *initial optimism* [45], which assigns large initial values to each $\hat{Q}(x, u)$, ensuring that each action is tried often enough to "drive down" its $\hat{Q}$ value to near its true value, and ϵ-greedy policies [33] that select the current best action with probability ϵ and choose randomly otherwise. An in-depth study of this topic is given in Reference 46.

Temporal credit assignment. While the update rule (13.34) is guaranteed to converge, in practice this can require a large number of trials to reach a good estimate for Q. The issue is that of *credit assignment*: When a reward is received, to which actions do we attribute it? In (13.34), credit is assigned only to the action immediately preceding the reward. This is troubling because if a large reward (say, obtaining a Ph.D.) occurs, credit for this reward will initially only be granted to the action that immediately led to this reward (finishing a dissertation) and not to any earlier actions that made the reward possible (enrolling in graduate school). Only after earning many Ph.Ds (!) will the influence of this reward propagate backward to have an influence on the decision to enroll in graduate school.

To combat this problem, more aggressive credit-assignment schemes have been developed that endeavor to squeeze more out of each action taken by the decision maker. One of the most effective techniques is to maintain an *eligibility trace* — a list of recent state–action pairs along with a weight for each. Eligibility traces are so-named because they determine which $\hat{Q}(x, u)$ values are eligible to be updated after the next action. Recently visited states are marked as eligible. After receiving a new cost l, we perform an update similar to (13.34) for each eligible state–action pair. As time passes, the weight of each eligible state decays (reducing the amount of change in its $\hat{Q}$ on subsequent iterations) until it is finally removed from the eligible list.

13.2.6.3 Additional decision makers

Everything up to this point has focused on a single decision maker interacting with an uncertain environment. Uncertainty is modeled as a "game against nature." A broad class of generalizations can be made if we allow additional decision makers in the system, each with its own independent goals. This is the realm of *game theory* [47, 48], which is concerned with the general situation of multiple decision makers interacting in some way. The breadth and depth of game theory literature will force us to focus on just a few issues that are vital for planning in the presence of other decision makers.

Suppose we have n decision makers (not counting nature). At each stage, decision maker i will select an action from an action set U^i that has some influence on the resulting state. This means we must extend the state transition

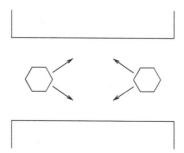

FIGURE 13.6   An illustration of the need for mixed strategies. The left robot attempts to pass through a corridor while the right robot attempts to block its progress. Each must independently decide whether to move to the left or right. If the left robot plays a pure strategy, the right robot can take advantage and always get in the way. A mixed strategy that chooses each direction equally often will enable the left robot to escape.

function:

$$f : X \times U^1 \times \cdots \times U^n \times \Theta \to X \qquad (13.37)$$

Each decision maker also has its own cost functional L^i which depends on all n actions selected at each stage. It is assumed that each decision maker has complete knowledge of all of the L^i's. The special case where $n = 2$ and $L^1(\tilde{x}, \tilde{u}, \tilde{\theta}) + L^2(\tilde{x}, \tilde{u}, \tilde{\theta}) = 0$ is unsurprisingly called a *zero-sum game*. This corresponds to the situation where two players are in direct competition for some limited resource.

What does a plan look like when there are multiple decision makers? The deterministic policies considered so far are no longer adequate. Figure 13.6 illustrates a very simple problem that requires a *mixed strategy* that selects actions at random according to some distribution. By contrast, the deterministic strategies we studied for the single-agent case are also called *pure strategies*.

With a single decision maker, we defined optimality in terms of the expected or worst-case cost. When there are multiple decision makers, optimality is usually defined in terms of *regret*, which is a measure of how much a decision maker could have improved his reward if he had known what actions the other players would take. For a single stage game in which the actions selected are $u^1, \ldots, u^n$, the regret R^i for player i is

$$R^i = L^i(u^1, \ldots, u^n) - \max_{u' \in U^i} L^i(u^1, \ldots, u', \ldots, u^n) \qquad (13.38)$$

For two-player zero-sum games, a pair of policies for which $R^1 = R^2 = 0$ is called a *saddle point*. A fundamental result in game theory is that if mixed strategies are allowed, then a saddle point will always exist. For nonzero-sum games and those with multiple players, the idea of a saddle points can be generalized to *Nash equilibria*, which are also based on the idea of eliminating regret.

13.3 PLANNING UNDER SENSING UNCERTAINTY

In this section we address the planning problem in which the knowledge of the robot's current state is limited, or not available at all. This accounts for the cases when the robot's sensors do not uniquely determine the current state of the robot (sensing uncertainty) and when the robot's control is not perfect (prediction uncertainty).

One common approach is to make an estimation of the current state, with all the information available, and determine some bound for the state uncertainty. Then the uncertainty may be *ignored*, and the algorithms of the previous sections may be applied. However, the state estimation may be completely avoided in the computation of a plan, that is, the robot may be able to reach/achieve its goal without ever determining its current state. This gives rise to the study of the *information space*, which will be the main topic of this section.

Information spaces have appeared throughout the robotics literature in many forms and under many different names. Information space concepts arise in maze searching [49], preimage planning [50], error detection and recovery [51], manipulation [52–55], bug algorithms [56, 57], gap navigation trees [58, 59], perceptual kinematic maps [60], perceptual equivalence classes and information invariants [61, 62], sensor-based planning [63], searching unknown dynamic environments, D^* [64], pursuit-evasion [65–69], probabilistic navigation [70], Bayesian localization and SLAM [71, 72], and guaranteed localization [73–75], and topological maps [76], to cite just a few examples.

In general, the robot can gather information about the state from the following sources:

- *Initial conditions.* Information the robot has about the task before the first sensing measurement is taken or the first action is performed. The particular initial condition for a planning problem, denoted by η_0, can have several forms:
  1. *Known state.* The initial state $x_1 \in X$ is given. Uncertainty appears when nature interferes with the state transition equation.
  2. *Nondeterministic.* A set $X_1 \subset X$ is given. The initial state is known to lie within X_1.
  3. *Probabilistic.* A probability distribution $P(x_1)$ over X is given.

- *Sensor observations.* Online measurements of the state are made. In general they do not give all the information of the state, either because some state variable cannot be measured (a sensor for it is not available), or due to limitations in the sensor construction, sensor resolution, disturbances due to noise, etc.
- *Previous actions.* The record of the actions may provide the robot with useful information. For example, under the assumption of perfect control, if the previous action was to move to the east, the current state is more to the east as the previous state, although neither the previous nor the current state is known.
- *Available actions.* The state may be inferred from knowledge of what actions are available to the robot.

13.3.1 Discrete State Spaces

We first describe the information space when X, the state space, is finite or countably infinite. The new element for computing a plan is that the robot does not have a complete knowledge of the current state, but it can measure it in some way through observations. Because of this, we begin our discussions with modeling the robot's *sensors*.

13.3.1.1 Sensors

A sensor is a device that provides some measurement of the current state. When the robot performs a sensing in the environment, the sensor *maps* the state space into the observation space Y. The observation space is the set of all possible readings of the sensor, giving "hints" of the current state. This is different from the case presented in Section 13.2.1.1, in which the observation only gave hints of the possible action that nature would take. The sensor mapping, denoted by h, takes several forms:

- *State sensor mapping.* Given a state $x \in X$, the observation $y = h(x) \in Y$ is completely determined.
- *State-nature sensor mapping.* Nature is allowed to interfere with the sensor measurements. Let $\Psi(x)$ denote a finite set of *nature sensing actions*, defined for each $x \in X$. The mapping produces an observation $y = h(x, \psi)$ for every $x \in X$ and for every $\psi \in \Psi(x)$. As with Θ in Section 13.2.1, the particular ψ chosen by nature is assumed to be unknown.
- *History-based sensor mapping.* This case is similar to the last one, but the observation may depend on previous states. If the plan is in stage k, the observation is $y = h_k(x_1, x_2, \ldots, x_k, \psi_k)$. In this case $\psi_k \in \Psi_k$ is the particular sensing action chosen by nature.

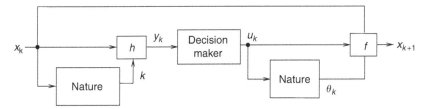

FIGURE 13.7 Planning with sensing uncertainty. Note that the decision maker does not have direct access to the state.

Adding sensing uncertainty to the model of Figure 13.1, yields the model presented in Figure 13.7. The decision maker does not have direct access to the state, which can only be measured through sensors.

13.3.1.2 Definition of the information space

Let X, U, and f follow the same definitions as in Section 13.2.2. If the plan is at stage k, we want to determine which information is available to the robot, either from the new observations, or the accumulation of previous information. It is assumed that the robot keeps a record of each of the observations made. Thus, the *observation history*, $\tilde{y} = (y_1, y_2, \ldots, y_k)$, is the ordered sequence of observations up to state k. Similarly, the *action history*, $\tilde{u} = (u_1, u_2, \ldots, u_{k-1})$, is the record of the actions taken. It runs until stage $k - 1$, because action u_{k-1} is applied in state x_{k-1}, to yield the current state x_k, where the observation y_k is made. Remember that η_0 denotes the initial condition. The *information state* at state k is defined as

$$\eta_k = (\eta_0, \tilde{u}_{k-1}, \tilde{y}_k) \qquad (13.39)$$

that is, the initial condition together with the history. Alternatively, an information state can be expressed recursively as

$$\eta_k = (\eta_{k-1}, u_{k-1}, y_k) \qquad (13.40)$$

since the difference between the previous and the current information state consists of the new observation made and the new action taken.

The set of all possible information states η_i for $1 \leq i \leq k$, is called the *information space*, $\mathcal{I}$. Similar to the case of prediction uncertainty, presented in Section 13.2.2, a plan in the information state is defined as a mapping π, but in this case using the information space. This yields $\pi : \mathcal{I} \to U$. The components of a planning problem for information spaces on countable state spaces are:

- A nonempty *state space*, X, which is either finite or countably infinite.
- A finite *action space*, U. It is assumed that U contains the special termination action u_F.
- A finite *nature action space*, $\Theta(x, u)$ for each $x \in X$ and $u \in U$.
- A *state transition equation*, f, that produces a state, $f(x, u, \theta)$ for every $x \in X$, $u \in U$, and $\theta \in \Theta(x, u)$.
- A finite or countably infinite *observation space*, Y.
- A finite *nature observation action space*, $\Psi(x)$ for each $x \in X$.
- A *sensor mapping*, h.
- An *initial condition*, η_0.
- A *goal set*, $X_G \subseteq X$.
- A real-valued additive cost functional L, which may be applied to any state-action history, $(\tilde{x}_{K+1}, \tilde{u}_K)$, to yield

$$L(\tilde{x}_{K+1}, \tilde{u}_K) = \sum_{k=1}^{K} l(x_k, u_k) + l_F(x_{K+1}) \qquad (13.41)$$

If the termination action, u_F, is applied at some stage k, then for all $i \geq k$, $u_i = u_F$, $x_i = x_k$, and $l(x_i, u_F) = 0$ if $x_i \in X_G$, or ∞ otherwise.

As before, the cost functional $L(\tilde{x}, \tilde{u})$ allows the evaluation of the quality of a plan. Since there is uncertainty in the state prediction and in the sensing, we can use either worst-case or expected-case analysis for evaluating plans. If $\mathcal{H}(\pi, \eta_0)$ denotes the set of all possible state–action histories given the plan π from the initial condition, the cost of the plan with worst-case analysis is

$$G_\pi = \max_{(\tilde{x}, \tilde{u}) \in \mathcal{H}(\pi, \eta_0)} L(\tilde{x}_{K+1}, \tilde{u}_K) \qquad (13.42)$$

If a probabilistic model of the uncertainty is known, the expected cost of a plan is

$$G_\pi = E_{\mathcal{H}(\pi, \eta_0)} L(\tilde{x}_{K+1}, \tilde{u}_K) \qquad (13.43)$$

13.3.2 Deriving Information States

In its original definition, the information space seems unmanageable. In fact, it only seems useful for planning problems where the number of states is very small, since the history representing an information state grows linearly with the number of stages. The main idea here is to map the original information space into a smaller space, ensuring that when a successful plan exists over the original space, a plan will exist also in the smaller space. As expected, in the general case, the smaller space will present plans that are feasible, but may not be optimal in the original space. For most of the planning problems asking for a feasible plan is already a challenging task.

In general, let $\kappa : \mathcal{I} \rightarrow \mathcal{I}^\circ$ denote a subjective mapping from an information space $\mathcal{I}$ to a *derived information space*, $\mathcal{I}^\circ$. Ideally, $\mathcal{I}^\circ$ should be as small as possible while ensuring that solutions to the planning problem exist. While the design of the mapping κ may take advantage of specific planning problem characteristics, we next present two general approaches to derive information states for $\mathcal{I}^\circ$.

Nondeterministic-derived information states. The first method we discuss is based on the inferences that can be done given an information state. If the information state η_k is available, it is possible to compute the set $X_k(\eta_k)$ in which the actual x_k is known to lie. The set $X_k(\eta_k)$ is called a *derived information state*. To compute the derived information state, we have to infer over the observations and actions performed. For the observations, we can define

$$H(y) = \{x | y = h(x, \psi), \text{for } \psi \in \Psi(x)\} \qquad (13.44)$$

that is, the set of all possible states the robot may be in, given an observation. The set $H(y)$ is called the *preimage* of y. Similarly, if we let the actions available depend on the current state, the robot can determine a set of states V where it may be, by computing

$$V(U_k) = \{x' | U_k = U(x') \text{ for } x' \in X\} \qquad (13.45)$$

in which U_k are the actions available at stage k. The current state then lies in the set $H \cap V$. Note, however, that it can be assumed that the robot has some kind of sensor that detects which kind of actions are available. This reduces the computation of V and H into only the computation of H. Thus, we will discuss only the case when U will be fixed for all $x \in X$.

From the state transition equation, it is possible to know which states may be reached if action u is applied at state x. Let F be this set, formally defined as

$$F(x, u) = \{x' \in X | \exists \theta \in \Theta(x, u) \text{ for which } x' = f(x, u, \theta)\} \qquad (13.46)$$

Using F and H, we next present how to compute the derived information state, $X_k(\eta_k)$, for any state k, using induction. Note that F and H eliminate the direct appearance of nature actions. The base case ($k = 1$) of the induction is

$$X_1 = \eta_0 \cap H(y_1) \qquad (13.47)$$

This first step consists only of making consistent the initial condition with the first observation. Now assume inductively that $X_k(\eta_k) \subseteq X$ is available, and $X_{k+1}(\eta_{k+1})$ should be computed. First note that $\eta_{k+1} = (\eta_k, u_k, y_{k+1})$, and the new information is provided only by u_k and y_{k+1}. From (13.44), the state is anywhere in $H(y_{k+1})$. On the other hand, if x_k was known, after applying u_k, the state lies somewhere in $F(x_k, u_k)$. Since x_k is unknown, but it is known that $x_k \in X_k(\eta_k)$, the new derived information state is

$$X_{k+1}(\eta_k, u_k, y_{k+1}) = \bigcup_{x_k \in X_k(\eta_k)} F(x_k, u_k) \cap H(y_{k+1}) \qquad (13.48)$$

Given that the derived information state is always a subset of X, the derived information space can be defined as $\mathcal{I}^\circ = 2^X$. Note that if X is finite, $\mathcal{I}^\circ$ is also finite, which makes it preferable if the number of stages is much larger than the size of X.

Probabilistic-derived information states. As before, we will compute derived information states, but assuming that nature is modeled probabilistically. Nature is also assumed to follow a Markov model, in which its actions depend only on the current state, as opposed to actions or state histories. Thus, a derived information state becomes a conditional probability distribution. The set functions H and F become $P(x_k|y_k)$ and $P(x_{k+1}|x_k, u_k)$, respectively. To compute $P(x_k|y_k)$ Bayes rule is applied as:

$$P(x_k \cap y_k) = P(x_k|y_k)P(y_k) = P(y_k|x_k)P(x_k) \qquad (13.49)$$

Solving for $P(x_k|y_k)$ yields

$$P(x_k|y_k) = \frac{P(y_k|x_k)P(x_k)}{P(y_k)} = \frac{P(y_k|x_k)P(x_k)}{\sum_{x_k \in X} P(y_k|x_k)P(x_k)} \qquad (13.50)$$

Bayes' rule requires the knowledge of $P(x_k)$ and $P(y_k|x_k)$. The prior $P(x_k)$ will be replaced later by a derived information state, while the probability $P(y_k|x_k)$ is easily computed as

$$P(y_k|x_k) = \sum_{\psi \in \Psi(x_k):y_k=h(x_k,\psi)} P(\psi|x_k) \qquad (13.51)$$

Since each information state is a probability distribution over X, it can be written as $P(x_k|\eta_k)$, if it is derived from η_k. As before, derived information states can be computed inductively. For the base case ($k = 1$) we have $\eta_0 = P(x_1)$ and the first observation y_1. Together they determine $P(x_1|y_1)$ as

$$P(x_1|\eta_1) = P(x_1|y_1) = \frac{P(y_1|x_1)P(x_1)}{\sum_{x_1 \in X} P(y_1|x_1)P(x_1)} \qquad (13.52)$$

Assuming inductively that $P(x_k|\eta_k)$ has been computed, $P(x_{k+1}|\eta_{k+1})$ has to be determined. Once again the derived information state can be written as $P(x_{k+1}|\eta_k, u_k, y_{k+1})$. Considering first the effect of u_k, note that

$$P(x_{k+1}|\eta_k, x_k, u_k) = P(x_{k+1}|x_k, u_k) \qquad (13.53)$$

because η_k contains no additional information regarding the prediction of x_{k+1} when x_k is given. To eliminate x_k from $P(x_{k+1}|x_k, u_k)$ marginalization is used, giving the derived information state

$$P(x_{k+1}|\eta_k, u_k) = \sum_{x_k \in X} P(x_{k+1}|x_k, u_k, \eta_k)P(x_k|\eta_k)$$

$$= \sum_{x_k \in X} P(x_{k+1}|x_k, u_k)P(x_k|\eta_k) \qquad (13.54)$$

The next step is to take into account the observation, y_{k+1}. From (13.50), k is replaced with $k+1$ and $P(x_k)$ is replaced with the information accumulated, to give

$$P(x_{k+1}| y_{k+1}, \eta_k, u_k) = \frac{P(y_{k+1}|x_{k+1}, \eta_k, u_k)P(x_{k+1}|\eta_k, u_k)}{\sum_{x_{k+1} \in X} P(y_{k+1}|x_{k+1}, \eta_k, u_k)P(x_{k+1}|\eta_k, u_k)}$$
$$(13.55)$$

The expression for $P(x_{k+1}|\eta_k, u_k)$ was given in (13.54). To calculate $P(y_{k+1}|x_{k+1}, \eta_k, u_k)$ note that

$$P(y_{k+1}|x_{k+1}, \eta_k, u_k) = P(y_{k+1}|x_{k+1}) \qquad (13.56)$$

because the observation depends only on the state.[2] Since $P(y_{k+1}|x_{k+1})$ is given as part of the sensor model, we are finished deriving the computation of $P(x_{k+1}|\eta_{k+1})$ from $P(x_k|\eta_k)$.

In this case, the derived information space is the set of all probability distributions over X. Thus, the planningproblem can be expressed again

---

[2] Here we are assuming that the sensor mapping does not depend on the history.

entirely in terms of the derived information space. A goal region can be specified as constraints on the probabilities. For example, for some particular $x \in X$, the goal might be to reach any derived information state for which $P(x|\eta_k) > 0.9$.

Let $n = |X|$. It is possible to embed $\mathcal{I}^\circ$ in $\mathbb{R}^n$ with each state $x \in X$ representing a vertex of a $(n-1)$-simplex. The coordinates of each vertex are expressed using probabilities $(p_1, p_1, \ldots, p_n)$ as barycentric coordinates. Here p_i is the probability of being in state x_i. Since $p_1 + \cdots + p_n = 1$, the vertices of the simplex (i.e., $(1, 0, \ldots, 0), (0, 1, \ldots, 0), \ldots, (0, 0, \ldots, 1)$) correspond to the cases when the state is completely known. A planning problem of this kind is known as a *Partial Observable Decision Process* (POMDP).

Efficient solutions to POMDPs form an active area in the research community [77, 78]. The problem is clearly very difficult, since the dimension of the space grows linearly with the number of states. However, the method of value iteration, presented in Section 13.2.3.1 can be applied. Let $\vec{x} \in \mathcal{I}_t$ be a derived information state. A worst-case analysis yields a cost functional of

$$\vec{l}(\vec{x}_k, u_k) = \max_{x_k \in X_k(\eta_k)} l(x_k, u_k) \tag{13.57}$$

and

$$\vec{l}_F(\vec{x}_F) = \max_{x_F \in X_F(\eta_F)} l_F(x_F) \tag{13.58}$$

Thus, the dynamic programming recursion is similar to the one presented in Section 13.2.3.1, but using derived information states:

$$G_k^*(\vec{x}_k) = \min_{u_k \in U} \left\{ \vec{l}(\vec{x}_k, u_k) + \sum_{\vec{x}_{k+1} \in \mathcal{I}_t} G_{k+1}^*(\vec{x}_{k+1}) P(\vec{x}_{k+1}|\vec{x}_k, u_k) \right\} \tag{13.59}$$

Note that the set of observations and nature actions is finite, since $\mathcal{I}^\circ$ is finite. This implies that $P(\vec{x}_{k+1}|\vec{x}_k, \vec{u}_k)$ is only an approximation distributed over a finite set of points of $\mathcal{I}_t$. The space $\mathcal{I}^\circ$ is a continuous space which usually requires the specification of a probability density function.

A policy can be found by approximating with a grid in the $(n-1)$-simplex, and using interpolation for evaluating points not in the grid [79, 80]. This method will be described in more detail for information spaces when the state space is continuous.

13.3.3 Continuous State Spaces

Until now, we have described information spaces when the underlying state space X is countable. Now we consider the case when X is a continuous space.

13.3.3.1 Sensors

As expected, the catalog of sensors is richer in the continuous case. Some models of sensors are:

- *Linear sensing.* It is assumed that $Y = X$. Thus, an *identity sensor* can be defined in which $y = h(x)$ makes the state immediately known. If there is a bound r in the error of the measurement, the state lies in the ball of radius r centered at y. This error is also commonly modeled with a probability distribution (i.e., a Gaussian).
- *Projection.* In this model, the dimension of the observation space, n_y, is smaller than the dimension of the state space. Either the observations ignore coordinates of X (i.e., a gyroscope gives orientation, but ignores position), or X is embedded in a smaller dimensional space (i.e., a photograph takes $X \subset \mathbb{R}^3$ into $\mathbb{R}^2$).
- *Landmark sensor.* A landmark sensor detects specific identifiable features in the environment. In its more abstract form, it detects specific points in the space (i.e., goal points or regions).

Specific sensors, such as an *odometry sensor*, which gives an estimation of the distance traveled, can be defined in terms of a projection sensor modeled together with a history-based sensor mapping. In recent years, *depth sensors* have been widely used in mobile robotics. This type of sensors gives measurements of the spatial distribution and shape of the obstacles in the environment. This accounts for sensors such as the *sonar*, or the *laser range finder*. Each sensor has an *upper range*. Obstacles farther from the sensor than this range cannot be detected. As the range is decreased, the sensor becomes a *proximity* sensor, and in the limit case it becomes a *contact* sensor. Note that the physical implementation may vary widely here. While an acoustic sonar measure time-of-flight of a high frequency sound, the contact sensor may be a device that makes a reading when it is *pushed*, thus indicating that distance is equal to 0.

13.3.3.2 Discrete-stage information spaces

The simpler case corresponds to a plan with discrete stages, and many of the concepts for discrete spaces, at least at first glance, are the same as their continuous counterparts. Let the state space $X \subset \mathbb{R}^m$ be an n-dimensional manifold.[3] The *observation space* $Y \subseteq \mathbb{R}^m$ is now an n_y-dimensional manifold, for $n_y \leq m$. Also, let $U \subseteq \mathbb{R}^m$ be an n_u-dimensional manifold for $n_u < m$.

Given that the time is discrete, the concepts presented for discrete spaces in Section 13.3.1.2 remain the same, but taking into account the fact that the variables are continuous.

13.3.3.3 Continuous-time information spaces

Most of the definitions presented in Section 13.3.3.2 remain the same when we consider a continuum of stages. Thus, X, Y, $\Psi(x)$, and $\Theta(x, u)$ are defined as before. However, the state transition equation now takes the form

$$\frac{\partial x}{\partial t} = \dot{x} = f(x, u, \theta) \tag{13.60}$$

for $x \in X$, $u \in U$, and $\theta \in \Theta(x, u)$. This means that the nature actions $\Theta(x, u)$ should be expressed in terms of velocities. Also, in the discrete case, an information state was expressed in terms of history sequences, but in the continuous case, histories become a function of time. Thus, $\tilde{y}_t : [0, t) \rightarrow Y$ $\tilde{u}_F : [0, t) \rightarrow U$, and $\tilde{x}_F : [0, t) \rightarrow X$ are the *observation history*, *action history*, and *state history*, respectively, up to time t.

The sensor mappings are now expressed with:

1. *State-sensor mapping.* $y(t) = h(x(t))$
2. *State-nature mapping.* $y(t) = h(x(t), \psi(t))$
3. *History-based sensor mapping.* $y(t) = h(\tilde{x}_F, \psi(t))$

Note that $\tilde{x}$ is usually the solution of a differential equation.

The *information state at time* t becomes

$$\eta_t = (\eta_0, \tilde{u}_t, \tilde{y}_t) \tag{13.61}$$

which has the same form and meaning as its discrete counterpart, but in continuous time. The set of all possible η_t is the *information space at time* t, $\mathcal{I}_t$. Since each $\eta_t \in \mathcal{I}_t$ is a function of time, $\mathcal{I}_t$ is a space of functions. Combining all the information spaces up to time $T \in [0, \infty)$, a single information space $\mathcal{I}$ is obtained as

$$\mathcal{I} = \bigcup_{t \in T} \mathcal{I}_t \tag{13.62}$$

---

[3] For readers unfamiliar with the term, an n-manifold is a space that locally looks like $\mathbb{R}^n$. Our everyday notion of a surface corresponds to a 2-manifold as a subset of $\mathbb{R}^3$ (see Reference 81).

To evaluate the quality of a plan, a new cost functional should be defined. Let L denote a real-valued, additive cost functional, which may be applied to any state–action history, $(\tilde{x}_t, \tilde{u}_t)$, defined as

$$L(\tilde{x}_t, \tilde{u}_t) = \int_0^{t'} l(x(t'), u(t')) \mathrm{d}t' + l_F(x(t)) \qquad (13.63)$$

in which $l(x(t'), u(t'))$ is the instantaneous cost, and $l_F(x(t))$ is a final cost.

13.3.4 Examples of Planning in the Information Space

13.3.4.1 Moving in an L-shaped corridor

This idealized example, which appeared originally in Reference 23, is intended to illustrate the issues that arise in selecting an appropriate map κ for derived information states. The state space, X, for the example shown in Figure 13.8 has 19 states, each of which corresponds to a location on one of the white tiles. For convenience, let each state be denoted by (i, j). There are 10 *bottom states*, denoted by $(1, 1), (2, 1), \ldots, (10, 1)$, and 10 *left states*, denoted by $(1, 1), (1, 2), \ldots, (1, 10)$. Since $(1, 1)$ is both a bottom state and a left state, it will be called the *corner state*.

It is assumed for this problem that there are no sensor observations. Nature, however, interferes with the state transitions, which leads to a form of nondeterministic uncertainty. If we try to apply an action that takes one step, nature

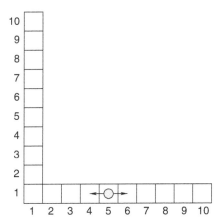

FIGURE 13.8   An example that involves 19 states. There are no sensor observations; however, actions can be chosen that enable the state to be estimated. The example provides an illustration of collapsing the information space.

may cause two or three steps to be taken, if possible. This can be modeled as follows. Let $U = \{(1,0), (-1,0), (0,1), (0,-1)\}$ and let $\Theta = \{1,2,3\}$. The state transition equation is defined as $f(x,u,\theta) = x + \theta u$, unless it is impossible to move to the required location, in which case $f(x,u,\theta) = x$. For example, if $x = (5,1)$, $u = (-1,0)$, and $\theta = 2$, then the resulting next state is $(5,1) + 2(-1,0) = (3,1)$.

Since there are no sensor observations, the information state at stage k is

$$\eta_k = (u_1, \ldots, u_{k-1}) \tag{13.64}$$

Now use the derived information space, $\mathcal{I}^\circ = 2^X$. The initial state, $x_1 = (10,1)$ is given, which means that the initial information state, η_1, is $\{(10,1)\}$. The goal is to arrive at the information state, $\{(1,10)\}$, which means that the task is to design a plan that moves from the lower right to the upper left.

With perfect information, this would be trivial; however, without sensors the uncertainty may grow very quickly. For example, after applying the action $u_1 = (-1,0)$ from the initial state, the derived information state becomes $\{(7,1), (8,1), (9,1)\}$. After $u_2 = (-1,0)$ it becomes $\{(4,1), \ldots, (8,1)\}$. A nice feature of this problem, however, is that uncertainty can be reduced without sensing. Suppose that for 100 stages, we continue to apply $u_k = (-1,0)$. What is the resulting information state? As the corner state is approached, the uncertainty is reduced because the state cannot be further changed by nature. It is known that each action, $u_k = (-1,0)$, decreases the X coordinate by at least one each time. Therefore, after nine or more stages, it is known that $\eta_k = \{(1,1)\}$. Once this is known, then the action $(0,1)$ can be applied. This will again increase uncertainty as the state moves through the set of left states. If $(0,1)$ is applied nine or more times, then it is known for certain that $x_k = (1,10)$, which is the required goal state.

A successful plan has now been obtained: apply $(-1,0)$ for nine stages, then apply $(0,1)$ for nine stages. Recall from Section 13.3.1.2 that a strategy is generally specified as $\pi : \mathcal{I} \to U$; however, for this example, it appears that only a sequence of actions is needed. The actions do not depend on the information state. Why did this happen? If no observations are obtained during execution, then there is no way to use feedback. There is nothing to learn by executing the plan. In general, for problems that involve no sensors and a fixed initial information state, a *path* in the information space can be derived from a plan. It is somewhat strange that this path is completely predictable, even though the original problem may involve substantial uncertainties. We always know precisely what will happen in terms of the information states if there are no sensors and the initial condition is fixed.

To make the situation more interesting, assume that any subset of X could be used as the initial condition. In this case, a plan $\pi : \mathcal{I} \to U$ must be formulated to solve the problem. From each initial information state η, a path in $\mathcal{I}$ can still be computed from π. Specifying a plan over all of $\mathcal{I}$ appears complicated, which motivates the next consideration.

The ideas from Section 13.3.2 can be applied here to collapse the information down from 2^{19} (over half of a billion) to 19 derived information states. The mapping $\kappa : \mathcal{I} \rightarrow \mathcal{I}^\circ$ must be constructed. We first make a naive attempt to collapse the information state down to only three states. Let $\mathcal{I}^\circ = \{g, l, a\}$, in which g denotes "goal," l denotes "left," and a denotes "any." The mapping is

$$\kappa(\eta) = \begin{cases} g & \text{if } \eta = \{(1, 10)\} \\ l & \text{if } \eta \text{ is a subset of the set of left states} \\ a & \text{otherwise} \end{cases} \qquad (13.65)$$

It might seem that this derived information space will lead to a very compact plan for solving the problem. Based on the successful plan described so far, the plan on $\mathcal{I}^\circ$ can be defined as $\pi(g) = u_F$, $\pi(l) = (0, 1)$, and $\pi(a) = (-1, 0)$. What is wrong with this? Suppose that the initial state is $(10, 1)$. There is no way to require that $u_k = (-1, 0)$ be applied nine times to reach the l state. If $(-1, 0)$ is applied to the a state, then it is not possible to determine when the transition to l should occur.

Now consider a different derived information space. Suppose that there are 19 derived information states, which includes g as defined previously, l_j for $1 \leq j \leq 9$, and a_i for $2 \leq i \leq 10$. The mapping κ is defined as $\kappa(\eta) = g$ if $\eta = \{(1, 10)\}$. Otherwise, $\kappa(\eta) = l_i$, for the smallest value of i such that η is a subset of $\{(1, i), \ldots, (1, 10)\}$. If there is no such value for i, then $\kappa(\eta) = a_i$, for the smallest value of i such that η is a subset of $\{(1, 1), \ldots, (1, 10), (2, 1), \ldots, (i, 1)\}$. Now the plan may be defined as $\pi(g) = u_F$, $\pi(l_i) = (0, 1)$, and $\pi(a_i) = (-1, 0)$. Although it might not appear to be any better than the plan obtained from collapsing $\mathcal{I}^\circ$ to three states, the important difference is that the correct information state transitions occur. For example, if $u_k = (-1, 0)$ is applied at a_5, then a_4 is obtained. If $u = (-1, 0)$ is applied at a_2, then l_1 is obtained. From there, $u = (0, 1)$ is applied to yield l_2. These actions can be repeated until eventually l_9 and g are reached.

13.3.4.2 The Kalman filter

When the transition function f, and the sensor mapping h are both linear functions, and nature actions, θ and ψ, can be modeled as Gaussian, the derived information states will follow a Gaussian distribution too. These assumptions are reasonable in many mobile robotics contexts. In this case, a mapping $\kappa : \mathcal{I} \rightarrow \mathcal{I}^\circ$, in which $\mathcal{I}^\circ$ is the space of all Gaussians, will collapse $\mathcal{I}$ without any loss of information. This is referred to as a *linear-Gaussian* model, which is the basis for the most common approach for collapsing $\mathcal{I}$, the *Kalman filter*.

Each Gaussian is specified by an n-dimensional mean vector μ, and an $n \times n$ symmetric covariance matrix, Σ.

Since the Kalman filter relies on linear models, f can be written as

$$x_{k+1} = A_k x_k + B_k u_k + G_k \theta_k \tag{13.66}$$

in which A_k, B_k, and G_k are real-valued matrices of appropriate dimensions. The subscript k is used because the Kalman filter works even if f is different in every stage. Similarly, the sensor mapping becomes

$$y_k = C_k x_k + H_k \psi_k \tag{13.67}$$

Since an information state $P(x_k | \eta_k) \in \mathcal{I}^{\circ}$ is represented by its mean vector and its covariance matrix, our goal here is to compute μ_k and Σ_k at stage k. We next give the update expressions, omitting their derivation, that can be found in many textbooks on stochastic control (i.e., [82]). Given the initial conditions μ_0 and Σ_0, we have

$$\Sigma'_{k+1} = A_k \Sigma_k A_k^T + G_k \Sigma_\theta G_k^T \tag{13.68}$$

$$\Sigma_{k+1} = (I - L_{k+1} C_{k+1}) \Sigma'_{k+1} \tag{13.69}$$

$$\mu_{k+1} = A_k \mu_k + L_{k+1}(y_{k+1} - C_{k+1} A_k \mu_k) \tag{13.70}$$

with

$$\Sigma_{k+1} = (I - L_{k+1} C_{k+1}) \Sigma'_{k+1} \tag{13.71}$$

The expression for L_k (substitute $k + 1$ for k to obtain L_{k+1}) is

$$L_k = \Sigma'_k C_k^T [C_k \Sigma'_k C_k^T + H_k \Sigma_\psi H_k]^{-1} \tag{13.72}$$

When nature is not Gaussian, or the transition equation is not linear, the derived information states density can be approximated using a grid, with numerical integration between the grid points. Let $S \subset X$ be the set of points in the grid. In the initial step, $P(s)$ is computed from $p(x)$ by numerically evaluating the integrals of $p(x_1)$ over the Voronoi region of each sample. Now suppose that $P(s_k | \eta_k)$ has been computed over S_k, and the task is to compute $P(s_{k+1} | \eta_{k+1})$ given u_k and y_{k+1}.

Considering only u_k, $P(s_{k+1} | s_k, u_k)$ approximates $p(x_{k+1} | x_k, u_k)$ when computed in the manner described above. At this point the densities needed have been approximated by discrete distributions.

The resulting distribution is $P(s_{k+1} | \eta_k, u_k)$, and the effect of y_{k+1} in $p(x_{k+1} | y_{k+1})$ can be computed approximately by $P(s_{k+1} | y_{k+1})$ using the grid samples. The resulting distribution, $P(s_{k+1}|\eta_{k+1})$ represents the approximate derived information state. It turns out that the Voronoi regions over the samples do not even need to be carefully considered. One can work directly with a collection of samples randomly drawn from the initial probability density $p(x_1)$. The general method is referred to as *particle filtering*, and has yielded good performance in applications to experimental mobile robotics [83].

13.3.4.3 Sensorless manipulation

Imagine a planning problem in which the robot does not have any sensors, so that there are no observations at all. Moreover, the initial condition is unknown. Is it still possible to compute a plan to reach a goal state? As we will explore in the next example, in some problems knowing only the action history is enough to compute a successful plan.

In the context of manufacturing, a *part* may need to have a specific orientation before being assembled with other components. In a sensorless setting, a robot, in this case a robotic arm with a gripper, needs to orient a part without any feedback [84, 85]. The part is modeled as a convex polygon. Its initial orientation is unknown; the goal is to bring the part to a known orientation, up to symmetry. The manipulation process is shown in Figure 13.9. The part moves on the conveyor toward a fence, against which it comes to rest after possibly rotating to reach a stable orientation. The robotic arm grasps the part, changes its orientation, and drops it up again in the conveyor. This process is repeated until the part achieves the desired orientation against the fence.

The natural state space for this problem is S^1, corresponding to the orientation of the part. At each step, the robotic arm rotates the part through some angle, so the action space is likewise S^1. These continuous spaces need to be transformed into finite sets. The key of the transformation is to identify *critical events* which partition the space into equivalence classes, then plan over this

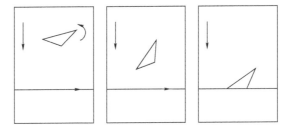

FIGURE 13.9   Overhead overview of a part on the conveyor. The conveyor moves downward. The robot picks up a part and rotates it through a chosen angle before placing it on the conveyor. The part then drifts on the conveyor into contact with the fence, possibly rotating compliantly as it comes to rest.

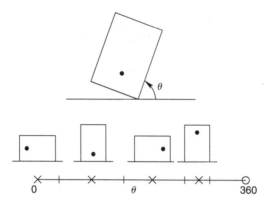

FIGURE 13.10 Effects of rotation actions on a rectangular part. The action space is divided into four equivalence classes according to the resulting state. The crosses mark a representative action from each class.

set of equivalence classes rather than the full space. These critical events are problem specific. In the case of the part orienting, the critical events in action space are orientation angles such that for a given information state, rotations either greater or less than these angles will reach different information states.

Thus, the state space X is chosen as the set of all the stable orientations of the part when it is lying statically on the fence. Since the part is polygonal, the size of X is bounded above by the number of edges in the part. Using the concepts presented in Section 13.3.2, the derived information space is $\mathcal{I}^\circ = 2^X$. The initial derived information state consists of all possible stable orientations (i.e., $\eta_0 = X$), since the part orientation is initially unknown. The action set is the range of rotation angles available to the gripper, partitioned into intervals of rotations that lead to identical resulting information states. The effects of a specific rotation action on a rectangular part are shown in Figure 13.10. The critical events in the continuous action set for an information state with two states are shown in Figure 13.11.

The objective is to find a sequence of actions such that the derived information state at the final stage corresponds to a single possible orientation of the part. Once one orientation is uniquely identified, the robotic arm may perform an additional rotation to achieve any desired goal orientation. With the finite action set, a directed graph can be constructed whose nodes are information states and whose edges are transitions resulting from the discrete action set. Standard graph searching techniques can be used to search for a directed path to a singleton information state. This path in the collapsed information space graph constitutes a plan for eliminating uncertainty in the part's orientation.

The reason that successful planning is still possible starting from total uncertainty and without sensor feedback is that some actions in this information space

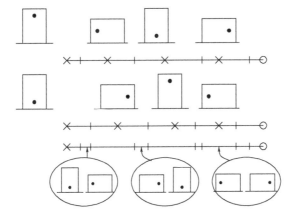

FIGURE 13.11   Critical events for an information state with two states.

have a *conformant* property, in which the same resulting state can be reached by the same action from many different initial states. By selecting conformant actions, uncertainty can be reduced. The same principle is applied in the context of mobile robot localization with extremely limited sensing in Reference 74.

13.4 CONCLUSION AND BIBLIOGRAPHICAL REMARKS

Our presentation has been tightly constrained by space limitations. There are a number of books to which we refer the reader for elaboration. Treatments of decision theory in general appear in References 4–6. Bertsekas [24] covers much of the same material as the present chapter and is well-stocked with examples. Part III of Reference 23 is also quite similar in content but is much more detailed. A general treatment of the infinite horizon case is given in Reference 43. Sutton and Barto [46] is the definitive introduction to reinforcement learning. Ghallib et al. [86] consider planning with primarily logic-based representations. Russell and Norvig [87] cover planning under (mainly probabilistic) uncertainty from an artificial intelligence perspective. Some recent papers on decision-theoretic planning are collected in Reference 88. Game theory is addressed in greater detail in References 47, 48, and 89.

REFERENCES

1. K. Arrow and L. Hurwicz. An optimality criterion for decision-making under ignorance. In C. F. Carter and J. L. Ford (eds), *Uncertainty and Expectation in economics*, Basil Blackwell and Mott Ltd., Oxford, UK, 1972.
2. A. Wald. *Statistical Decision Functions*. John Wiley & Sons, New York, 1950.

3. J. Von Neumann and O. Morgenstern. *Theory of Games and Economic Behavior.* Princeton University Press, Princeton, NJ, 1944.
4. J. O. Berger. *Statistical Decision Theory.* Springer-Verlag, Berlin, 1980.
5. M. H. DeGroot. *Optimal Statistical Decisions.* McGraw Hill, New York, NY, 1970.
6. P. A. Devijver and J. Kittler. *Pattern Recognition: A Statistical Approach.* Prentice-Hall Publications, Englewood Cliffs, NJ, 1982.
7. R. O. Duda, P. E. Hart, and D. G. Stork. *Pattern Classification*, 2nd ed., Wiley, New York, 2000.
8. T. Mitchell. *Machine Learning.* McGraw-Hill, New York, 1997.
9. S. Barbera, P. J. Hammond, and C. Seidl (eds). *Handbook of Utility Theory, Volume 1: Principles.* Kluwer, Dordrecht, 1998.
10. C. P. Robert. *The Bayesian Choice*, 2nd ed. Springer-Verlag, Berlin, 2001.
11. K. Z. Haigh and M. M. Veloso. High-level planning and low-level execution: towards a complete robotic agent. In W. Lewis Johnson (ed.), In *Proceedings of First International Conference on Autonomous Agents*, pp. 363–370, Marina del Rey, CA, February 1997. ACM Press, New York, NY.
12. I. A. Nesnas, A. Wright, M. Bajracharya, R. Simmons, and T. Estlin. CLARAty and challenges of developing interoperable robotic software. *Invited to International Conference on Intelligent Robots and Systems (IROS)*, Nevada, October 2003.
13. I. A. Nesnas, A. Wright, M. Bajracharya, R. Simmons, T. Estlin, and Won Soo Kim. CLARAty: an architecture for reusable robotic software. In *SPIE Aerosense Conference*, Orlando, FL, April 2003.
14. R. Simmons, R. Goodwin, K. Z. Haigh, S. Koenig, and J. O'Sullivan. A layered architecture for office delivery robots. In W. Lewis Johnson (ed.), *Proceedings of First International Conference on Autonomous Agents*, pp. 245–252, Marina del Rey, CA, February 1997. ACM Press, New York, NY.
15. R. Simmons, D. Goldberg, A. Goode, M. Montemerlo, N. Roy, B. Sellner, C. Urmson, A. Schultz, M. Abramson, W. Adams, A. Atrash, M. Bugajska, M. Coblenz, M. MacMahon, D. Perzanowski, I. Horswill, R. Zubek, D. Kortenkamp, B. Wolfe, T. Milam, and B. Maxwell. Grace: an autonomous robot for the AAAI robot challenge. *AI Magazine*, Vol. 24, 51–72, Summer 2003.
16. C. Urmson, R. Simmons, and I. Nesnas. A generic framework for robotic navigation. In *Proceedings of the IEEE Aerospace Conference*, Montana, March 2003.
17. R. E. Bellman. *Dynamic Programming.* Princeton University Press, Princeton, NJ, 1957.
18. R. Howard. *Dynamic Programming and Markov Processes.* MIT Press, Cambridge, MA, 1960.
19. G. H. Golub and C. F. Van Loan. *Matrix Computations (3rd ed).* Johns Hopkins University Press, Baltimore, MD, 1996.
20. A. G. Barto, S. J. Bradtke, and S. P. Singh. Real-time learning and control using asynchronous dynamic programming. *Artificial Intelligence*, 72: 81–138, 1995.
21. D. P. Bertsekas. Distributed dynamic programming. *IEEE Transactions on Automatic Control*, 27: 610–616, 1982.

22. D. P. Bertsekas. Distributed asynchronous computation of fixed points. *Mathematical Programming*, 27: 107–120, 1983.

23. S. M. LaValle. *Planning Algorithms*. Cambridge University Press, London, New York, 2006. To appear in print. Available online at http://msl.cs.uiuc.edu/planning/.

24. D. P. Bertsekas. *Dynamic Programming and Optimal Control*, 2nd ed., vol. 1. Athena Scientific, Belmont, MA, 2001.

25. D. H. Lehmer. Mathematical methods in large-scale computing units. In *Proceedings of 2nd Symposium on Large-Scale Digital Computing Machinery*, pp. 141–146. Harvard University Press, Cambridge, MA, 1951.

26. H. Niederreiter. Society for Industrial and Applied Mathematics, *Random Number Generation and Quasi-Monte-Carlo Methods*, Philadelphia, USA, 1992.

27. A. G. Sukharev. Optimal strategies of the search for an extremum. *U.S.S.R. Computational Mathematics and Mathematical Physics*, 11: 910–924, 1971. Translated from Russian, *Zh. Vychisl. Mat. i Mat. Fiz.*, 11: 910–924, 1971.

28. J. Matousek. *Geometric Discrepancy*. Springer-Verlag, Berlin, 1999.

29. I. H. Sloan and S. Joe. *Lattice Methods for Multiple Integration*. Oxford Science Publications, Englewood Cliffs, NJ, 1990.

30. X. Wang and F. J. Hickernell. An historical overview of lattice point sets. In K.-T. Fang, F. J. Hickernell, and H. Niederreiter (eds), *Monte Carlo and Quasi-Monte Carlo Methods 2000*, pp. 158–167. Springer-Verlag, Berlin, 2002.

31. S. Davies. Multidimensional triangulation and interpolation for reinforcement learning. In *Advances in Neural Information Processing Systems*, Nashua, NH, 1996.

32. D. P. Bertsekas and J. Tsitsiklis. *Neuro-Dynamic Programming*. Athena Scientific, Kiekrz, Poland, 1996.

33. C. J. C. H. Watkins. *Learning from Delayed Rewards*. PhD thesis, Cambridge University, Cambridge, England, 1989.

34. P. Konkimalla and S. M. LaValle. Efficient computation of optimal navigation functions for nonholonomic planning. In *Proceedings of First IEEE International Workshop on Robot Motion and Control*, pp. 187–192, 1999.

35. S. M. LaValle and P. Konkimalla. Algorithms for computing numerical optimal feedback motion strategies. *International Journal of Robotics Research*, 20: 729–752, 2001.

36. D. Moore. *Simplical Mesh Generation with Applications*. PhD thesis, Cornell University, Ithaca, NY, 1992.

37. R. Munos and A. Moore. Barycentric interpolators for continuous space and time reinforcement learning. In M.S. Keams, S.A. Solla, and D.A. Cohn (eds.,) *Neural Information Processing Systems*, vol. 11, pp. 1024–1030. MIT Press, Cambridge, MA, 1998.

38. R. E. Bellman and S. E. Dreyfus. *Applied Dynamic Programming*. Princeton University Press, Princeton, NJ, 1962.

39. D. P. Bertsekas. *Dynamic Programming: Deterministic and Stochastic Models*. Prentice-Hall, Englewood Cliffs, NJ, 1987.

40. R. E. Larson. A survey of dynamic programming computational procedures. *IEEE Transactions on Automation Control*, 12: 767–774, 1967.

41. R. E. Larson and J. L. Casti. *Principles of Dynamic Programming, Part II.* Dekker, New York, NY, 1982.

42. L. P. Kaelbling, M. L. Littman, and A. Moore. Reinforcement learning: a survey. *Journal of Artificial Intelligence Research*, 4: 237–285, 1996.

43. D. P. Bertsekas. *Dynamic Programming and Optimal Control*, 2nd ed., vol. 2. Athena Scientific, Belmont, MA, 2001.

44. C. J. C. H. Watkins and P. Dayan. Q-learning. *Machine Learning*, 8: 279–292, 1992.

45. R. I. Brafman and M. Tennenholtz. R-max — a general polynomial time algorithm for near-optimal reinforcement learning. *Journal of Machinery Learning Research*, 3: 213–231, 2003.

46. R. S. Sutton and A. G. Barto. *Introduction to Reinforcement Learning*. MIT Press, Cambridge, MA, 1998.

47. T. Başar and G. J. Olsder. *Dynamic Noncooperative Game Theory*. Academic Press, London, 1982.

48. G. Owen. *Game Theory*. Academic Press, New York, NY, 1982.

49. M. Blum and D. Kozen. On the power of the compass (or, why mazes are easier to search than graphs). In *Proceedings of Annual Symposium on Foundations of Computer Science*, pp. 132–142, 1978.

50. T. Lozano-Pérez, M. T. Mason, and R. H. Taylor. Automatic synthesis of fine-motion strategies for robots. *International Journal of Robotics Research*, 3: 3–24, 1984.

51. B. R. Donald. *Error Detection and Recovery for Robot Motion Planning with Uncertainty*. PhD thesis, Massachusetts Institute of Technology, Cambridge, MA, 1987.

52. S. Akella, W. H. Huang, K. M. Lynch, and M. T. Mason. Sensorless parts feeding with a one joint robot. In J.-P. Laumond and M. Overmars (eds), *Algorithms for Robotic Motion and Manipulation*, pp. 229–237. A K Peters, Wellesley, MA, 1997.

53. M. A. Erdmann and M. T. Mason. An exploration of sensorless manipulation. *IEEE Transactions on Robotics and Automation*, 4: 369–379, 1988.

54. K. Y. Goldberg and M. T. Mason. Bayesian grasping. In *IEEE International Conference on Robotics and Automation*, 1990.

55. R. H. Taylor, M. T. Mason, and K. Y. Goldberg. Sensor-based manipulation planning as a game with nature. In *Fourth International Symposium on Robotics Research*, pp. 421–429, 1987.

56. I. Kamon, E. Rivlin, and E. Rimon. Range-sensor based navigation in three dimensions. In *IEEE International Conference on Robotics and Automation*, 1999.

57. V. J. Lumelsky and A. A. Stepanov. Path planning strategies for a point mobile automaton moving amidst unknown obstacles of arbitrary shape. *Algorithmica*, 2: 403–430, 1987.

58. L. Guilamo, B. Tovar, and S. M. LaValle. Pursuit-evasion in an unknown environment using gap navigation graphs. In *Proceedings of IEEE/RSJ International Conference on Intelligent Robots and Systems*, 2004.

59. B. Tovar, S. M. LaValle, and R. Murrieta. Optimal navigation and object finding without geometric maps or localization. In *Proceedings of IEEE International Conference on Robotics and Automation*, pp. 464–470, 2003.

60. J. Y. Herve', P. Cucka, and R. Sharma. Qualitative visual control of a robot manipulator. In *Image Understanding Workshop*, pp. 895–908, 1990.

61. B. R. Donald. On information invariants in robotics. *Artificial Intelligence*, 72: 217–304, 1995.

62. B. R. Donald and J. Jennings. Sensor interpretation and task-directed planning using perceptual equivalence classes. In *IEEE International Conference on Robotics and Automation*, pp. 190–197, Sacramento, CA, April 1991.

63. H. Choset and J. Burdick. Sensor based planning, part I: the generalized Voronoi graph. In *IEEE International Conference on Robotics and Automation*, pp. 1649–1655, 1995.

64. A. Stentz. Optimal and efficient path planning for partially-known environments. In *IEEE International Conference on Robotics and Automation*, pp. 3310–3317, 1994.

65. L. J. Guibas, J.-C. Latombe, S. M. LaValle, D. Lin, and R. Motwani. Visibility-based pursuit-evasion in a polygonal environment. *International Journal of Computational Geometry and Applications*, 9: 471–494, 1999.

66. S.-M. Park, J.-H. Lee, and K.-Y. Chwa. Visibility-based pursuit-evasion in a polygonal region by a searcher. Technical Report CS/TR-2001-161, KAIST, Department of Computer Science, Korea, January 2001.

67. T. D. Parsons. Pursuit-evasion in a graph. In Y. Alavi and D. R. Lick (eds), *Theory and Application of Graphs*, pp. 426–441. Springer-Verlag, Berlin, 1976.

68. I. Suzuki and M. Yamashita. Searching for a mobile intruder in a polygonal region. *SIAM Journal of Computing*, 21: 863–888, 1992.

69. Y. Yavin and M. Pachter. *Pursuit-Evasion Differential Games*. Pergamon Press, Oxford, England, 1987.

70. R. Simmons and S. Koenig. Probabilistic navigation in partially observable environments. In *Proceedings of International Joint Conference on Artificial Intellegence*, 1995.

71. D. J. Austin and P. Jensfelt. Using multiple Gaussian hypotheses to represent probability distributions for mobile robot localization. In *IEEE International Conference on Robotics and Automation*, pp. 1036–1041, 2000.

72. D. Fox, W. Burgard, S. Thrun, and A. B. Cremers. Position estimation for mobile robots in dynamic environments. In *Proceedings of American Association of Artificial Intellegence*, 1998.

73. G. Dudek, K. Romanik, and S. Whitesides. Localizing a robot with minimum travel. In *ACM-SIAM Symposium on Discrete Algorithms*, pp. 437–446, 1995.

74. Jason M. O'Kane and Steven M. LaValle. Almost-sensorless localization. In *IEEE International Conference on Robotics and Automation*, 2005.

75. M. Rao, G. Dudek, and S. Whitesides. Randomized algorithms for minimum distance localization. In *Proceedings Workshop on Algorithmic Foundations of Robotics*, pp. 265–280, 2004.

76. E. Remolina and B. Kuipers. A logical account of causal and topological maps. In *Proceedings of the Seventeenth International Joint Conference on Artificial Intelligence (IJCAI-01)*, pp. 5–11, 2001.

77. M. Littman, L. Kaelbling, and A. Cassandra. Planning and acting in partially observable stochastic domains. *Artificial Intelligence*, 101: 99–134, 1998.

78. N. Zhang and W. Lin. A model approximation scheme for planning in partially observable stochastic domains. *Journal of Artificial Intelligence Research*, 7: 199–230, 1997.

79. M. Hauskrecht. Value functions approximations for partially observable Markov decision processes. *JAIR*, 13: 33–94, 1994.

80. M. Hauskrecht. Incremental methods for computing bounds in partially observable Markov decision processes. In *14th National Conference on Artificial Intelligence*, Providence, Rhode Island, 1997. AAAI Press/MIT Press.

81. J. G. Hocking and G. S. Young. *Topology*. Dover, New York, NY, 1988.

82. H. Kwakernaak and R. Sivan. *Linear Optimal Control Systems*. Wiley, New York, NY, 1972.

83. D. Fox, S. Thrun, W. Burgard, and F. Dellaert. Particle filters for mobile robot localization. In A. Doucet, N. de Freitas, and N. Gordon (eds), *Sequential Monte Carlo Methods in Practice*, pp. 470–498. Springer, Amsterdam, New York, 2001.

84. S. Akella and M. Mason. Using partial sensor information to orient parts. *IJRR*, 18: 963–997, 1999.

85. K. Y. Goldberg. Orienting polygonal parts without sensors. *Algorithmica*, 10: 201–225, 1993.

86. M. Ghallib, D. Nau, and P. Traverso. *Automated Planning: Theory and Practice*. Elsevier, Amsterdam, New York, 2004.

87. S. Russell and P. Norvig. *Artificial Intelligence: A Modern Approach*, 2nd ed. Pearson Education, Inc., Upper Saddle River, NJ, 2003.

88. G. Della Riccia, D. Dubois, R. Kruse, and H.-J. Lenz (eds). *Planning Based on Decision Theory*. Springer-Verlag, Vienna, 2003.

89. P. D. Straffin. *Game Theory and Strategy*. Mathematical Association of America, Washington, DC, 1993.

Biographies

Jason M. O'Kane is a Ph.D. candidate in the Department of Computer Science at the University of Illinois at Urbana-Champaign. In 2001, he earned a B.S. degree from Taylor University in Upland, Indiana. In 2005, he earned a M.S. degree from the University of Illinois at Urbana-Champaign. His research is in geometric algorithms for robotics.

Benjamín Tovar earned the B.S. degree in electrical engineering from ITESM at Mexico City, Mexico, in 2000, and the M.S. in electrical engineering from University of Illinois, Urbana-Champaign, in 2004. Currently (2005) he is pursuing the Ph.D. degree in computer science at the University of

Illinois. Prior to his M.S. he worked as a research assistant at Mobile Robotics Laboratory at ITESM Mexico City. He is mainly interested in motion planning, visibility-based tasks, and minimal sensing for robotics.

Peng Cheng is a Ph.D. candidate in the Department of Computer Science at University of Illinois at Urbana-Champaign. He earned his B.E. and M.E. degrees in electrical engineering from Tsinghua University, Beijing, China in 1996 and 1999, respectively. In 2001, he graduated with a M.S. degree in computer science from Iowa State University. He has worked in the areas of motion planning and robotics, especially on problems with real or virtual robotic systems under differential constraints, which include how to represent, characterize, plan, and control the motion of these systems.

Steven M. LaValle is an associate professor in the Department of Computer Science at the University of Illinois at Urbana-Champaign. He earned his Ph.D. in electrical engineering in 1995 from the University of Illinois. From 1995 to 1997, he was a postdoctoral researcher and lecturer in the Department of Computer Science at Stanford. From 1997 to 2001, he was an assistant professor at Iowa State University. In 1999, he received the CAREER award from the U.S. to National Science Foundation. He has published in the areas of robotics, computational geometry, artificial intelligence, computational biology, computer vision, and control theory. He recently authored an online book, *Planning Algorithms*, which will be published by Cambridge University Press.

14 Behavior-Based Coordination in Multi-Robot Systems

Chris Jones and Maja J. Matarić

Contents

The successful deployment of a multi-robot system (MRS) requires an effective method of coordination to mediate the interactions among the robots and between the robots and the task environment in order for a given system-level task to be performed. The design of coordination mechanisms has received increasing attention in recent years and has included investigations into a wide variety of coordination mechanisms. A popular and successful framework for the control of robots in coordinated MRS is *behavior-based control* [1,2]. Behavior-based control is a methodology in which robots are controlled through the principled integration of a set of interacting behaviors (e.g., wall following, collision avoidance, landmark recognition, etc.) in order to achieve desired system-level behavior. This chapter describes, through explanation, discussion of demonstrated simulated and physical mobile robots, and formal design and analysis, the range and capabilities of behavior-based control applied to multi-robot coordination.

We begin by providing a brief overview of single-robot control philosophies and architectures, including behavior-based control, in Section 14.1. In Section 14.2 we move from single robots to MRS and discuss the additional challenges this transition entails. In Section 14.3 we use empirical case studies to discuss and demonstrate three important ways in which robots can interact, and thus coordinate, their behavior. In Section 14.4 we discuss formal approaches to the design and analysis of MRS that are of fundamental importance if the full potential of MRS is to be achieved. Finally, in Section 14.5 we briefly discuss the future of coordinated behavior-based MRS and conclude the chapter.

14.1 OVERVIEW OF ROBOT CONTROL ARCHITECTURES

In this section, we briefly discuss the most popular approaches and techniques for the control of a single robot. In Section 14.1.1 we proceed with the fundamental principle of this chapter, the control of multiple robots, and how it is related to, and different from, the control of a single robot.

14.1.1 Single Robot Control

We define *robot control* as the process of mapping a robot's sensory information into actions in the real world. We do not consider entities that make no use of sensory information in control decisions as robots, nor do we consider entities that do not perform actions as robots, because neither category is truly interacting in the real world. Any robot must, in one manner or another, use incoming sensory information to make decisions about what actions to execute. There are a number of control philosophies dictating how this mapping from sensory information to actions should occur, each with its advantages and disadvantages.

A continuum of approaches to robot control can be described as a spectrum spanning from deliberative to reactive control.

The *deliberative* approach to robot control is usually computationally intensive due to the use of explicit reasoning or planning using symbolic representations and world models [3]. For the reasoning processes to be effective, complete and accurate models of the world are required. In domains where such models are difficult to obtain, such as in dynamic and fast-changing environments or situations with significant uncertainty in the robot's sensing and action, it may be impossible for the robot to act in an appropriate or timely manner using deliberative control [3,4].

In contrast to deliberative control, the *reactive* approach to robot control is characterized by a tight coupling of sensing to action, typically involving no intervening reasoning [5,6]. Reactive control does not require the acquisition or maintenance of world models, as it does not rely on the types of complex reasoning processes utilized in deliberative control. Rather, simple rule-based methods involving a minimal amount of computation, internal representations, or knowledge of the world are typically used. This makes reactive control especially well suited to dynamic and unstructured worlds where having access to a world model is not a realistic option. Furthermore, the minimal amount of computation involved means reactive systems are able to respond in a timely manner to rapidly changing dynamics.

A middle ground between deliberative and reactive philosophies is found in *hybrid* control, exemplified by three-layered architectures [7,8]. In this approach, a single controller includes both reactive and deliberative components. The reactive part of the controller handles low-level control issues requiring fast response time, such as local obstacle avoidance. The deliberative part of the controller handles high-level issues on a longer time-scale, such as global path planning. A necessary third component of hybrid controllers is a middle layer that interfaces the reactive and deliberative components. Three-layered architectures aim to harness the best of reactive controllers in the form of dynamic and time-responsive control and the best of deliberative controllers in the form of globally efficient actions over a long time-scale. However, there are complex issues involved in interfacing these fundamentally differing components and the manner in which their functionality should be partitioned is not yet well understood.

Behavior-based control, described in detail in Section 14.1.2, offers an alternative to hybrid control. It can also include both deliberative and reactive components, but unlike hybrid control, it is composed of a set of independent modular components that are executed in parallel [1,2].

The presented spectrum of control approaches is continuous and a precise classification of a specific controller on the continuum may be difficult. The distinction between deliberative and reactive control, and hybrid and behavior-based control is often a matter of degree, based on the amount of computation performed and the response time of the system to relevant changes in the world. In a specific domain, the choice of controller is dependent on many factors, including how responsive the robot must be to changes in the world, how accessible a world model is, and what level of efficiency or optimality is required.

14.1.2 Behavior-Based Control

The control methodology we focus on in this chapter is behavior-based (BB) control. The BB approach to robot control must not be classified as strictly deliberative or reactive, as it can, and in many cases is, both. However, BB control is most closely identified (often incorrectly so) with the reactive side of the control spectrum, because primary importance is placed on maintaining a tight, real-time coupling between sensing and action [7,8].

Fundamentally, a BB controller is composed of a set of modular components, called behaviors, which are executed in parallel. A *behavior* is a control law that clusters a set of constraints in order to achieve and maintain a goal [1,2]. Each behavior receives inputs from sensors or other behaviors or both and provides outputs to the robot's actuators or to other behaviors. For example, an obstacle avoidance behavior might send a command to the robot's wheels to turn left or right if the robot's sensors detect that the robot is moving directly toward an obstacle. There is no centralized world representation or state in a BB system. Instead, individual behaviors and networks of behaviors maintain any models or state information.

Many different behaviors may independently receive input from the same sensors and output action commands to the same actuators. The issue of choosing a particular action given inputs from potentially multiple sensors and behaviors is called *action selection* [10]. One of the well-known mechanism for action selection is the use of a predefined behavior hierarchy, as in the Subsumption Architecture [9], in which commands from the highest-ranking active behavior are sent to the actuator and all others are ignored. (Note, however, that the Subsumption Architecture has most commonly been used in the context of reactive and not BB systems.) Numerous principled as well as ad hoc methods for addressing the action selection problem have been developed and demonstrated on robotic systems. These include varieties of command fusion [11] and spreading of activation [12], among many others. For a comprehensive survey on action selection mechanisms, see Reference 15.

Behavior-based systems are varied, but there are two fundamental tenets all BB systems inherently adhere to (1) the robot is embodied and (2) the robot is situated. A robot is *embodied* in the sense that it has a physical body and its behavior is limited by physical realities, uncertainties, and consequences of its actions, all of which may be hard to predict or simulate. A robot is *situated* in the sense that it is immersed in the real world and acts directly on the sensory information received from that world, not on abstract or processed representations of the world.

Behavior-based control makes no assumptions on the availability of a complete world model; therefore, it is uncommon for a BB controller to perform extensive computation or reasoning relying on such a model. Instead, BB controllers maintain a tight coupling of sensing and action, allowing them to act in a timely manner in response to dynamic and fast-changing worlds. However, BB systems have also demonstrated elegant use of distributed representations enabling robot mapping and task learning [14–16].

This section has discussed approaches and philosophies to the control of a single robot, with a focus on the BB approach. In Section 14.2, the scope is expanded to consider the control of a coordinated group of multiple robots.

14.2 FROM SINGLE ROBOT CONTROL TO MULTI-ROBOT CONTROL

In this section we discuss the advantages and additional issues involved in the control of MRS as compared to the single-robot systems (SRS) discussed in Section 14.1.1. An *MRS* is a system composed of multiple, interacting robots. The study of MRS has received increased attention in recent years. This is not surprising, as continually improving robustness, availability, and cost-effectiveness of robotics technology has made the deployment of MRS consisting of increasingly larger numbers of robots possible. With the growing interest in MRS comes the expectation that, at least in some important respects, multiple robots will be superior to a single robot in achieving a given task. In this section we outline the benefits of a MRS over a SRS and introduce issues involved in MRS control and how they are similar and different to those of SRS control.

This chapter is focused on distributed MRS in which each robot operates independently under local sensing and control. *Distributed MRS* stand in contrast to *centralized MRS*, in which each robot's actions are not completely determined locally, as they may be determined by an outside entity, such as another robot or by any type of external command. In distributed MRS, each robot must make its own control decisions based only on limited, local, and noisy sensor information. We limit our consideration in this chapter to distributed MRS because they are the most appropriate for study with regard to systems that are scalable and capable of performing in uncertain and unstructured real-world environments where uncertainties are inherent in the sensing and action of each robot. Furthermore, this chapter is centered on achieving system-level coordination in a distributed BB MRS. Strictly speaking, the issues in a centralized MRS are more akin to a scheduling or optimal assignment and less of a problem of coordination in a distributed system.

14.2.1 Advantages and Challenges of Multi-Robot Systems

Potential advantages of MRS over SRS include a reduction in total system cost by utilizing multiple simple and cheap robots as opposed to a single complex and expensive robot. Also, multiple robots can increase system flexibility and robustness by taking advantage of inherent parallelism and redundancy. Furthermore, the inherent complexity of some task environments may require the use of multiple robots, as the necessary capabilities or resource requirements are too substantial to be met by a single robot.

However, the utilization of MRS poses potential disadvantages and additional challenges that must be addressed if MRS are to present a viable and effective alternative to SRS. A poorly designed MRS, with individual robots working toward opposing goals, can be less effective than a carefully designed SRS. A paramount challenge in the design of effective MRS is managing the complexity introduced by multiple, interacting robots. As such, in most cases just taking a suitable SRS solution and scaling it up to multiple robots is not adequate.

14.2.2 Necessity of Coordination in MRS

In order to maximize the effectiveness of a MRS, the robots' actions must be spatio-temporally coordinated and directed toward the achievement of a given system-level task or goal. Just having robots interact is not sufficient in itself to produce interesting or practical system-level coordinated behavior. The design of MRS can be quite challenging because unexpected system-level behaviors may emerge due to unanticipated ramifications of the robots' local interactions. In order for the interacting robots to produce coherent task-directed behavior, there must be some overarching coordination mechanism that spatio-temporally organizes the interactions in a manner appropriate for the task.

The design of such coordination mechanisms can be difficult; nonetheless, many elegant handcrafted distributed MRS have been demonstrated, both in simulation and on physical robots [17–19]. The methods by which these systems have achieved task-directed coordination are diverse and the possibilities are seemingly limited only by the ingenuity of the designer. From a few robots performing a manipulation task [20,21], to tens of robots exploring a large indoor area [22,23], to potentially thousands of ecosystem monitoring nanorobots [24,25], as the number of robots in the system increases, so does the necessity and importance of coordination. Section 14.3 examines mechanisms by which system-level coordination can be successfully achieved in a MRS.

14.3 FROM LOCAL INTERACTIONS TO GLOBAL COORDINATION

Given the importance of coordination in a MRS, we now address the issue of how to organize the robots' local interactions in a coherent manner in order to achieve system-level coordination. There are many mechanisms by which the interactions can be organized. We classify them into three broad and often overlapping classes: interaction through the environment, interaction through sensing, and interaction through communication. These classes are not mutually exclusive because MRS can, and often do, simultaneously utilize mechanisms from any or all of these classes to achieve system-level coordinated behavior.

In the following sections, we describe each of these interaction classes in detail. Through the discussion of empirical case studies we demonstrate how each type of interaction can be used to achieve system-level coordination in a MRS.

14.3.1 Interaction through the Environment

The first mechanism for interaction is through the robots' shared environment. This form of interaction is *indirect* in that it consists of no explicit communication or physical interaction between robots. Instead, the environment itself is used as a medium of indirect communication. This is a powerful approach that can be utilized by very simple robots with no capability for complex reasoning or direct communication.

An example of interaction through the environment is demonstrated in *stigmergy*, a form of interaction employed by a variety of insect societies. Originally introduced in the biological sciences to explain some aspects of social insect nest-building behavior, stigmergy is defined as the process by which the coordination of tasks and the regulation of construction do not depend directly on the workers, but on the constructions themselves [26]. This concept was first used to describe the nest-building behavior of termites and ants [27]. It was shown that coordination of building activity in a termite colony was not inherent in the termites themselves. Instead, the coordination mechanisms were found to be regulated by the task environment, in that case the growing nest structure. A location on the growing nest stimulates a termite's building behavior, thereby transforming the local nest structure, which in turn stimulates additional building behavior of the same or another termite.

Through the careful design of robot sensing, actuation, and control features, it is possible to utilize the concept of stigmergy in task-directed MRS. This powerful mechanism of coordination is attractive as it typically requires minimal capabilities of the individual robots. The robots do not require direct communication, unique recognition of other robots, or even distinguishing other robots from miscellaneous objects in the environment, or the performance of computationally intensive reasoning or planning.

Stigmergy, and more generally interaction through the environment, has been successfully demonstrated as a mechanism to coordinate robot actions in a number of MRS. It has been demonstrated in an object manipulation domain [28] in which a large box was transported to a goal location through the coordinated pushing actions of a group of robots. There was no globally agreed upon plan as to how or over what trajectory the box should be moved; however, each robot could indirectly sense the pushing actions of other robots through the motions of the box itself. Through simple rules, each robot decided whether to push the box or move to another location based on the motions of the box itself. As a large enough number of robots pushed in compatible directions, the box moved, which in turn encouraged other robots to push in the same direction.

Other examples of the use of stigmergy in MRS include distributed construction in which a given structure was built in a specified construction sequence [29]. The individual robots were not capable of explicit communication and executed simple rule-based controllers in which local sensory information was directly mapped to construction actions. The construction actions of one robot altered the environment, and therefore the subsequent sensory information available for it and all other robots. This new sensory information then activated future construction actions. In Section 14.3.2, we discuss in detail how the concept of stigmergy was utilized in a MRS object clustering task domain [26].

14.3.2   Interaction through the Environment Case Study: Object Clustering

We now describe an empirical case study in an object clustering task domain for which interaction through the environment was used to achieve system-level coordination. The clustering task domain requires a group of objects, originally uniformly positioned in an enclosed environment, to be repositioned by a group of robots into a single dense cluster of objects. There is no a priori target location for the cluster in the environment. Rather, the position of the cluster is to be determined dynamically at the time of task execution.

The particular approach to the object clustering task we describe here is from work presented in Reference 26. There, the robots performing the task were extremely simple, capable only of picking up and transporting and dropping a single object at a time. The robots had very limited local sensing and no explicit communication, memory of past actions, or recognition of other robots. Even with these highly limited capabilities, a homogeneous MRS composed of such robots was shown to be capable of successfully and robustly performing the object clustering task.

The robots in this task domain were able to coherently achieve system-level coordination in the formation of a single cluster. The mechanism by which they achieved coordination was an example of interaction through the environment. The robots communicated through their individual placement of objects over time, thus modifying the task environment, and thereby indirectly influencing the future object-placement behaviors of other robots and themselves. The location of the final cluster was not determined through explicit communication, negotiation, or planning on the part of the robots. Rather, it was determined through a symmetry break in the initially uniform distribution of objects. Once a small cluster began to form, it was likely to grow larger. During the early stages of task execution, several clusters were likely to be formed. However, over time, a single large cluster resulted.

The robots in this work were designed in a manner that carefully exploited the physical dynamics of interaction between the robots and their environment. Their hardware and rules were tuned so as to be probabilistically more likely to pick up an object that is not physically proximate to other objects (thus conserving clusters), to not drop objects near boundaries (thus avoiding hard-to-find objects), and to be probabilistically more likely to deposit an object near other objects (thereby building up clusters). Together, their properties resulted in a form of positive feedback in which the larger a cluster of objects became the more likely it was to grow even larger.

Similar approaches employing stigmergy were also demonstrated in the physical segregation and sorting of a collection of object classes. Additional studies with physical robots have been conducted and, by making various changes in the robots and the task environment, is has been demonstrated that one can influence the location of the final cluster by initializing the initial distribution of objects in a nonuniform manner [26].

Given this specific example of system-level coordination achieved through the use of interaction through the environment, in the following section we move on to the next method of organizing the robot's interactions: interaction through sensing.

14.3.3 Interaction through Sensing

The second mechanism for interaction among robots is through sensing. As described in Reference 19, interaction through sensing "refers to local interactions that occur between robots as a result of sensing one another, but without explicit communication." As with interaction through the environment, interaction through sensing is also *indirect* as there is no explicit communication between robots; however, it requires each robot to be able to distinguish other robots from miscellaneous objects in the environment. In some instances, each robot may be required to uniquely identify all other robots, or classes of other robots. In other instances, it may only be necessary to simply distinguish robots from other objects in the environment.

Interaction through sensing can be used by a robot to model the behavior of other robots or to determine what another robot is doing in order to make decisions and respond appropriately. For example, flocking birds use sensing to monitor the actions of other birds in their vicinity to make local corrections to their own motion. It has been shown that effective flocking results from quite simple local rules followed by each bird responding to the direction and speed of the local neighbors [30].

In the following section, we describe a case study in a formation marching domain in which interaction through sensing is used to achieve coordinated group behavior. Other domains in which interaction through sensing has been utilized in MRS include flocking [31], in which each robot adjusts it motions according to the motions of locally observed robots. Through this process, the robots can be made to move as a coherent flock through an obstacle-laden and dynamic environment. Interaction through sensing has also been demonstrated in an adaptive division of labor domain [32]. In that domain, each robot dynamically changes the task it is executing based on the observed actions of other robots and the observed availability of tasks in the environment. Through this process, the group of robots coherently divides the labor of the robots appropriately across a set of available tasks.

14.3.4 Interaction through Sensing Case Study: Formation Marching

In this section, we describe an empirical case study of a formation marching task domain for which interaction through sensing was used to achieve system-level coordination. The formation marching task domain requires a group of robots to achieve and maintain relative positions to one another as the group moves through the environment in a global formation. Each robot in the MRS operates under local sensing and control and is not aware of global information such as all other robot's positions and headings. In some environments, the formation may need to be perturbed in order for the group to move through a constrained passage or around obstacles. In such cases, the formation needs to correctly realign after the perturbation.

The approach to formation marching described here was presented in Reference 35. The general idea of the approach is that every robot in the MRS positioned itself relative to a designated neighbor robot. This neighbor robot, in turn, positioned itself relative to its own designated neighbor robot. As all robots are only concerned with their relative positions with respect to their neighbor robot, no robot is aware of, or needs to be aware of, the global positions and headings of all robots in the formation. Each robot only needs to be capable of determining the distance and heading to its neighbor. The global geometry of the formation was then determined through the defined chain of neighbors.

A "leader" robot has no neighbors and independently determines the speed and heading of the entire formation. Therefore, as the leader robot moves forward, the robot(s) that had the leader as their neighbor also move forward. This forward motion propagates down the chain of designated neighbors, causing the entire formation to move.

The formation could be dynamically changed by altering the structure of the local neighbor relationships. For example, if the desired formation is a line, each robot may be designated a neighbor robot to its left or right for which it desires to stay next to in order to maintain the line formation. If a cue is given to all robots to change to a diamond formation, each robot may follow a new neighbor at a different relative position and the line formation would then be dynamically changed to a diamond.

In the following section we move to the next method of organizing the robot's interactions: interaction through communication.

14.3.5 Interaction through Communication

The third mechanism for interaction among robots is through explicit communication. Unlike the first two forms of interaction, described earlier, which were indirect, in interaction through communication robots may communicate with others directly. Such *robot-directed* communication can be used to request information or action from other robots or to respond to received requests.

Communication in physical robotics is not free or reliable and can be constrained by limited bandwidth and range, and unpredictable interference. When utilizing it, one must consider how and toward what end it is used. In some domains, such as the Internet, communication is reliable and of unlimited range; however, in physical robot systems, communication range and reliability are important factors in system design [2,34].

There are many types of communication. Communication could be direct from one robot to another, direct from one robot to a class of other robots, or broadcast from one robot to all others. Furthermore, the communication protocol can range from simple protocol-less schemes to a complex negotiation-based and communication-intensive schemes. The information encoded in a communication may be state information contained by the communicating robot, a command to one or more other robots, or a request for additional information from other robots, etc.

Communications may be task-related rather than robot-directed, in which case it is made available to all (or a subset) of the robots in the MRS. A common task-related communication scheme is *publish/subscribe messaging*. In publish/subscribe messaging, subscribing robots request to receive certain categories of messages, and publishing robots supply messages to all appropriate subscribers.

In the next section, we describe a case study of the effective use of interaction through communication.

14.3.6 Interaction through Communication Case Study: Multiple Target Tracking

The case study on interaction through communication in this section is focused on the use of explicit communication in a multi-target tracking task as discussed in Reference 37. In multi-target tracking, the goal is to have a set of robots with limited sensing ranges position and orient themselves such that they are able to acquire and track multiple objects moving through their environment. The locations, trajectories, and number of targets are not known a priori. These difficulties are compounded in a distributed MRS, where the system must determine which robot(s) should monitor which target(s). Robots redundantly tracking the same target may be wasting resources and letting another target remain untracked. In this domain, explicit communication between the robots has been shown to be capable of effectively achieving system-level coordination.

In the implementation described in Reference 37, each robot had a limited sensing and communication range. Communication was used by each robot to transmit the position and velocities of all targets within its sensing range to all other robots within its communication range. This simple communication scheme involved no handshaking or negotiation.

Each robot was constantly evaluating the importance of its current tracking activities and possible changes in position that could increase the importance of its tracking activities. Communication was used to allow each robot to keep a local map of target movements within the robot's communication range but outside its sensing range. As a result, the group as a whole effectively tracked a maximum number of targets with a minimum number of available robots.

This demonstration of the use of interaction through communication concludes the discussion of MRS coordination mechanisms. As was mentioned earlier, any given MRS is likely to use any or all of the three mechanisms in varying degrees to achieve system-level coordination. Through an improved understanding of each of these mechanisms of coordination, one is better positioned to design a MRS utilizing the most appropriate combination of mechanisms for achieving a given task. In the next section we provide a discussion on formal methods for the design and analysis of MRS that can provide a principled foundation upon which to base such design decisions.

14.4 FORMAL DESIGN AND ANALYSIS OF MRS

The design of coordination mechanisms for MRS has proven to be a difficult problem. In the last decade, the design of a variety of such mechanisms over a wide range of task domains has been studied [17,18]. Although the literature highlights some elegant solutions, they are generally domain-specific and provide only indirect insights into important questions such as how appropriate a given coordination mechanism is for a particular domain, what performance characteristics one should expect from it, how it is related to other coordination mechanisms, and how one can modify it to improve system performance. These questions must be answered in a principled manner before one can quickly and efficiently produce an effective MRS for a new task domain. To fully utilize the power and potential of MRS and to move the design process closer to a science, principled design tools and methodologies are needed for establish a solid foundation upon which to construct increasingly capable, robust, and efficient MRS.

The design of an effective task-directed MRS is often difficult because there is a lack of understanding of the relationship between different design options and resulting task performance. In the common trial-and-error design process, the designer constructs an MRS and then tries it out either in simulation or on physical robots. Either way, the process is resource-intensive. Ideally, the designer should be equipped with an analytical tool for the analysis of a potential MRS design. Such a tool would allow for efficient evaluation of different design options and thus result in more effective and optimized MRS designs.

The BB paradigm for multi-robot control is popular in MRS because it is robust to the dynamic interactions inherent in any MRS. Any MRS represents a highly nonlinear system in which the actions of one robot are affected by the actions of all other robots. This makes any control approach that relies on complex reasoning or planning ineffective because it is intractable to accurately predict future states of a nontrivial MRS. For this reason, BB control is frequently used in MRS. The simplicity of the individual robots also confers the advantage of making the external analysis of predicted system performance on a given task feasible.

In the remainder of this section we discuss a variety of approaches to the analysis and synthesis of MRS.

14.4.1 Analysis of MRS Using Macroscopic Models

Macroscopic models reason about the system-level MRS behavior without explicit consideration of each individual robot in the system. As such, macroscopic models are generally more scalable and efficient in the calculation of system-level behaviors even as the studied MRS consists of increasingly larger numbers of robots.

A macroscopic mathematical MRS model has been demonstrated in a foraging task domain [36]. The model was used to study the effects of interference between robots, the results of which could be used to modify individual robot control or determine the optimal density of robots in order to maximize task performance. A macroscopic analytical model has been applied to the study of the dynamics of collective behavior in a collaborative stick-pulling domain using a series of coupled differential Equations [37].

A general macroscopic model for the study of adaptive multi-agent systems was presented in Reference 38 and was applied to the analysis of a multi-robot adaptive task allocation domain that was also addressed experimentally in Reference 32. In this work, the robots constituting the MRS maintain a limited amount of persistent internal state to represent a short history of past events but do not explicitly communicate with other robots.

14.4.2 Analysis of MRS Using Microscopic Models

In contrast, microscopic modeling approaches directly consider each robot in the system and may model individual robot interactions with other robots and with the task environment in arbitrary detail, including simulating the exact behavior of each robot. However, most microscopic approaches model the behavior of each robot as a series of stochastic events. Typically, the individual robot controllers are abstracted to some degree and exact robot trajectories or interactions are not directly considered.

A microscopic probabilistic modeling methodology for the study of collective robot behavior in a clustering task domain was presented in Reference 41. The model was validated through a largely quantitative agreement in the prediction of the evolution of cluster sizes with embodied simulation experiments and with real-robot experiments. The effectiveness and accuracy of microscopic and macroscopic modeling techniques compared to real robot experiments and embodied simulations was discussed in Reference 42. Furthermore, a time-discrete, incremental methodology for modeling the dynamics of coordination in a distributed manipulation task domain was presented in Reference 43.

14.4.3 Principled Synthesis of MRS Controllers

One step beyond methodologies for the formal analysis of a given MRS design lie formal methodologies for the synthesis of MRS controllers. *Synthesis* is the process of constructing an MRS controller that meets design requirements such as achieving the desired level of task performance while meeting constraints imposed by limited robot capabilities. Being able to define a task domain and then have a formal method that designs the MRS to accomplish the task while meeting the specified performance criteria is one of the long-term goals of the MRS community.

An important piece of work in the formal design of coordinated MRS was the development of information invariants, which aimed to define the information requirements of a given task and ways in which those requirements could be satisfied in a robot controller [42]. Information invariants put the design of SRS and MRS on a formal footing and began to identify how various robot sensors, actuators, and control strategies could be used to satisfy task requirements. Furthermore, the work attempted to show how these features were related and how one or more of these features could be formally described in terms of a set of other features. The concept of information invariants was experimentally studied in a distributed manipulation task domain [43] and was extended through the definition of equivalence classes among task definitions and robot capabilities to assist in the choice of appropriate controller class in a given domain [44].

There has also been significant progress in the design of a formal design methodology based on a MRS formalism that provides a principled framework for formally defining and reasoning about concepts relevant to MRS: the world, task definition, and capabilities of the robots themselves, including action selection, sensing, maintenance of local and persistent internal state, and broadcast communication from one robot to all other robots [45]. Based on this formalism, the methodology utilizes an integrated set of MRS synthesis and analysis methods. The methodology includes a suite of systematic MRS synthesis methods, each of which takes as input the formal definitions of the world, task, and robots *sans* controller and outputs a robot controller designed through a logic-induced procedure. Each of the synthesis methods is independent and produces a coordinated MRS through the use of a unique set of coordination mechanisms, including the use of internal state [46], inter-robot communication [32], or selection of deterministic and probabilistic action. Complimentary to the synthesis methods, this methodology incorporates both macroscopic [45] and microscopic MRS modeling approaches. Together, the synthesis and analysis methods provide more than just pragmatic design tools. A defining feature of this design methodology is the integrated nature of the controller synthesis and analysis methods. The fact that they are integrated allows for the capability to automatically and iteratively synthesize and analyze a large set of possible designs, thereby resulting in more optimal solutions and an improved understanding of the space of possible designs. This principled approach to MRS controller design has been demonstrated in a sequentially constrained multi-robot construction task domain [32,45,46].

A theoretical framework for the design of control algorithms in a multi-robot object clustering task domain has been developed [47]. Issues addressed in this formalism include how to design control algorithms that result in a single final cluster, multiple clusters, and how to control the variance in cluster sizes.

Alternative approaches to the synthesis of MRS controllers can be found in evolutionary methods [48] and learning methods [31,49]. There also exist a number of MRS design environments, control architectures, and programming languages which assist in the design of coordinated MRS [50–52].

14.5 CONCLUSIONS AND THE FUTURE OF MULTI-ROBOT SYSTEMS

Behavior-based control has been a popular paradigm of choice in the control of MRS. The BB control methodology represents a robust and effective way to control individual as well as multiple robots. In an MRS, the task environment is inherently dynamic and nonlinear as a result of the numerous types of interactions between the individual robots and between the robots and the task environment. This makes complex control strategies relying on accurate world models to perform computationally complex reasoning or planning ineffective. BB control provides a tight coupling between sensing and action and does not rely on the acquisition of such world models. As such it is a very effective control methodology in the dynamic and unstructured environments in which MRS inherently operate.

Behavior-based MRS have been empirically demonstrated in a diverse array of task domains — from foraging, to object clustering, to distributed manipulation, to construction. Each of these task domains requires some overarching mechanism by which to coordinate the interactions of the individual robots such that the resulting system-level behavior is appropriate for the task. We have described and illustrated three different mechanisms to achieve this coordinated behavior: interaction through the environment, interaction through sensing, and interaction through communication. Each provides a coordination scheme capable of organizing the individual robot's behaviors toward system-level goals.

Another advantage of BB MRS is their amenability to formal analysis and synthesis. Due to their rather straightforward and direct coupling of sensing to action, formal methods of synthesis and analysis become tractable and effective in producing and predicting the system-level behavior of a BB MRS.

The future possibilities and potentials of BB MRS are seemingly unlimited. As technology continues to improve and the nature and implications of different strategies for coordination are better understood, more task domains will become valid candidates for the application of MRS solutions.

REFERENCES

1. Matarić, M. (1997). Behavior-Based Control: Examples from Navigation, Learning, and Group Behavior. Hexmoor, H., Horswill, I., and Kortenkamp, D. (eds), *Journal of Experimental and Theoretical Artificial Intelligence*, special issue on Software Architectures for Physical Agents 9: 323–336.

2. Arkin, R. (1998). *Behavior-Based Robotics.* Cambridge, MA: MIT Press.
3. Rosenchein, S. and Kaelbling, L. (1995). *A Situated View of Representation and Control.* Special Issue on Computational Research on Interaction and Agency, Amsterdam, New York: Elsevier Science, pp. 515–540.
4. Brooks, R. (1990). Elephants Don't Play Chess. *Designing Autonomous Agents* 6. Cambridge, MA: MIT Press, pp. 3–15.
5. Brooks, R. (1991). Intelligence Without Representation. *Artificial Intelligence* 47: 139–160.
6. Brooks, R. and Connell, J. (1986). Asynchronous Distributed Control Systems for a Mobile Robot. In *Proceedings of SPIE's Cambridge Symposium on Optical and Optoelectronic Engineering.* Cambridge, MA, pp. 77–84.
7. Connell, J. (1992). SSS: A Hybrid Architecture Applied to Robot Navigation. In *Proceedings of IEEE Conference on Robotics and Automation*, Nice, France, pp. 2719–2724.
8. Gat, E. (1998). On Three-Layer Architectures. In D. Kortenkamp, R. P. Bonnasso, and R. Murphy (eds), *Artificial Intelligence and Mobile Robotics*, MIT, AAAI Press, Cambridge, MA, pp. 195–210.
9. Brooks, R. (1986). A Robust Layered Control System for a Mobile Robot. *IEEE Journal of Robotics and Automation* 2: 14–23.
10. Pirjanian, P. (2000). Multiple Objective Behavior-based Control. *Robotics and Autonomous Systems* 31: 53–60.
11. Payton, D., Keirsey, D., Kimble, D., Krozel, J., and Rosenblatt, J. (1992). Do Whatever Works: A Robust Approach to Fault-Tolerant Autonomous Control. *Applied Intelligence* 2: 225–250.
12. Maes, P. and Brooks, R. (1990). Learning to Coordinate Behaviors. In *Proceedings of the American Association of Artificial Intelligence (AAAI-91)*, Boston, MA, pp. 796–802.
13. Pirjanian, P. (1999). Behavior Coordination Mechanisms — State-of-the-Art. University of Southern California Institute of Robotics and Intelligent Systems, IRIS Technical Report IRIS-99-375.
14. Matarić, M. (1992). Integration of Representation Into Goal-Driven Behavior-Based Robots. *IEEE Transactions on Robotics and Automation* 8: 304–312.
15. Nicolescu, M. and Matarić, M. (2001). Experience-Based Representation Construction: Learning from Human and Robot Teachers. In *Proceedings of IEEE/RSJ International Conference on Robots and Systems (IROS-01).* Maui, Hawaii, pp. 740–745.
16. Nicolescu, M. and Matarić, M. (2000). Extending Behavior-Based Systems Capabilities Using an Abstract Behavior Representation. University of Southern California Institute for Robotics and Intelligent Systems, IRIS Technical Report IRIS-00-389.
17. Cao, Y., Fukunaga, A., and Kahng, A. (1997). Cooperative Mobile Robotics: Antecedents and Directions. *Autonomous Robots* 4: 7–27.
18. Dudek, G., Jenkin, M., and Milios, E. (2002). A Taxonomy of Multirobot Systems. In Balch, T. and Parker, L. (eds), *Robot Teams: From Diversity to Polymorphism*, A.K. Peters, Natick, MA: pp. 3–22.

19. Balch, T. and Parker, L. (eds) (2002). *Robot Teams: From Diversity to Polymorphism*. A.K. Peters, Natick, Massachusetts.

20. Bohringer, K., Brown, R., Donald, B., and Jennings, J. (1997). Distributed Robotic Manipulation: Experiments in Minimalism. In Khatib, O. (ed.) *Experimental Robotics IV*, Lecture Notes in Control and Information Sciences 223, pp. 11–25. Berlin: Springer-Verlag.

21. Gerkey, B. and Matarić, M. (2004). A Formal Analysis and Taxonomy of Task Allocation in Multi-Robot Systems. *International Journal of Robotics Research* 23: 939–954.

22. Howard, A., Parker, L., and Sukhatme, G. (2004). The SDR Experience: Experiments with a Large-Scale Heterogenous Mobile Robot Team. In *9th International Symposium on Experimental Robotics*. Singapore, June 18–20.

23. Konolige, K., Fox, D., Ortiz, C., Agno, A., Eriksen, M., Limketkai, B., Ko, J., Morisset, B., Schulz, D., Stewart, B., and Vincent, R. (2004). Centibots: Very Large Scale Distributed Robotic Teams. In *9th International Symposium on Experimental Robotics*. Singapore, June 18–20.

24. Rahimi, M., Pon, R., Kaiser, W., Sukhatme, G., Estrin, D., and Srivastava, M. (2004). Adaptive Sampling for Environmental Robotics. In *Proceedings of the International Conference on Robotics and Automation*, pp. 3537–3544. New Orleans, LA.

25. Zhang, B., Sukhatme. G., and Requicha, A. (2004). Adaptive Sampling for Marine Microorganism Monitoring. To appear in *Proceedings of IEEE/RSJ International Conference on Intelligent Robots and Systems (IROS)*. Sendai, Japan, October 30–September 2.

26. Holland, O. and Melhuish, C. (1999). Stigmergy, Self-Organization, and Sorting in Collective Robotics. *Artificial Life* 5: 173–202.

27. Franks, N. and Deneubourg, J.-L. (1997) Self-Organising Nest Construction in Ants: Individual Worker Behavior and the Nest's Dynamics. *Animal Behaviour* 54: 779–796.

28. Kube, C. and Zhang, H. (1996). The Use of Perceptual Cues in Multi-Robot Box-Pushing. In *Proceedings of IEEE International Conference on Robotics and Automation (ICRA-96)*, pp. 2085–2090, Minneapolis, MN.

29. Jones, C. and Matarić, M. (2004). Automatic Synthesis of Communication-Based Coordinated Multi-Robot Systems. In *Proceedings of IEEE/RSJ International Conference on Intelligent Robots and Systems (IROS)*. Sendai, Japan, October 30–September 2.

30. Reynolds, C. (1987). Flocks, Herds, and Schools: A Distributed Behavior Model. *Computer Graphics* 21: 25–34.

31. Matarić, M. (1995). Designing and Understanding Adaptive Group Behavior. *Adaptive Behavior* 4: 51–80.

32. Jones, C. and Matarić, M. (2003). Adaptive Division of Labor in Large-Scale Minimalist Multi-Robot Systems. In *Proceedings of the IEEE/RSJ International Conference on Robotics and Intelligent Systems (IROS)*. Las Vegas, NV, pp. 1969–1974.

33. Fredslund, J. and Matarić, M. (2002). A General, Local Algorithm for Robot Formations. *IEEE Transactions on Robotics and Automation, Special Issue on Multi-Robot Systems* 18: 837–846.

34. Gerkey, B. and Matarić, M. (2001). Principled Communication for Dynamic Multi-Robot Task Allocation. In *Proceedings of International Symposium on Experimental Robotics 2000*. Waikiki, Hawaii, pp. 341–352.

35. Parker, L. (1997). Cooperative Motion Control for Multi-Target Observation. In Proceedings of IEEE/RSJ International Conference on Intelligent Robots and Systems, Grenoble, France, pp. 1591–1598.

36. Lerman, K. and Galstyan, A. (2002). Mathematical Model of Foraging in a Group of Robots: Effects of Interference. *Autonomous Robots* 13: 127–141.

37. Lerman, K., Galstyan, A., Martinoli, A., and Ijspeert, A. (2001). A Macroscopic Analytical Model of Collaboration in Distributed Robotic Systems. *Artificial Life* 7: 375–393.

38. Lerman, K. and Galstyan, A. (2003). Macroscopic Analysis of Adaptive Task Allocation in Robots. In *Proceedings of the IEEE/RSJ International Conference on Intelligent Robots and Systems*. Las Vegas, NV, pp. 1951–1956.

39. Martinoli, A., Ijspeert, A., and Mondada, F. (1999). Understanding Collective Aggregation Mechanisms: From Probabilistic Modeling to Experiments with Real Robots. *Robotics and Autonomous Systems* 29: 51–63.

40. Martinoli, A. and Easton, K. (2002). Modeling Swarm Robotic Systems. In Siciliano, B. and Dario, P. (eds), *Experimental Robotics VIII*. Heidelberg: Springer-Verlag, pp. 297–306.

41. Martinoli, A., Easton, K., and Agassounon, W. (2004). Modeling Swarm Robotic Systems: A Case Study in Collaborative Distributed Manipulation. *Special issue on Experimental Robotics, International Journal of Robotics Research* 23: 415–436.

42. Donald, B. (1995). Information Invariants in Robotics. *Artificial Intelligence* 72: 217–304.

43. Donald, B., Jennings, J., and Rus, D. (1995). Information Invariants for Distributed Manipulation. In Goldberg, K., Halperin, D., Latombe, J.-C., and Wilson, R. (eds), *International Workshop on the Algorithmic Foundations of Robotics*. Boston, MA, pp. 431–459.

44. Parker, L. (1998). Toward the Automated Synthesis of Cooperative Mobile Robot Teams. In *Proceedings of SPIE Mobile Robots XIII*, vol. 3525. Boston, MA, pp. 82–93.

45. Jones, C. and Matarić, M. (2004). Synthesis and Analysis of Non-reactive Controllers for Multi-robot Sequential Task Domains. In *Proceedings of the International Symposium on Experimental Robotics*. Singapore, June 18–20.

46. Jones, C. and Matarić, M. (2004). The Use of Internal State in Multi-Robot Coordination. In *Proceedings of the Hawaii International Conference on Computer Sciences (HICCS)*. Waikiki, Hawaii, pp. 27–32.

47. Zhang, A., Chung, M., Lee, B., Cho, R., Kazadi, S., and Vishwanath, R. (2002). Variance in Converging Puck Cluster Sizes. In *Proceedings of the First International Joint Conference on Autonomous Agents and Multiagent Systems (AAMAS)*. Bologna, Italy, pp. 209–217.

48. Floreano, D. (1993). Patterns of Interactions in Shared Environments. In *Proceedings of the Second European Conference on Artificial Life*. Brussels, Belgium, pp. 347–366.

49. Parker, L. (1998). Alliance: An Architecture for Fault-Tolerant Multi-Robot Cooperation. *Transactions on Robotics and Automation* 14: 220–240.
50. Matarić, M. (1995). Issues and Approaches in the Design of Collective Autonomous Agents. *Robotics and Autonomous Systems* 16: 321–331.
51. Arkin, R. and Balch, T. (1997). Aura: Principles and Practice in Review. *Journal of Experimental and Theoretical Artificial Intelligence* 9: 175–189.
52. Alur, R., Grosu, R., Hur, Y., Kumar, V., and Lee, I. (2000). Modular Specification of Hybrid Systems in CHARON. In *Proceedings of the 3rd International Workshop on Hybrid Systems: Computation and Control*, Pittsburgh, PA, pp. 6–19.

BIOGRAPHIES

Chris Jones is a robotics researcher at iRobot Corporation in Burlington, MA. Dr. Jones earned his Ph.D. (2005) and M.S. (2003) in computer science from the University of Southern California and his B.S. (1999) in Computer Engineering from Texas A&M University. Prior to joining the Research Group at iRobot, he has been involved in robotics research and development in such places as the Center for Robotics and Embedded Systems at the University of Southern California, the Artificial Intelligence Lab at the University of Zurich, the Intelligent Systems and Robotics Center at Sandia National Laboratories, and the Robotics Research Lab at Texas A&M University. His interests include mobile robot autonomy, the coordination of multiple robots, and human-robot interaction.

Maja J. Matarić is an associate professor in the Computer Science Department and Neuroscience Program at the University of Southern California, founding director of USC's interdisciplinary Center for Robotics and Embedded Systems (CRES), and co-director of the USC Robotics Research Lab. Prof. Matarić earned her Ph.D. in computer science and artificial intelligence from MIT in 1994, M.S. in computer science from MIT in 1990, and B.S. in computer science from the University of Kansas in 1987. She is a recipient of the Okawa Foundation Award, the USC Viterbi School of Engineering Service Award, the NSF Career Award, the MIT TR100 Innovation Award, the IEEE Robotics and Automation Society Early Career Award, the USC Viterbi School of Engineering Junior Research Award, and the USC Provost's Center for Interdisciplinary Research Fellowship, and is featured in the Emmy award-nominated documentary movie about scientists, *"Me & Isaac Newton."* She is an associate editor of three major journals: *International Journal of Autonomous Agents and Multi-Agent Systems, International Journal of Humanoid Robotics*, and *Adaptive Behavior*. She has published over 30 journal articles, 17 book chapters, 4 edited volumes, 94 conference papers, and 23 workshop papers, and has two books in the works with MIT Press. She is active in educational outreach and is collaborating with K-12 teachers to develop hands-on robotics curricula for students at all levels as tools for promoting science and engineering topics and recruiting women and under-represented students. Her Interaction Lab pursues research aimed at endowing robots with the ability to help people through assistive interaction.

V

System Integration and Applications

Each component described in the preceding parts of the book plays an important role in the operation of an intelligent and autonomous system. Despite their individual importance, we should note that an intelligent autonomous system cannot be fully realized if any one of the components is lacking, or if they are not properly integrated.

System integration is the glue that brings the components together into a cohesive structure, and is therefore an important part of complex systems. The three chapters in the last part of the book examine the issues that exist in system integration from different points of view. Each chapter also presents case studies of intelligent systems currently being implemented in various applications, ranging from consumer products, to automotives, to military vehicles.

Chapter 15 discusses and analyzes the challenges of system integration for complex autonomous systems, with emphasis on consumer robotics. The system integration problem is presented as the optimization of three related but conflicting measures — performance, complexity, and price — and how these metrics influence the design of consumer robotic systems. The chapter focuses on the use of software architecture for the integration of the various components of an autonomous robot, and explores the characteristics and requirements for the design of such a software architecture while keeping in mind the conflicting metrics. The Evolution Robotics Software Platform (ERSP) is presented in detail as a software architecture that is able to fulfill the requirements demanded for system integration for commercial robots. The effectiveness of the ERSP was evaluated via two case studies involving the SONY AIBO ERS-7 and the development of a robotic vacuum cleaner.

Automotive systems and autonomous highways are treated in detail in Chapter 16, the second chapter of this part. The chapter describes the problems and motivation for automating highways and automobiles, such as issues with safety, congestion, and pollution. It offers an interesting insight into the future of automobiles, reviewing the slew of current technologies in hardware and sensing that will possibly make autonomous automobiles a reality in the near future. This chapter thus focuses upon the hardware requirements of autonomous automobiles and evaluates the degree to which these requirements are met by today's technologies. The reader is given a comprehensive overview of technologies in sensing (e.g., vision and GPS), actuators, and vehicular control. Interesting examples and case studies are included to illustrate the implementation and extent of the success of these technologies in state-of-the art systems, such as the Intelligent Multimode Transit System (IMTS) from Toyota.

Finally, Chapter 17 presents the 4D/RCS architecture that represents a comprehensive methodology for integrating the components within autonomous robots and also on a larger scale of cooperating robots. This integrates the numerous components contributing to autonomy into a coherent whole that is capable of fully utilizing the functionalities of each module to realize a truly intelligent system. The hierarchical and modular structure of the 4D/RCS architecture facilitates decentralized decision making by lower-level nodes and also the use of various levels of abstraction of available knowledge such that each node only maintains knowledge at the level of abstraction that is required. This allows the decomposition of high-level task descriptions as they propagate through the hierarchy and translates into more specific actions at the lower levels of the structure. The chapter also describes the successful implementation of the 4D/RCS architecture in the form of the AL2 architecture for teams of Unmanned Ground Vehicles, as well as in the U.S. Army Demo III Experimental Unmanned Vehicle (XUV) project.

This part of the book thus concludes and unifies the themes and component modules of autonomous systems presented in the earlier parts. We hope that it offers readers an insight into how the individual modules may be integrated and implemented successfully in real systems at different levels — our very first tentative steps into a world where autonomous systems coexist and participate seamlessly in the daily operations of their human counterparts.

15 Integration for Complex Consumer Robotic Systems: Case Studies and Analysis

Mario E. Munich, James P. Ostrowski, and Paolo Pirjanian

CONTENTS

15.1 INTRODUCTION

The field of robotics, as it continues to grow and diversify, will invariably be faced with the challenge of engineering increasingly complex systems. The complexity arises in trying to solve the multi-faceted difficulties of placing increasing levels of autonomy in robotic systems that must navigate and interact with the real world. Add to that the economic constraints placed on consumer robotics, such as service or entertainment robots, and the scientific and engineering challenges continue to multiply. Our goal in this chapter is to provide an overview of, as well as to discuss and analyze through case studies, many of the important issues encountered in the integration of complex, autonomous consumer robotic systems.

The role of system integration in developing complex systems is an area that is often underemphasized and undervalued. The reasons for this can be as varied and complex as the systems themselves. In academic research, it is very nearly impossible to get "credit" for the integration component of developing robotic systems — it is considered to be a necessary evil that must be endured to verify experimentally the underlying scientific contributions. It is thus not a goal in itself, but simply a means to an end. Research in the field of system integration for complex systems is often centered around best practices, since it can be difficult to mathematically formalize the underlying principles. Furthermore, by its very nature system integration is a cross-disciplinary endeavor. The system integrator often must strike a balancing act across engineering disciplines, having to accommodate pulls from electrical engineering, mechanical engineering, and computer science.

At the same time, the integration efforts that must be accomplished in order to develop autonomous robotic systems, especially in the consumer space, provide many unique and decidedly engaging challenges. Academic research into systems science and engineering has provided a number of steps forward in this area, with recent emphasis in the analysis of complex systems and the development of modular, open architectures for robotics. In several areas, such as the newly emerging fields of mechatronics and microelectromechanical systems (MEMS), we have seen a clear awareness of the need to bridge multiple traditional disciplines in order to address the complexity in robotic systems. And, in the private sector there are many examples demonstrating an appreciation for the importance of system integration, as seen, for example, in the strides made in the automotive and aerospace industries in building complex and highly reliable electromechanical systems and their use of modular and standardized components and interfaces.

In this chapter, we analyze the role of system integration in building complex autonomous systems in the area of consumer robotics. Our primary goal is to define the main elements necessary to develop a systematic methodology for integration. We phrase this problem loosely in terms of a multi-objective optimization problem that tries to balance conflicting goals of performance and system complexity, with the additional challenge in the consumer space of having to balance these against the eventual retail price of the robot.

After providing a brief background in Section 15.2 and discussing related work in Section 15.3, we present our approach to integration of complex autonomous systems in Sections 15.4 and 15.5. In Sections 15.6 and 15.7, we study these issues in more detail by analyzing the role of software architecture on system integration. Finally, we validate and discuss these premises using two cases studies: an entertainment robot discussed in Section 15.8 and a robotic vacuum cleaner described in Section 15.9. These case studies are driven by our experience in the consumer robotics market; some examples of commercial robots that employ components developed at Evolution Robotics are shown in Figure 15.1. The first robot is the AIBO™, an entertainment robot manufactured by Sony. The second robot is the eVac™, a robotic vacuum cleaner built by Sharper Image. The third robot is the ER2, a companion robot developed at Evolution Robotics. Our experience in developing and integrating such autonomous consumer robots helps motivate the current exposition.

15.2 BACKGROUND

What does integration mean and entail? What is integration for autonomous robots? What are the challenges of integration for complex autonomous systems? These are just a few of the many questions that one needs to consider when studying integration for complex systems. An important goal of studying integration is to understand and describe the principles that govern integration and which hopefully can be applied to support the design and integration of any autonomous system. One major premise for accomplishing this goal is that it is possible to obtain such indepth understanding of integration principles and techniques that can be generalized across most autonomous systems. The dilemma, however, is that (a) it is hard to enumerate every possible robot and (b) each robot can be significantly different in hardware and software. How do we define, then, an integration methodology or tools for robotics in general?

Part of the answer to this question is that we need to focus on the areas of integration that are, or that we believe will be, common across most robots. For example, most mobile robots will require obstacle avoidance, localization, and other navigational capabilities. Furthermore, one has to consider the variety of implementations of each individual component and hope that it is possible

(a)   (c)

(b)

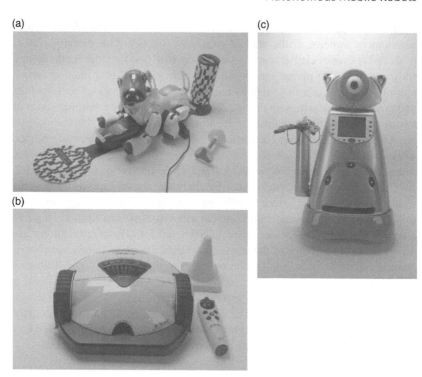

FIGURE 15.1   Autonomous robots built using Evolution components: (a) the Sony AIBO ERS-7M2, (b) the Sharper Image E-Vac, and (c) the ER2.

to have a common way of describing its interface and interaction with other system modules.

Following this line of thought one may argue that describing common building blocks is a necessary and essential part of a system integration framework. The integration framework will describe how each system component interfaces and interacts with other system components. Thus, in a sense, the framework defines and constrains how system components can be integrated according to some philosophy, methodology, rules, or guidelines. In addition, the integration framework will impose a certain definition and interpretation on the concept of a component and types of components. It will also provide guidelines, and where possible or practical, constraints, on the design of the system and how these components are to be interconnected to satisfy the system-level goals. These are issues that in the robotics community are considered as part of a *system architecture*.

In trying to establish a common integration framework and system architecture for robotics, there are many practical issues that arise due to the nature of the problem. For example, some of the characteristics that must be addressed include how to provide platform independence, scalability across various hardware and applications, expandability, and so on. These characteristics will be discussed in detail in Section 15.6.

15.3 RELATED WORK

In the 1960s, the AI (artificial intelligence) and robotics communities developed *symbolic planners*, such as STRIPS used to control Shakey the SRI robot [1]. Later it was realized that pure planning approaches, also known as "sense-plan-act" architectures, suffer when faced with the dynamics and uncertainty of the real world.

Realizing the limitations of planning systems, a new approach was taken in the mid-1980s which can be viewed as the *deliberative approach*. The deliberative architectures, such as NASA's NASREM [2], were characterized by having a hierarchical control structure where higher-level modules provided goals for lower levels. However, these systems also relied heavily on symbolic representations and hence suffered from similar problems as symbolic planners.

In the late 1980s, Brooks proposed a complete departure from using planners and symbolic representations. His Subsumption Architecture [3] relied on reactive modules that implement robot competencies by reacting to sensory data without much processing. Brooks demonstrated that subsumptive robots could react to real time events in the environment and exhibit very robust behaviors. The reactive approach later evolved into the behavior-based approach [4], where robot control is distributed among goal-oriented modules known as behaviors.

The late 1980s and the 1990s saw new approaches which attempted to combine the best of deliberative and reactive approaches into the hybrid or three-layer architecture [5]. The hybrid, deliberative-reactive, architectures generally consist of a reactive executive that deals with real-time responses to dynamic events, a high-level deliberator that reasons about long-term goals, and a mediator that coordinates the interaction between the two layers. Examples of hybrid systems include the task control architecture [6], ATLANTIS [7], and three-tier architecture. For a detailed overview of architectures, see Reference 8.

Even today there are several world-wide efforts trying to develop a common software control architecture. Since NASREM, NASA has initiated two significant efforts for developing a common, software control architecture for robotics. These efforts are mission data systems (MDS) and CLARAty [9] (coupled-layer architecture for robot autonomy). Other examples of proposed architectures include OROCOS [10], an international effort in Europe aiming at developing a common architecture and PLAYER [11], a modular architecture for distributed hardware access being developed at USC. There are also several efforts from commercial companies working on common architectures, including Saphira and ARIA from ActivMedia, OPEN-R/SDE from Sony, RoboStudio from NEC, and more.

15.4 SYSTEM INTEGRATION AS A MULTI-OBJECTIVE OPTIMIZATION PROBLEM

In order to initiate a discussion of system integration for autonomous consumer robotic systems, a useful place to start is to understand the implications of each of those terms. By studying "robotic" systems, there is a built-in level of *complexity* and integration required in dealing with a system generally possessing electrical and mechanical hardware as well as sophisticated software. Folding in "autonomy" implies a level of *functionality* and *performance* that goes beyond simple, reprogrammable pick-and-place-type robots. Finally, introducing a "consumer" aspect adds both additional requirements in terms of the expected *performance* of the system and in terms of the acceptable *price* at which the device will be purchased.

As we seek to understand the important factors that should be considered when building a systematic methodology for integration of complex robotic systems, we therefore focus on the impact of three critical measurement areas: complexity, performance, and price.

15.4.1 Complexity

Managing complexity is one of the most critical aspects of integration for robotic systems. One avenue is towards keeping very focused, simple devices. This trend is partially illustrated by the wave of robotic vacuum cleaners currently in the market, which for the most part focus on a single, well-defined task. Some of these devices have even raised debates as to whether they should be categorized as "robotic"; this debate notwithstanding, many of the current robotic vacuum cleaners represent relatively complex electromechanical systems with significant integration challenges. And it is also clear that the success of these products and the growth of the consumer robotics market will lead to demands for new products that in turn require increased complexity. The trick for the system integrator, of course, is how to manage system complexity to make it acceptable.

For robotic systems, unfortunately it can be difficult to provide precise measures of complexity. For the current context, we will generally think of two types of complexity: component level and task or system level complexity. In both cases, we tend to focus on two types of measures, namely, quantity or variety. For example, at the component level, we tend to measure complexity in the number of components, such as the number of lines of code, mechanical elements, or behavioral modules. We also measure component complexity in terms of the variety. For instance, the variety of electromechanical elements, such as different sensors and actuators, can have a significant impact on the complexity of the system integration effort. Likewise, system level complexity can be measured in the number and variety of the tasks that the robotic system is expected to accomplish. An underlying goal of robotics is to develop systems that can perform a wide variety of difficult and challenging tasks. The reality of modern robotics, and especially commercial robotics, however, is that tradeoffs must be made to reduce the system level complexity in order to make the integration effort tractable and the cost of the system reasonable.

15.4.2 Performance

The fundamental measure of performance for a system is the degree to which it satisfies the requirements of the task (or tasks) for which it was designed. This is generally laid out in the design specification for the system, and can vary greatly from robot to robot. Because of this variety, there is very little that can be done here to address the role of system integration in satisfying this measure of performance. There are, however, additional measures of performance that can be viewed more generically and used to evaluate the tradeoffs made in system integration.

One of the main themes of this book is autonomy for robotic systems. Autonomy is an important aspect of consumer robotic systems, and can largely be measured by how well the system can handle itself independent of (or with minimal) human intervention. One way to measure this is thus to analyze the frequency of, the time between, and the magnitude of human interventions.

Our focus in this chapter on the consumer space drives additional functional performance requirements motivated by the need to work in real-world environments and with high reliability. Performance in this context implies measuring additional factors such as robustness over time to user and environmental perturbations, as well as the ability to monitor, identify, and resolve problem states and malfunctions. These can be difficult to quantify but are important performance characteristics to balance against the complexity of the development and integration.

15.4.3 Price

In contrast with much of the work in academic research, where the cost of the system is a very minor part of the equation, we want to highlight here the very significant role of, and the interesting challenges that can be introduced by considering, the price of the robot. We have chosen to call this price instead of cost, for a number of reasons, but mostly to emphasize that the aspect that matters most in a consumer space is generally what the final price charged to the customer will be. Of course, the price is driven by many costs, such as development costs, part costs, marketing costs, etc.

It is important to note, too, that just as with the author who receives only a small royalty on a seemingly expensive book (many in the academic community can relate to this!), the budget for hardware and software components represents only a small portion of the overall price of the product. There are many other costs associated with bringing a consumer product to market, such as manufacturing costs, licensing fees (external or internal to cover development costs), marketing costs, shipping costs, and maintenance and repair costs. With these many additional costs, it is important to keep in mind that the parts cost will generally only represent a small portion — between *one-third* and *one-fifth* — of the actual price charged for the product.

It is very important, therefore, to realize that tradeoffs made in parts cost may sometimes appear in other areas of the final price of the consumer robot. For example, choosing longer life, but more expensive parts, can reduce the overall maintenance and support costs and reduce the rates of returns. And of course there is the cost of system integration itself, which is often directly impacted by the cost tradeoffs made in other areas!

15.5   TRADEOFFS AND CHALLENGES IN INTEGRATION OF A COMPLEX AUTONOMOUS SYSTEM

15.5.1   Component Simplicity vs. System Complexity

As a general rule, system design, and hence the integration effort that is involved, strives to strike a balance between keeping individual components simple and well-understood, and composing many such elements in a complex system to achieve high-levels of performance. In this section, we discuss some of the areas in which this balance is achieved.

15.5.1.1   Hardware vs. software

A very natural process in achieving component simplicity is that the design effort tends to compartmentalize. Most often for robotic systems, this tends to happen across hardware and software boundaries, though in large projects the segmentation may occur even within those domains. The end result is a "pitch it over the fence" mentality, where each camp hands off their work to the other, and blames the other for the problems that ensue.

The role of system integration must be to tear down the fences — when possible, before they are even built. One way to do this, which we return to repeatedly, is through well-defined interfaces and specifications. These can be used to provide "contracts" between groups working on independent subsystems. This provides a means of resolving conflicts by having prearranged agreements and definitions. It also often provides a surprising mechanism for opening dialog across groups when problems reveal flaws or gaps in the original specifications. It is important to view the occurrences of shortfalls of interface specifications not as a failure of the system or any particular individual, but as an opportunity for increased understanding between two groups and for resolution of conflicting concepts that were not fully understood when they were first agreed upon.

This relates to another mechanism for tearing down the fences between specialized subgroups: education and communication. Trying to design hardware without a basic understanding of how it will be utilized by software algorithms often leads to critical losses in development time and system performance. Similar results arise from a lack of understanding in software design of the practical limitations and failure modes of hardware components. Gaining exposure and insight into orthogonal areas of the design process enables a quicker development cycle. And just as importantly, it can help protect a complex system from catastrophic failures due to incompatible design and implementation efforts or the misuse of specialized components.

15.5.1.2   Generalization vs. specialization

Another tradeoff that must be considered in balancing this optimization problem is between generalized, modular components and dedicated, specialized components. On one hand, modular, reusable components have the decided advantage that they reduce the work (and thus the costs) associated with designing new components and can generally leverage testing and experience gained through their use in previous products. On the other hand, dedicated components can be more streamlined and thus cost-efficient, as well as have more targeted suitability and higher performance for the given application.

Generally the goal of reusable components is easy to justify when working within a product family, where variations are small and reuse is natural. More challenging is balancing the tradeoffs when working across a variety of products, where the variations in price and performance can vary significantly. For example, can one expect a sensor or algorithm designed for indoor use to translate to outdoors, or for a module that runs on a Pentium processor to scale down to an embedded PIC?

One additional balance that must be struck is the issue of what we call "paying it forward." That is, when is it worth investing in development costs up front that will lead to reduced product costs down the line? This issue is one that is felt most acutely in the commercial sector, where development funding can be a precious commodity, but also where the fate of the product very often revolves around its cost. There are many examples in which dedicated hardware (including embedded components, customized chips and boards, and ASIC's) have been developed and leveraged to provide a key piece of functionality at a much reduced cost. Some examples include dedicated color processing boards for vision, low-level sensing and avoidance motor controller boards, and speech synthesis and recognition chips. This dilemma is almost always decided by balancing the cost of developing the specialized hardware vs. the expected return on investment derived from the lower final cost. However, this requires an especially good sense (or very lucky guess!) of how well the product will do in the market.

15.5.1.3 Abstraction and aggregation

Along similar lines of balancing generalization vs. specialization, an important mechanism for increasing performance while limiting complexity growth is through the use of abstraction and aggregation. By this, we refer to collecting (aggregating) basic components that work in concert together to form higher-level (abstract) performance modules. In doing so, one can reduce the integration effort from dealing with many independent low-level modules to piecing together the actions represented by only a few aggregated ones.

The underlying notion we represent pictorially in Figure 15.2. In this figure, let us think, for example, of the performance being measured by the number and diversity of the tasks that must be performed. The abstraction level, on the other hand, can be thought of as representing the unit complexity of implementing and integrating each of these tasks, for example, the number of modules needed to complete the task. Thus, greater abstraction represents an easier integration task. As the required performance levels increase, if the unit complexity remains the same then the overall integration complexity grows in proportion to the performance. On the other hand, if, as shown in Figure 15.2, we can reduce the unit complexity by aggregating the solution into higher-level representations — effectively "preintegrating" low-level modules into higher-level aggregates — then the overall integration challenge can be made to remain relatively constant.

Now, the skeptic would say this is just smoke and mirrors — the integration complexity is just hidden in the aggregates, but still remains. While this is

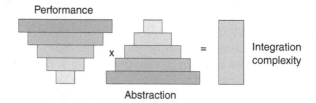

FIGURE 15.2 Using increased levels of abstraction and aggregation to provide consistent integration complexity under increasing performance requirements.

partially true, there is a powerful tool here for the system integrator to utilize through compartmentalizing the complexity by breaking up the integration task into focused composition of low-level modules into higher-level aggregates and the separate task of integrating these abstract aggregate modules. We will return to this issue in Section 15.9 as we discuss integration for an intelligent vacuum cleaner.

15.5.2 Testing, Testing, Testing. . .

Testing is a fundamental component of system integration that cannot be under-emphasized, and yet a balance must be struck between deep and thorough testing and the ability to move a product to market quickly. Good testing can be hard to achieve [12], and this is particularly true in developing autonomous robotic systems that must survive the unexpected nature of the real world.

At the core of any robotic system are its components, and as such they must be tested thoroughly to ensure they live up to their stated functionality. Component-level testing (in software, these are often referred to as "unit" tests) ensures this, while at the same time helps to enforce the proper use of interfaces and design specifications.

Upon integration of complete systems, there is a different level of testing that focuses on the functionality of the system. Depending on the particular features of the system under testing, it might be possible to run system-level tests with zero or very little user intervention. For example, the performance of a recognition system can be tested using a set of predefined images. However, in the case of many robotic systems, manual testing is usually required. Reliability and robustness of single components and entire systems need additional testing that focuses on performance achievement over long periods of time or a variety of conditions.

In the end, one must always understand that autonomous systems, especially in the consumer space, will face unique settings and unintended uses that could not have been predicted by the designer. For this reason, a final and equally critical step in testing is building in and testing internal self-monitoring of the autonomous robotic system. Such "health monitoring" is often not necessary in lab settings where human intervention is convenient and expected. However, with increasing levels of autonomy and the need to work in a consumer's environment, the expectation must be that the robot itself be able to identify and diagnose potential problems.

Just as with testing during development, self-monitoring can be performed at both the component and the system level. Component testing, for example, can involve checking a "heartbeat" signal from individual components or identifying problematic states, such as power surges or unreasonable fluctuation of sensor values. System-level testing may involve determining "stuck" states of the robot in which the robot is not able to satisfactorily perform its actions, or could involve identifying potential danger areas, such as those that might cause harm to the user or to the robot. As we discuss below, we feel that it is important that these types of monitoring capabilities be built into the overall system architecture, rather than being seen as add-ons to individual elements.

Before leaving this topic, we also would like to highlight an interesting technique found in the best practices of testing component-level software. This involves designing, and to the extent possible implementing, the tests *before* implementing the given system itself. Although at first blush this can seem counter-intuitive — how can one build a test for something that has yet to be built? — it turns out to be a very powerful tool in understanding the assumptions and the implications of the specifications governing the design. It provides an additional tool for the system integrator to insert himself into the process at an early stage, rather than becoming involved only once it is time to put the finished pieces together.

15.6 Integration through Architecture

Most often system integration is viewed as the *glue* that holds together the different, and sometimes conflicting, aspects of the product. For example, it involves tuning the software to work with a particular hardware platform; modifying individual components, such as sensors, to function in a particular environment; or even shaping the user interface and documentation to accommodate a particular target end user.

While these aspects are important, we also want to emphasize that system integration should play a prominent role as the *skeleton* of the product, providing a framework and a source of commonality of purpose in the product development. Thus, the integration impacts the development work from the beginning as a guiding force, rather than as a final step to patch things together.

To further investigate this important role of system integration, we focus now on the role of software architecture in providing a partial framework for system integration. This choice of focus is necessary to limit the scope of this chapter. However, we wish to stress that some of the most important aspects of system integration for robotic systems lie in understanding and shaping the interplay between hardware and software. Many of the aspects of the software architecture presented begin to blur the line distinguishing hardware from software, and in many ways we view this as a partial, foundational step toward the more general notion of a *system architecture*.

In describing the way the software architecture plays a role in system integration, we highlight lessons learned from the significant research literature on this topic, and draw motivation from the constraints and requirements of the service robotics sector. The software architecture chosen for developing an autonomous robotic system must provide an appropriate abstraction of the hardware components and a solid paradigm for decision making and action under uncertainty and in real-world applications. By real world we mean unstructured, usually cluttered, dynamic, and typically unknown environments. Further, the software architecture must enable the programmer or system engineer to rapidly develop and customize the control software toward a specific target, whether it is a vacuum cleaner, an entertainment toy, or a home companion. Each target application area will in general possess different and possibly conflicting requirements and constraints — thus tradeoffs must be made to realize the final implementation of the architecture.

In a long-term view, the architecture used for each target would have an implementation adapted for that particular application domain. This would embody the "niche finding" property referred to by Arkin [13], whereby architectures must find their place in a competitive world by being adapted to their particular domain area, or niche. For instance, robots in the entertainment sector would require the architecture to support powerful models of personality and emotion, which most likely is an unnecessary capability for vacuum cleaning robots. We view this as an example of the tradeoffs made between generalization and specialization — the overall philosophy and design of the architecture acts as general set of principles to guide the integration, while the implementation of the architecture becomes a specialized embodiment suited to a specific domain. This ability to tailor the architecture to satisfy a particular commercial sector can be critically important, since the computational platform for robots ranging from industrial robots to toys can vary significantly from Pentium CPUs with gigahertz clock speeds and gigabytes of memory down to embedded CPUs running at tens of megahertz with only kilobytes of RAM. This highlights the need to balance the component cost against the required performance of the autonomous system.

The following list outlines the main characteristics and requirements for the design and implementation of a software architecture targeted toward system integration for commercial robotics. It is clear that not all of the characteristics can be satisfied simultaneously — some are even contradictory — but our goal is to seek balances that optimize the tradeoffs, especially in relation to optimizing the metrics of performance, complexity, and price.

Modularity. The architecture should include support and guidelines for a modular code structure that allows only the applicable portions of the code to be installed, executed, or updated. This includes providing the support infrastructure for easily composing the modules to form a seamlessly integrated system.

Code reusability. This implies that modules can be reused in a variety of applications and should support working across different configurations, for example, as sensor type and placement is modified.

Portability and platform independence. In many cases, it is desirable to use the software developed for an application across different hardware (e.g., CPUs, robots) and software (e.g., operating systems) platforms. On the hardware side, this means that the software should be easily configurable for different robot configurations, such as sensor type and layout, motor type, and overall mobility. It is desirable that the software can be easily ported to run on different CPU architectures and different operating systems, such as Linux, Windows™, or MacOS™. In the commercial sector, an important aspect of this is to support implementations that can run on embedded microprocessors with limited memory and processing power.

Scalability. It should be easy to expand the system by adding new software modules and hardware components. Also, the overhead of supporting modular components should increase reasonably with the scale of the system.

Lightweight. The modularity and flexibility provided by the architecture should not introduce significant overhead to the system. This is clearly a conflicting goal with many of the other characteristics; however, the architecture must provide the right balance between generality and efficiency. This is especially important for commercial robotics, where computational overhead directly impacts cost.

Open and flexible. The system should enable access to the implementation of the architecture through a well-established *application program interface* (API), and should provide flexibility in allowing customizations that respect the overall architecture design.

Dynamic reconfigurability. It should allow for dynamic reconfiguration of the system, including adding, removing, upgrading, or reconnecting components to the system. This includes the infrastructure for maintaining and updating the system over time.

Ease of integration with external applications. Although this can be difficult to measure, the goal should be to provide convenient and flexible mechanisms to integrate the modules developed under this architecture with external libraries and applications, for example, customized software developed by third parties.

Networking support. As networking infrastructure becomes more and more commonplace, it is important that the architecture properly support working across Ethernet networks. Furthermore, remote process control and shared data across networks is desirable.

Fault monitoring. The system should support component-level determination of task success or failure, along with mechanisms for handling failures at different levels of the architecture. It should also provide self-monitoring through online evaluation of its state, as well as satisfaction of its task objectives.

Testing infrastructure. The architecture should support the ability to test each component, as well as provide system-level testing facilities, both at the development and product stages.

Reactive and deliberative. The system should provide tight perception-action feedback loops to react promptly to unexpected situations as well as higher-level planning for efficient use of resources over longer time frames. Plans should guide, not control, reactive components.

In addition, we feel strongly that the *tools* provided to support system integration can be just as important as the components that are provided by the architecture. These include, for example, tools for rapidly building applications, configuring the system, debugging during development, or visualizing the system and analyzing its performance. And of course, simulation environments play an ever-expanding role in supporting development and testing. However, while simulations can enable rapid debugging of new code, a hard-learned lesson (sometimes often repeated!) in system integration for complex autonomous robotic systems is how dramatically inadequate simulation environments can be in replicating the real world. We suggest, without proof, that this inadequacy is directly related to the fragile nature of complex systems. Developing in a simulated environment allows one to respond to many of the situations envisioned by the creator of the simulator. But the real world provides infinitely more variety in seeking out the subtle failure modes of the system.

Finally, we also note that beyond support tools, it is important that an architecture provides a framework and guidelines to support and instruct code development. This is an example of the *skeleton* referred to earlier that provides a basis upon which the development and integration can be centered.

15.7   A SOFTWARE ARCHITECTURE FOR CONSUMER ROBOTIC SYSTEMS

In this section, we introduce an implementation of an architecture that attempts to satisfy the above mentioned design considerations. The *Evolution Robotics Software Platform* (ERSP) provides basic components and tools for rapid development, prototyping, and integration of robotics applications. The software architecture it provides, called the Evolution Robotics Software Architecture (ERSA), is the underlying infrastructure and one of the main components of ERSP.

Figure 15.3 shows a diagram of the software structure and the relationship among the software, operating system, and applications. There are five main blocks in the diagram — three of them, *Applications*, *OS and drivers*, and *3rd Party Software*, correspond to components that are external to ERSP. The other two blocks correspond to subsets of ERSP: the core libraries (left-hand-side block) and the implementation libraries (center block). We focus on the *Architecture* (ERSA) portion of the core libraries. The other core libraries provide system-level infrastructure for developing robotic applications — essentially, the "muscle" attached to the architectural skeleton.

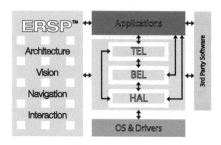

FIGURE 15.3   ERSP structure and relation to application development.

The *Vision* component includes algorithms for color segmentation and tracking, optical flow computation, and a *Visual Pattern Recognition* (ViPR™) module [14]. The *Navigation* component includes exploration, mapping, obstacle avoidance, path planning, and *visual Simultaneous Localization and Mapping* (vSLAM™) [15–17]. The *Human–Robot Interaction* module includes speech recognition, speech synthesis, and tools for building GUIs.

The Evolution Robotics Software Architecture provides a set of interfaces (APIs) for integration of the different software modules and with the robot hardware. ERSA allows for building task-achieving modules that make decisions and control the robot, for orchestrating coordination and execution of these modules, and for controlling access to system resources. ERSA is composed of three layers, the *Hardware Abstraction Layer* (HAL), the *Behavior Execution Layer* (BEL), and the *Task Execution Layer* (TEL).

The architecture corresponds to a mixed architecture in which the two first layers follow a behavior-based philosophy [3,4] and the third layer incorporates a deliberative stage for planning and sequencing [5]. The first layer, HAL, provides *interfaces* to the hardware devices and low-level operating system (OS) dependencies. This layer assures portability of ERSA and application programs across robots and computing environments. It also enables rapid configuration of the software to support new robot platforms or sensor layouts. The second layer, BEL, provides infrastructure for development of modular robotic competencies, known as *behaviors*, for achieving tasks with a tight feedback loop such as following a trajectory, tracking a person, avoiding an object, etc. Behaviors are the basic, reusable building blocks on which robotic applications are built. BEL also provides techniques for coordination of the activities of behaviors, for conflict resolution (action-selection mechanisms), and for resource scheduling. Finally, the third layer, TEL, provides infrastructure for developing *event-driven tasks* along with mechanisms for the coordination of task executions. Tasks can run in sequence or in parallel, and execution of tasks can be triggered by user-defined events.

While behaviors are highly reactive, and are appropriate for creating robust control loops, tasks are a way to express higher-level execution knowledge and to coordinate the actions of behaviors. Tasks can run asynchronously using event triggers or synchronously with other tasks. Time-critical modules such as obstacle avoidance are typically implemented in the BEL, while tasks implement skills that are not required to run at a fixed execution rate. Behaviors tend to be synchronous and highly data driven. TEL is more appropriate to deal with complex control flow which depends on context and certain conditions that can arise asynchronously.

15.7.1 ERSP in the Role of the System Integration Architecture?

This section evaluates the design of ERSP in relation to the fulfillment of the requirements postulated in Section 15.6.

Modularity. ERSP was designed so that the interdependency between components has been reduced to a minimum, with functional units grouped into individual libraries. The infrastructure for the three layers of ERSA (HAL, BEL, TEL) is implemented in three, separate libraries. In addition to ERSA, two libraries are devoted to vision components (one for basic vision primitives and the other for ViPR) and one library implements all navigation capabilities (vSLAM, exploration, path planning, etc.). These libraries constitute the "core" of ERSP as shown in Figure 15.3. A different set of libraries represent the "implementation" of basic ERSP APIs for particular cases, for example, `libevoviavoice` corresponds to the implementation of the `ISpeechRecognizer` and `ISpeechTTS` interfaces for the case of the ViaVoice™ speech recognition and text-to-speech engines. The modularity of ERSP allows its users to decide which functionality (and correspondingly which subset of the libraries) to integrate in a particular application.

Portability and platform independence. The HAL is the interface between robotic applications and the underlying hardware. HAL software controls the robot's interactions with the physical world and even with low-level OS dependencies. As such, it plays a critical role in the integration of the hardware and software components in the system. The use of a HAL has proven to be a boon to our integration efforts, particularly in porting modules across different robot platforms and OSs.

The HAL does this by *abstracting* away the details of particular hardware devices and platform-specific ways of interacting with hardware or other resources. We define a *resource* to be a physical device, connection point, or any other means through which the software interacts with the external environment. Resources include sensors and actuators, network interfaces, microphones and speech recognition systems, or a battery. We have also extended the notion of resources to include fundamental computational units that operate on sensory data. For example, both vSLAM and ViPR can be accessed directly as resources. This level of abstraction helps to maintain a constant level of integration complexity as discussed earlier in Section 15.5.

The software module that provides access to a resource, often through appropriate OS or other library calls, is termed a *resource* or *device driver*. The description of the resources and the corresponding drivers are managed through configuration files based on the *eXtensible Markup Language* (XML). The use of resource configuration files allows configuration changes to be made on the fly, which enhances one's ability to test the impact of design changes without having to recompile any code.

Code reusability. To protect higher-level modules from low-level dependencies, HAL provides a number of well-defined *interfaces* for interacting with a variety of robotic devices. These interfaces are a set of public, C++ abstract classes. Again, we emphasize the role of abstraction in reducing the complexity of the integration task. The use of abstraction found in object oriented programming has proven to be a powerful tool in providing platform independence and portability, as the abstract classes buffer the user from the details of a given implementation. They also facilitate the interaction with resources using real-world concepts and units. The particular driver implementation of these interfaces is determined at run-time, based on the set of hardware or other resources currently being used.

For example, the HAL provides an `IRangeSensor` resource interface with methods that determine the distance to an obstacle. In addition, the `IRangeSensor` has knowledge about the uncertainty associated with its measurements. An obstacle avoidance algorithm can use `IRangeSensor` to determine the position of obstacles to avoid. You can implement `IRangeSensor` for IR sensors and sonar sensors. At the application level, you work only with `IRangeSensor` and do not have to worry about any device-specific details such as converting the IR or sonar readings into proper distances. The device-specific properties of the sensor are specified in an XML-based resource schema file. For an `IRangeSensor`, these might include the calibration curve that maps raw sensor values to distances, as well as parameters describing the sensor measurement uncertainty. HAL selects the proper driver to handle these details, based on the sensor type(s) that are installed on the current robot. This way, algorithms can be developed generically to work with a variety of robotics platforms. The use of interfaces for isolation of implementation details is also employed in the case of OS-dependent constructs like multi-threading, synchronization, and file handling.

Scalability. As depicted in Figure 15.2, one of the design criteria for the architecture was the capability of providing increased levels of abstraction that encapsulate increasing performance requirements in order to keep a consistent integration complexity. ERSP includes two different solutions to this problem, one in the behavior layer and the other in the task layer. The application designer has the ability to choose which of the two approaches is best suited for his/her application.

The BEL provides the infrastructure for aggregating behaviors into a single, meta behavior. These behavior aggregates implement the original behavior interface and can be used as any other behavior in applications. Aggregates play a crucial part in solving the problem of scalability of behavior networks, making the networks manageable as the number of behaviors components grows. As discussed in Section 15.5.1.3, this aids the system integrator by keeping the integration complexity relatively constant. We return to this issue in Section 15.9.

The TEL incorporates the infrastructure for linking behaviors and tasks. TEL allows the definition of *task primitives*, which act essentially as wrappers around behavior networks, making them look and act like individual tasks. Other tasks can then use the task primitive exactly in the same way that they use any other task. At its core, then, a primitive is an XML file describing the connections between each behavior. The task then provides connections to the inputs of desired behaviors, and can read data from outputs of the behaviors within the network. The task primitive can then connect incoming events to inputs on behavior ports, and can route outgoing data from the behavior network to trigger events. Finally, the task primitive must handle any initialization that occurs when it is started, and any cleanup from being terminated.

Lightweight. Each layer of ERSP has been designed to add a minimum overhead to applications. HAL, BEL, TEL have been implemented following a similar model in which a manager (the *Resource Manager* for HAL, the *Behavior Manager* for BEL, and the *Task Manager* for TEL) is responsible for managing (instantiation, execution, shutdown, and clean-up) of the appropriate software modules (*resource or device drivers* for HAL, *behaviors* for BEL, and *tasks* for TEL). Depending on the nature of the application, the user decides which layers of ERSP would be needed. The run-time overhead of ERSP corresponds to the manager(s) needed to run the desired layers. Tests of variation in performance in terms of increase in memory consumption and decrease in speed have shown a negligible hit by using the manager(s).

Open and flexible. ERSP is built on a set of well-defined APIs. Resource drivers follow the `IResource` interface, behaviors implement the `IBehavior` interface, and behavior networks implement the `IBehaviorNetwork` interface. ERSP provides a basic implementation of these three interface classes for users' convenience. However, the system would still work properly if a different implementation of these classes is used or even if a different implementation of the `ResourceManager` or of the `BehaviorManager` is employed. In contrast with HAL and BEL, TEL has not been fully completed and lacks the flexibility of handling alternative implementations of their main components. Nevertheless, TEL allows users to create their own tasks and primitives, and run the tasks in serial, parallel, and mixed modes.

Dynamic reconfigurability. A resource must be located, instantiated, and activated before it becomes available to the system. After it is no longer needed, or the system shuts down, the resource must be released and its memory reallocated. For this purpose, we utilize a *Resource Manager* that is responsible for managing the system resources across their life cycle. The Resource Manager loads the resources based on the information provided by the resource configuration file. Resources specified in the resource configuration file are available on a need-to-use basis. If a particular portion of an application requires a set of resources at a certain point in time, the resources will be activated by the Resource Manager and provided to the application. Once the resources are not needed, the Resource Manager will deactivate them while keeping them in the list of available resources for future use. One drawback of the current implementation, however, is that dynamic and system-wide changes in the resource configuration file can only be effected by a complete shutdown and deallocation of resources, followed by a reload of the resource configuration file. Similarly, updates to the system require that the system be restarted — there is no infrastructure within ERSP for internally updating libraries.

BEL and TEL have extensive support for dynamic reconfiguration of behavior networks and tasks. The Behavior Manager is capable of dynamically loading and shutting down behavior networks. Multiple networks can be run in parallel and at different rates. Each behavior library provides a "factory" function for each behavior type, which the Behavior Manager can call to create a new instance of that type. The Behavior Manager also reads parameters from the XML-based behavior network, which can be used at run-time to override default parameters for each behavior. Behaviors can also be disabled from execution by using gating behaviors that disable data transfer to the input ports of the behavior to be disabled. These gating behaviors are used in the case in which portions of a behavior network should be enabled/disabled depending on the mode of operation of the application.

Ease of integration with external applications. ERSP does not have any particular mechanism for integration of third party software and applications but rather allows for incorporating them in any of the three layers of the architecture. One particular example is the integration of third party speech recognition (ASR) and text-to-speech (TTS) engines to applications built with ERSP. HAL was the layer used for the integration of these engines since it was beneficial in many counts: the input from a single ASR (and the output to a single TTS) was accessible to (from) a multiplicity of modules, a well-defined interface isolated higher layers from the type of engines that were used (we had implementations for the IBM ViaVoice ASR and TTS engines, the Microsoft ASR and TTS engines, the ATT Naturally Speaking TTS engine, and the Sphinx ASR engine).

Networking support. ERSP provides support for networking using sockets. Several applications have been written in a client-server mode that uses the networking support. For example, the application presented in Section 15.9 has been split into two portions: one runs on the robot (that has the size of a super-sized robotic vacuum cleaner) and the other runs on a client computer. The client application is a GUI-based application that controls the actions of the robot and collects information on the state of the robot. The client also collects debug information that helps developers correct problems in the behavior of the robot.

Networking support is provided within behavior networks with the *Malleable Behavior*. This type of behavior opens a socket connection on a given port and handles data communication encapsulated in an XML text format. One current limitation of ERSP is that it only supports the transfer of simple data structures, for example, strings and arrays. Transfer of more complex data structures, such as images, is handled by customized behaviors.

Networking support in ERSP is still lacking some important functionality such as support for multi-robot coordination, distributed computation across different robots and/or processors, and transparent data sharing across networks.

Fault monitoring. The Resource Manager is also a natural vehicle for monitoring the health of the low-level hardware components, and can provide a central source of information about these components. The Behavior Manager collects the information on the state of the behaviors and can alert the system upon failure. From the coding standpoint, ERSP has been implemented in C++ without C++ exceptions, but rather providing result codes that define the state of execution of the code. Clearly defined coding rules that enforce checks for unexpected results provide a highly reliable and robust implementation of the components of ERSP. However, ERSP lacks internal support for a heartbeat monitoring system that can assess the state of the robot at any time and take appropriate measures upon failure.

Testing infrastructure. This is one of the characteristics that has the least support in ERSP. Component-level testing infrastructure for unit-tests is provided in ERSP. Functional testing of components can be achieved with different applications suites like the one described in Section 15.8. However, general, system-level testing facilities have not yet been implemented.

Reactive and deliberative. The layered design of ERSP tries to address the need for an architecture that is both reactive and deliberative. As mentioned earlier, the two first layers follow a behavior-based philosophy [3,4] and the third layer incorporates a deliberative stage for planning and sequencing [5].

While behaviors are highly reactive, and are appropriate for creating robust control loops, tasks express higher-level execution knowledge and coordinate the actions of multiple behaviors or behavior networks. For example, an action that is best written as a behavior would be a robot using vision to approach a recognized object. An action that is more appropriate for a task, on the other hand, would be a robot navigating to the kitchen, finding a bottle of beer, and picking it up.

In addition to enabling event-driven, task-oriented processes, TEL has the added advantages of providing *familiarity* and *ease of scripting*. Defining tasks is similar to writing standard C++ functions, not writing finite state machines. Also, creating an interface with most scripting languages, including Python, is simple, allowing one to make use of the power of high-level scripting languages, while controlling tasks in a natural way. By sequencing and combining tasks using functional composition, you can create a flexible plan for a robot to execute, while writing code in a traditional procedural style. TEL support for parallel execution, task communication, and synchronization allows for the development of plans that can be successfully executed in a dynamic environment. TEL also provides a high level, task-oriented interface to the BEL.

15.7.2 Development Tools

The development of robotic applications is complicated by the number of elements that compose a robot. ERSP provides a number of tools, modules, and a framework aimed toward *easing application development* efforts.

The XML files are used for configuration of robots, for description of the internals of resources and behaviors, and for detailing behavior networks. The use of a single, XML-based resource configuration file enhances *usability* by concentrating all system- and platform-dependent changes to a single file. Modifications of the number, type, or location of sensors or actuators thus involve only a single modification of the resource configuration file, not a recompilation of the application. Settings for resource and behavior parameters are also described with XML files, allowing for quick modification of parameters while tuning applications. From the user standpoint, quick and easy access to parameters is key for *fast debugging* of algorithms.

An important *development tool* provided in ERSP is the Behavior Composer. This graphical tool allows the user to create behavior networks with a drag-and-drop procedure. A behavior is just a box that has inputs, outputs, and parameters. The behavior composer allows one to place behaviors in a network, to select the flow of data by drawing connections between the input/output ports of the behaviors, and to modify parameters with a property editor. The network is saved in XML format and ready to be executed. In addition, the Behavior Composer enforces *type safety* for data flowing between behaviors, by only allowing connections between ports of compatible semantic type.

Figure 15.4 shows a behavior network for an application in which the robot drives around avoiding obstacles while responding to user commands recognized with an ASR engine. The application switches states depending on the input command and provides a response with a TTS engine.

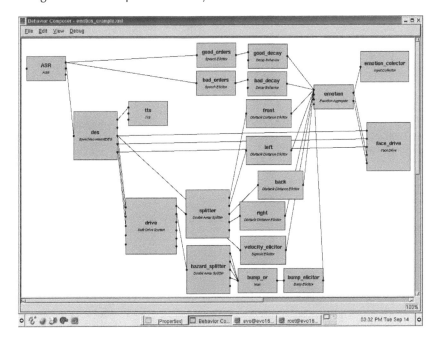

FIGURE 15.4   Behavior composer. ERSP includes a GUI tool for ease of development of applications (behavior networks) at the BEL layer.

The system that drives the robot and avoids obstacles is encapsulated in the `SafeDriveSystem` behavior. This behavior is a Behavior Aggregate composed of approximately 15 sub-behaviors.

15.8   Case Study 1: Sony AIBO

The first example that tests the described design methodology corresponds to the integration of the ViPR module of ERSP into the latest software release for the Sony AIBO™ (ERS-7™). ViPR supports AIBO's robust self-charging capabilities, which enhances the autonomy of the robot, by allowing the robot to reliably find its charging station (see Figure 15.1). ViPR also supports reliable command-and-control human-robotic interaction by using a set of cards which the user shows to the robot in order to initiate commands (see Figure 15.5). The AIBO represents perhaps the most complex consumer robot available on the market, and embodies many of the tradeoffs between performance, complexity, and price that have been described earlier.

This project presented several challenges, namely, the extraction of a single module of ERSP, cross-compilation of the software for a completely new OS

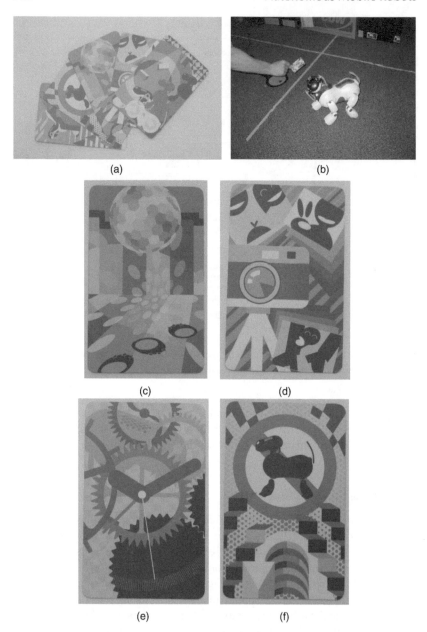

(a)   (b)

(c)   (d)

(e)   (f)

FIGURE 15.5   Examples of cards used to command the Sony AIBO. (a) Deck of cards and (b) AIBO looking at a card. Detailed views (c)–(f) of four of these cards: the first one instructs the robot to "dance," the second one to "take a picture," the third one to "set the alarm clock," and the last one to "turn around".

and microprocessor, and the achievement of a set of computational and memory requirements. In fact, the module was required to be compiled with two different compilers, gcc and Green Hills, posing additional compatibility challenges due to the differences between the implementations of the "standard" libraries.

The modular design of ERSP allowed for a simple integration of ViPR into the Sony AIBO: just extract the ViPR libraries from ERSP and custom compile them for AIBO's proprietary OS, called Aperios™, and for a 64-bit RISC processor (ERSP was used only on ×86 platforms before this project).

The ERSP achieves OS and platform independence by using a set of abstraction functions that isolates the ERSP libraries from the actual implementation and function names of the underlying standard C-library. For example, the ViPR libraries use `ersp_stat()` for getting the status of a file and this function properly points to `stat()` in Linux, to `_stat()` in Windows, and to `OFS::stat()` in the Green Hills version of Aperios. Another example is the case of `ersp_time()` that is used to obtain the time in seconds. This function points to `time()` in both Linux and Windows, but it had no implementation in either the gcc or the Green Hills versions of Aperios™. This is one of the few cases in which we had to add a piece of code to ERSP in order to implement `ersp_time()` using the `GetSystemTime()` function of Aperios. This platform independence satisfied an important criterion of the software architecture discussed earlier, and accelerated development time by allowing us to spend most of the development time working on the platform of our choice.

Applications developed for AIBO are based on a proprietary software architecture developed by Sony called Open-R. The application is basically composed of a set of modules that are loaded at run time according to a list of modules and interconnections among them. Similar to the Behavior Manager in ERSP, modules are executed by a software manager and data transference between modules is achieved with an event system. This modularity plays a significant role in maintaining a tractable level of complexity, even as the system performance level is increased. However, it does not fully support in an easy way the notion of complexity reduction through abstraction illustrated in Figure 15.2. ViPR was integrated and linked into a module that received input images from the camera and provided output recognition events. The public ViPR API was used to develop this module and no modifications of ERSP were required for integration. This helped manage the cost implication of using ViPR, by reducing development costs, since existing, standardized components could be used, and by limiting the cost to only a module of ERSP, rather than the entire platform. In addition, the well-defined API made possible the integration of ViPR between two teams separated both by distance (United States and Japan) and sometimes language.

Testing and quality assurance (QA) presented some of the most challenging parts of the project. The project had a set of requirements stated in terms of recognition performance, computational performance, memory and CPU usage, and overall quality of the code. We had to verify that we had achieved those requirements without knowing the environment and the conditions in which AIBO would be deployed. This "generalization" requirement is quite difficult to test since there are infinitely many possibilities. Therefore, Sony provided us with a "nominal" test set of images that would be challenging enough to cover a range of reasonable usability conditions. This test set was used to test computational performance of ViPR in addition to recognition performance.

The overall quality of the code was tested using a set of internal tests for detection of memory leaks and successful handling of conditions that might create exceptions. We developed an application suite that was run for many days (and even weeks) at a time to verify stability of the code. One of the applications had AIBO performing recognition of the pole located in the back of the charging station. This application was executed for long periods of time in a location whose lighting was changing during the day. The recognition results were captured in order to verify the stability of recognition and of the calculation of the pose of the robot. A separate set of tests was used to evaluate the load of the CPU when ViPR was run. This wide range of tests gave a means by which to certify the performance of the code before it was ever delivered. This had the effect of reducing the integration time, and hence the cost of the project, while ensuring the software met the desired performance levels.

15.9 CASE STUDY 2: AUTONOMOUS CAPABILITIES FOR VACUUM CLEANING

In this section, we examine the evolution of a robotic system as it progressed toward a consumer product. Our goal is to highlight some of the critical challenges that arose during this process. For example, this development path followed an interesting cycle when viewed from the metrics of performance, complexity, and price. The initial development cycle targeted increasing levels of performance, especially in terms of autonomy and robustness, while attempting to maintain a manageable level of complexity. The resulting solution was infeasible in the consumer space, due to its price. What followed was a gradual shift toward squeezing down the price, while attempting to maintain as much performance as possible.

Early on in the development of this robot, a main focus was to enable improved autonomy through improved awareness of the robot's location. *Simultaneous Localization and Mapping* (SLAM) is one of the most fundamental, yet most challenging, problems in mobile robotics. To achieve full autonomy a robot must possess the ability to explore its environment without user intervention, build a reliable map, and localize itself in the map. In particular, if *global positioning sensor* (GPS) data and external beacons are unavailable, the robot must somehow, by itself, determine what are appropriate reference points on which to build a map.

Unfortunately, the only existing solutions to the problem at the time required either that the robot have access to GPS (not available in indoor consumer environments); to markers or beacons manually placed throughout the environment; or to expensive hardware, such as laser range finders, that enabled solving the SLAM problem. Motivated by the need for a low-cost, flexible alternative, Evolution Robotics developed the first *visual Simultaneous Localization and Mapping* (vSLAM) algorithm [15,16]. The algorithm is vision- and odometry-based, and enables low-cost navigation in cluttered and populated environments. No initial map is required, and the algorithm satisfactorily handles dynamic changes in the environment, for example, lighting changes, moving objects, or people. At a system level, vSLAM uses inputs to the system of odometry data and images, and outputs the robot pose and an abstract vSLAM map.

In contrast to previously proposed algorithms, the vSLAM system generates, detects, and estimates the relative pose to a landmark utilizing a single camera. By using a localization scheme with a particle filter [18] and an adaptive mixed proposal distribution [19,20], vSLAM enables navigation with good accuracy in a large variety of real-world environments. The adaptive mixed proposal distribution also enables vSLAM to recover from "kidnapping" scenarios; that is, situations where the robot is lifted up and moved without being notified. Indeed, an important aspect of the system is the ability for other modules to monitor and detect when the system has been kidnapped. This satisfies the health monitoring characteristic of the system architecture. As an example of its usage, mapping algorithms must use this information in order to avoid incorrectly updating the map based on invalid position information, and planning algorithms are invoked to recover from situations in which the robot becomes lost.

SLAM in general, and vSLAM in particular, provides a fundamental component of localization and mapping that is important for any complex autonomous robotic system. After having implemented vSLAM, we decided to focus on specific applications that could utilize such autonomous capabilities. At the same time, the appearance of robotic vacuum cleaners such as the Roomba™ from iRobot, the Trilobite™ from Electrolux, the RC3000™ from Karcher (see Figure 15.6), and others, gave birth to the robotic vacuum cleaner market.

None of these robotic vacuum cleaners had localization capabilities, but instead achieved spatial coverage using a series of heuristics and algorithms that moved the robot in pseudo-random patterns. Thus, it seemed that there was a clear fit between this newly emerging market of robotic vacuum cleaners and the localization solution provided by vSLAM. Localization would enable systematic floor coverage for efficiency (cleaning speed and energy utilization)

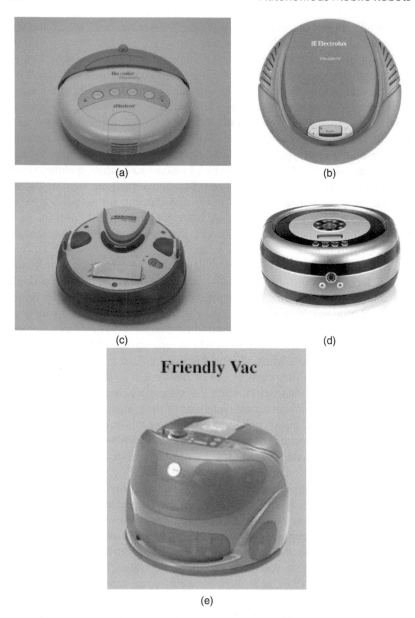

(a)   (b)

(c)   (d)

(e)

FIGURE 15.6   Robotic vacuum cleaners. (a) Roomba from iRobot, (b) Trilobite from Electrolux, (c) RC3000 from Karcher, (d) VC-RP30W from Samsung, (e) the Friendly Vac from Friendly Robotics.

FIGURE 15.7   R3R robot from evolution robotics.

and completeness (maximum area covered), as well as autonomous homing and docking for self-charging. One crucial element was missing: a suitable robotic platform for demonstrating the capabilities of vSLAM to the target market. For this reason, we developed our own robot for this purpose, the Evolution Robotics R3R shown in Figure 15.7. The goal in developing the R3R was to demonstrate, in a robot of form factor similar to actual vacuum cleaner robots, the many advanced capabilities that would be useful for vacuum cleaning, including vSLAM, autonomous exploration, automatic map generation, floor coverage, self-docking, obstacle avoidance, and path planning.

All the robotic vacuum cleaners shown in Figure 15.6 have a circular design since this shape allows for minimal risk of hitting and scratching furniture and getting entangled in dangling pieces of cloth. Following the same principles, the R3R robot was designed with a circular shape that was about 30 cm in diameter (similar to the diameter of a Roomba). The height and weight of the robot, at about 30 cm and 10 kg, respectively, were much greater than desired. This was mainly driven by the design choice of the computational unit; in order to avoid the challenges and risks of integrating into an embedded system, we chose to use an off-the-shelf minibook PC with a 2 GHz Pentium 4 CPU. This design choice led to the need to use a set of four high capacity, flat Li-ion batteries, higher-power actuators, and a much larger form factor than desired. This also led to additional hardware challenges when trying to design the suspension to support a heavy, wheel-centered, differential drive robot with a high center of gravity.

In retrospect, we feel the choice was still a good one, since we were able to directly leverage many advantages of the Pentium implementation of the software architecture, but there were clearly additional challenges introduced by this choice. In effect, our optimization choice was to sacrifice component cost to reduce complexity and development time and cost. Performance was enhanced in many ways, for example, by having significant processing power to support many modules executing parallel tasks, but was also adversely impacted by the difficulties in form factor mentioned earlier. We return to the issue of cost reduction in Section 15.9.2.

The R3R robot shown in Figure 15.7 corresponds to the final prototype design; however, there were many intermediate versions of the robot during the development of this prototype. Software was being developed along with the hardware, so there was a need to allow for quick reconfiguration of the hardware, minimally affecting the software development. The HAL layer of ERSP provided such capability: changes in the hardware configuration just needed to be reflected in the XML file that described the robot, the resource-config.xml, allowing for applications being run without having to recompile any code. In the case in which new sensors were added to the R3R or when a particular sensor type was changed, there was a need for the development of a new driver for the sensor. This resource driver needed to follow a predefined software interface that was used by applications at the HAL and TEL layers. The net result was that applications could be run just by adding the new sensor library and modifying the resource-config.xml, without the need for recompilation of the main application code. The hardware isolation provided by the HAL layer had a very important impact on development productivity since the integration of hardware modifications caused almost no disturbance to the software development process.

Figure 15.8 presents the main software modules of the system. The figure also shows the various pieces of data that are interchanged between the modules. The Exploration, Sweeping, and Docking modules also represent different mutually exclusive states of operation of the robot. The Path Planning, vSLAM, and Obstacle Mapping are functional modules that serve each of the different states of the robot. The functionality of the modules is the following (numbers in parentheses represent the number of modules comprising each higher-level module):

vSLAM (2). This module performs SLAM based on visual and odometry inputs. It creates a map of visual landmarks, computes visual measurements, and estimates the pose of the robot based on the odometry readings and the visual measurements.

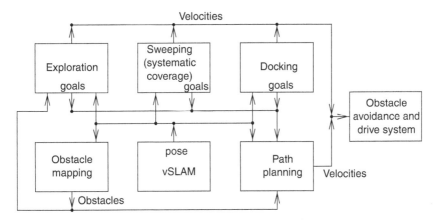

FIGURE 15.8 Software modules comprising the demo system.

Path planning (5). This determines the set of motions needed to move the robot to a given location, taking into consideration the current position on the map and the set of obstacles that have been found. It outputs velocities that command the Drive System.

Obstacle mapping (3). This keeps a record of the locations in which obstacles have been detected. It outputs an occupancy grid that has been corrected by taking into consideration the current pose of the robot.

Exploration (4). This module decides the places that the robot has to visit based upon the robot's pose, the location of the obstacles, and history of locations that the robot has visited. It outputs either goal locations for the Path Planner or velocities for the Drive System.

Sweeping (systematic coverage) (4). This prepares a set of trajectories that maximizes the coverage of the area to clean. It determines target paths to be followed and chooses where to sweep based on a map of the area that has been covered. In addition, the coverage algorithms have specially adapted routines to guarantee that the robot cleaned around the boundaries of obstacles in order to maximize coverage.

Docking (5). This module manages the return of the robot to the charging station from its current position. The docking process has two stages: long-range approach and short-range docking. In the first stage, the docking module provides a goal location to the Path Planner based on the current pose of the robot. In the second stage, a finer grained, short-range approach to the charging station and self-docking is controlled directly by this module. ViPR is used in this final approach in order to compute the pose of the robot with respect to a planar landmark that has been attached to the charging station, and a set of velocities is provided to the Drive System.

Obstacle avoidance and drive system (15+). This module senses and reacts to obstacles while driving the robot at the given input velocities. It includes sensors for obstacle detection, actuators that drive the motors, and a set of action selection mechanisms that allow for balancing the two (possibly) opposing goals of driving at the desired velocities while avoiding obstacles.

As can be seen from the above, each of the high-level aggregates is generally comprised of three or four lower-level modules (although some of the complexity is hidden in powerful library calls, such as is the case with vSLAM and Obstacle Mapping). Using this abstraction has been a valuable component of reducing some of the integration complexity in putting together such a complex system (as shown in Figure 15.2).

It is also interesting to look in more detail at the Obstacle Avoidance and Drive System module, which has the largest number of modules and uses several levels of abstraction. Figure 15.9 shows the internal components of this module. The application was developed using the BEL layer of ERSP; therefore, all the blocks shown on the figure represent behaviors. The center block is called `SafeDriveSystem` and corresponds to an aggregate behavior that delivers all the functionality of the avoidance and drive module. This meta-behavior is comprised of two different aggregates, the `SensorAggregate` and the `AvoidanceAggregate`. The first of these behaviors is responsible for instantiating one behavior per sensor included in the resource-config.xml file and collecting sensor information into a set of data that is used by the `AvoidanceAggregate`. This second aggregate is responsible for taking the desired velocities and the information collected from the sensors, evaluating the best possible way of achieving the goals while avoiding obstacles, and then driving the motors. `SensorAggregate` is itself composed of three aggregates that handle each type of sensor present in the R3R: a `SensorRing` with IR sensors (or for that matter, laser or sonar sensors) for measuring distance to obstacles, `BumpSensorRing`, and `StairSensorRing` with IR sensors for measuring hazards such as stairs. We have also shown an expanded view of the modules that further comprise the `SensorRing`. Similarly, the `AvoidanceAggregate` contains many modules to make decisions based on the sensory information, and includes an internal aggregate, `OccupancyGridAggregate`, that creates a local occupancy grid to provide memory to the avoidance system.

Figure 15.9 is also a good example of the pyramidal structure that was described in Figure 15.2, in using aggregation and abstraction to improve performance without significant increases in integration complexity. All the modules presented in Figure 15.8 follow a similar implementation that use the tools provided in ERSP to keep the complexity of the integration of the system at reasonable levels when adding more functional modules to the overall application.

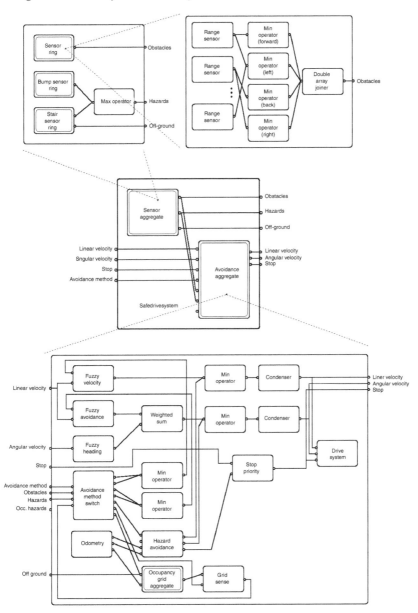

FIGURE 15.9 Internals of safedrivesystem.

Having shown the complexity of the system, testing becomes a major challenge due to the number of components and the internal complexity of each of the algorithms. In fact, the challenge is not only testing, but also debugging since tracing a problem to find its root cause may become an overwhelming task. Fortunately, ERSP's modular design makes it simple for users and developers to focus on testing individual components well before integration and system testing. The behavior networks that compose the aggregates could be tested individually and independently before integration. Algorithm prototyping and replacement can be done by just simply replacing behaviors in the corresponding behavior network.

15.9.1 Lessons Learned

Before turning to the work performed in reducing costs via an embedded design, we briefly capture a few lessons learned and challenges that still exist in using ERSA and developing such a system.

One of the most significant design challenges is how to design general, modular, reusable behavioral components that have a performance comparable to those that are tailored for a specific robot hardware or user environment. For example, there are different design choices, in both hardware and software, to be made when using IR sensors vs. a laser range finder vs. sonar sensors. Similarly, an omnidirectional robot will utilize different strategies in obstacle avoidance, path planning, and docking than a differential drive robot. Even in the case of two robots with the same types of sensors and drive system, the placement of the sensors and the form factor of the robot can motivate very different solutions that are hard to generalize into a single control behavior. In approaching this problem, we generally attempted to develop behaviors that could work as broadly as possible, but remained willing to develop specialized behaviors when the performance could not be achieved otherwise. Doing the former generally leads to an increased complexity within a given behavior, but lowers future development costs and simplifies maintenance; doing the latter has the potential to improve performance, but makes maintaining the code more difficult, as the component tailored to each configuration must be maintained separately.

Another area that posed challenges to us was the lack of shared, global data space available within ERSA. In order to have different behaviors communicate data, they must be passed over ports; for tasks, data must be passed as events. Having a central, "blackboard" repository for shared data would enable developers of different modules to more readily communicate. However, there is the trade-off of increased complexity as one must properly handle synchronization of data (reading and writing), ensuring that it is not stale, and enforcing data types.

One final area that posed a challenge was integrating system aspects that must operate at different execution rates. For example, low-level obstacle avoidance must execute within milliseconds, while vSLAM and planning need only operate at rates on the order of seconds. The current implementation of ERSA allows for multiple behavior networks to run at different rates; however, there is no easy mechanism for communicating between these networks. This is an area for future development. Additionally, there are no real-time guarantees provided by Linux or ERSA, so CPU intensive tasks, such as vSLAM, can slow down time critical tasks such as stair avoidance to unacceptable levels. For this reason, we chose to use hardware acceleration (e.g., embedded microcontrollers) whenever possible. For example, we off-loaded low-level hazard and obstacle avoidance to the hardware controller board, which meant we could easily maintain sensing rates without any additional costs (except for development costs). In the next section, we describe in more detail another example of reducing hardware costs by implementing selected, critical algorithms in embedded hardware.

15.9.2 Embedded Implementation of vSLAM

The demonstration of vSLAM and the algorithms for path planning, exploration, coverage, and self-docking based on the R3R were an important first step into prototyping the features of the robotic vacuum cleaner of the future. However, this demonstration was not sufficient to make a compelling business case for integrating these technologies into actual robotic vacuum cleaner products. The main hesitation from potential manufacturers stemmed from our use of a 2 GHz Pentium 4 CPU as the computation unit. For this reason, we developed an embedded version of the R3R to demonstrate a robot that would resemble an actual product computational-wise.

We thus embarked on a project to develop a version of the R3R based around a low-cost embedded CPU. The goal for this project was to maintain roughly the same level of performance, while reducing the component price by over an order of magnitude (roughly $1500 down to $100). This reduction in cost was obtained at the expense of an increase in the complexity of the design and implementation of the system; however, we utilized many techniques described earlier to bring the complexity roughly in line with the original robot. The embedded R3R required a hardware-software codesign that involved not only the selection of the appropriate embedded CPU(s), but also the partition of the functionality among the computational units and the adaptation of the software to handle such partition. Many of the requirements for the software architecture postulated in Section 15.6 were crucial in maintaining the complexity of the embedded R3R at the same level as the original R3R from the software point of view. Modularity of the design allowed for swappable components that moved from a pure software implementation to a hardware or a mixed implementation. Dynamic reconfiguration enabled using the software or the hardware version of a component depending on the operating conditions and on the CPU(s) load.

The development of the embedded R3R was an iterative process that involved choosing the embedded processor, evaluating the performance of the system with this processor, profiling the code, and selecting which functionality to off-load to hardware. One main criteria was used in this selection: achievement of the desired performance at the required cost. Following this criteria we selected two modules to be implemented in DSP coprocessors. The first module was the vSLAM frontend, where ViPR recognition is performed, that was the major consumer of CPU cycles. The second module was the low-level sensing/actuation module (drive system and obstacle avoidance) that required a higher-performance rate than what was achievable with the selected embedded processor.

Figure 15.10 presents the internals of the vSLAM frontend. The first two modules were part of ViPR, that performed feature extraction from images and that computed an affine match between the landmark and the images. The third module calculated a full 3D match to the landmark and estimated the pose of the robot in order to create a visual measurement. The affine matching module of ViPR was composed of two steps, a lookup of features in a k–d tree and an evaluation of match candidates and an affine matching computation. The feature extraction module and the k–d tree lookup accounted for about 70% of the computation of the frontend. Most of the calculations in these two modules used integer operations; therefore, we implemented them in a low-cost DSP (as shown in Figure 15.10).

An interesting side benefit of moving to an embedded processor was the ability to reverse the vicious cycle of power consumption and battery weight. Removing the need to feed a power-hungry Pentium processor allowed us to reduce the battery load. This, in conjunction with a lighter processor board, made the overall system lighter, which further extended the battery life.

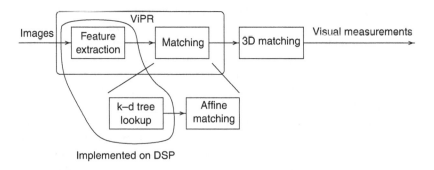

FIGURE 15.10   Internals of the vSLAM frontend.

In the end, the embedded R3R could easily be run five times as long as the original!

15.10 Conclusions

In this chapter, we have discussed and analyzed some of the key challenges and constraints that arise when developing a complex autonomous robotic system for the consumer market. We propose analyzing the integration challenge along three primary system-level measures: performance, complexity, and price. These measures drive design and integration decisions and lead one toward a series of secondary constraints and criteria that help guide the integration process, including modularity, scalability, interface design and abstraction, portability, and testing capabilities. We discuss these ideas in the context of a software architecture designed to support autonomous consumer robotic systems, and analyze the utility of these ideas using two case studies of robots designed and built for the consumer market.

References

1. R. E. Fikes and N. J. Nilsson. Strips: a new approach to the application of theorem proving to problem solving. *Artificial Intelligence*, 5: 189–208, 1971.
2. J. Albus et al. NASA/NBS standard reference model for telerobot control system architecture (NASREM). Technical Note 1235, NIST, Gaithersburg, MD, July 1987.
3. R. A. Brooks. A robust layered control system for a mobile robot. *IEEE Journal of Robotics and Automation*, RA-2:14-23, 1986.
4. R. C. Arkin. *Behavior-Based Robotics*. MIT Press, Cambridge, MA, 1998.
5. E. Gat. Three-layer architectures. In R. D. Kortenkamp, R. P. Bonasso, and R. Murphy (eds), *Artificial Intelligence and Mobile Robots*. AAAI/MIT Press, Cambridge, MA, 1998.
6. Reid G. Simmons. Structured control for autonomous robots. *IEEE Transactions on Robotics and Automation*, 10: 34–43, 1994.
7. E. Gat. Integrating planning and reacting in a heterogeneous asynchronous architecture for controlling real-world mobile robots. *AAAI-92 Robot Navigation*, pp. 809–815, 1992.
8. P. Pirjanian. Behavior coordination mechanisms — state-of-the-art. Technical Report IRIS-99-375, Institute for Robotics and Intelligent Systems, University of Southern California, October 1999.
9. I. A. Nesnas, A. Wright, M. Bajracharya, R. Simmons, W. S. Kim, and T. Estlin. CLARAty: An architecture for reusable robotic software. In *SPIE Aerosense Conference*, Orlando, Florida, 2003.
10. W. Li, H. Christensen, A. Oreback, and D. Chen. An architecture for indoor navigation. In *Proceedings of International Conference on Robotics and Automation*, pp. 1783–1793, New Orleans, April 2004.

11. R. Vaughan, B. Gerkey, and A. Howard. On device abstractions for portable, resuable robot code. In *Proceedings of the IEEE/RSJ International Conference on Intelligent Robots and Systems*, Las Vegas, NV, October 2003.

12. Keith Stobie. Too darned big to test. *ACM Queue*, 3, 2005. http://www.acmqueue.com/modules.

13. R. C. Arkin. Just what is a robot architecture anyway? Turing equivalency versus organizing principles. In *AAAI Spring Symposium: Lessons Learned from Implemented Software Architectures for Physical Agents*, 1995.

14. N. Karlsson, M. E. Munich, L. Goncalves, J. Ostrowski, E. Di Bernardo, and P. Pirjanian. Core technologies for service robotics. In *Proceedings of the International Conference on Intelligent Robots and Systems (IROS)*, Sendai, Japan, October 2004.

15. L. Goncalves, E. Di Bernardo, D. Benson, M. Svedman, J. Ostrowski, N. Karlsson, and P. Pirjanian. A visual frontend for simultaneous localization and mapping. In *Proceedings of International Conference on Robotics and Automation (ICRA)*, Barcelona, Spain, 2005.

16. N. Karlsson, E. Di Bernardo, J. Ostrowski, L. Goncalves, P. Pirjanian, and M. E. Munich. The vSLAM algorithm for robust localization and mapping. In *Proceedings of International Conference on Robotics and Automation (ICRA)*, Barcelona, Spain, 2005.

17. N. Karlsson, L. Goncalves, M. E. Munich, and P. Pirjanian. The vSLAM algorithm for navigation in natural environments. *Korean Robotics Society Review*, 2: 51–67, 2005.

18. A. Doucet, N. de Freitas, and N. Gordon (eds). *Sequential Monte Carlo Methods in Practice*. Springer-Verlag, New York, 2001.

19. D. Fox, S. Thrun, W. Burgard, and F. Dellaert. Particle filters for mobile robot localization. In A. Doucet, N. de Freitas, and N. Gordon (eds). *Sequential Monte Carlo Methods in Practice*, chapter 19. Springer-Verlag, New York, 2001.

20. D. V. Hinkley. *Bootstrap Methods and their Application*. Cambridge Series in Statistical and Probabilistic Methematics, 1997.

Biographies

Mario E. Munich is vice president of software and senior research scientist at Evolution Robotics, with expertise in computer vision, speech recognition, machine learning, and mobile robotics. He holds a degree of electronic engineer (with honors) from the National University of Rosario, Argentina and an M.S. and Ph.D. in electrical engineering from the California Institute of Technology. His research interests include computer vision, human–computer interfaces, pattern recognition, machine learning, affective computing, robotics, and embedded systems.

James P. Ostrowski is senior research scientist at Evolution Robotics, with expertise in mobile robotics, nonlinear dynamics, and vision-based control.

He holds a B.Sc. in electrical engineering from Brown university and an M.S. and Ph.D. in Mechanical engineering from the California Institute of Technology. From 1996 to 2002, he was a faculty in the Department of Mechanical Engineering at the University of Pennsylvania, where he was a member of the General Robotics, Sensing, Automation, and Perception (GRASP) Laboratory and is currently an adjunct associate professor. He is a recipient of the National Science Foundation CAREER Award for young faculty and the Ford Motor Foundation Award for Faculty Advising, and was formerly an associate editor for the *IEEE Transactions on Robotics and Automation*.

Paolo Pirjanian is the chief scientist and general manager for Robotics and Vision division at Evolution Robotics, Inc and heads the R&D efforts in developing core technologies for consumer and commercial robotics products. His main work is focused around vision-based and optical navigation, control architectures, and human–robot interaction. Paolo has served as a part-time lecturer at the Computer Science department of University of Southern California. Prior to joining Evolution Robotics, Paolo initiated several research thrusts on multi-robot system for space exploration at the Jet Propulsion Laboratory, NASA. He received his Ph.D. degree in robotics from Aalborg University, Denmark where he also served as a research professor. He received an Outstanding Technical Leadership Award from JPL, NASA for technical leadership in developing advanced robot architectures for planetary outposts in 2001. He received the IEEE RAS Industry Early Career Award in 2004. Paolo is the U.S. Chair of the IEEE Industrial Activities Committee and serves on the editorial boards of *Journal of Autonomous Robots and Intelligent Service Robots*.

16 Automotive Systems/Robotic Vehicles

Michel R. Parent and Stéphane R. Petti

CONTENTS

16.1 Introduction

16.1.1 A Key Product of the 20th Century

The automobile has been one of the most important products of the 20th century. It has generated an enormous industry and has given individuals a freedom of movement, which has completely changed our ways of living. Indeed, the automobile has been the key factor to a large change in the way our urban societies are structured. If we look at the change in the population of a large city such as London in the last half of the 20th century (Table 16.1) we see that a large number of inhabitants have moved from the center of the city to the periphery, creating the concept of the suburbs. This shift of population has been driven mostly by the availability of the mobility offered by the automobile and the desire to live in a house with a garden. Together with this shift of population, new structures have emerged with decentralized organizations for work, shopping, and entertainment and as a consequence, a decrease in the interest for going to town centers. We have now reached a situation where the access to a private automobile is synonymous to freedom and this has led to car densities between

TABLE 16.1
London Population

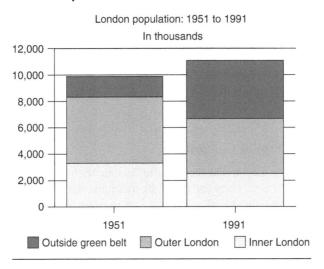

London population: 1951 to 1991
In thousands

Legend: ■ Outside green belt ▦ Outer London ☐ Inner London

500 and 800 cars per 1000 inhabitants in most industrialized countries and a strong desire to reach these levels in other countries. These densities mean that nowadays, almost anyone within the proper age limit has access to a car for moving around and those not of age, must rely on the "soccer moms" as they are called in America.

16.1.2 Problems with Safety

Such enormous development has obviously brought some problems to our societies. The first problem is concerned with safety. It is estimated that 1 million people die every year in traffic-related accidents worldwide. This is a catastrophe of bigger magnitude than all armed conflict, which has happened in the past. The most advanced countries have been able to drastically reduce the number of fatalities through better vehicles (better handling, better braking, and better passive safety) and through better infrastructures (modern freeways are ten times safer than regular roads). However, these improvements seem to reach a limit in terms of the number of deaths per million passengers-kilometers, even in industrialized countries. Indeed, the problem of the automobile is that it is inherently dangerous for the driver, the passengers, and the other road users. It goes at such speeds that very slight errors can have catastrophic results. And human errors can happen to any type of driver. The simplest and most common error is distraction. A few seconds of inattention are sufficient to collide with a pedestrian crossing the street, with a car, which has stopped ahead, or to go off the road. Another human error which do happen is the mishandling of an emergency situation. A large percentage of drivers will take improper action in such situations and incur an accident, which could be avoided. The only solution to solve these problems of human errors is to remove the driver from the control loop. An interim step is to assist the driver to warn him/her in case of potential danger (e.g., in the case of excessive speed before a dangerous bend or when a car is present in the blind spot while changing lane), or to take over the control in emergency situations (e.g., emergency braking in case of impending collision).

16.1.3   Problems of Congestion

The success of the automobile also leads to the saturation of the infrastructures, in particular, in cities. Each car needs a certain amount of space in order to operate. The normal width of roads is 3.5 m in order to accommodate various types of vehicles and steering imprecision. Spacing between vehicles has also to be kept at a minimum to prevent collisions during decelerations (this depends essentially on the driver's reaction time). This spacing is obviously very dependent on the speed and it is usually recommended that it should be at least equal to 1.5 times the speed in meter/second. This is equivalent to a time gap of 1.5 sec. This spacing leads to a maximum throughput of about 2,200 cars per hour per lane. This is not much if we consider that a suburban train can carry about 60,000 passengers per hour on an infrastructure of similar dimension. Furthermore, such high-density car traffic often leads to a flow breakdown (stop and go traffic) and to accidents both of which drastically decrease the capacity. The solution to these problems lies once again in the removal of the driver from the control loop to improve the lateral guidance (and hence reduce the width of lanes) and the longitudinal control (with possible time gaps of around 0.3 sec) while maintaining the traffic safety. Such techniques of automatic driving could multiply by a factor of ten the throughput of the infrastructures.

Another problem is that of congestion for parking. Each individual vehicle is used for a very small percentage of its life. Most of the time, it occupies space very unproductively. Typically, a car will need about 10 m^2 if parked at the curb but this space is very limited in cities and cannot accommodate all the cars of the residents and visitors. In parking lots, each car will need four times this amount in order to have access to each individual slot. These spaces can reach a high price in dense cities and the cost of parking (if paid by the owner) can be a strong deterrent for going to or living in these places. The solution for reducing the cost of parking lies in the automatization of the parking in order to obtain higher-car densities, in the reduction of car size for cities, and in the sharing of cars (see Figure 16.1 and Section 16.1.5 on car-sharing).

FIGURE 16.1 Car-sharing from Toyota.

16.1.4 Problems with Emissions and Nuisances

The large development of the car and truck population has also led to critical problems of noise and pollution at the local level as well as greenhouse gas emission at the global level. Although the car manufacturers have been able to drastically reduce the local pollutants, the noise in cities is now perceived as the major problem by the inhabitants. At the global level, the generation of CO_2 through the use of fossil fuels is also considered as a major problem, which will require drastic steps such as the limit in the use of cars which generate CO_2 above a certain level. In the long term, this will lead the industry to offer vehicles running on different types of energy or to new forms of transportation systems.

Already, some cities have taken such steps to allow only zero emitting vehicles (ZEV) in some zones, either for passenger transport or for freight transport (this latter being responsible for a major part of the nuisances). The European project ELCIDIS has experimented with such concepts for freight with distribution platforms located outside the cities so that freight is transferred to electric van for distribution in the cities (Figure 16.2).

Another path for reducing the amount of emissions and noise generated in cities while reducing the space needed for private vehicles is to change the balance between individual vehicles and public transport. In general (and if the occupation factor is high enough), public transport is much more energy efficient (and space efficient) than individual vehicles. Furthermore, public vehicles can use electricity and hence reduce noise and local pollution. Therefore, if we could change the balance between private transport and public transport (Table 16.2), we could drastically reduce the negatives effects of transport in cities. Several steps are now being taken in this direction in particular

FIGURE 16.2   Electric vans for freight distribution.

TABLE 16.2
Surface Transport Modal Split

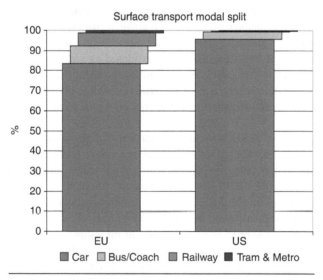

with more space dedicated to public transport with right of ways. New vehicles are also designed to improve the comfort and the efficiency of public transport, in particular with electronic guidance of vehicles, which can behave as a light rail with regular road infrastructure (see Figure 16.3 and Figure 16.4).

FIGURE 16.3 Civis.

FIGURE 16.4 Phileas.

16.1.5 Car-Sharing and Cybercars

In order to take into account the problems presented earlier while preserving the convenience and flexibility of individual vehicles, a solution is being developed around the concept of public individual vehicles. In this scheme, subscribers to a mobility service can "borrow" a vehicle from a fleet at various points in the city and either return it to the same location, or, in some systems, return it to any other authorized location. This system, which has initially been tested in the 1960s without much success, is now getting quite popular in countries like Switzerland, Germany, and Canada. The development of these car-sharing schemes relies mostly on new communication and localization technologies for a better management of the fleet and easier reservation by the users (mostly through Internet and mobile phones). More than a hundred thousand of such vehicles are already in operation in various cities.

The major interest of car-sharing is to encourage the users to take public transport, or to walk when possible, and to take an individual vehicle only when this is a better alternative. Another interest is to be able to select a vehicle better adapted to the city. In several car-sharing schemes (Praxitele, Liselec, ICVS,...), electric or hybrid vehicles are offered which solve the problems of local pollutions and noise while improving the safety (it has been demonstrated that electric vehicles have less accidents).

The problems of car-sharing organizations are, however, the difficulty of a good return on investment at this time due to the lack of demand for each vehicle. The operators must make sure that the vehicles are available where and when needed but this is quite difficult since the number of pick-up points must be limited for management reasons and redistribution techniques. The solution to this problem lies with the automatic movement of the vehicle in order to make them available where and when needed. This solution, which originated in the Praxitele project, has been tested by Honda in Motegui in 1998 and is being further developed in the European CyberCars Project (www.cybercars.org) (see Figure 16.5).

16.1.6 The Future of the Automobile

The automobile has certainly been a major step in the development of our advanced societies and it has now become a necessity for its sustainability. Those who do not have access to an individual vehicle (because of age, physical abilities, or financial situation) have a hard time getting the same freedom of movement as those who have the access and therefore have less opportunities for work, shopping, entertainment, etc. The automobile has become a "must" for any citizen and this need is now occurring throughout the world.

Although this need is creating an enormous industry and therefore an improvement in the global economy, there are serious doubts concerning the negative factors such as lack of safety and emissions. It is certainly not sustainable for our planet to support billions of automobiles running in the same way as today (the current automobile population is around 800 million at the beginning of the 21st century with extreme growth in China and India).

One way to limit these negative impacts is to move from an industry of products to an industry of service where anyone would have access to mobility in the most cost-efficient way. With this service industry, each customer would have a choice of transportation mode. For those trips occurring in high-demand time and zone, mass transport would be more efficient and hence cheaper.

FIGURE 16.5   Cybercars in Amsterdam-Floriades.

At other times, an individual transport would be better. Sometimes, for a single trip, two or more modes would be the best. Of course, both mass transport and individual transport would be offered by companies in the most cost-efficient way and in accordance with local regulations. In this context, automated driving will be developed because of decrease in cost of operation and improvement in safety (this is already occurring in certain transport modes such as the metros). This approach toward automation follows the development of robotics in the factory, basically for the same reasons (cost) but also for improved safety and meeting regulations (in particular noise and emissions).

Obviously this move toward fully automatic vehicles will not happen everywhere at the same time. There are two trends toward this future. One is driving assistance which is spreading quite rapidly since the late 1990s with numerous techniques which have appeared in recent high-end private and commercial vehicles (buses and trucks). The other is with the arrival of people-movers based on automated guided vehicles in specific locations and on dedicated tracks (protected or not). It can be forecast that in the next 10 years, these two trends will merge with individual vehicles with dual mode capabilities: manual (assisted) driving on regular roads and fully automatic driving on dedicated zones where no (or few) manual vehicles will be allowed, ensuring therefore a smooth and safe operation of the automated ones. This type of vehicle will be perfect for the implementation of mobility services with vehicles which can be called on demand (perhaps through mobile phone) when and where needed. With the development of such zones, new dedicated infra-structures will be built specifically for these vehicles to move automatically and at high speed from one automated zone to the next. This is the most realistic path for the "automated highways" to happen.

The rest of this chapter will describe the techniques already developed for the sensors, for the actuators and their controllers, and for the control algorithms in these two approaches which transform the old automobiles of the 20th century into more or less autonomous robots. We will present the robotic techniques already in use for applications such as drivers' assistance and fully autonomous vehicles already on the road. Indeed, the new automobile with its sensing capab-ilities, acting capabilities, and control techniques are becoming robots and will use many past and present developments from the robotics research community.

16.2 THE AUTOMOTIVE SENSORS

In order for the modern vehicles to become more independent from the driver in their behavior and to assist or completely replace him/her, it must acquire information from its environment as in any robotics system. This is why the new vehicles must be equipped with sensors, which will feed the actuators through the control algorithms. Presently, a rather large number of sensors are already available or soon will be for the automotive market. This section reviews such sensors.

16.2.1 Ultrasound Sensors

These sensors are the simplest and least expensive available at this time. Ultrasound sensors operate by emitting a cone-shaped ultrasonic wave (pressure wave) through an ultrasonic (electrostatic or piezoelectric) transducer and receiving its echo. Within a stipulated distance range, the incoming echo is checked and the time taken for the sound to travel the distance is determined from which the distance to the object is calculated. If the distance between the sensor and the objects is too small, the echo arrives before the ultrasonic trans-ducer has reached steady state and is ready to receive. Thus, objects in this dead band cannot be detected reliably. Usually, in order to have wide-area detection, a set of few sensors is used for measuring their orientation [1] or position [2]. In this case, neighboring ultrasonic sensors can influence each other mutually at an extent that is generally only determined experimentally. One solution is to synchronize the output waves. Ultrasonic sensors are today widely spread in the automotive industry for a few applications, the most common being the back maneuvers and parking assist systems. These sensors are cheap, light, small, and low-power consuming. Their main drawbacks are the short range (only a few meters), their poor angular precision (10 to 30° depending on the sensors), and their sensitivity to wind, humidity, and object shape or orientation. Some developments are under way to have ultrasonic sensors with high-angular resolution through the use of an array of receivers but the range will always remain a problem.

16.2.2 Inertial Sensors

These sensors are not truly environment sensors but proprioceptive sensors in the sense of robotics. However, they are now widely used for stability functions and for navigation in the automotive industry. Thanks to new electronic technologies (in particular, MicroElectroMechanical Systems or MEMS), they can be very inexpensive.

Traditionally, a full inertial measurement unit comprises six sensors allowing measurement over the six degrees of freedom of a vehicle, namely three orientations (roll, pitch, and yaw) and three accelerations. However, the vehicle being restricted to operate on a known surface (usually the horizontal plane), a six axis inertial navigation system is not needed and in fact, often only the angular rotation around the vertical axis and the longitudinal acceleration are of importance to estimate the position. The accelerations are measured using accelerometers and permit to retrieve the vehicle displacements over time by double integration. Unfortunately this causes the drift errors intrinsic to these proprioceptive sensors to increase at a square rate of the distance.

16.2.2.1 Gyroscopes — gyrometers

The spinning mass gyroscope is the classical gyro that has a mass spinning steadily with free movable axis (so called gimbal). When the gyro is tilted, gyroscopic effect causes precession (motion orthogonal to the direction tilt sense) on the rotating mass axis, hence letting you know the angle moved.

More precise, without the moving parts causing friction (therefore inherent drift), the optical *gyrometer* (measuring the rotational speed instead of the angle) has been developed over the past decade, based on the Sagnac effect. When two light beams propagate in opposite directions around a common path, they experience a relative phase-shift depending upon the rotation rate of the plane of the path. The actual heading or direction is obtained by integrating the output. The ring laser gyroscope (RLG) is the first type of these sensors. The input laser beam is split into two beams that travel the same path in a prism but in opposite directions (one clockwise and the other counter-clockwise). The beams are recombined and sent to the output detector. In the absence of rotation, the path lengths will be the same. If the apparatus rotates, there will be a difference in the path lengths traveled by the two beams, resulting in a net phase difference and destructive interference. The net signal will vary in amplitude depending on the phase shift providing a measurement of the rotation rate. In the case of the fiber-optic gyroscope (FOG), the phase difference is detected by interfering the two beams outside the path. The FOG being a simpler device than RLG, it is currently receiving more attention due to its potential to achieve the required performance at a lower cost. Also attractive, thanks to a good immunity to magnetic fields, they can be found (Hitachi) in automotive navigation system, used in conjunction with GPS (discussed later in Section 16.2.6.1) and in several other systems (cleaning robots, unmanned dump trucks, devices for route surveying, and mapping).

Vibrating gyroscope is the third type of gyroscope. A vibrating element (vibrating resonator), when rotated, is subjected to the Coriolis effect that causes secondary vibration orthogonal to the original vibrating direction. By sensing the secondary vibration, the rate of turn can be detected. The vibration is often exerted and detected by means of the piezoelectric effect. This type of gyro is suitable for mass production and almost free of maintenance.

Recently, monolithic integration of MEMS with driving, controlling, and signal processing electronics makes possible new generation of smaller, cheaper sensors. MEMS gyros use capacitive silicon-sensing elements coupled with stationary silicon beams attached to the substrate and measure the Coriolis-induced displacement of the resonating mass and its frame.

16.2.2.2 Accelerometers

Accelerometers detect the motion of an object by means of instrumented spring-mass "seismic" structures. Under acceleration, a force acts on the inertial mass causing a displacement of the (silicon) moving structure with the fixed frame. The analog output is inferred by transduction based on piezoelectric effect of quartz and special ceramics, piezoresistive or capacitive measurement principles. The last two types have in recent years been produced, as gyroscopes, using the silicon micromachining technology (MEMS). Capacitive-based MEMS accelerometers are actually preferred over piezoelectric accelerometers in many cases because they generally offer higher sensitivity and better resolution. The same is true for MEMS vibrating beam accelerometers. Also MEMS accelerometers usually do not have the problem that piezoelectric accelerometers have with low-frequency components. Compared to piezo, the latter produces a higher-level output signal for increased noise immunity.

16.2.3 Laser Detection and Ranging

Light Detection and Ranging (LIDAR) are devices consisting of a photon source, often a laser for Laser Detection and Ranging (LADAR), a photon detection system, a timing circuit, and optics for both the source and receiver.

The interest of LADAR stems from the natural three-dimensional (3D) spatial data it produces, defined by its spherical coordinates (r, θ, φ) mapped to a two-dimensional (2D) matrix [11]. The distance r from the device to targets struck by the emitted photons is measured by the time-of-flight (TOF) divided by the speed-of-light. A device that proceeds in a single-shot measurement is usually referred to as laser rangefinders. LADAR, on the other hand, is generally assumed to generate a 2D or 3D range image.

16.2.3.1 Range measurement

The simplest method is the direct TOF technique where a pulse is sent and its echo is received. The shortest pulse source currently used in an operational LADAR device is 250 psec. The Chirped Pulse method is less sensible to noise, thanks to the pulse that is coded using pseudorandom techniques similarly to radars (see Section 16.2.4).

A second type of method is based on continuous sinusoidal signal (CW — Continuous Wave). An unmodulated CW source is suitable for velocity measurements but is incapable of measuring range. The phase-based AM-CW, aimed at improving accuracy over direct TOF through the use of phase detection, is based on the source modulation with a fixed sinusoidal frequency, f. A phase shift of $\Delta\phi = 2\pi f (2d/c)$ will be observed between the transmitted and received signal. Therefore, the object distance is given by $d = (\Delta\phi c / 4\pi f)$. Another way of getting range information using CW is by generating a chirp waveform in the frequency domain (FMCW — Frequency modulated continuous wave). The modulation might be linear or sinusoidal. The methods are practically difficult to implement in a stable, linear system. (One approach is to thermally control the laser cavity; as the cavity expands and contracts under thermal excitation the coherent wavelength of the laser changes or changes the length of the cavity mechanically using extremely fast piezoelectric actuators.) The major sources of error are from the precision with which the initial laser pulse is generated (CW, pulsed, or chirped), the nature, and ambiguity of the light detection. However, a precision of a few centimeters can usually be obtained, which is sufficient for most applications.

16.2.3.2 Azimuth measurement

Scanning methods are the most widely used to built a LADAR data frame by illuminating each pixel in a range image, retrieving range and angular information (azimuth and tilt for 3D LADAR). Frequently there is a trade-off between speed and accuracy. LADAR frames can be created by scanning high-resolution laser rangers in which a single degree-of-freedom laser rangefinder is mechanically swept over the scene using either encoder-equipped pan/tilt servos or a rotating mirror (polygonal scanners, galvanometric scanners) combined

FIGURE 16.6   ALASCA ladar (IBEO AS).

with either a pan or tilt servo. Specific LADARs for the automotive industry are now under development. The ALASCA sensor (IBEO AS) [12] uses a rotating mirror and a combination of a single-illuminating laser with four receivers, which allows for a vertical field of view of several degrees (see Figure 16.6).

The major problem with scanning comes from the cost and reliability (over the lifetime of a vehicle) of the mechanical components. Microfabrication may provide the way forward. With reduced size comes reduced inertia, which in turn permits higher performance. Micromirror arrays could prove to be a very useful technology for the control of the resolution of LADAR sensors. Micromirrors also can act as a distributed scanner that generates a large number of microbeams that can scan the workspace from different angles and positions.

Another alternative to scanning is the use of Focal Plane Array (FPA) which is a 2D "chip" in which individually addressable photo sensitive "pixels" can be accessed. Early FPA detectors were developed as, first, infrared imagers and later as FLIR (Forward Looking InfraRed) detectors, largely for military purposes. For measuring range, additional electronics must be added to an FPA in the form of timing circuitry, a cost that tends to limit the size of the array. This circuitry must fit behind each pixel in the array and usually causes the pixel size to be large (e.g., relative to the pixel sizes in digital cameras) (see Figure 16.7). New work in FPA design shows promise for the resolution (now at 124×160) as well as for the level of miniaturization needed to improve range resolution and speed since most of the effort is being directed at acquiring the LADAR frames in real-time (Flash LADAR).

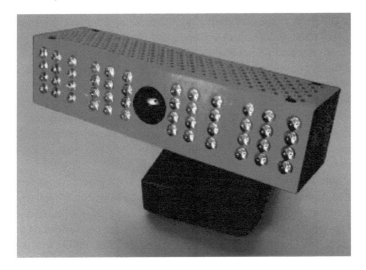

FIGURE 16.7 Prototype 3D camera.

16.2.4 Radar

The Radar (Radio Detection and Ranging) is based upon propagation of a signal, a high-frequency electromagnetic wave, emitted from the sensor. This signal reflects from any obstacle on its path and the sensor will receive in turn this echo, a signal that will be processed and from which range, azimuth, and velocity of the obstacle (or target) can be determined. The relative velocity of the target is measured using the Doppler effect (a wave reflecting on a moving object of a speed difference with the sensor of V_r, has its frequency modified by a value of $\Delta f = 2V_r f / c$), there are however different ways to measure the range and the azimuth angle, depending mostly upon the required accuracy and therefore resolution [4].

16.2.4.1 Range measurement

The pulse radar sends a signal and immediately listens to its echo, which permits the sensor to have a single emitter/receiver, reducing the cost. In order to increase the resolution, the signal has to be short and might lack in strength, which causes difficulty to extract the echo from the noise. The signal strength is indeed proportional to $1/d^4$ (with d the distance). The sensitivity is usually improved by using pseudorandom modulation technique over the pulse length (PseudoNoise Phase Shift Keying, PN-Frequency Hopping, PN-Pulse Position Modulation) in order to send a specific coded signal, a chirp pulse, the reception of which is based on the emitted signal characteristics and become much more reliable. The shortcomings of coding the signal is that for a few centimeters resolution, very short signals must be sent (state of the art of 300 psec) requiring a high-spectral resolution (Ultra Wide Band — UWB) making the measurement of the echo amplitude and phase difficult and impossible, thus making the measurement of Doppler effect, that is, the obstacle velocity impossible. Pulse Doppler radars exist and are based on a specific signal processing based on the demodulation of the received signal with the source frequency. These radars are, however, more sensitive to interferences and present a substantial blind zone in front of the sensor (time for the pulse emission — a few nanoseconds for these radars).

Continuous wave radars on the other hand, send a continuous frequency modulated signal and simultaneous reception. The received microwave is delayed by an amount that is retrieved by comparing the current emitted frequency signal with the received one. In case the wave was reflected by a moving obstacle, the Doppler shift is added and inserts an ambiguity that is solved, thanks to the modulation type. The frequency is modulated using double ramps (Linear FM) of steps (Frequency Shift Keying — FSK) or by a combination of it [3]. Thanks to this technique, the range and velocity estimations of the target can be made, based on Fast-Fourier Transform (FFT) algorithms or possibly on high-resolution spectral modeling [5]. FMCW has many advantages when operating at the high frequencies used for automotive radar (76 to 77 GHz). At these frequencies, achieving a well-controlled pulse transmission for a pulse radar is difficult. FMCW offers straightforward integration gain through using a low-detection bandwidth to compensate for the low-transmit power that can be achieved. FMCW also offers very short-range capability as receiver recovery time, necessary for a pulse radar, is not required.

16.2.4.2 Azimuth measurement

Usually, the angular information does not strongly depend upon the wave shape but on the antenna. The gain of the antenna is proportional to the surface of the antenna and inversely proportional to the square of the wavelength. Therefore, in order to have a high-gain antenna, for applications (like automotive) where space is an issue, high-frequency microwave would be preferred. Angular information might be determined by triangulation in case two (high-range resolution) radars are used. Another way is to mechanically scan the antenna. It allows a large detecting area, but a usually poor-velocity resolution due to the limited available time of analysis at each angular position. Commutation of receiving signal over different antennas is another technique but the most common for high-angular resolution requirements is the monopulse method. Over a specific detecting area, the reflected signals from the same target are received on two shifted antennas as two beams having different directions and allow to precisely retrieve the angular position of the obstacle.

16.2.4.3 Automotive radars

The 77 GHz radar for the automotive industry is commercially available since early 1999 [6], based on conventional RF-components. A 79 GHz band with 4 GHz bandwidth should be made available for short range radar applications within EU member states and European conference of Postal and Telecommunication Administration (CEPT) countries by 2005. However, the cost of these systems will limit their use for long-range and high-accuracy (at least longitudinally) application such as ACC (adaptive cruise control) or emergency braking. The 24 GHz radar is under development and should address the applications for short range and low cost [7,9,10]. However, at this time, the legislation in Europe does not allow wide use of this frequency.

16.2.5 Vision Sensor

16.2.5.1 Principles

The previous sensors provide high-spatial resolution mainly by determining the range value to the target at the expense of a lower definition due to the single array information (besides next generation LADAR). Vision sensors offer a 2D array of up to a million pixels with a large field of view, the range detection and the angular field of view depending upon the optics (lens and focal). This rich source of information allows for much smarter, though involved, applications, for example, classification and recognition of object. Wide-luminance range sensors are provided by CMOS imagers with non-linear luminance response and will certainly replace current charge coupled device (CCD) imagers. In fact, CMOS devices can offer several advantages over CCD-based imagers, including enhanced functionalities with individual pixel signal processing, lower power consumption, and lower cost. Within the several application fields in which the advanced signal processing capabilities of CMOS imagers are useful, the automotive one is currently the object of many research activities and offer wide market opportunities. System on chip incorporates major IC components on a single chip. High-temporal dynamics [14] and a fast read out resolve fast movement. Thus, the digital image sequence is passed to an evaluation unit that performs appropriate signal processing to extract the desired output.

16.2.5.2 Specificities of automotive applications

The design of vision sensors for automotive vision has to take into account specific constraints [13,15]. When most of the current vision systems use full color, accurate image reproduction, and photographic or video aspect ratios, operating over a visible light spectrum of 0.45 to 0.7 μm and realized with

CCD or CMOS sensors, most automotive applications are monochrome and may involve variable aspect ratios, pixel size, and pitch. Automotive imagers must collect light at very low levels, and yet still resolve objects in direct sunlight. Another constraint concerns practical resolution requirements. The automotive vision system will be used as sensor input for computational vision that will perform higher-level task by extracting features from the imager and its pixels information in order to perform lane recognition, passenger occupancy detection, forward vehicle, or pedestrian detection [21,22]. The last constraint is generic for any automotive system: provide features with customary automotive durability, at an affordable cost.

16.2.5.3 Stereovision systems

The ranging capability of active sensors (radars, ladars) is an extremely rich data source. A stereo-based vision system can similarly provide a direct absolute measurements of the scene [18]. Computing depth from two images is a computationally intensive task. It involves finding for every pixel in the left image the corresponding pixel in the right image. Correct corresponding pixel is defined as the pixel representing the same physical point in the scene. The distance between two corresponding pixels in image coordinates is called the disparity and is inversely proportional to distance. In other words, the nearer a point is to the sensor, the more it will appear to shift between left and right views. Stereo depth computation, in particular, has many advantages over other 3D sensing methods. First, stereo is a passive sensing method. As we have seen, active sensors rely on the projection of some signal and often pose high power requirements or safety issues under certain operating conditions. They are also detectable — an issue in security or defence applications. Second, stereo sensing provides a color or monochrome image, which is exactly (inherently) registered to the depth image. This image is valuable in image analysis, either using traditional 2D methods, or novel methods that combine color and depth image data. Third, the operating-range functions of lens field-of view, lens separation, and image size are flexible. Fourth, stereo sensors have no moving parts, an advantage for reliability. Computation relates to the frame rate, which needs to be high with low latency and remains an issue for many applications. In safety applications such as airbag deployment, the 3D position of vehicle occupants must be understood to determine whether an airbag can be safely deployed — a decision that must be made within tens of milliseconds (Siemens VDO, Delphi). For vehicle tracking applications, by means of 3D information, it is easy to distinguish vehicles and shadows on the ground, vehicles and reflections on the road, or detect overlapping vehicles (Omron Corp.). However, a key problem of these stereovision systems is a correct calibration [16,17]. In all applications where not only recognition is

Figure 16.8   Lane departure warning system.

important, but a correct localization in world coordinates is essential, a precise mapping between image pixels and world coordinates becomes mandatory. This correspondence may vary during system operations due to many reasons; in automotive applications, vehicle movements and drifts due to sudden vibrations may change the position and orientation of the cameras, making this mapping less reliable as the trip proceeds.

Automotive application using cameras have started to appear on the market by the end of the 1990s. The first commercial application has been the lane departure warning, not a truly robotic application but a first step in this direction. This system (see Figure 16.8), which warns the driver when the vehicle tends to drift away from its lane, is now often used in trucks. The first application, which put some control of the vehicle, was the introduction of lane keeping assistant (Nissan Cima in 2002) where a single camera helps the driver to stay in the middle of the lane by applying a corrective torque on the steering wheel.

16.2.5.4 Future vision processors

The complexity of computational vision encouraged the introduction of dedicated ASICs aimed at developing hardware-based vision processing functions [19]. As innovative approaches, an Embedded Perception Processor, inspired by the physiological mechanisms of human vision, has been developed based on a "Perception Paradigm" modeling the visual perception capabilities of the human brain (BEV S.A.) [20]. Contrasting with the usual "DSP Paradigm" using expensive image processing techniques, this processor is an electronic modeling of a spatio-temporal neuron, which is the basic building block of the perception processor allowing real-time analyses of successive frames of video and the determination of the speed, direction, hue, luminance, and saturation of each pixel. As a result, the processor (Generic Visual Perception Processor [GVPP]) is a single-credit size chip able to detect the presence of objects in a motion video signal, and then to locate and track those objects as they move in real time. Extremely fast, a single GVPP chip is capable of performing up to 20 billion operations per second.

16.2.6 Global Navigation Satellite-Based System

The Global Positioning System, global navigation the first satellite-based system (GNSS) launched by the U.S. Department of Defense in 1980 has found a lot of crucial civil application and is nowadays a key technology for most of navigation and positioning solutions [23,24]. The goal of the GNSS is to build a system that provides information on time, position, and velocity everywhere on the planet.

16.2.6.1 Global positioning system

A constellation of 24 satellites evenly spaced, placed at 20.200 km altitude, in circular 12 h orbits and inclined at 55° to the equatorial plane provide, at a reasonable cost, an earth wide coverage. Since its start, new generation of satellites have succeeded. The atomic clocks control all inboard signals very precisely, thanks to (for the BlockII) two rubidium and two cesium clocks, which gives for each satellite four time standards. As we will see in the following sections, these clocks are the heart of the global positioning system (GPS). The satellites transmit two microwave carrier signals derived from the fundamental L-Band frequency, the L1 frequency (1575.42 MHz) and the L2 frequency (1227.60 MHz). On these carriers, two main binary codes are shifted. First of all, the most used information, the C/A Code (Coarse Acquisition), is a repeating 1 MHz Pseudo Random Noise (PRN) Code modulated upon L1 only, which delivers an effective length of 300 m. There is a different C/A code PRN for each satellite, which is often also used to identify the satellites. The C/A code is also designated as the Standard Positioning Service (SPS). Second, the P-Code (Precise) is a 10 MHz PRN code that modulates both the L1 and L2 carrier phases. It has been reserved for U.S. military and other author ized users. Besides, a data message is modulated onto the L1 carrier at 50 Hz providing status information, satellite clock bias corrections, and ephemeredes (orbits).

In order to keep civilians from using the GPS, two techniques have been implemented in order to add bias error to the signal. The Selective Availability (S/A) operating on the C/A Code and accounting for observed variation of amplitude of 50 m, is obtained by dithering the fundamental frequency of the satellite clock. The S/A has been turned off on May 2, 2000 [28]. Furthermore, with the Anti-Spoofing (AS) turned on, the P-Code is encrypted into the Y-Code in order to limit the access to authorized users only (Precise Positioning Service).

16.2.6.2 GLONASS

The GLONASS satellite navigation system is deployed by the Russian Federation, owned and operated by the Russian defense department from 1982. It is equivalent to GPS with a few particularities. The 24 satellites are distributed over three orbital planes instead of six, which allows quicker reorganization of constellation in case of a satellite failure and a better coverage on the polar region. They transmit the same PRN-code at different frequencies $L1 = 1602$ MHz $+ n \times 0.5625$ MHz, where "n" is frequency channel number ($n = 0, 1, 2, \ldots$) for each satellite, which provides more robust resistance to interferences and jamming. The signals from the Glonass system are not degraded. In addition to the Channel of Standard Accuracy (CSA) that provides a horizontal accuracy of 60 m (at 0.997 probability) a Channel of High Accuracy (CHA) will be accessible to authorized users. The current status is ten orbiting satellites. However, outages have been observed so the usable constellation is not always ten satellites but rather nine or eight.

16.2.6.3 GALILEO

The European Union intends for the Galileo system to provide four navigation services and one search and rescue (SAR) service. The primary signals of Galileo are intended to provide an "Open Service" (OS) of a high quality, consisting of six different navigation signals on three carrier frequencies. OS performance will at least equal that expected from the "follow-on" generation (Block IIF) of GPS satellites scheduled to begin launching in 2005 and the future GPS III system architecture currently being investigated. The GPS IIF/III satellites will offer wideband signals on three civil (open) frequencies: one high-chipping rate signal (L5 centered at 1176.45 MHz) and two low-chipping rate signals (L1 at 1575.42 MHz, L2 at 1227.60 MHz). Moreover, the GPS modernization program will offer additional civil and military code structures on L2.

16.2.6.4 GPS receiver-based localization

GPS receivers are usually small electronic devices that offer different reception capabilities. The choice is usually based upon the absolute positioning accuracy vs. the price of the receiver. Small civil SPS receivers can be purchased for under $200 and some can accept differential corrections. Receivers that can store files for post-processing with base station files cost more ($2000 to 5000). Receivers that can act as DGPS reference receivers (computing and providing correction data) and carrier phase tracking receivers (and two are often required) can cost many thousands of dollars ($5,000 to 40,000). Military PPS receivers may cost more or be difficult to obtain. Other costs include the cost of multiple receivers when needed, post-processing software, and the cost of specially trained personnel.

16.2.6.5 Basic principle

The basic principle to answer the question of positioning relies on the basic geometry tool of triangulation, which is the capability to find an intersection point from three different spheres. Assuming that we see from a point of the earth three satellites and that we know their exact position, by calculating the distance from this point to each satellite, that is, the time travelled by a coded electromagnetic wave from the satellite to our position times the speed of light, we are able to solve for our position, that is, longitude, latitude, and altitude. Therefore, in order to have a precise positioning information, it is mandatory to have (1) a precise information on the satellites position within an Earth Centered Fixed Frame (ECFF) for instance and (2) an exact measurement of the time of travel of this specific coded signal that is emitted by the satellite. At the speed of light, an error of 1 msec in measuring the travel time would result in an error of a about 300 km! Kepler's law gives precise orbits for each satellite, where they are positioned when they are launched. The orbits information are communicated to the receivers by means of the navigation message and their related "ephemeris" errors resulting from deviation from the nominal orbit are calculated from the base stations, transmitted to the satellite in order for it to send this information to the user on earth, the GPS receiver, such that it can take this error into account. The second condition, however, is much harder to satisfy because of two main factors. First of all, a GPS receiver embeds a very cheap clock with respect to the $50K to 100K satellite atomic clock. The synchronization of the satellite clock and the GPS receiver clock then becomes mandatory though a complicated task and will require a fourth satellite. Furthermore, the electromagnetic wave signal sent by the satellite does not always travel at the speed of light, especially within the atmosphere and specifically through the ionosphere (at an altitude of 50 km to 500 km), where the speed will depend upon the chemical composition of the environment and the thickness of the layer that is traveled.

16.2.6.6 Code range positioning

The basic position calculation can be performed by a multichannel (in order to receive signals from different satellites at the same time) single L1 frequency receiver. Assuming C/A PRN codes are generated precisely at the same time at the satellite and the receiver, the code is shifted within the receiver in order to match with the code sent. The time required to shift the signal will give the time of traveling and will be used to estimate the distance to the satellite or its pseudorange. The offset between the clocks will then be calculated in order to account for this error. In order to do so, the measurement of a fourth satellite is made. This measurement will not coincide with the already defined point if there is a clock offset. Since this offset modifies all calculation, it is modified as to find the proper value for a single intersection point with the four different measurement, which gives the clock offset. For simple receiver, the ionospheric refraction error is modeled within the receiver software and is taken into account for the final calculation. The typical accuracy value with S/A turned off for these systems is about 10 m at the 95% probability level. Dual frequency receivers (on L1 and L2) have the advantage to be able to define more precisely the ionospheric error. The refraction depends indeed upon the frequency, a lower frequency (L2) get more refracted, that is, is more delayed than higher frequency. This information is used by smart receivers to get exact ionospheric errors. Unfortunately this requires a very sophisticated receiver since only the military has access to the signals on the L2 carrier. Civilian companies have worked around this problem with some tricky strategies where the receiver operates in a codeless or quasi-codeless mode. Unfortunately they are very secret. Of course P-Code receivers do not present only this advantage, since they have access to an extra code, demodulated on L1 and L2 they can obtain a very good accuracy.

16.2.6.7 Code phase differential GPS

When the S/A was still turned on and the precision of the GPS did not meet many application requirements, the development of a technique came up, based on the use of (at least) two receivers, one of which, the reference, is supposed to be located at a precisely known stationary position and the other one the (moving) receiver for which the position is to be determined. The reference frame calculates the errors on the measured pseudoranges since it knows its position and transmits it to the second receiver in order to improve its position calculation. Such a system presents, however, some shortcomings for a few applications. In order for the error information to be valid it is supposed that the two receivers monitor the same satellites and receive the signals in similar conditions which means that the roving receiver cannot go far from the base station (100 km). Furthermore, a radio link must be established between the two receivers so that the base can send correct information to the roving one. Aside from the basic calculation, using the available information in a smart way, the GPS has also been designed to do new tricky calculations that have risen during the last decade.

16.2.6.8 Augmented differential GPS

A precision of a few meters might be sufficient for many applications; however, for a global use, differential GPS using earth-based reference as seen previously, is not practical as it would require thousands of reference bases. The aviation community first had in mind to establish a system that could improve the integrity of the GPS, as this issue is crucial in avionics. When a GPS is not working correctly, the delay between the time it is informed and, in turn, the user is informed through its system status is too big. So the Federal Aviation Administration (FAA) got the idea that they could set up their own monitoring system that would respond much quicker. In fact, they figured they could park a geosynchronous satellite somewhere over the United States that would instantly alert the aircraft when there was a problem. Then they reasoned that they could transmit this information right on a GPS channel, so that the aircraft could receive it on their GPS receivers and would not need any additional radios. The additional benefits are to have an extra satellite always on view and to use the earth-based reference to establish a consistent continental-wide correction map that can be informed to the user via the satellite in order to correct the pseudoranges and get a national differential GPS accuracy positioning, for free! A satellite-based augmentation system (SBAS) has already been deployed on the North American continent (WAAS for the United States and CWAAS for the Canada) as well as on Japan (MSAS) and is under deployement over Europe through the EGNOS system which has started its operation phase in July 2005 [25].

16.2.6.9 Internet-based differential GPS

Recently introduced, the main idea of this technique is to establish a wide area-consistent correction map to transmit it to the receiver through the Internet [26,27]. Any system connected on the Internet via a standard connection or wireless, using wireless LAN, or 2.5 (GPRS) or 3G (UMTS) telecommunication compliant system might then have access to this data. This service could have the advantage to be more request-specific instead of broadcasting all information, it could provide the one needed by the receiver and furthermore it could bring an earth-wide solution as information could be gathered from the entire planet and transmitted to the entire planet.

16.2.6.10 Carrier phase differential GPS

Some applications require even more accuracy, down to the decimeter or even centimeter. The problem with code phase receivers is that they compare and match signals (C/A PRN Code) at a cycle width of a microsecond, which represents the speed of light 300 m. Code matching techniques have achieved very good quality of the order of 1 to 2%, which still represents a few meters. The carrier frequency on the other hand has a cycle rate of over a gigahertz, which gives a wavelength of 19 cm for L1 and 24 cm for L2. A few percent matching precision on a phase matching would result in accuracy down to the centimeter. This technique in fact has been widely used by geographic surveyors for static positioning resulting in accuracy down to millimeters. The main difficulty with this technique is to track the number of phases. It might be possible to know exactly where in a specific phase a matching is occurring. But if the number of phase cannot be recalled, the precision becomes completely uncertain. The kinematic method is a technique used to solve this "integer ambiguity" by means of a lot of sampling and by defining a relative distance, a baseline vector, to a known base that observes simultaneously the same carrier phases. In case the roving receiver moves, the ambiguity must be solved on the fly (OTF), which is much more involved. Current systems that make such calculation are very expensive. They are based on very precise DGPS receivers that have accuracy of the order of the meter. Over this tolerance, the ambiguity is much faster to resolve. Corrections and base carrier phase information are sent via a radio link from the base to the roving receiver in real time and has led to the real time kinematic technique (RTK). Precision down to a few centimeters can be achieved after the phase has been locked. However, the tracking of the phase is very sensible to perturbations like multipath errors, hidden base, and requires at least five visible satellites.

16.2.6.11 Sensor fusion for improved localization

Autonomous robots rely on different sensors in order to obtain a coherent representation of the world state. As sensors exhibit different properties and react differently according to a specific environment, it might be interesting to handle these sensors' discrepancies in a consistent way and combine the different sources of sensory information into one representational format in order to obtain the most accurate estimate of the dynamic system states. GPS position information is corrupted or even absent within enclosed environment whereas proprioceptive sensors like gyros or accelerometers must account for drift. The combination of these sources of information, however, might provide a more consistent information mutually benefitting each other in each instance. There are different formalisms for sensor fusion, for global localization the most used being the Kalman filter [29], fuzzy inference, or neural networks-based algorithms or more recently using Bayesian inference [30] or Dempster Shafer inference [31]. Recent work is even aiming at adding vision sensor through operators that have smart detection features in order to achieve a localization precision of a few centimeters.

16.3 Automotive Actuators

Until until very recently, the control of an automobile was performed exclusively through mechanical impediments linking the driver to the throttle for the engine, to the brake pads for braking, to the gearbox for changing gears, to the clutch to disconnect the engine from the transmission, and to the front wheels for steering. In the first half of the 20th century, great developments took place to reduce the efforts of the driver with power assistance for steering and for braking and with the replacement of mechanical gearbox and clutch by hydraulic automatic transmission therefore simplifying greatly the task of controlling the acceleration. However, until very recently, the driver was still in total control of the three basic functions : steering, braking, and acceleration even if this means poor efficiency and sometimes loss of control of the vehicle itself.

16.3.1 Power Train Actuators

For a very long time the control of the engine torque was performed exclusively through a mechanical link between the pedal and a throttle, that is, a valve which blocks more or less the entrance of air into the cylinders. In turn, the amount of air which entered the carburetor, determined through variously complex mechanisms (depending on the performances), the amount of gas which was mixed with the air. Modern engines have a different approach and independently control the amount of intake air through an electronically controlled throttle (the valve is operated by an electric motor) and the amount of gas through electronically controlled injectors resulting in a fine control of the torque while minimizing the emissions.

The engine torque (and furthermore, the torque available at the wheel which controls the vehicle acceleration) is also dependant on the rotation speed of the engine which in turn depends on the gear ratio. This gear ratio can be changed manually through the gear box and through the clutch operated with a pedal. This combination of clutch and gearbox can be replaced by a hydraulic gearbox with mechanical or nowadays electronic selection of the gear. Some vehicles also use continuously changing gear ratios (the CVT or Continuously Variable Transmission) through a system of two variable diameter pulleys and a belt (see Figure 16.9). Modern CVT offer full electronic control of the transmission ratio. Another approach for a fine control of the torque available at the wheel while minimizing the emissions is the "robotized gearbox." In this approach, an electronic controller (sometimes replaced by buttons or levers operated by the driver as in Formula 1 racing) pilots an actuator which operates the clutch and another actuator which changes the gears.

Modern electronic controllers nowadays tend to integrate the control of the engine with the control of the gears in order to have the best control of the power train with minimum emissions. Most of the time, the driver is still in charge of the input to these controllers through the accelerator pedal (which sends its signals to the controllers), but in drivers' assistance systems such as ACC (Adaptive Cruise Control) where the speed is dependant on information coming from a radar sensor, or in ISA (Intelligent Speed Adaptation) where speed is limited according to the location of the vehicle, this input can be overridden or completely unnecessary.

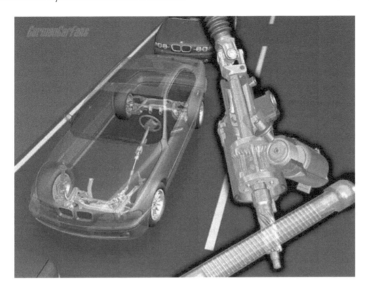

FIGURE 16.9   Active steering (BMW).

16.3.2 Brake Actuators

The first actuator which came (at least partially) under the control of electronics was the brake. In October 1978, after 14 years of development from Teldix GmbH and then Bosh which took over (and in close cooperation with Daimler which deposited the brand name ABS), luxury models from Daimler and then from BMW were equipped with an electronic control of the brake which took over some of the control from the driver in order to limit the pressure on the hydraulic circuit of each wheel in order to prevent wheel locking. This was the first robotic system introduced in a vehicle with sensors (the vehicle speed and the wheel speed), information processing, and actuation in the form of a reduction in hydraulic pressure. It is still amazing that such a complex system with about 140 components in its first generation and the possibility of catastrophic failure (complete loss of braking) was introduced in an industry known for its conservativeness. However, the gains of such a system were so obvious that 25 years later, this equipment is becoming standard in all new vehicles sold in Europe. At the same time, the weight of the hydraulic unit has decreased from 6.3 to 1.2 kg, the number of components from 140 to 16, and the memory size of the control unit has increased from 2 to 128 kb.

In 1986, ABS was improved to include the control of slip during acceleration (TCS or Traction Control System, also called ASR), this time by braking the wheel from slipping too much and at the same time controlling the engine torque through the electric throttle and engine management unit. This was a major step since the control not only reduces the braking pressure but can also increase it leading the way to the full control of braking.

In 1995, a new function was introduced to control the steering through braking when the vehicle is skidding with ESP (Electronic Stability Program). By measuring the steering wheel angle and comparing it with the rotation speed of the vehicle measured by a solid state gyrometer, the electronics control the braking force on each side of the vehicle to prevent a loss of control (mismatch between steering angle and rotational speed). This function was mounted on all Mercedes Class A vehicles from the start in 1998.

Although it is now possible to control electronically the braking of the vehicle, the braking itself is still done through hydraulic circuits putting pressure on friction pads. In the future, these hydraulic circuits are bound to disappear with electrically actuated pads, which will greatly simplify the system by removing all the fluid pipes and the hydraulic control unit.

16.3.3 Steering Actuators

Steering is probably the most critical driver input in the control of the vehicle. Any failure of this component at high speed leads to a crash. This is probably why the legislation still imposes a direct mechanical link between the driver and the front wheels of the vehicle. This is why, for a very long time, the car manufacturers were only considering bringing some assistance in the steering effort through hydraulics circuits (power steering).

In the 1990s steering assistance through electric motors came into being. This change in technology is now spreading to most of the new models because of the flexibility of the control of this assistance which was too difficult (and expensive) to obtain in the hydraulic systems and because of lower cost. This electric steering assistance is based on a torque sensor which measures the torque applied on the steering wheel by the driver and then computes an assistance which can depend on the steering angle and on the vehicle speed. However, the torque can also depend on other sensors and can therefore lead to a total control of the vehicle with the driver not intervening on the steering. Such assistance is now appearing on some Japanese models with lane-keeping assistance (based on vision) and parking assistance (with the computation of a trajectory to reach a spot indicated by the driver on a control screen).

However, on these assistance systems, there is still a constant relationship between the steering wheel angle and the angle of the front wheels. Such a constant relationship has been removed with a revolutionary system available since 2004 on some BMW models called Active Steering. The core element of the Active Steering system is an override function provided by a planetary gearbox integrated into a split steering column. Acting through a self-inhibiting gear wheel, an electric motor intervenes as required in this planetary gearbox,

FIGURE 16.10   CVT transmission (Aisin AW).

either increasing or taking back the steering angle of the front wheels. Interacting with the power steering assistance, these two components adjust the
steering angle of the front wheels and the steering forces on the steering wheel
to the respective situation on the road and the driver's requirements. In technical
terms the various functions and benefits offered by Active Steering are based
on the principle of overlapping steering angles: an electromechanical adjuster
between the steering wheel and the steering gearbox adds an additional steering
angle to the angle predetermined by the driver (see Figure 16.10). This means
that the steering ratio can be adjusted according to road conditions but also that
a correction to the steering angle can be made without the driver noticing it.
This is what is done to improve further the ESP (Electronic Stability Program)
by acting not only on the brakes but also (and mostly) on the steering improving
therefore the stability of the vehicle in difficult situations such as in high-lateral
wind or slippery surfaces.

   However, this system might just be an intermediate step before the fully
independent steering with no mechanical connection between the driver and the
front wheels (drive-by-wire). Such systems have been tested many years ago and
are actually used in modified vehicles for handicapped drivers. They have not
been generalized because of the legislation which still insists on the mechanical

FIGURE 16.11   Side stick control of a truck (Daimler-Chrysler).

link, and because of the complexity of the system for safety reasons (need for fail-safe redundant elements). However, the R&D clearly goes into this direction with several steer-by-wire systems proposed by equipment manufacturers and tested on real vehicles such as trucks and light vehicles with either conventional steering wheels (but decoupled mechanically from the front wheels) or various forms of joysticks (see Figure 16.11).

16.4 VEHICLE CONTROL

Where robotics research has faced, for so many years, the tremendous difficulty of performing intelligent tasks, perception planning, or motion that seem so natural to human beings, the automotive industry has provided the driver, since many years, with added functions that are aimed at supporting him, not replacing him. This step-by-step automation process is the incremental approach. The first smart assistance has been provided to the driver by automated controls, through the anti-blocking system (ABS), then through electronically stability program (ESP) and now with new systems like adaptive cruise control (ACC). Precrash systems, ESP with steering correction, steering control at low speed, and then at high speed, and full control for collision avoidance will certainly be the future focus areas of the automobile industry. Furthermore, the driver's planning ability as well is increased (road planning thanks to navigation systems, and parking maneuvers assistance systems) but in the future complete trajectory planning and assistance might came into being until we reach the goal which was set in the early 1990s in America with the Automated Highway System (AHS).

16.4.1 Longitudinal Control

16.4.1.1 Adaptive cruise control

The adaptive cruise control (ACC) has been the first commercial application of intelligent technology with a perception of the environment and an action on the throttle and/or the brakes. The principle of an ACC is to adjust automatically the speed of the car to ensure that a constant headway is maintained between vehicles. It operates usually between 30 and 180 km/h like a standard cruise control by controlling the accelerator thus maintaining a preset speed, but is also able to brake (decelerate) the vehicle in order to maintain a safe distance from the vehicle in front. The usual upper limit is a 2 sec headway distance starting at minimum of 1 sec.

An ACC system has to accomplish the following main tasks [32]: the object detection performed by means of exteroceptive range sensors is in charge of the detection and characterization of the objects (type/speed/range/azimuth). Most of the systems that are on the market use either millimeter radar technology or lidar technology (see sensor Section 16.2.3 and Section 16.2.4). Earlier radar-based systems suffered from poor field of view (FOV) and a near cut-off range (5 m), which limited their use in dense or slow traffic. On curves, the beam width necessitates some steering in order not to lose the target, which was not possible. Object tracking processes the data and group the collective detected objects into distinct targets, according to similar attributes (i.e., distance and relative velocity) by mean of specific filters. Kalman filters have been used in many applications and are essentially single object trackers due to the unimodality of the Gaussian distribution that is assumed. Multiple object trackers require multiple Kalman, Bayesian [33], and particle filters [34]. Some difficulties remain, for instance, the selection and deselection of a car when changing lane remains a difficult situation in terms of driver's acceptance as the system will always react slower than a human driver would, due to the different filters and the anticipation the driver has while overtaking. Some recent work has been presented in order to fuse the radar or lidar information with a video-based sensor in order to improve the tracking algorithm's robustness and reliability.

Path estimation by mean of wheel sensors (from ABS) or a yaw rate sensor estimates the roadway curvature estimation in order to identify the target with respect to the roadway curvature. Furthermore, sensing limitations occurring by obstruction (weather conditions, high border roads) or on tops of hills and bottom of valleys seriously challenge the ACC systems. Advanced systems have been presented using further information from GPS-based navigation systems, providing advance information on road topology.

One of the most popular traditional control techniques is PID (pro-portional/integral/derivative) control. In spite of its simplicity and many advantages, this type of controllers presents some shortcomings for vehicle control (longitudinal or lateral). When the control process is highly nonlinear, it is required to retune the controller parameters in order to keep the desired performance. The gain scheduling technique, which consists in embedding a table of PID control parameters, is widely used. For problems associated with noise and for which the process has time varying parameters, it is desirable to tune the parameters online, obtaining a self-tuning PID controller. Adaptive controllers popular from the 1980s for nonlinear process control often com-bine conventional control law, self-tuning, and neural networks for nonlinear parameter estimation. Acceleration controllers have been designed using fuzzy controllers and successfully implemented though they might appear difficult to tune. Sliding mode control approaches have being used as well. The controller is designed by combining constant rate control law and various switching surface. As is well known, the complex structure of controller will lead to the increase of computation time in the real-time control, and there may be a problem that the integral term of distance error employed in switching surface will attenuate the convergence rate of sliding mode control. The regulation speed controllers usu-ally use feedback or feedforward linearization and have to face the problem of varying road inclination and vehicle load that have to be estimated. For comfort reasons, limits must be set on acceleration and jerk furthering the complexity of the control algorithm.

However, ACC is only a first step toward Stop and Go and accident avoidance systems that will require to have solved braking control issues.

16.4.1.2 Precrash system/automatic emergency braking

Along with this ACC system, long studies have been conducted for the development of an active safety system to prevent collisions, using intelli-gent technologies, without any major satisfactory result up to now. Precrash system appears as a natural evolution of the comfort ACC system to safety aiming at avoiding or at least mitigating a collision impact. This function relies heavily on the detection sensors that measure the distance from which time to collision analysis is performed [8]. A commercialized system [35] (Toyota Motor Corp.) is based on a new radar sensor using electronic steering (see Figure 16.12). This unique phased array radar uses FMCW signal and allows from very short to long-range detection aiming at measuring distance for the ACC system as well as for the safety critical precrash system. For this applic-ation, some other proprioceptive sensors provide information about the host

FIGURE 16.12   Millimeter wave radar for precrash system (Denso).

vehicle kinetics. When the vehicle detects an oncoming vehicle, the precrash system control unit determines whether or not this vehicle is an obstacle based on time to collision analysis. In case the host driver or the coming vehicle begins a steering avoidance maneuver, the safety features are not activated, otherwise the system will be activated. When activated, that is, unavoidable collision is detected, the system operates the seatbelt motor and retracts seatbelts so that occupants are restrained immediately before collision (it is possible to decrease the chest deceleration by 3 to 5 G and the chest deflection by 3 to 5 mm in the event of a 55 kph collision), and increases hydraulic pressure of the brake system in accordance with the driver's braking force in order to assist and make more efficient its braking. Autonomous braking systems are under investigation, in case of an emergency situation only. An Automatic Emergency Braking (AEB) system is presented in Reference 36 (IBEO AS) with a single newly developed lidar approach. This system consists in a warning phase and a brake activation phase. AEB is an active safety application which overrules the driver in case of an unavoidable crash by an immediate full breaking. The crash is unavoidable, if an obstacle is in the driving path, the braking distance is longer than the distance to the obstacle, and no escape route exists to pass the obstacle. In Reference 37 (DaimlerChrysler) a similar system for heavy vehicles is presented in order to prevent or mitigate rear end collision of other vehicles. Usually, as soon as the sensor system recognizes a critical distance situation, the system will prefill the brake system and cause the vehicle to achieve full braking performance much earlier in the braking process. This results either in the complete avoidance of an accident or at least in reducing the impact energy by up to 50% (when braking is performed 1 sec earlier). Particularly for small vehicles with short crash zones, this additional safety function represents a beneficial enhancement of the safety equipment. However, in order to have a reliable system, most of the approaches use sensor fusion. In Reference 38 a Bayesian network-based platform is presented for sensor fusion.

16.4.1.3 Stop and go

As a next step, ACC with Full Speed Range Function, namely the Stop and Go function for low-speed area, will help the driver to keep a safe distance with the preceding vehicle at low speed. This application will assist the driver in traffic jams on highways and in urban areas and will react under speed of 40 kph on moving and stopped vehicles. The requirement of this function is a robust selection of the relevant targets in front of the vehicle from 0 to 50 m. An ACC sensor does not satisfy the requirements due to cut-off range and blind spots. Long-range ACC sensor in connection with a special near-range smart radar sensor might be a solution. Short-range radar, Optical, ultrasound, or fusion of these sensors [41] are suitable. The vehicle will be slowed down to a standstill if needed instead of turning off at 30 km/h as is presently the case. The possible traffic events that have to be correctly interpreted by the environment sensor system for highest safety are extremely complex, specially the detection of fixed obstacles without any false alarm. If the vehicle has come to a standstill, another ACC comfort function is activated: the braking system will maintain the brake pressure in order to reliably hold the car even on a slope without the driver having to apply the brakes. The control scheme of this function is different than for the ACC system, and the stopping phase is a particularly tedious issue. Fuzzy controllers have been studied [39] as well as adaptive controllers [40]. The first commercialized Stop and Go system has been launched in March 2004 for the Japanese market (Toyota Motor Corp.). The new system keeps track of the preceding vehicle at speeds of 30 km/h or lower. When the preceding vehicle stops, the system provides visual and audio warnings urging the driver to apply the brakes. If the driver does not respond in time, the system slows the vehicle to a complete stop. It thus assists the driver in stop-and-go traffic by reducing pedal work. The key to the new system lies in a broader-range laser sensor attached on the center of the front bumper for detecting vehicles ahead and improved recognition capabilities, as well as the use of a high-performance braking system that operates smoothly in low-speed ranges.

16.4.2 Lateral Control

Current systems operating a lateral control on a vehicle are aimed at keeping the vehicle from inadvertently drifting out of the lane. Using a CMOS Camera and an image processing algorithm, the system registers the position of the lane in relation to the vehicle. The "Lane Departure Warning" (LDW) system is based on a system approach, which recognizes features such as marking lines on the road with the aid of image-providing sensors. Lane recognition is one of the first image-processing applications in vehicles. This system recognises the position of the vehicle in relation to the lane markings and compares this position with the driver's intentions (which can be determined on the basis of changes in the steering angle, activation of the flasher unit, and brake pedal), can provide additional support for the driver and warn him in critical situations. In order to guarantee successful image processing in a wide variety of lighting and weather conditions, sturdy, model-based image-processing operators determine the position of the lane marking in advance at different distances. Special features such as filter lanes and motorway exits are recognized as such, in order to avoid misinterpretations. The distance between the vehicle and the lane marking thus determined is used to issue an acoustic or haptic warning to the driver, if necessary. One of the most frequent causes of accidents — unintentional drifting out of lane can thus be prevented. Possible warning alerts can be a trembling in the steering wheel, a vibrating seat, or a virtual washboard sound (a noise people recognize as generated by driving over a lane marker). As a next step, the system becomes an active lane-keeping assistant (LK), through an intervention in the steering and actively supports the driver in keeping the vehicle to the lane. However, for commercialized systems, the driver always retains the driving initiative, meaning that though he can feel the recommended steering reaction as a gentle movement of the steering wheel, his own decision takes priority at all times. One of the commonly used model for lateral control takes into account the slipping angle of the front wheels as well as the wind force [42]. Different controllers have been studied, conventional PID or adaptive, Optimal (LQ), or even fuzzy controllers. Fuzzy controllers often exhibit good properties, but the lack of theoretical foundation makes it difficult to give robustness guarantee required for such a system as an active lane-keeping system. Optimal controllers and adaptive controllers exhibit the best properties usually for this application, yet are sometimes difficult to implement, depending upon the model and the assumptions.

16.4.3 Full Vehicle Control

Some systems are already moving to a higher level of automated control in the commercial domain. Automated Bus Rapid Transit (ABRT) combines the service quality of rail transit with the flexibility of buses. A BRT system can include off-vehicle fare collection, rapid passenger loading, high-tech vehicles, dedicated lanes, modern stations, extended green lights at intersections, and more frequent service. BRT can be less expensive to develop than fixed-rail transit systems. Besides, BRT service can be tailored to serve busy urban

FIGURE 16.13   IMTS (Toyota).

corridors by using high-capacity vehicles, frequent service, and parallel local and express routes. By adding the driving automation on a BRT, the system can be made more efficient and safer, as it is already the case with automated metros.

A recent BRT from Toyota, the Intelligent Multimode Transit System (IMTS) [43] consists of vehicles navigated and controlled by magnetic markers imbedded in the middle of their dedicated roads (see Figure 16.13). The markers are embedded in the center of the track in intervals ranging between 1 and 2 m. The onboard sensor is fitted under the front axle. Therefore, the magnetic marker system contains information of both the lateral displacement of the sensor center from a magnetic marker and also the accumulative number of markers passed from the initial position. When an IMTS vehicle passes through a special marker, whose magnetic field is different from normal ones, the passing signal resets the counter to zero. The allocation of the markers along the track is planned in advance so that the counter number corresponds to the actual position along the track. All IMTS vehicles have numerical tables containing information on curvature, slope, speed restriction, and so on at every point along the track they are moving on. In this way, information on the appropriate steering angle and speed limit is available without continuous vehicle-to-road communication. Model-based control algorithm ensures easy implementation and has good control performance in general. A simple vehicle-dynamics model is commonly available and it is easy also to include

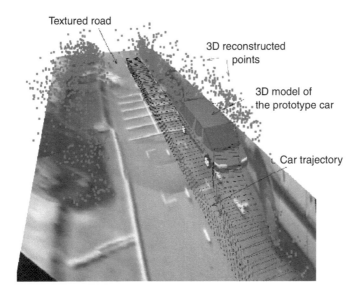

Textured road

3D reconstructed
points

3D model of
the prototype car

Car trajectory

FIGURE 16.14   3D parking assistance.

the dynamics model as well as a kinematical model describing the geometrical relation between a vehicle configuration and the track topology. However, state variable estimation, using a Kalman filter-based estimator, is necessary to apply the LQ control algorithm to the steering control as there are no sensor data available on a yaw angle. The yaw angle is that of the body direction to the tangent of the track. The platoon running function (three electronically linked vehicles run in file formation at uniform speeds) of the IMTS consists in precisely controlling the speed of all the vehicles in the platoon to be the same at all times. Since distance sensors, such as millimeter-wave radars have sensing delay, the IMTS buses use vehicle-to-vehicle communication devices to synchronize speeds.

The ABRT technology can also be found with smaller vehicles now called the cybercars for on demand door-to-door operation. These vehicles have been put in operation for the first time at Schipohl airport in December 1997 and have been the object of intense development in the last 5 years with the Cybercars project (www.cybercars.org). The long-term future of these cybercars may be with the development of dual mode vehicles and in particular for car-sharing operations with an automatic mode for operation in city centers (restricted to this type of vehicles) and a driver operated (with assistance) mode for regular infrastructures. The automobile industry is already looking at the development of such vehicles.

Already, some fully automated features are being introduced on production cars. An Intelligent Parking Assist system (IPA) has been recently introduced on

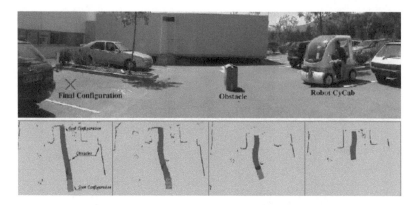

FIGURE 16.15   Autonomous navigation for a car-like robot.

FIGURE 16.16   Arrival of the winner at the DARPA Grand Challenge.

the market (Aisin), offering the ability for the vehicle to be parked without the driver using the steering wheel. The technology that automatically directs the vehicle uses an electrically operated power steering system and monitoring technology to take the vehicle to a targeted parking position defined by the driver. The latter interacts with the system by braking as needed, and by viewing images on a dashboard screen from a rear-of-vehicle positioned camera. Installed on hybrid cars, it benefits the electric motor at low speed, in reverse, that moves the car. In the future, IPA might use 3D imaging [44] (see Figure 16.14).

Future vehicles will exhibit an increasing number of drivers' assistance functions. Among them, tedious tasks will be replaced by autonomous maneuvers. Overtaking maneuvers as well as parking maneuvers [45,46] have already been studied and implemented on prototype platforms. However, planning a trajectory for a car while avoiding moving and static obstacles (see Figure 16.15) remains a challenge that is attracting a lot of attention within the robotic research community [47–49]. In 2004 and 2005, the DARPA Grand Challenge has brought together a large number of these researchers to demonstrate the feasibility of these techniques and in October 2005, several vehicles have succeeded in completing a difficult course of 132 miles in the desert in totally autonomous mode with several difficult obstacles (see Figure 16.16). However, there remains many issues to be solved before such systems become sufficiently robust and reliable for the public.

REFERENCES

1. Shoval, S. and Borenstein, J., Measurement of Angular Position of a Mobile Robot Using Ultrasonic Sensors, in *Proceedings of ANS Conference on Robotics and Remote Systems*, Pittsburgh, April 1999.
2. Fox, D. et al., Position Estimation for Mobile Robots in Dynamic Environments, In *Proceedings of the 15th National Conference on Artificial Intelligence*, Madison, Wisconsin, 1998.
3. Meinecke, M. M. and Rohling, H., Combination of FMCW and FSK Modulation Principles for Automotive Radar Systems, in *German Radar Symposium*, Berlin, 2000.
4. Honma, S. and Uehara, N., Millimeter-Wave Radar Technology for Automotive Application, Technical Reports, Mitsubishi Electric Advance, June 2001.
5. Chabert, M. et al., On the Use of High Resolution Spectral Analysis Methods in Radar Automotive, in *Workshop on Intelligent Transportation*, Germany, 2004.
6. Hoess, A. et al., Design and Realization of a Novel, Synchronized 77 GHz Radar Network for Automotive Use, in *IMS Workshop on Circuit and Antenna Technologies for Automotive Radars*, Seattle, USA, June 3, 2002.
7. Rollmann, G. and Schmid, V., 24 GHz Short Range Radar — Potential Benefits and Regulatory Issues, in *Workshop on Intelligent Transportation*, Germany, 2004.
8. Meinecke, M. M. and To, T. B., A 24 GHz Radar Based Automotive Precrash System, in *Workshop on Intelligent Transportation*, Germany, 2004.
9. Klotz, M., An Automotive Short Range High Resolution Pulse Radar Network. PhD thesis, Technische Universitat Hamburg, Hamburg, 2002.
10. Klotz, M. and Rohling, H., A high range resolution radar system for parking aid applications, in *Proceedings of 5th International Conference on Radar Systems*, Brest, France, May 1999.

11. Evans, D., Performance Analysis of Next-Generation LADAR for Manufacturing, NISTIR 7117, Building and Fire Research Laboratory National Institute of Standards and Technology, Gaithersburg, Maryland 20899, May 2004.

12. Thomas, K., Laserscanner for Automotive Applications, Germany, in *Workshop on Intelligent Transportation*, Germany, 2004.

13. Schauerte, J. et al., A 360 × 226 Pixel CMOS Imager Chip Optimized for Automotive Vision Applications, Frank, Delphi Research Labs, SAE technical paper 2001-01-0317.

14. Battiato, S., Castorina, A., and Mancuso, M., High dynamic range imaging for digital still camera: an overview, *Journal of Electronic Imaging*, 12, 459, 2003.

15. Muramatsu, S. et al., Automotive Vision Platform Equipped with Dedicated Image Processor for Multiple applications, SAE technical paper 2004-01-0179.

16. Broggi, A., Bertozzi, M., and Fascioli, A., Self-Calibration of a Stereo Vision System for Automotive Applications, in *Proceedings of IEEE International Conference on Robotics and Automation*, Seoul, Korea, 2001.

17. Zhang, Z. et al., A Robust Technique for Matching Two Uncalibrated Images Through the Recovery of the Unkown Epipolar Geometry, Research Report INRIA RR-2273, May 1994,

18. Faugeras, O., 3D Reconstruction of Urban Scenes from Sequences of Images, Research Report INRIA RR-2572, June 1995.

19. IST-2001-34410 CAMELLIA — Core for Ambient and Mobile intELLigent Imaging Applications, DELIVERABLE 3.1 — Report on Computer Vision Algorithms, July 2003.

20. URL: http://www.bev.lu

21. Chapuis, R. et al., Accurate Vision Based Road Tracker, in *Proceedings of Intelligent Vehicle Symposium*, Versailles, 2002.

22. Saito, A., Kimachi, M., and Ogata, S., Silhouette Vision: New Video Vehicle Detection Field Proven Robust and Accurate, in *Proceedings of 6th World Congress on Intelligent Transport Systems*, Toronto, November 1999.

23. Hofman-Wellenhof, B., *GPS Theory and Practice*, Lichtenegger H. Wien (ed.), New York, Springer-Verlag, 1994.

24. Dana, P., Global Positioning System Overview, Department of Geography, University of Texas at Austin, The Geographer's Craft Project Notes, 1994.

25. Gauthier, L. et al., EGNOS: The First Step of the European Contribution to the Global Navigation Satellite System, ESA Bulletin No. 105, February 2001.

26. Muellerschoen, R. J. et al., An Internet-Based Global Differential GPS System, in *ION National Technical Meeting*, Anaheim, CA, January 2000.

27. Torán-Martí, F., Ventura-Traveset, J., and de Mateo, J. C., Internet-Based Satellite Navigation Receivers using EGNOS: the ESA SISNET Project, in *NAVITEC Conference*, Netherlands, December 2001.

28. Van Dyke, K. L., The World After Selective Availability: Benefits to GPS Integrity, in *Position Location and Navigation Symposium*, San Diego, CA, USA, March 2000.

29. Roumeliotis, S. I. and Bekey, G. A., An Extended Kalman Filter for Frequent Local and Infrequent Global Sensor Data Fusion, in *SPIE International*

   Symposium on Intelligent Systems and Advanced Manufacturing, Pittsburgh, Pennsylvania, USA, October 1997.

30. Coue, C., Pradalier, C., and Laugier, C., Bayesian Programming for Multi-Target Tracking: An Automotive Application, in *International Conference on Field and Service Robotics*, Lake Yamanaka (Japan), 2003.

31. El Najjar, M. E. and Bonnifait, Ph., A Roadmap Matching Method for Precise Vehicle Localization using Belief Theory and Kalman Filtering, in *Proceedings of 11th International Conference on Advances in Robotics IEEE*, 2003.

32. Widmann, G. R. et al., Comparison of Lidar-Based and Radar-Based Adaptive Cruise Control Systems, SAE technical paper 2000-01-0345.

33. Coué, C. et al., Using Bayesian Programming for Multi-Sensor Multi-Target Tracking in Automotive Applications, in *Proceedings of the IEEE International Conference on Robotics and Automation*, Taipei (TW), September 2003.

34. Khan, Z., Balch, T., and Dellaert, F., An MCMC-Based Particle Filter for Tracking Multiple Interacting Targets, College of Computing, Georgia Institute of Technology, Atlanta, GA Technical Report number GIT-GVU-03-35, October 2003.

35. Tokoro, S. et al., Electronically Scanned Millimeter Wave Radar for Pre-Crash Safety and Adaptive Cruise Control System, in *Proceedings of Intelligent Vehicle Symposium*, Columbus, Ohio (US), 2003.

36. Dittmer, M., Automatic Emergency Breaking, in *Workshop on Intelligent Transportation*, Hamburg, 2004.

37. Schaefers, L., PROTECTOR — Emergency Braking System for Heavy Trucks, in *Proceedings of Intelligent Transportation Systems World Conference*, Madrid, 2003.

38. Kawasaki, N. and Kiencke, U., Standard Platform for Sensor Fusion on Advanced Driver Assistance System using Bayesian Network, in *Proceedings of Intelligent Vehicle Symposium*, Parma (IT), 2004.

39. Eizad, Z. and Vlacic, L., A Control Algorithm and Vehicle Model for Stop and Go Cruise Control, in *Proceedings of Intelligent Vehicle Symposium*, Parma (IT), 2004.

40. Yi, K. and Moon, I., A Driver-Adaptive Stop-and-Go Cruise Control Strategy, in *Proceedings of Intelligent Vehicle Symposium*, Parma (IT), 2004.

41. PellKofer, M. et al., Sensor Data Fusion for a Stop and Go Driving Assistant, in *Proceedings of Intelligent Transportation Systems World Conference*, Madrid, 2003.

42. Chaib, S., Netto, M. S., and Mammar, S., H-Infinity, Adaptive, PID and Fuzzy Control: A Comparison of Controllers for Vehicle Lane Keeping, in *Proceedings of Intelligent Vehicle Symposium*, Parma (IT), 2004.

43. Suzuki, T., Aso, M., and Shida, M., An Intelligent Multimode Transit System — A Fusion of Fully-Automated Platooning Buses with Existing Railroad Infrastructure, in *Proceedings of Intelligent Transportation Systems World Conference*, Madrid, 2003.

44. Bendahan, R. et al., 3D Vision System for Vehicles, in *Proceedings of Intelligent Transportation Systems World Conference*, Madrid, 2003.

45. Paromtchik, I., Planning Control Commands to Assist in Car Maneuvers, in *Proceedings of International Conference on Advanced Robotics*, Coimbra, 2003.

46. Paromtchik, I. E., Garnier, Ph., and Laugier, C., Autonomous Maneuvers of a Nonholonomic Vehicle, in *Lecture Notes in Control and Information Sciences, Experimental Robotics*, A. Casals and A. T. Almeida (eds), London, Springer-Verlag, p. 277, 1998.

47. Lefebvre, O., Lamiraux, F., and Pradalier, C., Obstacles Avoidance for Car-Like Robots, Integration and Experimentation on Two Robots, *in Proceedings of International Conference on Robotics and Automation*, Taipei, 2004.

48. Petti, S. and Fraichard, Th., Safe Motion Planning within Dynamic Environment, in *Proceedings of IEEE/RSJ International Conference on Intelligent Robots and Systems*, Edmonton (CA), 2005.

49. Kelly, A. and Nagy, B., Reactive nonholonomic trajectory generation via parametric optimal control, *International Journal of Robotics Research*, 22, 7–8, July–August, 2003.

BIOGRAPHIES

Michel R. Parent is currently the program manager at INRIA of the R&D team on advanced road transport (IMARA research group). This group focuses on research and development of new forms of road transport, in particular on fully automated vehicles (the cybercars).

Michel Parent has spent half of his time in research and academia at such places as Stanford University and MIT in the United States and INRIA in France, and the other half in the robotics industry. He is the author of several books on robotics, vision, and intelligent vehicles, and numerous publications and patents. He was the coordinator of the European Project CyberCars between 2001 and 2004.

He has an engineering degree from the French Aeronautics School (ENSAE), a Masters degree in Operation Research, and a Ph.D. in Computer Science, both from Case Western Reserve University, United States.

Stéphane R. Petti started a Ph.D. in 2003 at the Mines of Paris and INRIA in the field of robotics applied to intelligent vehicles, in collaboration with Aisin AW Co.

He entered the automotive industry in 1999, after a first experience in a fluid simulation software company. He spent 5 years with Aisin AW Europe (Belgium) as a development engineer for automatic transmission and automotive navigation systems' software.

Stéphane Petti graduated in mechanical engineering in 1997 (Affiliation of Technological University of Compiègne — France) and obtained a masters degree in mathematics in 1998 (University of Kansas — United States).

17 Intelligent Systems

Sesh Commuri, James S. Albus, and
Anthony Barbera

CONTENTS

Designing intelligent systems is a complex task requiring the integration of a diverse set of hardware and software components. This task is significantly more difficult if these systems are required to cooperate and function in a coordinated fashion. Intelligence can be defined as "the ability of a system to behave appropriately in an uncertain environment" where "appropriate behavior maximizes the likelihood of the system's success in achieving its goals [1]." Therefore, an intelligent system should be able to respond to sensory feedback at every level such that goals are achieved despite perturbations and unexpected feedback. Since intelligence responds to sensory feedback at all levels, overall effectiveness requires such "system intelligence" to be distributed in nature. Therefore, special attention must be paid to the architecture and design of such systems. Improper architecture could become a bottleneck for system performance and severely restrict the functionality of the system.

Engineering of intelligent systems should be built on scientific fundamentals and must provide designers with systematic methods to characterize the skills required for problem solving, define the behaviors that contribute to the success of the mission, represent the knowledge, and predict the future. The system must possess sensors to sense the environment and the means to effect changes in the surrounding environment. This must be done by the selection of appropriate behaviors while evaluating the cost and benefit of the action.

In this chapter, the implementation of distributed intelligence for teams of unmanned ground vehicles (UGV) is examined. The system requirements are analyzed and the implication of these requirements on hardware and software is discussed in Section 17.1. A brief background of existing solution methodologies from the literature is presented in Section 17.2. The 4D/(real-time control system) RCS architecture developed at the National Institute of Standards and Technology (NIST) and the implementation methodology are discussed in Section 17.3, and examples illustrating the implementation of RCS reference model architecture at NIST, the University of Oklahoma (OU), and Oklahoma State University (OSU) are presented in Section 17.4. Recent trends in the hardware–software codesign and hardware reconfiguration are presented and the direction of future research is addressed in Section 17.5. The conclusions are summarized in Section 17.6.

17.1 ARCHITECTURAL REQUIREMENTS FOR INTELLIGENT UGVs

An intelligent system functions by sensing the environment, perceiving and evaluating situations, modeling the world, and choosing behavior that is appropriate for the perceived situation. Perception involves recursive estimation to detect, track, measure, and classify objects, events, and situations. World modeling uses simulation and modeling techniques to generate expectations and predict results of hypothesized actions. Planning consists of a combination of case- and search-based techniques. Case-based methods may be used to limit range and resolution in the space of potential behaviors, and search-based methods used to optimize behavior within that limited space.

In general, an intelligent system has a hierarchical structure wherein global percepts are used to generate long-range plans at the higher levels, while simultaneously short-term actions are generated in response to local sensory information at lower levels. At each level, commands, goals, constraints, and priorities are passed downward from higher levels, while feedback from lower levels is filtered, generalized, classified, and used for planning and control.

Designing intelligent UGVs is substantially more complex than just the integration of smart components [2,3]. Embedding intelligence into systems requires a new design paradigm that takes into account the hardware and software complexities involved in the design of these systems [1,4]. The design must address issues such as:

- The ability of the team to negotiate obstacles, satisfy formation constraints like size and shape, and reach a goal destination.
- The ability to demonstrate desired group behaviors.
- The ability to dynamically dilate/contract the formation and the retasking of individual robots while in formation.
- The ability to share information between multiple robots for generation of multi-dimensional multi-resolutional maps etc.
- The efficiency of fault tolerance and learning algorithms.

The design must also address real-time and nonreal-time issues in sensor fusion, communication between network nodes, and the automation of the decision-making process.

17.2 BACKGROUND ON INTELLIGENT SYSTEMS

There are a number of architectures proposed for the development of intelligent systems. The application of Artificial Intelligence (AI) concepts to problem solving was examined in the development of SOAR architecture [5,6]. Swarming phenomena in nature also inspired the development of architectures that facilitated cooperative behavior in systems [7]. The need to implement fault tolerant systems motivated the development of behavior-based software architecture called Alliance [8]. Subsumption, another behavior-based architecture was proposed in References 9 and 10. While the SOAR architecture deals with the critical elements of intelligent systems like states, goals, plans, agents, behaviors, knowledge representation, it does not incorporate the concept of time, thus making its use in real-time robotic systems difficult [1]. Subsumption on the other hand, is primarily reactive in nature and does not model planning and problem solving capabilities that are crucial to the implementation of intelligent UGV systems.

Hybrid architectures embodying the benefits of both the real-time reactive and the long-term deliberative behaviors have been proposed for specific applications. For example, the AuRA [11,12] architecture embodies a high-level deliberative hierarchical planner with a low-level reactive controller based on schema theory. Control of a fleet of mobile robots was accomplished using the Multiple Resource Host Architecture (MRHA) where the focus was on planning paths and issues commands to the robots in the fleet. Atlantis, a three-level hierarchical architecture, was developed at JPL for the Mars rover project [13].

Many recent methods for controlling multi-robot systems focus on the software architecture of the system [14–16]. Efficient use of resources was addressed using centralized planning in the 3T architecture [17,18]. In contrast, the requirement for loss tolerance has led many researchers to consider distributed systems (e.g., [8,19–21]). More recently, attention has been given to designing architectures that combine the advantages of both centralized and distributed approaches (e.g., [21,22]). Distributed systems make possible the development of sophisticated systems with complex behaviors. In such systems, time-critical behaviors of the system can be implemented locally, while generalized system-level behaviors can be abstracted out and implemented on a central resource that communicates with all the distributed nodes in the system. Such implementations encourage modularity in the design and facilitate fault-tolerant design. Crucial in the design of such systems is the selection of the appropriate hardware and software components and the architecture for their integration.

Currently there exist a wide variety of intelligent system components. There are hundreds of perception algorithms, world modeling algorithms, representation schema, reasoning engines, decision theory formulae, expert systems tools, planning algorithms, hybrid systems, and control methods. While the component technologies have matured in the past few years, the design of intelligent systems is not a matter of simple system integration. Often, it is necessary to design these systems ground-up in order to meet the overall requirements [1,4]. In the next section, we will present a reference model architecture, called the 4D/RCS, and a methodology for building systems that comply with this architecture. This model architecture provides a formal method for integrating all of these components into a coherent whole that is able to exhibit intelligent behavior. The methodology presented provides the engineering discipline for designing, coding, testing, and upgrading the software embedded in the system, as well as engineering specifications for hardware configuration.

17.3   4D/RCS ARCHITECTURE AND METHODOLOGY

17.3.1   4D/RCS Architecture

Engineering of intelligent systems requires a reference model architecture, and a methodology for building systems that comply with that architecture. An architecture is a framework consisting of functional modules, interfaces, and data structures. A reference model architecture defines how the functional modules and data structures are integrated into subsystems and systems. The architecture represents a framework wherein issues such as the network connectivity, latency, bandwidth, reliability, and communication between modules that affect the system performance can be addressed. Therefore, the reference model architecture for intelligent systems can be viewed as providing an infrastructure for representing knowledge about the environment, the mission, tasks, plans, schedules, intentions, priorities, beliefs, and values. It also provides infrastructure for perception, attention, and cognition, including methods for reasoning, modeling, planning, and learning. For the architecture to be feasible, it should also define the required human interfaces, including displays and controls, simulation and training environments, and programming and debugging tools.

Real-Time Control System evolved from the bottom up as a real-time intelligent control system for real machines operating on real objects in the real world. Initially, RCS was proposed to address the real-time goal-directed control of sensory-interactive laboratory robots [23]. Since then, the architecture has been refined and applied to a variety of problems including intelligent manufacturing systems, industrial robotics, automated general mail facilities, automated stamp distribution systems, automated mining equipment, unmanned underwater vehicles, and unmanned ground vehicles [24,25]. The most recent version of RCS, that is, 4D/RCS, embeds elements of Dickmanns [26,27] 4D approach to machine vision within the RCS control architecture. 4D/RCS was designed for the U.S. Army Research Lab AUTONAV and Demo III Experimental Unmanned Vehicle programs [28] and has been adopted by the Army Future Combat System program for Autonomous Navigation Systems [1,29,30].

A block diagram of a 4D/RCS reference model architecture is shown in Figure 17.1. The 4D/RCS architecture consists of a multi-layered multi-resolutional hierarchy of computational nodes, each containing elements of sensory processing (SP), world modeling (WM), value judgment (VJ), behavior generation (BG), and a knowledge database (KD) (included in the WM in Figure 17.1). Each node in the architecture represents an operational unit in an organizational hierarchy.

Figure 17.2 shows a first level of detail in a typical 4D/RCS node. In each node, a behavior generation process accepts task commands with goals and parameters from a behavior generation process at the next higher level and issues commanded actions with subgoals and parameters to one or more behavior generation process at the next lower level. (Solid lines indicate normal data pathways. Dotted lines indicate channels by which an operator can peek at data or insert control commands whenever desired.) Figure 17.3 shows a second level of detail in 4D/RCS nodes. Each node contains both a deliberative and a reactive component. Bottom-up, each node closes a reactive control loop driven by feedback from sensors. Top-down, each node generates and executes plans designed to satisfy task goals, priorities, and constraints conveyed by commands from above. Within each node, deliberative plans are merged with reactive behaviors [30].

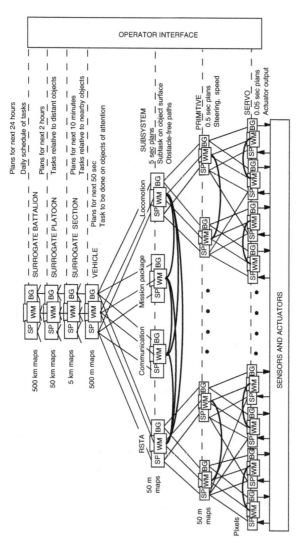

FIGURE 17.1 A 4D/RCS reference model architecture for autonomous ground vehicles. Processing nodes are organized such that the BG processes form a command tree. Information in the knowledge database is shared between WM processes in nodes above, below, and at the same level within the same subtree. On the right, are examples of the functional characteristics of the BG processes at each level. On the left, are examples of the scale of maps generated by the SP processes and populated by the WM in the KD at each level. Sensory data paths flowing up the SP hierarchy typically form a graph, not a tree. VJ processes are hidden behind WM processes in the diagram. A control loop may be closed at every node. An operator interface may provide input to, and obtain output from, processes in every node (Numerical values are representative examples only. Actual numbers depend on parameters of specific vehicle dynamics.)

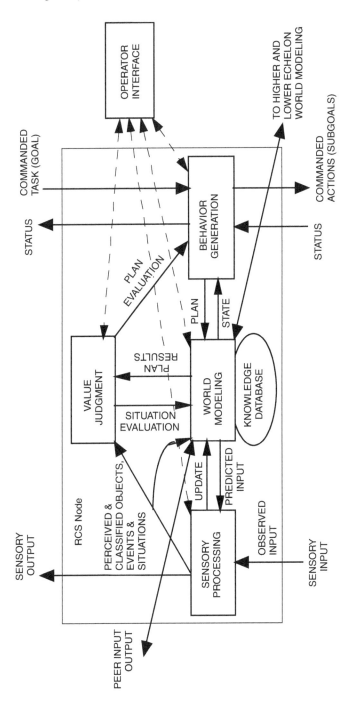

FIGURE 17.2 Internal structure of a 4D/RCS node. Behavior generation plans and executes actions. Sensory processing transforms sensor data into perceived and classified objects, events, and situations. World modeling maintains the knowledge database and generates predictions — both for SP recursive estimation and BG planning. Value judgment evaluates perceived objects, events, and situations for the world model and evaluates the predicted results of simulated plans for behavior generation.

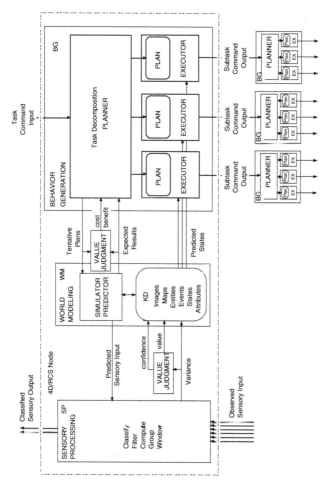

FIGURE 17.3 A typical 4D/RCS computational node. Task command input comes from a higher-level BG process in the 4D/RCS hierarchy. Each input task command is decomposed into a plan consisting of subtasks for subordinate BG processes. A WM process maintains a KD that is the BG unit's best estimate of the external world. A SP process operates on input from sensors by windowing (i.e., focusing attention), grouping, computing attributes, filtering, and recognizing entities, events, and situations. A VJ process evaluates expected results of tentative plans. A VJ process also assigns confidence and worth to entities, events, and situations entered into the KD.

Each BG process accepts tasks and plans and executes behavior designed to accomplish those tasks. The internal structure of the BG process consists of a planner and a set of executors (EX). At the upper right of Figure 17.3, task commands from a supervisor BG process are input. A planner module decomposes each task into a set of coordinated plans for subordinate BG processes. For each subordinate there is an executor that issues commands, monitors progress, and compensates for errors between desired plans and observed results. The executors use feedback to react quickly to emergency conditions with reflexive actions. Predictive capabilities provided by the WM may enable the executors to generate preemptive behavior.

Plans may be generated by any of a great variety of planning algorithms, for example, case-based reasoning, search-based optimization, or schema-based scripting. The RCS software engineering methodology has been developed on the strength of many different applications that have been implemented over the past 35 years using the RCS reference model architecture.

17.3.2 4D/RCS Methodology

The fundamental premise of the RCS software engineering methodology is that at each point in time, the task state (i.e., where the system is, where it is going, what it is doing, what the goal is, and what the constraints are) collectively define the requirements for all of the knowledge in the knowledge database (both procedural and declarative), and specifies the support processing required to acquire and maintain the knowledge database. In particular, the task state determines what needs to be sensed, what world objects, events, and situations need to be analyzed, what plans need to be generated, and what task knowledge is required to do so [31,32]. An example of the RCS methodology for designing a control system for a tactical behavior such as route reconnaissance is shown in Figure 17.4.

The RCS methodology consists of the following six steps:

Step 1 consists of an intensive analysis of domain knowledge derived from training manuals and subject matter experts. Scenarios are developed and analyzed for each task and subtask. The result of this step is a structuring of procedural knowledge into a task decomposition tree with simpler and simpler tasks at each echelon. At each echelon, a vocabulary of commands (i.e., action verbs with goal states, parameters, and constraints) is defined to evoke task behavior at each echelon.

Step 2 defines a hierarchical structure of organizational units that will execute the commands defined in step 1. For each unit, duties and responsibilities in response to each command are specified. This is analogous to establishing a work breakdown structure for a development

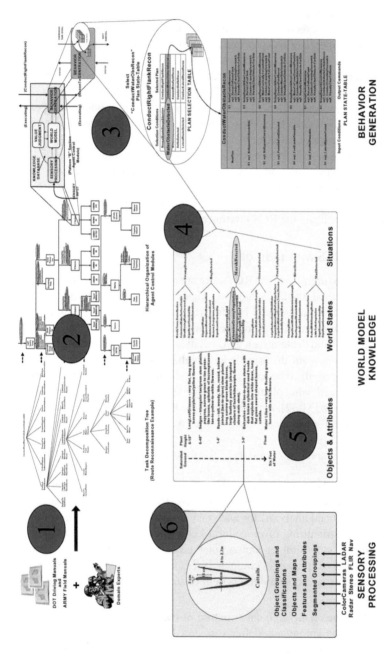

FIGURE 17.4 Six steps of the RCS software engineering methodology.

project, or defining an organizational structure for a business or military unit.

Step 3 specifies the processing that is triggered within each unit upon receipt of an input command. For each input command, a state-graph (or state-table, or extended finite state automaton) is selected that provides a plan (or procedure for making a plan) for accomplishing the commanded task. The input command thus selects (or causes to be generated) an appropriate behavior (which may be encoded as a state-table), the execution of which generates a series of output commands to units at the next lower echelon. The result of step 3 is that each organizational unit has, for each input command, a state-table of production rules that identify all the task branching conditions and specify the corresponding state transition and output command parameters. Task branching conditions may include the state of the task, the internal state of the vehicle, the state of objects of attention in the external world, and situational relationships between and among them.

Step 4 analyzes each of the branching conditions defined in step 3 to reveal dependencies on world states and situations. This step identifies the detailed relationships between entities, events, and states of the world that cause each state or situation to be true.

Step 5 identifies and names all of the world model entities and events along with their attributes and relationships that are relevant to detecting the world states and situations.

Step 6 uses the context of world states and situations to establish the distances, and timing requirements of various behaviors. From these, the resolution, speed, and stability requirements of sensors can be determined to enable the relevant entities, events, and situations to be measured and recognized. This then defines a set of requirements and/or specifications for sensor systems to support each subtask activity.

17.3.3 Representing Knowledge in 4D/RCS

The 4D/RCS architecture is designed to accommodate multiple types of knowledge representation formalisms, and provide an elegant way to integrate these formalisms into a common unifying framework. This section will describe the types of knowledge representations that have been researched or implemented within the 4D/RCS architecture for autonomous driving and the mechanisms that have been deployed to integrate them.

The hierarchical structure of 4D/RCS supports knowledge representation with different range and resolution, and different levels of abstraction, at the various echelons of control. It should be noted, however, that the 4D/RCS reference model actually contains three different hierarchies:

1. A task hierarchy consisting of a chain of command with echelons of control
2. A hierarchy of range and resolution of signals, images, and maps in space and time
3. A hierarchy of abstraction in representation of entities and events

Echelons of control are defined by decomposition of tasks into subtasks and the assignment of task skills and responsibilities to organizational units in a chain of command. Range and resolution of signals, images, and maps are defined by sampling interval and field of regard over space and time (e.g., pixel size and field of view in images, scale and size of maps, and sampling frequency of signals). Levels of abstraction are defined by grouping and segmentation algorithms that operate on the geometry of entities (e.g., points, lines, vertices, surfaces, objects, groups) and the duration of events (e.g., milliseconds, seconds, minutes, hours, days). These three hierarchies are related, but not congruent. For example, the range and resolution of maps are related to echelons of control by speed and size of the system being controlled. Resolution of images is related to spatial dimension by magnification. Resolution of maps is related to spatial dimension by scale. Pixels in images are related to pixels on maps by transformation of coordinates.

The RCS methodology begins with the task decomposition hierarchy that defines echelons of control. Task timing and system speed and size then determine range and resolution of images and maps. Levels of abstraction are determined by the logical requirements of task decomposition. Different system requirements will produce different relationships between these three hierarchical representations.

Typically, knowledge in 4D/RCS nodes at the lowest echelon of the control hierarchy consists of signals, images, and state variables, such as vehicle position, orientation, velocity, and acceleration; actuator positions, velocities, and forces; pressure sensor readings; position of switches and gearshift settings. Knowledge in nodes at the second echelon and above consists of map-based information, with decreasing resolution and increasing spatial extent at each higher echelon in the hierarchy. Maps are used to represent the size, shape, location, surface orientation, and roughness of terrain features and regions of interest. Knowledge in nodes at the third echelon and above contain both map-based representations and abstract data structures that represent named entities such as road edges and obstacle surfaces, along with their attributes and pointers that represent spatial and temporal relationships and class membership. Image and map representations are linked to abstract data structures by pointers that represent relationships such as "is_a" and "belongs_to." Back pointers indicate

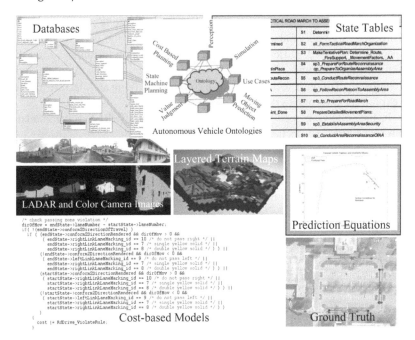

FIGURE 17.5   Types of knowledge representation in 4D/RCS.

where abstract entities are located in the map or image representations. Higher echelons represent information about the location, motion, and attributes of objects such as other vehicles, roads, intersections, traffic signals, landmarks, and terrain features such as buildings, roads, woods, fields, streams, fences, and ponds. The upper level echelons represent knowledge about groups of objects such as groups of people, groups of buildings, and road networks. Group attributes such as size, shape, and density are computed over the group. At each echelon, pointers define relationships between entities and events in situations.

Within each node, knowledge is stored within a knowledge database (KD) consisting of data structures that contain the static and dynamic information that collectively form a model of the world. Each node contains knowledge with the range, resolution, and level of abstraction required to support the behavior generation, sensory processing, and value judgment processes within that node. This includes a best estimate of the current state of the world relevant to the current task assigned to that node, plus world model parameters that define how the world state can be expected to evolve in the future under a variety of circumstances.

Figure 17.5 shows the many different types of knowledge representation formalisms that are currently being implemented within the 4D/RCS architecture to support autonomous driving. These formalisms range from iconic to symbolic and from procedural to declarative. Knowledge is captured in formalisms and at levels of abstraction that are suitable for the way in which it is expected to be used. Different knowledge representation techniques offer different advantages, and 4D/RCS is designed in such a way as to combine the strengths of all of these techniques into a common unifying architecture in order to exploit the advantages of each.

In the following sections, we will describe how 4D/RCS accommodates both procedural and declarative knowledge.

17.3.3.1 Procedural knowledge

Procedural knowledge is the knowledge of how to perform tasks. Procedural knowledge can be captured in task frames. A task frame is a data structure specifying all the knowledge necessary for accomplishing a task. A task frame is essentially a recipe consisting of a task name, a goal, a set of parameters, a list of materials, tools, and procedures, and set of instructions of how to accomplish a task. For each task that an RCS node is able to perform, there exists a task frame.

A task frame may include:

1. *Task name* (index into the library of tasks the RCS node can perform). The task name is a pointer or an address in a database where the task frame can be found.
2. *Task identifier* (unique identity for each task command). The task identifier provides a means for keeping track of tasks in a queue.
3. *Task goal* (a desired state to be achieved or maintained by the task). The task goal is the desired result of executing the task.
4. *Task goal time* (time at which the task goal should be achieved, or until which the goal state should be maintained).
5. *Task objects* (on which the task is to be performed). Examples of task objects include parts to be machined, features to be inspected, tools to be used, targets to be attacked, objects to be observed, sectors to be reconnoitred, vehicles to be driven, weapons or cameras to be pointed.
6. *Task parameters* (that specify, or modulate, how the task should be performed). Examples of task parameters are speed, force, priority, constraints, tolerance on goal position, tolerance on goal time, tolerance on path, coordination requirements, and level of aggressiveness.
7. *Agents* (that are responsible for executing the task). Agents are the subsystems and actuators that carry out the task.
8. *Task requirements* (tools needed, resources required, conditions that must obtain information needed). Tools may include instruments,

sensors, and actuators. Resources may include fuel and materials. Conditions may include temperature, pressure, weather, visibility, soil conditions, daylight, or darkness. Information needed may include the state and type of parts, tools, and equipment, or the state of a manufacturing process, or a description of an event or situation in the world.

9. *Task constraints* (upon the performance of the task). Task constraints may include speed limits, force limits, position limits, timing requirements, visibility requirements, tolerance, geographical boundaries, or requirements for cooperation with others.

10. *Task procedures* (plans for accomplishing the task, or procedures for generating plans). Plans may be prepared in advance and stored in a library, or they may be computed online in real-time. Task procedures may be simple strings of things to do, or may specify contingencies for what to do under various kinds of circumstances.

11. *Control laws and error correction procedures* (defining what action should be taken for various combinations of commands and feedback conditions). These typically are developed during system design, and may be refined through learning from experience.

Some of the slots in the task frame are filled by information from the command. Others are properties of the task itself and what is known about how to perform it. Still others are parameters that are supplied by the WM.

The task procedures (slot #10 in the task frame) consist either of plans, or planning procedures for generating plans.

Plans. In general, plans can be represented as state-tables (or state-graphs). State-tables and state-graphs are duals. The advantage of the state-graph representation is that behavior can easily be visualized. The advantage of the state-table representation is that state-tables can be directly executed by an extended finite-state automata so as to generate a sequence of output commands designed to accomplish the task goal within the constraints specified in the task frame. A state-table (or corresponding state-graph) may contain as many state-dependent branching conditions as necessary to cover the space of things that the system is capable of doing in response to situations represented in the system's world model.

Both state-graph and state-table representations can easily be changed by adding or modifying rules at any node in the state graph, or by adding nodes to the graph. This means that a system can learn and eventually optimize its performance. What is required for learning is for an expert critic to point out where in the state-graph the system should have performed differently, what piece of information in the system's world model should have been used to trigger the different behavior, and what the different behavior should have been. This is the type of information typically supplied to a human student by a human instructor or teacher.

Planning. Plans are the result of planning procedures. Planning may be performed off-line by human programmers, or in real-time by the intelligent machine itself or its human operator. In 4D/RCS, planning is distributed throughout the architecture. Each RCS node has its own autonomous planner, but is able to accept plans from an outside source such as a human operator.

There are two methods for generating plans that are currently being implemented in 4D/RCS.

Case-based planning. Case-based planning uses situation-action logic as the primary method of task decomposition. Case-based planning uses a library of plans represented as state-graphs (or state-tables) with at least one plan for each task command. There are typically two types of case-based planning: one that decomposes a task into job assignments for multiple agents, and a second that decomposes each agent's job assignment into a schedule that may or may not be coordinated with peer agents. Slots for parameters such as priorities, constraints, tolerances, modes, and speeds may be filled in at planning time.

Search-based planning. Search-based planning performs a search over the space of possible actions to find the "best" course of action to achieve the goal. Search-based planning may use a map or spatial graph with cost overlays to evaluate various possible paths through the map or graph. Typically, this requires an action model that predicts how each planned action will affect the system state at the appropriate level of resolution. Alternatively, search-based planning may use an inverse model that predicts what actions are required to generate a sequence of desired states. In either case, the planner generates a series of planned actions and resulting states that predict how the system is expected to behave in the real world environment. These simulated actions and predicted states can then be evaluated by a cost-function that takes into account constraints and priorities that are passed down from higher levels, as well as uncertainties and knowledge about the environment from sensors. A typical cost-function for an autonomous ground vehicle may take into account the cost of fuel or time, the difficulty of terrain, the estimated cost of collision with various types of objects, the risk of detection or attack by an enemy, and the benefit or payoff of achieving or maintaining a goal.

Search-based planning is a widely researched field. A variety of planning techniques have been reported in the literature. One algorithm that has proven particularly efficient is the incremental creation and evaluation of the planning graph [33]. This incremental approach reduces the number of planning-graph nodes that must be created and evaluated to find the cost-optimal path through the planning graph.

17.3.3.2 Declarative knowledge

Declarative knowledge is represented in a format that may be manipulated, decomposed, and analyzed by reasoning engines. Declarative knowledge may describe the size, shape, state (i.e., position, orientation, velocity), and class of entities. It may describe the start-time, duration, frequency, or temporal pattern of events. Declarative knowledge may also describe the spatial or temporal relationships that exist between and among entities and events in places and situations.

Declarative knowledge enables a system to know the current state of the environment and its own situation relative to other entities in the environment. Declarative knowledge enables a system to reason logically or mathematically to predict what will result from possible future actions and events. Two types of declarative knowledge that are captured within 4D/RCS are symbolic knowledge and iconic knowledge.

Symbolic Knowledge. Symbolic knowledge representations use abstract data structures to represent things (e.g., actions, entities, or events) in the world that can be referenced by name. Two types of symbolic representations that are being implemented within 4D/RCS are ontologies and relational databases.

Ontologies represent key concepts, their properties, their relationships, and their rules and constraints within a given domain. Two efforts have focused on the development of ontologies for autonomous navigation.

The first is an ontology of objects that may be encountered during on-road driving. This ontology will be used to estimate the damage that would be incurred by collisions with the different objects under a variety of conditions. Automated reasoning is used to estimate collision damage and compare it with the cost and risk of evasive action. This enables the real-time path planner to decide what is required to avoid the object, or whether it would be better to simply collide with the object rather than slam on the brakes or swerve violently to miss it. More information about this effort can be found in Reference 34.

The second is an ontology of tactical behaviors that is being implemented using the OWL-S specification (Web Ontology Language-Services) [35]. In this context, behaviors are actions that an autonomous vehicle would be expected to perform upon being confronted with a tactical situation. This ontology is stored within the 4D/RCS knowledge database, and the behaviors will be triggered when situations in the world are perceived to exist. More information about this effort can be found in Reference 36.

In addition to ontologies, relational databases have been developed to house symbolic information. For example, our intelligent vehicle world model contains a Road Network Database [37]. This database contains slots for detailed information about the roadway, such as where the road lies, rules dictating the traversal of intersections, lane markings, road barriers, road surface characteristics, etc. The purpose of the Road Network Database is to provide the data structures necessary to capture the information about road networks needed by a planner or control system to plan routes along the roadway at each level of abstraction.

Each echelon of planning requires knowledge at a different level of abstraction. To accommodate these requirements, the Road Network Database is hierarchically organized with each echelon in the road network hierarchy conceptually centered on the vehicle. At the lowest echelon, the Road Network Database represents information at a range and resolution that a low-level planner can use to plan trajectories to navigate a vehicle over the next few meters. At the highest echelon, the Road Network Database represents information at a range and resolution that a high-level planner can use to plan a trip across the country.

Iconic Knowledge. Iconic knowledge models objects and situations in space and time in a manner that directly represents spatial and temporal relationships (e.g., images, maps, and state trajectories.) Iconic representations typically use scalars, vectors, or arrays to represent things that can be measured (e.g., attributes) about the world. Iconic representations are typically referenced by location. The location of each element in an iconic representation often corresponds to (or projects onto) a dimension or location in physical space. For example, the location of a pixel in an image corresponds to a geometrical projection of the world onto the image, and vice versa. The location of an axon in a nerve bundle depends on the location of a tactile sensor on the skin. The contents of each element of the array may contain a Boolean or real number representing the value of a physical attribute such as light intensity, color, altitude, range, or density at that point in the array. The contents of each element may also contain numbers representing the values of spatial or temporal gradients of intensity, color, and range; or of image flow direction and magnitude. The contents of an element may also be a pointer to a symbolic data structure representing an entity (e.g., an edge, vertex, surface, object, or group) to which the pixel belongs. If the arrays of these various attributes and pointers are registered (as shown in Figure 17.6), the result is a three-dimensional (3D) matrix of attribute, entity, class, and worth images.

Iconic representations have scale, and are limited in range and resolution. Both images and maps have a finite number of pixel elements. Images have limited fields of view and maps have boundaries. Similarly, temporal events have a beginning and an end, and can only be sampled at a finite rate. Examples of iconic knowledge currently used within 4D/RCS include maps and images. Maps may be expressed in a variety of formats including survey and aerial maps, or digital terrain elevation databases (DTED) containing information about hydrology, ground cover, roads, bridges, streams, woods, and buildings. Images include video or LADAR images from cameras mounted on the ground vehicle. To be useful for path planning beyond line of sight, the information

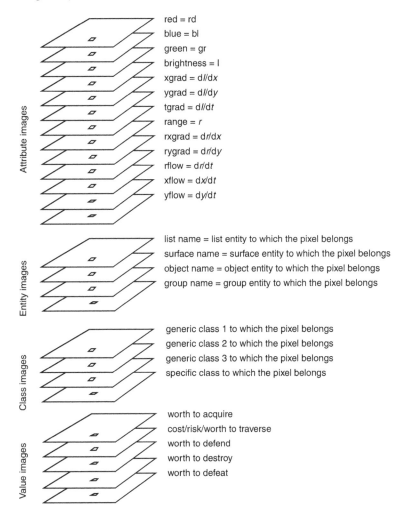

red = rd
blue = bl
green = gr
brightness = l
xgrad = d*l*/d*x*
ygrad = d*l*/d*y*
tgrad = d*l*/d*t*
range = *r*
rxgrad = d*r*/d*x*
rygrad = d*r*/d*y*
rflow = d*r*/d*t*
xflow = d*x*/d*t*
yflow = d*y*/d*t*

list name = list entity to which the pixel belongs
surface name = surface entity to which the pixel belongs
object name = object entity to which the pixel belongs
group name = group entity to which the pixel belongs

generic class 1 to which the pixel belongs
generic class 2 to which the pixel belongs
generic class 3 to which the pixel belongs
specific class to which the pixel belongs

worth to acquire
cost/risk/worth to traverse
worth to defend
worth to destroy
worth to defeat

FIGURE 17.6 Attribute, entity, class, and value images. These images are registered to form a 3D matrix such that each pixel has an attribute vector, a pointer to an entity frame, one or more pointers to the class(es) to which the pixel belongs, and values assigned to the region where the pixel is located.

gathered by sensors on the ground vehicle must be registered with a priori maps generated from external sources.

A hybrid iterative algorithm has been developed for registering 3D LADAR range images obtained from unmanned aerial vehicles with LADAR images obtained from unmanned ground vehicles [38]. Registration of the UGV LADAR to the aerial survey map minimizes the dependency on GPS for position estimation. This is important when GPS estimates are unreliable or unavailable.

Perception. Perception is the intelligent system's window into the world. Perception begins with sensing and ends with a world model that contains information that is relevant to the task at hand and adequate for whatever decision making and planning is required to generate successful behavior. In biological creatures, perception is a hierarchical process that starts with arrays of tactile sensors in the skin, arrays of photon sensors in the eyes, arrays of acoustic sensors in the ears, arrays of inertial sensors in the vestibular apparatus, arrays of proprioceptive sensors (that measure position, velocity, and force) in the muscles and joints, and a variety of internal sensors that measure chemical composition of the blood, pressure in the circulatory system, and several other modalities. In machine vision, image understanding begins with signals from one or more cameras and ends with a world model consisting of data structures that include a registered set of images and maps with labeled regions, or entities, that are linked to each other and to entity frames that contain entity attributes (e.g., size, shape, color, texture, temperature), state (e.g., position, orientation, velocity), class membership (e.g., road, lane marker, tree, vehicle, pedestrian, building), plus a set of pointers that define relationships among and between entities and events (e.g., situations).

It should be noted that, contrary to popular opinion, perception does not reduce a large amount of sensory data to a few symbolic variables that are then used to trigger the appropriate behavior. Instead, perception increases and enriches the sensory data by computing attributes and combining it with a priori information so that the world model contains much more information (not less) than what is contained in the sensory input.

Perception is an active, goal-driven process that focuses attention on those parts of the world that are important, and masks out (or assigns to the background) those that are irrelevant. Thus, perception does not treat all regions on the egosphere equally. The role of attention is to focus perceptual resources on what is important for achieving current and near future task goals.

In 4D/RCS, perception generates a hierarchy of image entities and entity frames. These are linked to a hierarchy of maps with differing range and resolution. It should be noted, however, that the hierarchy of range and resolution for maps is not parallel to the hierarchy of image entities and entity frames. The hierarchy of entities is generated by grouping and segmentation processes at each level of the SP hierarchy. The hierarchy of range and resolution of maps is specified by the planning horizon of the behavior generation processes in the BG hierarchy that use the maps. This can be seen in Figure 17.7, where distant objects in the image may appear only in the section echelon map, whereas close objects in the image appear magnified in the primitive echelon map.

There are five sensory processing steps at each level in the SP hierarchy in Figure 17.7.

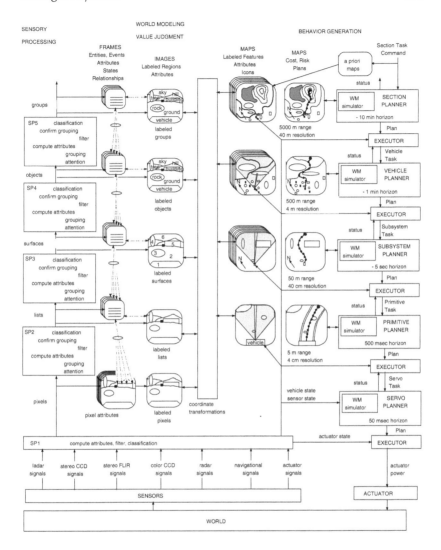

FIGURE 17.7 Five layers of the 4D/RCS architecture developed for the Demo III program. On the far-right are planner and executor modules. In the center-right are maps for representing terrain features, roads, bridges, vehicles, friendly/enemy positions, and the cost and risk of traversing various regions. On the far-left are five levels of sensory processing. At each level there are symbolic frames representing entities, segmented images with labeled regions, and links between the images and frames. In the center are coordinate transforms that use range information to assign pixels and regions in images on the left to regions in maps on the right. The maps have range and resolution defined by the planning horizon of the planners at each echelon. The images have resolution and field of view defined by the imaging optics.

The *first step* is to focus attention. At the lowest level in the SP hierarchy, focusing attention means pointing the high resolution part of the visual field toward those regions of the world that contain information important to the task. At higher levels, focusing attention means that SP computing resources are committed to regions in the image that are important to the task, while remaining regions are largely ignored.

The *second step* at each level is to group portions of the visual field that belong together into entities, and to segment each entity from the rest of the image. At the lowest level, grouping consists of integrating all the energy imaged on each single pixel. At higher levels, grouping pixels or entities according to gestalt heuristics such as proximity, similarity, contiguity, continuity, or symmetry. This step also establishes pointers between segmented regions in the image and entity frames that contain knowledge about the entities. Each grouping operation is a gestalt hypothesis.

The *third step* is to compute attributes and state of each entity, and store this information in the corresponding entity frame. Attributes may include size, shape, color, texture, and temperature. State includes position, orientation, and velocity.

The *fourth step* is recursive estimation on entity attributes to confirm or deny the gestalt hypothesis that created the entity in step two. Recursive estimation uses entity state and state prediction to track entities from one image to the next.

The *final step* is to compare confirmed entity attributes with attributes of class prototypes. When a match occurs, the entity can be assigned to the class. Once an entity has been classified, it inherits attributes of the class. There is a hierarchy of classes to which an entity may belong. For example, an entity may be classified as a geometrical object, as a tree, as an evergreen tree, as a spruce tree, and as a particular spruce tree. More computing resources are required to achieve more specific classifications. Thus, an intelligent system typically performs only the least specific classifications required to achieve the task.

17.4 EXPERIMENTAL RESULTS

In this section, the experimental validation of the RCS reference model architecture to the control of autonomous ground vehicles is presented. The RCS architecture is very flexible and can easily be adapted for control of multi-robot teams. This architecture is also well suited for incorporating learning algorithms at all levels of the hierarchy and to address the needs of multi-robot teams like dynamic sizing of the teams, retasking, etc. In the first example, the adaptation of 4D/RCS methodology by the research team at the OU and OSU for controlling teams of robots is addressed. In the second example, the validation is provided through the performance of the Army Demo III Experimental Unmanned Vehicle at the NIST.

17.4.1 Implementation of Reconfigurable UGV Teams at the OU

Conventional design solutions based on existing architectures cannot adequately address the demands of system intelligence. These requirements are exemplified in the control and coordination of multi-robot teams where the system needs to exhibit higher levels of fault tolerance, adaptability, and task and mission reconfigurability. In this example, we will demonstrate the flexibility of the 4D/RCS architecture for the implementation of reconfigurable UGV teams. Central to the implementation in this example is the architecture, called Adaptation and Learning at all Levels (AL^2) that is designed to exploit the recent advances in hardware and software reconfiguration. This work is motivated by the 4D/RCS architecture and is supported by a grant from the Army Research Office, U.S. Department of Defense.

17.4.1.1 AL^2 architecture

The implementation of intelligent UGVs at the OU and OSU is based on the 4D/RCS architecture and is influenced by the need for open architectures and standards [2,3,39]. This architecture called AL^2, is hierarchical in nature and allows for plug-and-play and fault tolerance at the lowest level and for learning and adaptive behaviors at the highest level. At each level of the hierarchy, each activity unit can understand and incorporate the adaptation and learning capabilities of the next lower level, which includes both those directly found in individual subunits and those resulting from the aggregation of units at recursively lower levels.

The AL^2 architecture integrates many of the features of previous architectures in the literature into a single framework that supports all of the system requirements. The research builds completely integrated teams of UGVs in which the teams and the network will adapt to the needs of the mission. In order to accomplish this goal, we propose a design methodology that enables us to design simple components whose performance can be rigorously analyzed. Complex hierarchical systems can then be constructed using these low-level building blocks. The architecture used is shown in Figure 17.8.

Conventional design process require the system functionality, and thereby the hardware and software resources to be fixed. Such design primarily addresses the compromise between the system performance and its cost and

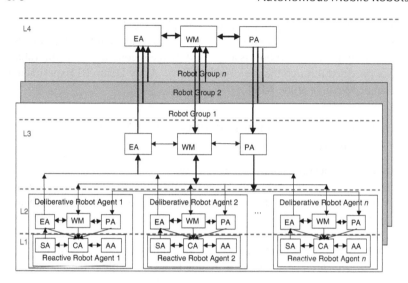

FIGURE 17.8 Architecture for adaptation and learning at all levels (AL^2).

flexibility [39,40]. The requirements of system intelligence like plug-and-play capability of sensors/actuators, fault tolerance, and retasking of robotic agents can be achieved only by planning redundancies in the hardware and reconfiguring the software during runtime. An alternate approach is to use reconfigurable hardware components, like Field Programmable Gate Arrays (FPGAs), to allow for both hardware and software reconfiguration [41,42]. By systematically partitioning the system, functionality requiring changing execution paths or those that impact performance can be assigned resources that allow both hardware and software reconfiguration. Successful implementation of this technique requires the development of new system architectures for Reconfigurable Computing. Reconfigurable Computing is the ability of the software to reach through to the hardware and alter the data path for execution thereby optimizing the performance. Our research indicates that the 4D/RCS reference model architecture is well suited for implementing reconfigurable designs for the control of UGV teams.

The architecture shown above, envisions one or more robotic agents working as a group. At the lowest level (L1), each robot agent has a control agent (CA), an actuator agent (AA), and a sensor agent (SA). The CA is responsible for attaining the commanded system performance at the lowest level. It can command the sensor agent to override its output values, recalibrate its signal, as well as perform rudimentary signal processing like filtering. The AA and SA have the lowest level of autonomy and are completely controlled by the CA. This level (L1) is characterized by stringent real-time requirements and deterministic behaviors. At a very fundamental level, this design is adequate for a robotic agent to function and perform repetitive tasks in a structured environment. Note that, because of our distributed communication infrastructure, the sensor, actuator, and control resources (and the corresponding subagents) for a single robot agent need not be present on the same physical platform. For example, a platform lacking a camera and image processing capabilities could still perform leader-following if the leader platform had a rear-facing or omni-directional camera (or other sensor) that could be used to sense the relative position of the follower platform.

In order to meet the requirements of fault tolerance, uncertainty in the system model and the environment, we use a distributed architecture wherein the higher layer (L2) incorporates elements that instil higher-level intelligence in the robot. In this layer, the sensory signals from layer L1 are processed by the Estimator Agent (EA). The output of the estimator is then used to modify/update the local representation of the WM and as input to the CA. The distributed intelligence paradigm that is proposed means that EA can now include algorithms for fault detection, dynamic sensor reconfiguration, and sensor fusion at the level of a deliberative robot agent. The WM entity in the robot agent maintains information about the environment that is necessary for the successful tasking of the robot. Typically, this would include local map information, friend/foe classification, targets and obstacles, etc. The planning agent (PA) utilizes the information from the local model of the world (WM) and the high-level task requirements to generate a plan that is communicated to the CA in layer L1. The PA implements algorithms for path planning, obstacle avoidance, optimization, etc., for an individual robot. Level L2 is characterized by increased autonomy and less stringent real-time requirements.

A team of robots may consist of a number of individual robot agents possibly with differing sensor/actuator suites and capabilities. The coordination between these agents is managed by the PA entity at the level of the robot group (L3). Information sharing between L2 entities is controlled by the entities in L3. This increases the security of the implementation because the L2 entities can function independently of each other, while still functioning in a coordinated manner. The primary function of the entities in layer 3 is to coordinate the working of the robot agents in the group. L3 handles all reassignments of tasks between different robot agents in L2. Introduction of new robot agents or sensor suites, etc., are the exclusive domain of L3. The outputs of all the EAs in layer L2 provide the input to the EA module in L3. Team-level sensor fusion amongst the different robotic agents is accomplished by the EA at L3. This EA module is used to update the WM in layer 3. This WM also manages the information sharing among the robot agents in L2. The PA in this layer does the task decomposition from the mission requirements and updates the individual PAs in L2. It is to be noted that the architecture specified is independent of hardware and software implementations and individual elements in L2. Layer 4 (L4) manages the coordination between groups of robot agents. The highest level of intelligence and autonomy and the lowest level of real-time criticality characterize L4. Dynamic reassignment of the responsibilities of each group is handled by L4.

It can be easily seen that this architecture enables the seamless sharing of resources across multiple robots and multiple teams. While the low-level fault accommodation algorithms help accommodate failures in individual robots, the ability to share resources across robots in a team, as well as across teams, provides a second sophisticated level of fault tolerance that will improve the overall success of the robotic mission. This architecture enables the development of groups of autonomous ground vehicles that are "intelligent." The architecture is flexible and is not dependent on the type of controllers or algorithms implemented in any given layer.

17.4.1.2 Hardware and software design methodology

In this section, the design methodology is introduced that enables the realization of the L1 layer of the AL^2 architecture. The discussion in the previous section leads one to the conclusion that while fault identification can be accomplished in the hardware or software, fault accommodation is done in the lowest level of the system. This is done by dynamically recreating data execution paths in the hardware that provide alternate paths for the control execution. Similarly, identification of plug-and-play components is done in software but the sensor circuitry is dynamically created in hardware. Thus by implementing the hierarchical architecture AL^2, the requirements of plug-and-play sensors and fault accommodation can be addressed by incorporating the ability to reconfigure the L1 layer of the system. The architecture also allows for the algorithms that incorporate learning and deliberative behaviors to be incorporated in L2 layer of the system and insulate those from changes at the L1 layer. Care however has to be exercised to ensure that the communication between L1 and L2 layers are not affected by the change.

The need to implement different hardware configurations at the lowest level of the system to meet the operational requirements for different tasks necessitates different types of hardware reconfiguration capabilities. These capabilities are summarized below [43,44].

Full vs. partial reconfiguration. Systems typically need to execute special tests at startup to verify proper system functioning. Once the startup tests are complete, the system can transition to the "run-time" mode. While it is easy to load test software to run system tests at startup, the tests that can be run are constrained by the hardware. By incorporating the ability to change the configuration of the FPGA device, the same hardware can be used for system tests at startup and then "fully" reconfigured for run-time operations. The ability to "fully" reconfigure the hardware is also essential to the retasking of the individual robot. When the robot is retasked, sensor and actuator configurations can be selected that adapt the robot for the specified task. Since the embedded hardware can be optimized for the specific task, the overall performance can be improved without an increase in system cost. Often, it is required to reroute the signals to accommodate for faults or add additional circuitry to handle signals from new sensors that come online. In such circumstances, unused portions of the FPGA can be configured to handle this requirement while the rest of the device is unaffected. Such reconfiguration, called partial reconfiguration, is essential to support retasking of individual robots, plug-and-play transducers, and for fault accommodation.

Static vs. dynamic reconfiguration. Static reconfiguration requires the system to be completely reconfigured before execution can begin. On the other hand, dynamic reconfiguration can take place while the system is under operation. However, care has to be exercised to prevent changing portions of the hardware during execution to prevent unforeseen outcomes. Dynamic reconfiguration is essential when it is not feasible to take the system off-line to implement changes. Depending on the system requirements, partial reconfiguration can be static or dynamic.

17.4.1.3 Design and implementation of UGV teams

The proposed AL^2 architecture is tested by implementing the L1 layer on the Xilinx's Virtex-II Pro platform. This platform was selected based on its capability in implementing reconfigurable architectures, and the excellent development tools and product support. The Virtex-II Pro XC2VP4 has a PowerPC core, 6768 logic cells, 504 KBits BRAM, 4 3.125 Gbps RocketIO transceivers, and 3.01 Mbits configuration space.

The Xilinx Virtex-II Pro device is a user programmable gate array with embedded PowerPC processor and embedded high-speed serial transceivers. The Xilinx Virtex architecture is coarse grained and consists of a number of basic cells called configurable logic blocks (CLBs). These logic blocks are arranged in rows and columns, with each CLB consisting of four logic cells arranged in two slices. Each CLB also contains logic that implements a four-input look up tables (LUTs) [41]. Each slice contains two function generators, two storage elements, arithmetic logic gates, large multiplexers, wide function capability, fast carry look ahead chain, and horizontal cascade chains. The function generators are configurable as four input look up tables (LUTs), sixteen bit shift registers, or as sixteen bit selective RAM memory. Each CLB also has fast interconnect and connects to a generalized routing matrix (GRM) to access general routing resources. The Virtex-II Pro has SelectIO-Ultra blocks (IOBs) that provide the interface between the package pins and the internal configurable logic. Active Interconnect Technology connects all these components together. The overall interconnection is hierarchical and is designed to support high-speed designs [45].

Reconfiguration Time:
The minimum unit of reconfiguration in Virtex 2 Pro is a "frame." It is proportional to the CLB width of the device. There are a total of 884 frames with 424 bytes per frame. Therefore with the system clock running at 100 MHz, the time required to configure a frame is 4 μsec, and the entire device can be reconfigured in 4 msec [46,47]. Such fast reconfiguration times make this platform ideal for implementing low-level controllers for intelligent systems.

(a) *Creation of Reconfigurable Resources*
The programmable elements in the Virtex-II Pro, including the routing resources, are controlled by values stored in the static memory cells. The device is configured by loading the bitstream into the internal configuration memory. These values can be reloaded to change the functions of the programmable elements. The Xilinx Virtex family of FPGAs supports both partial as well as dynamic reconfiguration. Partial reconfiguration can be achieved in one of the two ways, namely module-based partial reconfiguration and small-bit manipulations. In the module-based reconfiguration, the entire module can be reconfigured. The height of the reconfigurable module is the height of the device and the module can cover one or more columns. In small-bit manipulations, the reconfiguration is done by making a small change in the design, and then generating a bit-stream based only on the differences in the two designs. Switching the configuration from one implementation to another is easy and very quick. The process of implementing a general purpose IO is demonstrated in Figure 17.9. The design includes a PowerPC processor core connected to the high bandwidth processor local bus (PLB) and a bridge connecting the PLB and the on-chip peripheral bus (OPB). The required peripherals are just connected to the OPB and any external memory (BBRAM) can be accessed using the interface controller (IF_Controller). In the design example, a general purpose IO is selected. If subsequent reconfiguration of the system requires communication capabilities, say serial communication, then a different module can be generated with an UART device added to the OPB. Module-based reconfiguration will then result in enhanced system capability. Since the reconfiguration can be done in real time while the system is operational, system components can be added in real time to address changing needs during the retasking of the system.

(b) *Dynamic Fault Tolerance*
In the second design example, a PWM generator is implemented in the hardware to control the drive motors of the robot (Figure 17.10). Timer 1 is configured to generate the PWM signal while Timer 2 is configured in the "capture" mode to sense the feedback signal. Fault conditions are specified during the design process and the occurrence of a fault can be ascertained by the comparison of the PWM and the feedback signals. If a fault is detected during

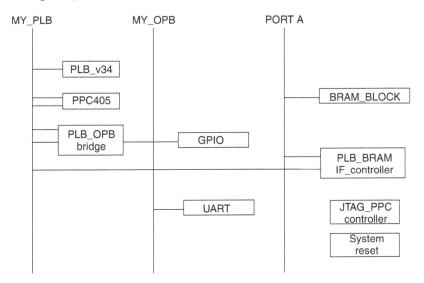

FIGURE 17.9 Implementation of general purpose IO and serial communication in Xilinx EDK.

operation, then a new timer (Timer 3) is created and the output of this timer is switched to the output pins. The changes involved in the reconfiguration between the two designs are relatively small and small-bit manipulation is ideal for this type of reconfiguration. The control cycle in this example was executed in real time with a sampling rate of 20 msec. The time for reconfiguration was about 16 μsec showing that dynamic fault accommodation is achieved in real time.

(c) *Built-In Self Test, System Retasking, and Fault Accommodation*
Built in Self Test and system retasking can be effectively implemented using "full reconfiguration" of the FPGA device. Here, a first configuration is loaded for self test and on successful completion a run-time configuration is loaded onto the FPGA device. Since the system can be optimized for every task separately, the overall performance is improved. Often, a system requires only a portion of its functionality to be changed, especially during fault recovery where there might be a need to reconfigure only the sensor module or simply bypass the sensor. This can be done using partial reconfiguration. Partial reconfiguration is implemented in the FPGA device where a portion of the circuitry is reconfigured while the rest of the device is unaffected and still in operation. Depending on the system requirements, partial reconfiguration can be accomplished by taking the system off-line (static reconfiguration) or while the system is under operation (dynamic reconfiguration). Dynamic reconfiguration

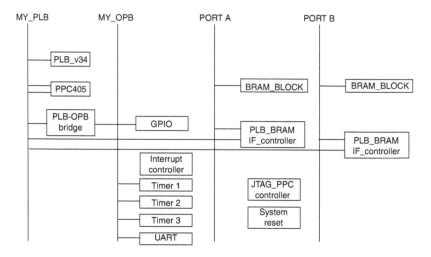

FIGURE 17.10 Implementation of PWM motor control with dynamic reconfiguration for fault accommodation.

is essential when it is not feasible to take the system off-line to implement changes.

(d) *Low Level Learning for Improving the Performance of the Controller*
Often, significant improvement in the performance can be achieved by compensating for nonlinearities like deadband and hysteresis in the drive mechanisms. These nonlinearities are dependent on the load and operating conditions. Effective compensation of these effects requires the implementation of adaptive algorithms. However, these algorithms are required only in those cases where the substantial degradation of the performance results in the absence of the compensation techniques. Further, since these algorithms can be implemented very efficiently in hardware, compensation for the nonlinear effects can be achieved dynamically on a case-by-case basis using dynamic, partial reconfiguration capability of the proposed architecture.

In the third design example, a neural network (NN) is implemented in the FPGA to compensate for actuator deadband (Figure 17.11). The controller continuously monitors the output and dynamically instantiates the NN in the FPGA when the performance degrades significantly due to load dependent deadband in the actuator dynamics. The NN is modeled and designed in Simulink using the Xilinx toolset provided by Mathworks Inc. (Figure 17.12). This design was successfully validated and a configuration bit stream generated. This model is loaded onto the FPGA for dynamic creation of the NN module in the hardware. When the control strategy is changed, it automatically results in the loading of the NN module in the FPGA and the associated software for execution in real time.

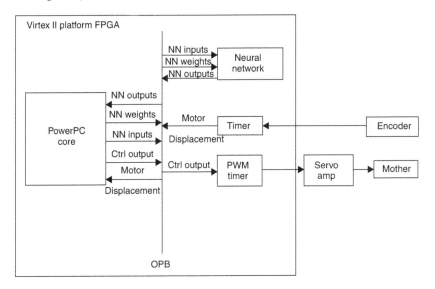

FIGURE 17.11   Neural network-based compensation of actuator nonlinearities.

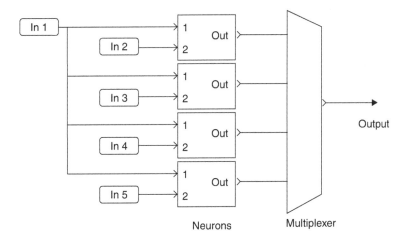

FIGURE 17.12   Implementation of a one-layer neural network in simulink using xilinx toolset.

(e) *Software Implementation for Reconfigurable Computing*

The controller, sensing, and actuation functions of the system are abstracted out and implemented as separate packages with well-defined interfaces. The controller communicates with the sensing module and the actuator module through the interface provided and is independent of the actual implementation within

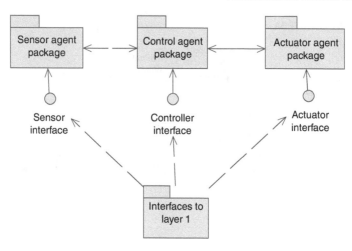

FIGURE 17.13   UML implementation of the layer 1 entities.

these modules. Providing standardized interfaces makes possible the implementation of distributed sensors and actuators. Further, by encapsulation of the sensing and actuation functions, system requirements such as plug-and-play of sensors, and fault accommodation can be achieved. Exposing the interface to these packages also aids in system diagnostics and testing of the low-level functionality.

The modeling of the sensors, actuators, and controller was done using Rational's Unified Modeling Language (UML 2.0) as shown in Figure 17.13 and Figure 17.14. The code was generated from these models for execution on the target platform.

The architecture developed was implemented in the development of a team of intelligent UGVs. The implementation is not constrained by the size of the team and the team can be dynamically created. The tasking of an individual member and the required sensor configuration can be dynamically performed at run-time. The implementation has been tested for scenarios that include wall-following, leader following modes of team behaviors. The research prototype of the reconfigurable robot is shown in Figure 17.15.

17.4.2 Demo III Experimental Unmanned Vehicle (XUV) Project at NIST

In this example, the experimental validation of the 4D/RCS architecture has been provided by the performance of the Army Research Lab Demo III experimental unmanned ground vehicle (XUV) shown in Figure 17.16. Four of these vehicles were put through an extended series of demonstrations and field tests during the fall and winter of 2002 to 2003.

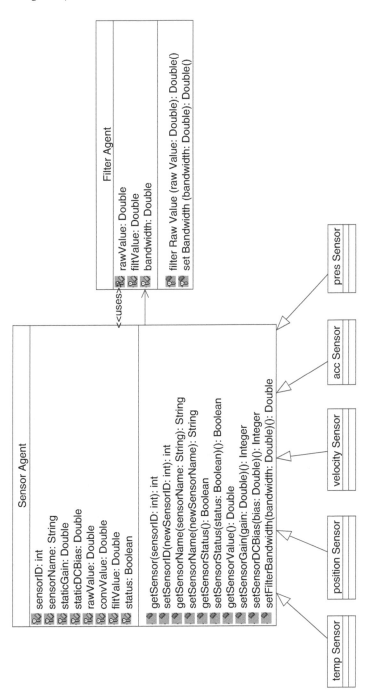

Figure 17.14   Generalization of the sensor class.

FIGURE 17.15   Prototype of a reconfigurable robot.

The XUVs were equipped with an inertial reference system, a commercial grade GPS receiver (accurate to about ±20 m), a LADAR camera with a frame rate of 10 frames per sec, and a variety of internal sensors. The LADAR had a field of view 90° wide and 20° high with resolution of about $\frac{1}{2}$° per pixel. It was mounted on a pan/tilt head that enabled it to look in the direction that it planned to drive. The LADAR was able to detect the ground out to a range of about 20 m, and detect vertical surfaces (such as trees) out to a range of about 60 m. Routes for XUV missions were laid out on a terrain map by trained Army scouts, and given to the XUVs in terms of GPS waypoints spaced more than 50 m apart.

The XUVs operated completely autonomously until they got into trouble and called for help. Typical reasons for calling for help were the XUV was unable to proceed because of some terrain condition or obstacle (such as soft sand on a steep slope, or dense woods), and was unable to find an acceptable path plan after several attempts at backing up and heading toward a different direction. At such a point, an operator was called in to teleoperate the vehicle out of difficulty. During these operations, data was collected on the cause of the difficulty, the type of operator intervention required to extract the XUV, the time required before the XUV could be returned to autonomous mode, and the work load on the operator.

FIGURE 17.16 The Army Demo III experimental unmanned vehicle. On the left-top is the LADAR. In the center-top is a Reconnissance Camera ball. On the right-top is a pan/tilt unit with a color stereo pair, a FLIR stereo pair, and a color high resolution monocular camera. The white panel in the center front is for a radar. The front bumper is instrumented to detect obstacles hidden in the weeds. (Photo courtesy of General Dynamics Robotic Systems.)

During three major experiments designed to determine the technology readiness of autonomous driving, the Demo III XUVs were driven a total of 550 km, over rough terrain (1) in the desert, (2) in the woods, (3) through rolling fields of weeds and tall grass, (4) on dirt roads and trails, and (5) through an urban environment with narrow streets cluttered with parked cars, dumpsters, culverts, telephone poles, and manikins. Tests were conducted under various conditions including night, day, clear weather, rain, and falling snow (see Figure 17.17). The unmanned vehicles operated without any operator assistance over 90% of both time and distance. A detailed report of these experiments has been published in Reference 48. High resolution LADAR ground truth data describing the terrain where the XUVs experienced difficulties was also gathered and analyzed [49].

It should be noted that the Demo III tests were performed in environments devoid of moving objects such as oncoming traffic, pedestrians, or other vehicles. The inclusion of moving objects in the world model, and the development of perception, world modeling, and planning algorithms for operating in the presence of moving objects is a topic of current research.

FIGURE 17.17   An Army Demo III experimental unmanned vehicle driving autonomously through the woods during a snow storm at Ft. Indiantown Gap, Pennsylvania, in January 2003.

17.5 CURRENT RESEARCH AND FUTURE DIRECTIONS

In this chapter, the engineering of Intelligent Systems was addressed through the development of a reference model architecture. The implementation of this architecture to the control of unmanned autonomous vehicles and teams of unmanned ground vehicles was then discussed.

Current research in our laboratories is focused on the following aspects of autonomous vehicle control:

1. Autonomous driving on normal roads and streets, for example, driving on country roads and city streets with oncoming traffic, negotiating intersections with traffic signals and pedestrians, and maneuvering in and out of parking spaces.
2. Autonomous tactical behaviors for teams of real and virtual autonomous military ground and air vehicles, for example, controlling the behavior of a platoon of scout vehicles consisting of ten unmanned ground vehicles and three unmanned air vehicles cooperating in the performance of a route reconnaissance mission prior to a troop echelon road march.

We believe that autonomous driving and the control of teams of intelligent autonomous vehicles are an excellent topic for future research for the following reasons:

First, it is a problem domain for which there is a large potential user base, both in the military and civilian sectors. This translates into research funding.

Second, it is a problem domain where physical actuators and power systems are readily available. Wheeled and tracked vehicle technology is mature, inexpensive, and widely available.

Third, it is a problem domain for which the technology is ready. The invention of real-time LADAR imaging makes it possible to capture the 3D geometry and dynamics of the world. This has broken the perception barrier. The continued exponential growth rate in computing power per dollar cost has brought the necessary computational power within the realm of economic viability. This has broken the cost barrier. Furthermore, reconfigurable computing, cognitive modeling, intelligent control theory, and software technology have advanced to the point where the engineering of intelligent systems is feasible. This has broken the technology barrier.

Finally, autonomous driving is a problem domain of fundamental scientific interest. Locomotion is perhaps the most basic of all behaviors in the biological world. Locomotion is essential to finding food and evading predators throughout the animal kingdom. The brains of all animate creatures have evolved under the pressures of natural selection in rewarding successful locomotion behavior. It is therefore, not unreasonable to suspect that building truly intelligent mobility systems will reveal fundamental new insights into the mysteries of how the mechanisms of brain give rise to the phenomena of intelligence, consciousness, and mind.

17.6 CONCLUSIONS

The current research in the engineering of intelligent systems has focused on the development of reference model architectures and implementation of features that instill intelligence in the overall system. These implementations have resulted in significant advances in the technology in the areas of perception, knowledge representation, planning, adaptation, learning, and control. Current research has also focused on utilizing recent advances in the reconfigurable computing technology to design systems that are intelligent and can configure themselves according to the needs of the mission. Recent implementations at the NIST and OU indicate that the 4D/RCS architecture is well suited for the implementation of Intelligent Systems and that the Intelligent UGVs are a good testbed for validating system autonomy, and intelligent behaviors.

In many ways, 4D/RCS is a superset of Soar, ACT-R, Dickmanns 4D approach, and even behaviorist architectures such as subsumption and its many derivatives. 4D/RCS incorporates and integrates many different and diverse concepts and approaches into a harmonious whole. It is hierarchical but distributed, deliberative yet reactive. It spans the space between the cognitive and reflexive, between planning and feedback control. It bridges the gap between spatial distances ranging from kilometers to millimeters, and between time intervals ranging from months to milliseconds. And it does so in small regular steps, each of which can be easily understood and readily accomplished through well-known computational processes.

Each organizational unit in 4D/RCS refines tasks with about an order of magnitude increase in detail, and an order of magnitude decrease in scale, both in time and space. At the upper levels, most of the computational power is spent on cognitive tasks, such as analyzing the past, understanding the present, and planning for the future. At the lower levels, most of the computational power is spent in motor control, and the early stages of perception.

However, at every level, the computational infrastructure is fundamentally the same (except for scale in time and space). Computational modules (that theoretically could be implemented as neural nets, or finite state automata, or production rules) accept inputs and produce outputs. Knowledge is represented in arrays, strings, pointers, frames, and rules. At various levels and in many different ways, computational modules process sensory data, model the world, and decompose high-level intentions into low-level actions. Within each module, this process is both limited in complexity and finite in scope. Perhaps most important, 4D/RCS makes the processes of intelligent behavior understandable in terms of computational theory. Thus, it can be engineered into practical machines.

We should note in closing that there remain many features of the 4D/RCS reference model architecture that have not yet been fully implemented in any application. However, enough of the 4D/RCS reference model has been implemented to demonstrate that the fundamental concept is valid and the more advanced features are feasible.

ACKNOWLEDGMENTS

The first author gratefully acknowledges the assistance of the Army Research Office, U.S. Department of Defense, in supporting this work through grant # DAAD 19-03-1-0142. The second and third authors gratefully acknowledge the support of the Army Research Lab, Aberdeen Proving Grounds, in supporting this work through Contract # MIPR4CARL80059, Charles Shoemaker, PM; and the support of the DARPA MARS Program through Contract # K300, Douglas Gage, PM.

REFERENCES

1. J. S. Albus and A. M. Meystel, *Engineering of Mind: An Introduction to the Science of Intelligent Systems*, Wiley Series on Intelligent Systems, 2000.

2. IEEE 1451.1, "Standard for smart transducer interface for sensors and actuators — network-capable application processor (NCAP) information model," 1999.

3. IEEE 1451.2, "Standard for a smart transducer interface for sensors and actuators — Transducer to microprocessor communication protocols and transducer electronic data sheet (TEDS) format," 1997.

4. J. S. Albus, "Features of intelligence required by unmanned ground vehicles," *Proceedings of the Performance Metrics for Intelligent Systems Workshop*, NIST, Gaithersburg, Maryland, 2000.

5. A. Newell, J. C. Shaw, and H. A. Simon, "Elements of a theory of human problem solving," *Psychological Review*, 65, 151–166, 1958.

6. P. S Rosenbloom, J. E. Laird, and A. Newell (eds), *The Soar Papers: Research on Integrated* Intelligence, MIT Press, Cambridge, MA, 1993.

7. M. Matarić, *Interaction and Intelligent Behavior*, Ph.D. thesis, Massachusetts Institute of Technology, 1994.

8. L. E. Parker, "ALLIANCE: an architecture for fault tolerant multi-robot cooperation," *IEEE Transactions on Robotics and Automation*, 14, 220–240, 1998.

9. R. A. Brooks, "A robust layered control system for a mobile robot," *IEEE Journal of Robotics and Automation*, 2, 14–23, 1986.

10. R. A. Brooks, *Cambrian Intelligence: The Early History of the New AI*, MIT Press, Cambridge, MA, 1999.

11. R. C. Arkin, "Integrating behavioral, perceptual, and world knowledge in reactive navigation," *Robotics and Autonomous Systems*, 6, 105–122, 1990.

12. R. C. Arkin, *Behavior Based Robotics*, MIT Press, Cambridge, MA, 1998.

13. E. Gat, "Integrating planning and reaction in a heterogenous asynchronous architecture for controlling real-world mobile robots," *Proceedings of the Tenth National Conference on Artificial Intelligence, San Jose, California*, AAAI Press, 1992.

14. Y. Cao, A. Fukunaga, and A. Kahng, "Cooperative mobile robotics: antecedents and directions," *Autonomous Robots*, 4, 7–27, 1997.

15. È. Coste-Marière and R. Simmons, "Architecture, the backbone of robotic systems," *Proceedings of the IEEE International Conference on Robotics and Automation*, San Francisco, California, USA, pp. 67–72, 2000.

16. G. Dudek, M. Jenkin, E. Milios, and D. Wilkes, "A taxonomy for multi-agent robotics," *Autonomous Robots*, 3, 375–397, 1996.

17. R. Peter Bonasso, R. J. Firby, E. Gat, D. Kortenkamp, David P. Miller, and Mark G. Slack, "Experiences with an architecture for intelligent, reactive agents." *Journal of Experimental and Theoretical Artificial Intelligence*, 9, 237–256, 1997.

18. D. Schreckenghost, P. Bonasso, D. Kortenkamp, and D. Ryan, "Three tier architecture for controlling space life support systems," *Proceedings of IEEE International Joint Symposia on Intelligence and Systems*, Rockville, Maryland, May 21–23, pp. 195–201, 1998.

19. R. Murphey and P. Pardalos (eds.), *Cooperative Control and Optimization*, Applied Optimization, vol. 66, Kluwer Academic Press, Dordrecht, 2002.

20. T. Balch and R. Arkin, "Behavior-based formation control for multirobot teams," *IEEE Transactions on Robotics and Automation*, 14, 926–939, 1998.

21. R. Simmons, T. Smith, M. B. Dias, D. Goldberg, D. Hershberger, A. Stentz, and R. Zlot, "A layered architecture for coordination of mobile robots," in *Multi-Robot Systems: From Swarms to Intelligent Automata: Proceedings from the 2002 NRL Workshop on Multi-Robot Systems*, A. Schultz and L. Parker (eds.), Kluwer, Dordrecht, pp. 103–112, 2002.

22. R. Volpe, I. Nesnas, T. Estlin, D. Mutz, R. Petras, and H. Das, "The CLARAty architecture for robotic autonomy," *Proceedings of IEEE Aerospace Conference*, pp. 121–132, 2001.

23. A. J. Barbera, J. S. Albus, and M. L. Fitzgerald, "Hierarchical control of mobile robots using microcomputers," *Proceedings of 9th International Symposium on Industrial Robots*, Washington D.C., March 13–15, 1979.

24. A. J. Barbera, M. L. Fitzgerald, J. S. Albus, and L. S. Haynes, "RCS — The NBS real-time control system," *Proceedings of the Robots 8 Conference and Exposition*, Detroit, MI, June 4–7, 1984.

25. J. S. Albus, "The NIST real-time control system (RCS): an approach to intelligent systems research," *Journal of Experimental and Theoretical Artificial Intelligence*, Special Issue, 157–174, 1997.

26. E. Dickmanns, "A general dynamic vision architecture for UGV and UAV," *Journal of Applied Intelligence*, 2, 251–270, 1992.

27. E. Dickmanns, "An expectation-based, multi-focal, saccadic (EMS) vision system for vehicle guidance," *Proceedings of the 9th International Symposium on Robotics Research*, Salt Lake City, UT, October 1999.

28. J. Albus, et al., "4D/RCS Version 2.0: A Reference Model Architecture for Unmanned Vehicle Systems," *NISTIR 6910, National Institute of Standards and Technology*, Gaithersburg, MD, 2002.

29. A. Barbera, J. Horst, C. Schlenoff, E. Wallace, and D. Aha, "Developing World Model Data Specifications as Metrics for Sensory Processing for On-Road Driving Tasks," *Proceedings of the 2003 PerMIS Workshop*, NIST Special Publication 990, Gaithersburg, MD, 2003.

30. J. S. Albus, "4D/RCS: a reference model architecture for intelligent unmanned ground vehicles," *SPIE AeroSense Conference 4715 on Unmanned Ground Vehicle Technology IV*, Orlando, FA, April 2–3, 2002.

31. A. Barbera, J. Albus, E. Messina, C. Schlenoff, and J. Horst, "How Task Analysis Can Be Used to Derive and Organize the Knowledge For the Control of Autonomous Vehicles," *Proceedings of the Knowledge Representation and Ontology for Autonomous Systems Workshop of the 2004 AAAI Spring Symposium*, Stanford University, Palo Alto, CA, 2004.

32. S. Balakirsky, *A Framework for Planning with Incrementally Created Graphs in Attributed Problem Spaces*, IOS Press, Berlin, Germany, 2003.

33. R. Provine, M. Uschold, S. Smith, S. Balakirsky, and S. Schlenoff, "Observations on the Use of Ontologies for Autonomous Vehicle Navigation Planning," *Knowledge Representation and Ontologies for Autonomous Systems*, 2004 AAAI Spring Symposium, pp. 98–102, 2004.

34. C. Schlenoff, R. Washington, and T. Barbera, "Experiences in Developing an Intelligent Ground Vehicle (IGV) Ontology in Protege," *Submitted to the 7th International Protege Conference*, Bethesda, MD, 2004.

35. The OWL Services Coalition, "OWL-S 1.0 Release," http://www.daml.org/services/owl-s/1.0/owl-s.pdf, 2003.

36. R. Madhavan, T. Hong, and E. Messina, "Temporal Range Registration for Unmanned Ground and Aerial Vehicles,"*Proceedings of the IEEE International Conference on Robotics and Automation (ICRA)*, New Orleans, LA, pp. 3180–3187, 2004.

37. C. Schlenoff, S. Balakirsky, T. Barbera, C. Scrapper, J. Ajot, E. Hui, and M. Paredes, "The NIST Road Network Database: Version 1.0," *National Institute of Standards and Technology (NIST)*, NISTIR-7136, Gaithersburg, Maryland, 2004.

38. R. Camden, B. Bodt, S. Schipani, J. Bornstein, R. Phelps, T. Runyon, F. French, and C. Shoemaker, Autonomous Mobility Technology Assessment: Interim Report — February 2003, *ARL-MR-565*, Army Research Laboratory, ATTN: AMSRL-WM-RP, Aberdeen Proving Ground MD 21005-5066, 2003.

39. F. M. Proctor, B. Damazo, C. Yang, and S. Frechette, "Open architectures for control," *National Institute on Standards and Technology*, Internal report, NISTIR-5307, Gaithersburg, Maryland, 1993.

40. L. Wills, S. Kannan, S. Sander, M. Guler, B. Heck, J. V. R. Prasad, D. Schrage, and G. Vachtsevanos, "An open platform for reconfigurable control," *IEEE Control Systems Magazine*, 21(3), 49–64, June 2001.

41. S. Donti, and R. L. Haggard, "A survey of dynamically reconfigurable FPGA devices," *Proceedings of IEEE*, 8, 422–426, 2003.

42. J. Harkin, T. M. McGinnity, and L. P. Maguire, "Partitioning methodology for dynamically reconfigurable embedded systems," *IEE Proceedings of Computers Digitial Technology*, 147, 391–396, 2000.

43. S. Commuri, R. Fierro, D. Hougen, and R. Muthuraman, "System intelligence requires distributed learning," *Proceedings of the IEEE International Symposium on Intelligent Control*, Taipei, Taiwan, pp. 67–72, 2004.

44. S. Commuri, Y. Li, D. Hougen, and R. Fierro, "Designing for system intelligence — a case study," *Proceedings of the 5th IFAC Symposium on Intelligent Autonomous Vehicles*, Lisbon, Portugal, July 2004.

45. *Virtex-II Pro: Platform FPGA Handbook*, Xilinx Inc., UG012, v 2.0, 2002.

46. Xapp 662: In-Circuit Partial Reconfiguration of Rocket IO Attributes v.1.1, Xilinx Inc.

47. B. Blodget, S. McMillan, and P. Lysaght, "A light weight approach for embedded reconfiguration of FPGAs", *Proceedings of the Design, Automation and Test in Europe Conference and Exhibitions*, Messe Munich, Germany, 3–7, March 2003, pp. 399–400.

48. C. Shoemaker, J. Bornstein, S. Myers, and B. Brendle, "Demo III: Department of Defense testbed for unmanned ground mobility," *SPIE Conference on Unmanned Ground Vehicle Technology, SPIE Vol. 3693*, Orlando, FA, April 1999.

49. C. Witzgall, G. S. Cheok, and D. E. Gilsinn, Terrain characterization from ground-based LADAR, *Proceedings of PerMIS '03 Workshop,* National Institute of Standards and Technology, Gaithersburg, MD 20899, 2003.

BIOGRAPHIES

Sesh Commuri is an associate professor in the School of Electrical and Computer Engineering at the University of Oklahoma in Norman. He earned his M.S. in electrical engineering from the Indian Institute of Technology, Kanpur, India in 1989 and the Ph.D in electrical engineering from the University of Texas at Arlington in 1996. Dr. Commuri has over 10 years experience working in the industry developing embedded systems for control applications. Dr. Commuri has been with the University of Oklahoma since January 2002. His research interests include embedded systems, reconfigurable architectures, mechatronics, wireless sensor networks, and the design and development of intelligent systems.

James S. Albus founded and led the Intelligent Systems Division at the National Institute of Standards and Technology for 20 years. He is currently a senior NIST fellow pursuing research in autonomous vehicles, perception, knowledge representation, and intelligent behavior. During the 1960s he designed electro-optical systems for more than 15 NASA spacecraft. During the 1970s, he developed a model of the cerebellum that is still widely used by neurophysiologists today. Based on that model, he invented the cerebellar model arithmetic controller (CMAC) neural net, and coinvented the real-time control system (RCS) that has been used for developing a wide variety of intelligent systems for both government and commercial applications. The most recent version of RCS has been chosen by the Army to provide autonomous navigation capabilities for all future combat system vehicles. He is the inventor of the NIST RoboCrane. Dr. Albus is the author or coauthor of five books and more than 150 scientific papers, journal articles, book chapters, and official government studies on intelligent systems and robotics. He has lectured extensively throughout the world.

Anthony Barbera is an electronics engineer in the Intelligent Systems Division at the National Institute of Standards and Technology, in Gaithesburg, Maryland.

Index